HANDBOOKS OF AMERICAN NATURAL HISTORY

ALBERT HAZEN WRIGHT, ADVISORY EDITOR

The Spider Book

BY JOHN HENRY COMSTOCK

WEB OF METARGIOPE TRIFASCIATA

The Spider Book

A MANUAL FOR THE STUDY OF THE SPIDERS
AND THEIR NEAR RELATIVES, THE SCORPIONS,
PSEUDOSCORPIONS, WHIP-SCORPIONS,
HARVESTMEN, AND OTHER MEMBERS OF
THE CLASS ARACHNIDA, FOUND IN
AMERICA NORTH OF MEXICO,
WITH ANALYTICAL KEYS FOR THEIR
CLASSIFICATION AND POPULAR ACCOUNTS
OF THEIR HABITS

By

John Henry Comstock

REVISED AND EDITED BY

W. J. Gertsch

American Museum of Natural History

Comstock Publishing Associates

A DIVISION OF

CORNELL UNIVERSITY PRESS

ITHACA AND LONDON

Copyright, 1912, 1940, by
DOUBLEDAY, DORAN & COMPANY, INC.

Copyright assigned 1948 to
COMSTOCK PUBLISHING COMPANY, INC.

Reissued 1948
Reissued 1965
Second printing 1967
Third printing 1971
Fourth printing 1975

All rights reserved. Except for brief quotations in a review, this book, or parts thereof, must not be reproduced in any form without permission in writing from the publisher. For information address Cornell University Press, 124 Roberts Place, Ithaca, New York 14850.

Published in the United Kingdom by Cornell University Press Ltd., 2-4 Brook Street, London W1Y 1AA

International Standard Book Number 0-8014-0084-8
Printed in the United States of America

PREFACE

THE SPIDER BOOK has served the needs of beginners and mature students of spiders in an admirable manner for more than twenty-five years. Appearing at a time when arachnology was the property of a few trained systematists, it opened the way to a new appreciation of spiders and their near relatives by the laity of amateurs, that assembly of individuals who yearn for a fuller knowledge of commonplace things. It brought together for the first time in concise form a wealth of information on the structure, habits and classification of the American arachnids. It aimed to correct many erroneous impressions about these common animals and to emphasize the interest and keen enjoyment in store for all who study these creatures. That The Spider Book has accomplished its purpose of popularizing the study of arachnids is evidenced by the fact that at present there are more serious students of spiders and a greater number of well informed amateurs than at any previous time in the history of American arachnology. To these many workers The Spider Book has been the Primer. No comparable work has appeared in any other language.

The Spider Book now has been out of print for a number of years. Its loss has been keenly felt by new students throughout the country who, deprived of the classic source of information on our fauna, have voiced repeatedly the wish for a reprint. Recognizing the desirability of again making this splendid work available to the public, the representatives of Doubleday, Doran, Inc. conferred with me on the feasibility of republication of The Spider Book, not as a mere reprint but in a revised form. Needless to say, I have considered it an honour to become identified even to a limited extent with a new edition of The Spider Book.

The keynote of the revision has been conservatism. To alter in any radical way the form or limits of the book was deemed unadvisable. Inasmuch as The Spider Book was intended as a popular introduction, it has seemed inexpedient to incorporate into the book controversial arrangement, debatable nomenclature, or radical departures from the present standard. The Chapters

Preface

which treat of the morphology and habits of spiders remain unchanged. Although there is much that could be added to them, they are still adequate in their present form. Numerous changes have been made in the sections on the classification of spiders and their relatives. They have not been made with the object of making The Spider Book complete in the sense that every species, or even every genus, is diagnosed. That was quite impossible in 1913 and is even less practicable today because of the enormous increase in the number of genera and species now known from the region covered by the book. All the recent orders of the Arachnida are now known to have representatives within our borders. An account of our only species of the Ricinulei has been included. Several families of spiders which have recently been shown to occur within the limits considered by the book have been treated in their proper places. Other families have been amplified or revised to a greater or lesser degree. Only those changes have been made which have seemed compatible with the purpose of the book as an introduction to fill the needs of beginning students on the arachnids.

The Spider Book will again fill the existing needs of student arachnologists for a summary of the biology and classification of spiders and the lesser orders of arachnids found in America north of Mexico. The appearance of a revised edition is an added compliment to the energy and accomplishment of Dr. Comstock who, in gaining an eminent position in the field of Entomology, still found time to give us an outstanding book on the Arachnida.

W. J. GERTSCH

CONTENTS

CHAPTER		PAGE
PREFACE BY W. J. GERTSCH		v
INTRODUCTION		ix
I.	SPIDERS AND THEIR NEAR RELATIVES	3
	I. The zoölogical position of the Arachnida	3
	II. The characters of the Arachnida	9
	III. The orders of the Arachnida	12
II.	THE EXTERNAL ANATOMY OF SPIDERS	95
III.	THE INTERNAL ANATOMY OF SPIDERS	137
IV.	THE LIFE OF SPIDERS	177
	I. Methods of study	177
	II. The development of spiders	182
	III. The food of spiders	185
	IV. Means by which spiders obtain their prey	186
	V. The silk of spiders	187
	VI. The types of webs of spiders	193
	VII. The building of an orb-web	196
	VIII. The nests of spiders	206
	IX. The pairing of spiders	207
	X. The motherhood of spiders	208
	XI. The venom of spiders	213
	XII. The aeronautic spiders	215
V.	THE ORDER ARANEIDA OR SPIDERS	218
VI.	THE SUPERFAMILY AVICULARIOIDEA OR TARANTULAS	222
VII.	THE SUPERFAMILY ARGIOPOIDEA OR TRUE SPIDERS	252

INTRODUCTION

OF ALL of our little neighbours of the fields there are none that are more universally shunned and feared than spiders, and few that deserve it less. There is a widespread belief that spiders are dangerous, that they are liable to bite, and that their bites are very venomous. Now this may be true of certain large species that live in hot countries; but the spiders of the temperate regions are practically harmless.

It is true, spiders bite and inject venom sufficient to kill an insect into the wounds made by their fangs. But they are exceedingly shy creatures, fearing man more than they are to be feared. If an observer will refrain from picking up a spider there is not the slightest danger of being bitten by one; and, excepting perhaps a single uncommon species, no spider is known in the northern United States whose bite would seriously affect a human being.

On the other hand, spiders are exceedingly interesting subjects for study; for some of the most remarkable exhibitions of instinctive powers are presented by them. What product of instinctive skill is more wonderful than the web of an orb-weaving spider!

He who loves the out-of-doors will find the interest incident to his walks greatly increased if he learns something of the habits of spiders. Their webs are to be found everywhere; and those of the different species differ greatly in structure, varying from an irregular, tangled maze, as that of the common domestic spider, to the wonderfully symmetrical nets of the garden-spiders.

If spiders did not occur in our fauna, and if the keepers of a zoölogical garden were to bring from some remote part of the world living examples of the little animals that spin from their bodies threads of silk of different kinds, some dry and inelastic, some viscid and elastic, and some, as the hackled band of *Filistata*, of wonderful complexity of structure, and with these threads construct snares of surprising regularity for trapping their prey, the presence of such marvellous animals would at-

Introduction

tract general attention, and we would make long journeys to see them. Fortunately, however, this marvel can be seen at home by any one that has eyes and will look.

While the web-building spiders are those that most often attract attention, there are many that differ greatly from these in habits; and the stratagems employed by these to escape their enemies and to obtain their prey, in many cases, are scarcely less wonderful than those of the web-building species. In a word, the abundance of spiders, the great variations in the habits of the different species, and the high development of instinctive powers of many of them, render them exceedingly available for purposes of study of animal behaviour, whether this study be pursued by the lover of nature for his own enjoyment or by the teacher who wishes to use it as a means of interesting young people in the world about us and in training their powers of observation.

The structure of spiders also offers attractive fields for study. This is especially true if attention be given to the correlation of structure and habits. As we find here the most elaborate spinning habits, so too we find here the most complicated organs for the production and manipulation of the silk which is spun. Some spiders are sedentary, either trapping their prey by snares or lying in ambush for it; others, like wolves, stalk their prey; some make use of any retreat that they find; while others dig tunnels in the earth; of the burrowing species, some merely strengthen the walls of their burrow with silk, leaving the entrance a simple opening into the earth; some build a watch-tower or turret about the entrance; and some close the entrance with a cunningly constructed hinged door. In each case the structure of the spider is specialized in a way that adapts it to its peculiar mode of life.

Many of these correlations can be readily seen by the comparatively untrained observer. The short and stout legs of the jumping spiders, the longer and more slender legs of the running species, the extra claw on each foot of those that cling to webs, the rakes of the cheliceræ or jaws of the burrowing tarantulas, the pearly lustre of the "night-eyes" of those that live in dark places, and the protective colours of many species are all easily observed adaptations to peculiar modes of life. And for the trained observers many problems in the morphology of these animals await solution.

Introduction

While the chief object of this book is to furnish an introduction to the study of the structure, classification, and habits of spiders, it has seemed wise to include in it accounts of the near relatives of spiders, of the other orders of the class Arachnida to which the spiders belong. Some of these, as the harvestmen, the mites, and the pseudoscorpions, are common in all parts of our country, and will be observed by all students of spiders. Others, as the scorpions and the whip-scorpions, abound in the warmer parts of our country and will be found by those who study there or who receive collections from the South. As no general work treating of the North American representatives of all of the orders of the Arachnida has been published, it is believed that the account of them given here will be a welcome addition to "The Spider Book."

THE SPIDER BOOK

CHAPTER I: SPIDERS AND THEIR NEAR RELATIVES

Class Arachnida

I.— THE ZOÖLOGICAL POSITION OF THE ARACHNIDA

Spiders, scorpions, harvestmen, mites, and certain other less familiar forms constitute a group of animals which is known to zoölogists as the Class *Arachnida* (A-rach′ni-da).

This is one of several classes of animals that agree in having the body composed of a series of more or less similar rings or segments, and in having some of these segments furnished with jointed legs (Fig. 1). All the animals possessing these characteristics are classed together as the *Arthropoda* (Ar-throp′o-da), which is one of the chief divisions or phyla of the animal kingdom.

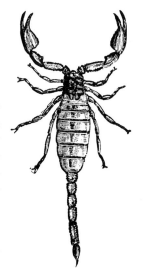

Fig. 1. A SCORPION, ONE OF THE ARTHROPODA

A similar segmented form of body is found among worms; but these are distinguished from the arthropods by the absence of legs. It should be remembered that many animals commonly called worms, as the tomato-worm, apple-worm, etc., are not true worms, but are the larvæ of insects, and have legs (Fig. 2); the angle-worm is the most familiar example of a true worm.

The Phylum Arthropoda is the largest of the phyla of the animal kingdom, including many more species than all the other phyla taken together. The more familiar of the classes included in it are the six mentioned below. Besides these there are other classes, represented by less familiar animals, which also have the body segmented and possess jointed legs. For an account of the classes of arthropods not mentioned here, and for discussions of the relations of the classes of arthropods to each other, the reader is referred to more general works on zoölogy.

I. Class CRUSTACEA (Crus-ta'ce-a).— The more familiar representatives of the Crustacea are the cray-fishes, the lobsters, the shrimps, and the crabs. Cray-fishes (Fig. 3) abound in our brooks, and are often improperly called crabs. The lobsters, the shrimps, and the true crabs live in the sea.

The Crustacea are essentially aquatic animals; a few of them live in damp places on land, but most of them live in the water. They breathe either through the general surface of the body or by means of gills; they are never furnished with tracheæ as are nearly all other arthropods.

The Crustacea differ also from all other arthropods in having two pairs of antennæ. In many of them the head and the thorax, the part of the body which bears the ambulatory legs, are united, forming a region known as the *cephalothorax;* the region behind the cephalothorax is the *abdomen.*

The examples named above are among the more conspicuous members of the class; but many other smaller forms abound both in the sea and in fresh water. Some of the more minute fresh-water forms are almost sure to occur in any fresh-water aquarium. In Fig. 4 are represented three of these, greatly enlarged.

Among the Crustacea that live in damp places on land are the sow-bugs, *Oniscidæ* (O-nis'ci-dæ). These frequently occur about water-soaked wood. One of them is represented in Fig. 5.

On the sea-coasts an immense number of species of Crustacea occur.

II. Class DIPLOPODA (Di-plop'o-da).—This class includes the millipedes. These are air-breathing arthropods in which the head is distinct, and the remaining segments of the body form a continuous region (Fig. 6). The most striking characteristic is the fact that most of the body-segments bear each two pairs of legs. As a rule, the body is not flattened as with the centipedes, and the antennæ are comparatively short and few-jointed.

The millipedes live in damp places and feed on decaying

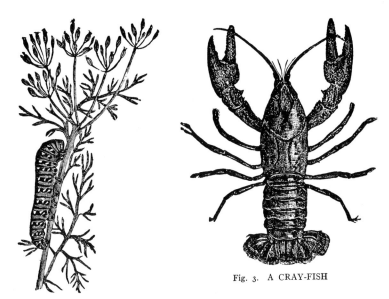

Fig. 2. THE "CARAWAY-WORM," THE LARVA OF A BUTTERFLY

Fig. 3. A CRAY-FISH

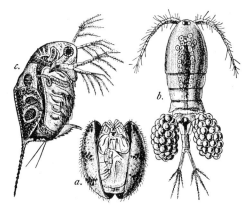

Fig. 4. CRUSTACEA
a, Cypridopsis   b, Cyclops   c, Daphnia

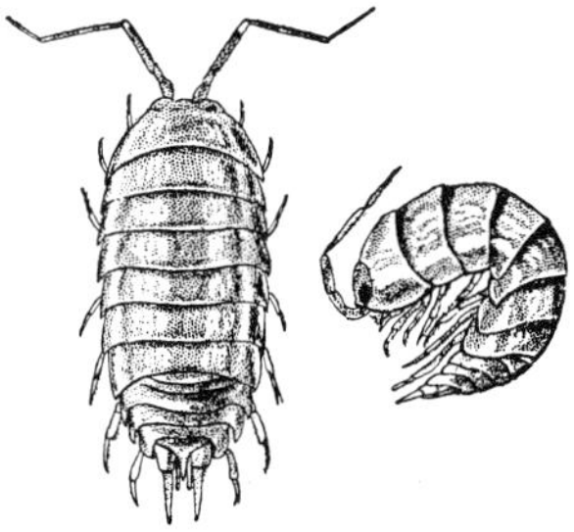

Fig. 5. A SOW-BUG (*Cylisticus convexus*) (after Sars)

Fig. 6. A MILLIPEDE

vegetable matter. They are harmless, except that occasionally they feed on growing plants.

III. Class CHILOPODA (Chi-lop'o-da).—The Chilopoda includes the centipedes. These, like the millipedes, are air-breathing and have an elongated body composed of similar segments (Fig. 7). They can be distinguished from the millipedes by the fact that each segment bears only a single pair of legs. The body is usually flattened, and the antennæ are usually long and many-jointed.

Fig. 7. A CENTIPEDE

The centipedes are predaceous, feeding on insects; they are common under stones and other objects lying on the ground. Many species are venomous. The poison glands open through the claws of the first pair of legs, which are bent forward so as to act with the mouth-parts. These creatures abound in all parts of the United States; those that are found in the North are comparatively small, and rarely, if ever, inflict serious injury to man; but the larger species, which occur in warmer regions, are said to be extremely venomous.

Formerly the millipedes and the centipedes were grouped together as the Class *Myriapoda* (Myr-i-ap'o-da); and this grouping is retained in many recent zoölogies.

IV. Class HEXAPODA (Hex-ap'o-da).— The class Hexapoda comprises the various orders of insects. The members of this class are air-breathing arthropods, with distinct head, thorax, and abdomen. They have one pair of antennæ, three pairs of legs, and usually one or two pairs of wings in the adult state (Fig. 8).

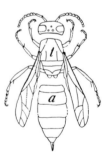

Fig. 8. AN INSECT; A WASP, WITH THE HEAD, THORAX AND ABDOMEN SEPARATED
t, thorax   *a*, abdomen

Among the more familiar examples of insects are grasshoppers, dragon-flies, butterflies, moths, and beetles. There are so many excellent popular works treating of insects that it is not necessary to dwell on this class here.

V. Class PALÆOSTRACHA (Pa-læ-os'tra-cha).—This class is composed almost entirely of extinct forms, there being living representatives of only a single order, the *Xiphosura* (Xiph-os-u'ra). And this order is nearly extinct; for of it there remains only the genus *Limulus* (Lim'u-lus), represented by only five known species.

The members of this genus are known as king-crabs or horseshoe-crabs; the former name is suggested by the great size of some of the species, the latter, by the shape of the cephalothorax (Fig. 9).

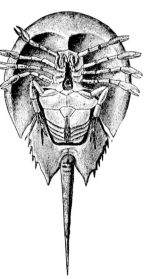

Fig. 9. A HORSESHOE-CRAB (*Limulus*) (after Leuckart)

The king-crabs are marine; they are found on our Atlantic Coast from Maine to Florida, in the West Indies, and on the eastern shores of Asia. They are found in from two to six fathoms of water on sandy and muddy shores; they burrow a short distance in the sand or mud and feed chiefly on worms. The single species of our coast is *Limulus polyphemus*.

In *Limulus* the body-segments are grouped in two regions, cephalothorax and abdomen; the abdomen is terminated by a long, strong spine. The cephalothorax bears six pairs of appendages; these correspond to the six pairs of appendages of Arachnida described later; and, as in the Arachnida, none of the appendages are jaw-like. The food is crushed by the basal part of the leg-like appendages, which are situated round the mouth. The abdomen also bears six pairs of appendages; these are plate-like and the members of each pair are united at the base.

The first pair of abdominal appendages form a nearly semicircular plate, which nearly covers the following appendages. On the posterior side of this plate are two openings, the outlets of the reproductive organs. Each of the following pairs of appendages bears a pair of gills. These are termed book-gills, each being composed of many thin plates, like the leaves of a book.

In its more general characteristics, *Limulus* is closely allied to the Arachnida, except that its respiration is aquatic. So close is this resemblance in structure to the Arachnida that many zoölogists, and among them some of those who have studied *Limulus* most carefully, regard the Xiphosura as an order of the Arachnida. A more conservative view is to regard the Palæostracha and the Arachnida as two closely allied but distinct classes.

The most familiar of the fossil representatives of the Palæostracha are the Trilobites.

VI. Class ARACHNIDA (A-rach'ni-da).—The members of this class are air-breathing arthropods, in which the head and thorax are usually grown together, forming a cephalothorax, which have four pairs of legs, and which apparently have no antennæ. The characteristics of the Arachnida are more fully discussed in the next section of this chapter.

TABLE OF CLASSES OF ARTHROPODS

The more striking of the distinguishing characters of the six classes of arthropods mentioned above can be stated in a tabular form as follows:

A. With two pairs of antennæ and at least five pairs of legs; respiration aquatic. CRUSTACEA
AA. With one pair of antennæ or apparently with none; respiration, except in *Limulus*, aerial. The number of legs varies from six to many.
  With one pair of feeler-like antennæ.
  C. With more than three pairs of legs, and without wings.
    D. With two pairs of legs on some of the body-segments. DIPLOPODA
    DD. With only one pair of legs on each segment. CHILOPODA
  CC. With only three pairs of legs, and usually with wings in the adult state. HEXAPODA
BB. Apparently without antennæ.
  C. Respiration aquatic. PALÆOSTRACHA
  CC. Respiration aerial. ARACHNIDA

II.—THE CHARACTERS OF THE ARACHNIDA

At first sight, the Arachnida appear to be distinguished from all other air-breathing arthropods by the absence of antennæ; for they have no feelers projecting forward from the head. But it has been found that a pair of nippers, the cheliceræ, with which the head is armed, correspond to the second antennæ of Crustacea. Vestiges of the first antennæ have been observed in the embryos of certain arachnids; but these appendages either disappear completely before birth or, according to the views of some writers, become consolidated to form the upper lip or rostrum. We can say, therefore, that in the Arachnida the first pair of antennæ are wanting as distinct appendages, and the second antennæ are modified so as to form prehensile organs.

Most Arachnida differ from all other air-breathing arthropods in having the segments of which the head and thorax are composed consolidated so as to form a single region, the cephalothorax; but in certain generalized forms the thoracic segments are more or less distinct. In most cases the cephalothorax and abdomen are distinct; but in the mites the entire body forms a single region.

In the scorpions and in some other forms the abdomen is divided into two portions; a broad preabdomen, and a slenderer tail-like division, the postabdomen.

Perhaps the most distinctive characteristic of the Arachnida is the fact that in this class the combined head and thorax bears only six pairs of appendages; this, however, is also true of the Palæostracha, which some writers class with the Arachnida. The nearest approach to this condition among air-breathing arthropods is found in the insects, where the head and thorax bear seven pairs of appendages.

Another striking characteristic of the Arachnida, which, however, is also possessed by the Palæostracha, is the absence of true jaws. In other arthropods one or more pairs of appendages are jaw-like in form and are used exclusively as jaws; but in the Arachnida the prey is crushed either by the prehensile antennæ alone (as in *Kœnenia*) or by these organs and other more or less leg-like appendages. The arachnids suck the blood of their victims by means of a sucking stomach; they crush their prey, but do not masticate it so as to swallow the solid parts.*

The Arachnida are air-breathing. Two forms of respiratory organs exist in this class: first, the book-lungs; and second, the tracheæ. Both of these are described in the chapter on the internal anatomy of spiders. In the mode of respiration, the Arachnida differ from the Palæostracha, with which they agree in the number of cephalothoracic appendages and in the absence of true jaws.

The reproductive organs open near the base of the abdomen, on the ventral side. In this respect the Arachnida resemble *Limulus*, the millipedes, and the Crustacea, and differ from the centipedes and insects, in which the reproductive organs open near the caudal end of the body.

The eyes are simple, that is, each eye is covered with a single cornea, not facetted like the compound eyes of insects. The number varies from two to twelve, and some species are eyeless.

The eyes vary greatly in their position. In some forms (Avicularioidea) they are borne on an ocular tubercle near the middle of the head. In this case there are two large round median eyes, one on each side of the median line; on each side of the median eyes, on the base of the tubercle, are three lateral eyes, which differ in size, shape, and appearance, both from the median eyes and from one another. This is probably the primitive arrangement of the eyes; but in most forms, the lateral eyes have wandered off laterally on to the cephalic lobes, which constitute the lateral portions of the head, and the optic tubercle is obliterated (Bernard '96).† A similar migration of the paired ocelli of insects has been pointed out by the writer (Comstock and Kochi '02). In some of the more generalized insects, the three ocelli are borne by the front; but in most insects they have wandered off from this sclerite into the epicranial suture or even on to the vertex, which is developed from the cephalic lobes of the embryo.

Of the six pairs of appendages borne by the cephalothorax, in most forms, the first two pairs are used for seizing and crushing

---

*For possible exceptions to this see accounts of the habits of the Solpugida and of the Phalangida, in the next chapter.

†The complete titles of works referred to in parentheses in the text are given in the bibliography at the end of the volume.

prey, and the last four, as legs; but there is considerable lack of uniformity in this respect. Owing to this lack of uniformity in the function of some of the appendages in different orders some writers designate the six pairs of appendages by numbers; but the older authors gave special names to the first and second pairs, and termed the last four pairs legs. The older nomenclature is so well established that it does not seem wise to attempt to change it, notwithstanding that in one order the second pair of appendages function as legs, and in another order, the third pair of appendages function as feelers.

The first pair of appendages are termed the *chelicerœ* (che-lic'e-ræ). They are situated in front of the mouth; all other appendages are postoral, except perhaps the second pair.

The second pair of appendages are the *pedipalps* (ped'i-palps). These are situated immediately behind the mouth, or one at each side of it. They are always more or less leg-like in form; and in the Microthelyphonida they function as legs.

Following the pedipalps are the four pairs of legs. In the Pedipalpida, the first pair of legs function as feelers.

Usually the basal parts of the pedipalps bear *masticatory ridges* for crushing the prey. Sometimes the crushing organ is a distinct sclerite borne by the basal segment of the pedipalp; it is then known as an *endite*. Masticatory ridges or endites are sometimes borne by the first or the first and second pairs of legs.

Fig. 10. TARSAL CLAWS OF A SPIDER

Fig. 11. A CHELA OF A PEDIPALP OF A SCORPION

The last segment of an appendage may bear claws, which resemble more or less the tarsal claws of insects. (Fig. 10.) In other cases the last segment itself is claw-like; in this case a pincer-

Spiders and Their Near Relatives

like organ is formed. Two types of pincers exist in the Arachnida. The first is known as a *chela;* an illustration of this is a great claw of scorpions. A chela is formed by the next to the last segment of an appendage being prolonged on one side so as to oppose the last segment (Fig. 11). An appendage bearing a chela is said to be *chelate*. In descriptions of chelate claws, the last segment is termed the movable claw of the chela; and the prolongation of the next to the last segment opposed to this, the fixed claw of the chela.

Another type of pincers is produced when the last segment of an appendage is claw-like and is folded back into a groove in the next to the last segment, like the blade of a pocket-knife into the handle (Fig. 12). An appendage bearing a pincer of this type may be said to be *uncate*.

Fig. 12. AN UNCATE CHELICERA

These two types of pincers are not distinct, intermediate forms being present. Such intermediate forms may be termed *semichelate;* the pedipalps of *Mastigoproctus* are illustrations of the semichelate type.

A detailed account of the structure of spiders is given in the following chapters; it is not necessary, therefore, to discuss further the anatomy of arachnids in this place.

III.—THE ORDERS OF THE ARACHNIDA

According to the classification that is most generally accepted, which excludes various groups of animals of uncertain position, the class Arachnida is divided into nine orders. All of them are represented within the area considered by this book and may be separated by the following table.

TABLE OF THE ORDERS OF THE ARACHNIDA

A. Abdomen distinctly segmented.
  B. Abdomen with a stout tail-like prolongation ending in a sting; pectines present beneath base of abdomen; pedipalps stout, chelate. (Scorpions) P. 21.   SCORPIONIDA

BB. Abdomen without a terminal sting, sometimes with a whip-like caudal appendage; pectines absent.
  C. Two or three posterior segments of the carapace free.
    D. No caudal appendage; racket organs beneath base of last pair of legs. (Solpugids) P. 32.   SOLPUGIDA
    DD. Caudal appendage present; without racket organs.
      E. Caudal appendage long, many-segmented. (Micro-whip-scorpions) P. 13.
          MICROTHELYPHONIDA
      EE. Caudal appendage short, at most three-segmented. (Family Schizomidæ) P. 16.
          PEDIPALPIDA
  CC. All segments of carapace fused together.
    D. Abdomen with a long, segmented whip; first legs long and slender. (Whip-scorpions of Family Thelyphonidæ) P. 16.   PEDIPALPIDA
    DD. Abdomen without a terminal whip.
      E. Cephalothorax with a movable hood (cucullus) in front. (Ricinulids) P. 48.   RICINULEI
      EE. Cephalothorax without movable hood in front.
        F. Abdomen narrowly joined to cephalothorax; first legs elongated. (Whip-scorpions of Family Tarantulidæ) P. 16.   PEDIPALPIDA
        FF. Abdomen broadly joined to cephalothorax.
          G. Palpi chelate. (Pseudoscorpions) P. 39.
            PSEUDOSCORPIONIDA
          GG. Palpi not chelate. (Harvestmen) P. 53.
            PHALANGIDA
AA. Abdomen unsegmented.
  B. Abdomen narrowly joined to cephalothorax; spinnerets on abdomen. (Spiders) P. 218.   ARANEIDA
  BB. Abdomen fused with the cephalothorax. (Mites and Ticks) P. 81.   ACARINA

Order MICROTHELYPHONIDA*

The Micro-Whip-scorpions

The tiny creatures constituting the order Microthelyphonida (Mi-cro-thel-y-phon'i-da) bear striking resemblance in the form

---

*This order was established under the above name by Grassi ('86). Later Thorell ('88) changed the name to *Palpigradi;* but this latter name cannot be justly accepted, although it is shorter and equally appropriate.

Spiders and Their Near Relatives

of the body to members of the Thelyphonidæ (p. 18); it was this that suggested the name of the order. They differ greatly from the whip-scorpions, however, both in size and in structure; the larger of the species found, as yet, measure less than one tenth of an inch in length, including the tail-like appendage.

The form of the body and the more general features of the appendages are shown in Fig. 13. The thorax appears to consist of only two segments when seen from above, the tergum of the segment bearing the fourth pair of appendages being a part of the carapace. On the ventral aspect of the cephalothorax there are four sternites; a larger one corresponding to the pedipalps and the first pair of legs, and three smaller ones corresponding to the second, third, and fourth pairs of legs respectively. The abdomen is distinctly segmented, and is terminated by a slender appendage consisting of about fifteen segments.

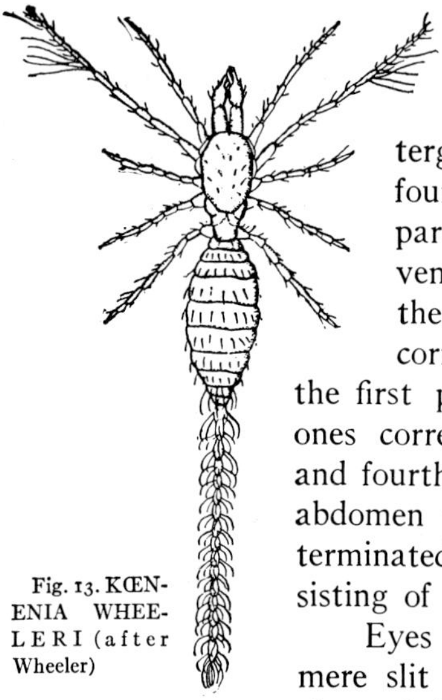

Fig. 13. KŒNENIA WHEELERI (after Wheeler)

Eyes are wanting. The mouth is a mere slit in an oral prominence, which is situated far forward so that it is partly between the bases of the cheliceræ; in other words, the mouth has not migrated so far back as in other arachnids. None of the appendages are furnished with either masticatory ridges or endites. This is the simplest oral apparatus found in the arachnida.

The cheliceræ are chelate, and are the only chelate appendages. The pedipalps are leg-like and are terminated by a pair of claws, as are the legs; or to express it differently, the pedipalps have not been modified into organs for some other function than locomotion, as is the case with the more specialized arachnids. It would be correct, therefore, to say that these creatures possess five pairs of legs; but to avoid confusion it is better to restrict the term legs to the last four pairs, as is done in describing other arachnids. Each pedipalp consists of nine segments.

The first pair of legs, using the term in the restricted sense just indicated, are the longest of the appendages, and resemble the corresponding appendages of the whip-scorpions in having the tarsi broken up into several segments, each of these legs consisting of twelve segments. The second and third pairs of legs are seven-jointed; the fourth, eight-jointed.

The respiratory organs, according to the investigation of Miss Rucker ('01), consist of three pairs of lung-sacs, which are situated in segments four, five and six of the abdomen, with their corresponding orifices on the ventral surface. These sacs are evidently evaginated through the internal blood-pressure, and invaginated by dorso-ventral muscles, a pair for each pair of sacs. When invaginated, the sacs appear like diminutive tracheæ if stretched by the retractor muscles; but if the latter become relaxed, allowing the sac to flatten dorso-ventrally and wrinkle, they appear like diminutive book-lungs.

The external reproductive organs are quite complicated; they are borne by the second and third abdominal segments.

A review of the order, which at present contains four genera belonging to a single family, was given by Roewer ('34).

Family KŒNENIIDÆ (Kœn-e-ni'i-dæ)

These tiny arachnids are widely distributed, species having been found in southern Europe, in Africa, Siam, Paraguay, Chile, Mexico, and Texas and California in the United States. In the genus *Prokœnenia* the abdomen is supplied with small, paired, ventral lung sacs on the fourth, fifth, and sixth sternites. In *Kœnenia* the ventral lung sacs are apparently absent.

The kœneniids live under stones in company with insects of the order Thysanura. The species are of a translucent white colour except the blades of the chelicerae, which have the yellow tint of thickened chitin. Excellent accounts of two of our species, *Prokœnenia (Kœnenia) wheeleri* and *Kœnenia florenciæ*, are given by Wheeler ('00) and Rucker ('01 and '03). A third species, *Prokœnenia calijornica*, was described by Silvestri ('13) from a single specimen taken in California. Additional species will probably be discovered when adequate collecting for these tiny creatures is done in our southern states.

Spiders and Their Near Relatives

Order PEDIPALPIDA*

The Whip-scorpions

These strange creatures are found in our country only in the extreme southern part, they being tropical animals; but they are distributed from the Atlantic to the Pacific. In their general form they bear some resemblance to scorpions; but they can be easily distinguished from scorpions by the form of the pedipalps, of the first pair of legs, and of the postabdomen.

The common name whip-scorpions was doubtless suggested by the slender caudal appendage of the Thelyphonidæ (Fig. 14); but it is almost as appropriate for the forms that lack this appendage, as they have the tarsi of the front legs broken up into many small segments, which gives this part of the leg a whiplash-like appearance (Fig. 18).

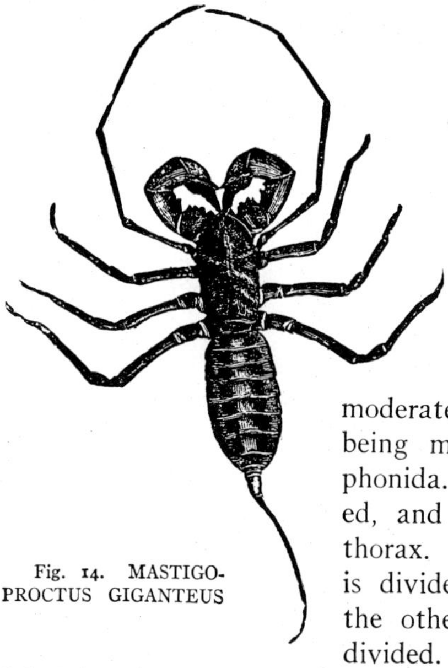

These arachnids are of moderate or large size, none of them being minute like the Microthelyphonida. The abdomen is segmented, and distinctly separate from the thorax. In one family the carapace is divided by transverse sutures; in the other two families it is not so divided. The chelicerae are two-jointed and uncate; that is the second segment is claw-like and folds back upon the end of the first segment. A remarkable feature, in some members of the order if not all, is that the chelicerae are attached to the head by a thin membrane in such a way as to be capable of being retracted into the head for a considerable distance (Laurie '94). The pedipalps are very stout and are six-jointed; they present the three types of claws, one in each of the

Fig. 14. MASTIGOPROCTUS GIGANTEUS

---
* The original form of this name was *Pedipalpi*, a family name proposed by Latreille (1806); and this form is still retained by many writers even though they rank the group as an order.

three families. The first pair of legs are elongated and are modified into feelers, the tarsus being divided into many segments, the last of which has a rounded tip instead of claws. The ambulatory organs are therefore reduced to three pairs. The respiratory organs are book-lungs; there are two pairs of these; they open on the posterior edge of the second and of the third abdominal segments.

A complete review of this order has been given recently by Werner ('35). The three families may be separated as follows:
A. Cephalothorax longer than broad, with nearly parallel sides.
   B. Carapace with two posterior segments free; caudal appendage short, at most three segmented. P. 17.
          SCHIZOMIDÆ
  BB. Carapace without transverse segmentation behind; caudal appendage long, many jointed. P. 18. THELYPHONIDÆ
AA. Cephalothorax broader than long, unsegmented, the sides strongly arched. P. 19.   TARANTULIDÆ

Family SCHIZOMIDÆ (Schiz-o'mi-dæ)

About thirty representatives of this family, previously known as the Schizonotidæ, have been described, chiefly from tropical and subtropical regions. They are regarded with great interest on account of the generalized condition of the carapace in which two of the posterior segments are present as free tergites. Eyes are wanting but certain pale areas of the cuticula probably indicate their former position. The first pair of legs is very long and slender. The palp is rather robust and often armed with characteristic spines and setæ. The caudal appendage is short, made up of one to three segments. In some males it is fused into a single rounded or elongated knob.

Fig. 15. TRITHYREUS CAMBRIDGEI (after Börner)

Specimens representing the two well known genera have been taken in the United States, but only a single species has been described up to the present time. In *Schizomus*, of which a few examples have been found in Texas, the posterior free segment of the cephalothorax is entire. In the other genus, *Trithyreus*, this segment is divided by a longitudinal suture.

A single species of this rare family has been described from the United States; this is *Trithyreus pentapeltis* (Tri-thyr'e-us pen-ta-pel'tis), which occurs in the desert regions of southern California. Only a few specimens have been taken; the largest of these measured less than one half inch in length. Figure 16 represents some of the details of structure of this species as given by Cook ('99).

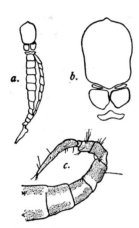

Fig. 16. TRITHYREUS PENTAPELTIS (after Cook). *a*, body without appendages; *b*, carapace more enlarged; *c*, caudal segments lateral view

Family THELYPHONIDAE (Thel-y-phon'i-dæ)

This family includes the tailed whip-scorpions, those in which the caudal end of the abdomen is furnished with a slender, many-jointed appendage (Fig. 14). In this family the carapace is not divided by transverse sutures. Eight eyes are present, two, the median eyes, near the centre of the front part of the head, and a group of three, the lateral eyes, on each side farther back. The pedipalps are stout and armed with tooth-like processes (Fig. 17). They are semichelate, the projecting process of the fifth segment being much smaller than the sixth segment which forms the other part of the chela; in the ordinary chelate type the opposite of this is the case. These pedipalps are remarkably developed for crushing the prey; they might be described as plurichelate; for in addition to the pincers formed by the fifth and sixth segments, another pair is formed by the fourth and fifth; and the second segment bears a prominent semicircular process, armed with strong teeth, which can be opposed to the third and fourth segments. The chief use of this process of the second segment is to be opposed to the corresponding process of the other

Fig. 17. THE PALPUS OF MASTIGOPROCTUS

Spiders and Their Near Relatives

pedipalp. The first segments of the pedipalps meet on the middle line on the ventral side of the head. The flexible tip of the first legs is composed of nine segments; and each of the six ambulatory legs is furnished with two tarsal claws.

This family is represented in the United States by only a single species, the giant whip-scorpion, *Mastigoproctus giganteus* (Mas-ti-go-proc′tus gi-gan-te′us). This species measures when full-grown from four to five inches in length. Figure 14 represents one less than natural size. In some parts of the South they bear the local name grampus, and are greatly feared on account of their supposed venomous powers; but it is probable that there is no foundation for this fear; for although it has been stated often that their bites are poisonous, I can find no direct evidence that this is true, and no poison glands have been found in this order.

This species burrows in sand under logs or other objects lying on the ground; it doubtless feeds on any insects that it can capture.

Family TARANTULIDÆ (Tar-an-tu′li-dæ)

This family includes the tailless whip-scorpions. These resemble the tailed whip-scorpions in the consolidated condition

Fig. 18. TARANTULA MARGINEMACULATA

of the carapace and in the possession of eight eyes. They differ in the absence of the caudal appendage and in having the abdomen joined to the thorax by a slender pedicel. The flexible tip of the first legs is very long and is composed of very many segments (Fig. 18).

This family is the Phrynidæ of some authors; it is represented in the United States by two genera, *Acanthophrynus* and *Tarantula*. These are separated as follows:

- A. Front border of the cephalothorax armed with long teeth; next to the last segment of the pedipalp with only one long spine on the upper inner margin. P. 20.
ACANTHOPHRYNUS
- AA. Front border of the cephalothorax unarmed or only denticulate; next to the last segment of the pedipalp with at least two long spines on the upper inner margin. P. 20.
TARANTULA

The genus *Acanthophrynus* includes only a single known species, *Acanthophrynus coronatus* (A-can-tho-phry′nus cor-o-na′tus); this is found in California and Mexico. This species is larger than those of the following genus, attaining a length of nearly two inches.

The genus *Tarantula* (Ta-ran′tu-la) is represented in the United States by three species. We have here an unfortunate instance in which the technical and popular uses of a name are very different. For although, according to the law of priority the generic name *Tarantula* must be applied to these animals, the popular name tarantula is irrevocably applied, in this country at least, to the large four-lunged spiders of the South and the Southwest.

Four species of *Tarantula* are known; these are separated by Kræpelin as follows:

- A. With two short spines between the longest spines of the tibia of the pedipalps. *T. fusimana*
- AA. With only one short spine between the longest spines of the tibia of the pedipalps.
  - B. The intermediate spine between the two principal spines of the pedipalps considerably longer than the spine immediately preceding the first of the two principal spines. *T. whitei*
  - BB. The intermediate spine shorter than the one preceding the first of the principal spines.
    - C. The principal spine on the upper edge of the next to the last segment of the pedipalps preceded by a distinct spine (Fig. 19). *T. palmata*

CC. The above-mentioned spine preceded by a short tubercle (Fig. 20). *T. marginemaculata*

Of these species *Tarantula palmata* alone has not been found in the United States. I found *T. marginemaculata* (Fig. 18) not uncommon near Miama, Fla., under logs. It is a very active creature.

According to several observers, the eggs of members of this genus are carried in a sac formed of a dark brown transparent material, containing some threads, and attached to the ventral surface of the abdomen (Laurie '94). It is supposed that the substance of which the egg-sac is composed exudes from openings in the first abdominal segment (Bernard '95); but our knowledge of the spinning glands of these animals is very incomplete.

Fig. 19

Fig. 20

Order SCORPIONIDA

The Scorpions

The order Scorpionida (Scor-pi-on'i-da) includes only the scorpions. Although these creatures do not live in the North, they have been pictured so often that their form is well-known. Their most striking features are the large size of the pedipalps which are furnished with very stout chelæ, and the division of the abdomen into two portions: a broad preabdomen, consisting of seven segments; and a slenderer tail-like division, the post-abdomen or cauda, consisting of five segments. At the end of the postabdomen there is a large poison-sting, which appears like a segment (Fig. 21).

The cephalothorax is compact and unsegmented; the abdomen is broadly joined to the thorax; the cheliceræ are chelate; the coxæ of the pedipalps are fitted for crushing the prey, which is seized by the large chelæ; the remaining four pairs of cephalothoracic appendages are fitted for

Fig. 21. A SCORPION

Spiders and Their Near Relatives

walking; they are seven-jointed, the tarsus consisting of three segments, and the patella being wanting; the legs are not chelate, but are furnished with tarsal claws. The coxæ of the first two pairs of ambulatory legs bear each an endite, which is directed toward the mouth, and doubtless helps hold the prey opposite the mouth (Fig. 22).

Fig. 22. VENTRAL ASPECT OF THE CEPHALOTHORAX AND TWO, THE THIRD AND FOURTH, ABDOMINAL SEGMENTS OF CENTRURUS GRACILIS

The cephalothorax bears a pair of eyes near the middle line, the median eyes, and on each side near the cephalo-lateral margin a group of from two to five, the lateral eyes. A few scorpions are blind.

Full-grown scorpions possess a pair of comb-like organs, the *pectines* (pec'ti-nes), on the lower side of the second abdominal segment (Fig. 22). The function of these organs is not yet known, but it has been suggested that it is tactile. "Pocock noticed that a scorpion which had walked over a portion of a cockroach far enough for the pectines to come in contact with it immediately backed and ate it."

Scorpions breathe by means of book-lungs, of which there are four pairs, opening on the lower side of the third to the sixth abdominal segments.

The sexes of scorpions differ in that the male has broader pincers and a longer postabdomen. Scorpions do not lay eggs, the young being developed within the mother; after the birth of the young, the mother carries them about with her for some time, attached by their pincers to all portions of her body.

Scorpions live in warm countries. They are common in the southern portion of the United States, but are not found in the North. They are nocturnal, remaining concealed during the day, but leaving their hiding places at dusk. When they run, the pedipalps are carried horizontally in front, and are used partly as feelers and partly as raptorial organs; and the postabdomen is bent upward over the back by some, others drag it along behind. They feed upon spiders and large insects, which they seize with the large chelæ of their pedipalps, and sting to death with their caudal poison-sting.

Scorpions are shy animals; when disturbed they attempt to run away and hide and do not sting unless molested.

The sting of a scorpion rarely if ever proves fatal to man, although the larger species, which occur in the Tropics, produce serious wounds. It is said that the best remedy for the sting of a scorpion is ammonia applied externally, and also administered in small doses internally.

Scorpions attain maturity slowly, Fabre ('07) who studied them in confinement, came to the conclusion that it required five years for them to reach their full size.

Fig. 23. CARAPACE OF A SCORPION

In the classification of scorpions the following special terms are used:

Upon the carapace, the dorsal covering of the cephalothorax, there are often more or less distinct keels; these are distinguished as the fore, middle, and hind, median and lateral keels respectively; these keels are connected with each other in various ways (Fig. 23). That part of the fore median keel which is above a median eye is termed the *superciliary ridge*.

Upon the postabdomen, or cauda, there are often two pairs of keels upon the dorsal aspect and two pairs on the ventral aspect of some or of all of the segments. The two pairs of keels nearest the middle line of the body are designated as the *dorsal median keels* and the *ventral median keels* respectively. Outside of these are the *dorsal lateral keels* and the *ventral lateral keels*. In addition to these eight principal keels *accessory keels* may be present.

Fig. 24. THE CHELA OF A SCORPION
ff, fixed finger
mf, movable finger

The large basal portion of the chela of a pedipalp is termed *the hand* (Fig. 24); and the apposed portions are distinguished as the *fixed finger* and the *movable finger* respectively, the movable finger being the last segment of the pedipalp.

Spiders and Their Near Relatives

The sclerite on the middle line of the body from which the combs arise is termed the *basal piece of the combs* (Fig. 25); the series of sclerites that form the front margin of a comb constitute the *marginal area;* next to these there is a series of sclerites constituting the *middle area;* the small sclerites between the middle area and the teeth are the *fulcra;* and the long slender appendages of the comb are the *teeth.*

Fig. 25. A COMB OF A SCORPION

The characters presented by the tarsi of the legs, and especially of the last pair of legs, are much used, as, for example, the presence or absence of tarsal spurs and the number of these spurs when present. Sometimes the last segment of a tarsus is prolonged above the claws forming a *dorsal lobe* (Fig. 26); and sometimes this segment is prolonged into a lobe on each side,

Fig. 26. TARSUS OF BUTHUS

the *lateral lobes* (Fig. 27); and there is often a more or less claw-like *empodium* below and between the claws (Fig. 26).

About a score of species of scorpions are known to occur in the United States; besides these, other Mexican species may be found in southern Texas. A general work on the scorpions of the world was published by Kræpelin in 1899 and a synopsis of our species was published by Banks in 1900. An interesting work on the habits of scorpions is that of Fabre in the ninth series of his *Souvenirs Entomologiques* ('07). Of more recent date a comprehensive review of the order has been published by Werner ('34).

Fig. 27. TARSUS OF PANDINUS

Five of the six families recognized by Werner are found in the United States. They may be separated by the following table.

TABLE OF FAMILIES OF SCORPIONS

A. Only one spur at the base of the last tarsal segment of the last pair of legs, and this is on the outside.
   B. With a spine beneath the sting. P. 28.   DIPLOCENTRIDÆ
   BB. Without a spine beneath the sting. P. 28.   SCORPIONIDÆ

AA. One or two spurs on each side at the base of the last tarsal segment of the last pair of legs.
  B. From three to five lateral eyes on each side.
    C. Sternum triangular (Fig. 22); usually a spine under the sting. P. 25. BUTHIDÆ
    CC. The lateral margins of the sternum nearly parallel (Fig. 34); sternum usually broader than long; no spine under the sting. P. 29. VEJOVIDÆ
  BB. Only two lateral eyes on each side. P. 29. CHACTIDÆ

Family BUTHIDÆ (Bu'thi-dæ)

The members of this family are most easily recognized by the form of the sternum, which is triangular in outline (Fig. 22). There are one or two spurs on each side at the base of the last tarsal segment of the last pair of legs. There are from three to five lateral eyes on each side. The hand of the chelæ is rounded and the fingers are long. The last segment of the tarsus of the legs is not terminated by lateral lobes. There is usually a spine under the sting.

This is a large family, containing more than 150 known species, representing 18 genera. Nearly one half of the species of scorpions occurring in our fauna belong to it. These represent four genera, which can be separated as follows:

A. A distinct tarsal spur at the distal end of the first tarsal segment of the third and fourth legs (Subfamily Buthinæ). P. 26. UROPLECTES
AA. No tarsal spur at the distal end of the first tarsal segment of the last pair of legs (Subfamily Centrurinæ).
  B. The oblique rows of teeth on the edge of the fingers of the chelæ have on each side a parallel row of minute teeth (Fig. 31). P. 27. CENTRUROIDES
  BB. The oblique rows of teeth on the edge of the finger of the chelæ not accompanied by parallel rows of minute teeth.
    C. The ends of the oblique rows of teeth on the fingers of the chelæ overlapping (Fig. 30). P. 27. TITYUS
    CC. The ends of the oblique rows of teeth on the fingers of the chelæ not overlapping but often connected in one direct line (Fig. 28). P. 26. ISOMETRUS

Spiders and Their Near Relatives

Genus UROPLECTES (U-ro-plec′tes)

This genus is the only representative in our fauna of the subfamily Buthinæ, which is characterized by the presence of a tarsal spur at the distal end of the first tarsal segment of the third and fourth legs. Of this genus the following species is the only one yet found within the limits of the United States.

The Mexican Uroplectes, *Uroplectes mexicanus* (U. mex-i-ca′nus).—This is a pale species. There is no spine under the sting; the teeth on the finger of the palpus are in many oblique rows, with stouter teeth at the end of each and to one side; there are from thirty to thirty-five teeth in the combs; and the keels on the under side of the last caudal segment are very strongly toothed. This species has been found in Texas and in California.

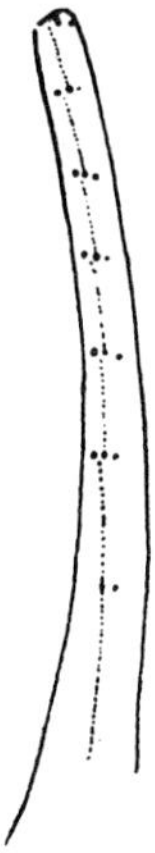

Genus ISOMETRUS (I-som′e-trus)

In this genus, there are only a few oblique rows of teeth on the edge of the fingers of the chelæ and these do not overlap (Fig. 28); there is a large spine under the sting (Fig. 29); and the abdomen has a single keel above. The following is our only species.

Fig. 28. FINGER OF ISOMETRUS MACULATUS

The spotted Isometrus, *Isometrus maculatus* (I. mac-u-la′tus). — This is a dirty yellow species marbled and flecked with black. The body is thin and slender. In the female the postabdomen is usually about as long as the rest of the body; in the male, it is often twice as long. The hand is long and thin, thinner than the tibia of the pedipalp; the finger is from one and a half to two times as long as the hand. The combs have from seventeen to nineteen teeth. The female grows to nearly two inches in length; the males, to nearly three inches.

Fig. 29. STING OF ISOMETRUS MACULATUS

This species is distributed throughout the tropical and subtropical regions of the world; in this country it is found in southern Florida, and California, and, probably, in the intermediate regions.

Genus TITYUS (Tit'y-us)

There are many overlapping oblique rows of teeth on the edge of the fingers of the chelae (Fig. 30). Two species have been found in the United States.

Tityus floridanus (T. flor-i-da'nus).—This is a reddish brown species nearly three inches in length, found at Key West. The sting is long and curved; the tooth beneath it is acute, but short.

Tityus tenuimanus (T. ten-u-i-man'us).— This is a yellowish brown species, about two inches in length, found in California. There is no spine beneath the sting.

Genus CENTRUROIDES (Cen-tru-roi'des)

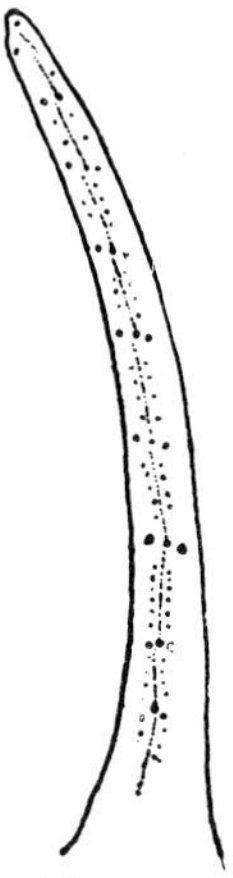

This name is now used for the familiar genus *Centrurus* which is unavailable. The oblique rows of teeth on the edge of the fingers of the chelae have on each side a parallel row of minute teeth (Fig. 31). There is a tooth on the lower margin of the fixed finger of the chelicera, and a spine under the sting may be either present or wanting. The following key will aid in the separation of most of our common species.

Fig. 30. FINGER OF TITYUS

Fig. 31. FINGER OF CENTRURUS

- A. Body nearly uniform dark reddish brown or black. Occurs in Florida and Texas. *Centruroides gracilis*
- AA. Body pale yellowish brown, the abdomen usually marked with longitudinal black stripes.
  - B. No spine beneath the sting. Occurs in California. *Centruroides exilicaudata*
  - BB. At least a small spine or tubercle beneath the sting.
    - C. Body uniform yellowish brown, unstriped. Occurs in Arizona. *Centruroides sculpturatus*
    - CC. Body with two dark stripes on a paler ground.
      - D. Triangle of median and lateral eyes very much darker than rest of carapace. Southern States, west into New Mexico. *Centruroides vittatus*
      - DD. Triangle not noticeably darker; appendages mottled. Occurs in Florida. *Centruroides hentzi*

Family DIPLOCENTRIDÆ (Dip-lo-cen′tri-dæ)

In this and the following family there is only one spur at the base of the last tarsal segment of the last pair of legs and this is on the outside. The lateral margins of the sternum are nearly parallel. There are no spurs at the end of the first tarsal segment of the third and fourth legs. There are usually three lateral eyes on each side but in the genus *Oeclus* there are only two. The sting (Fig. 32) is always provided with a blunt tubercle beneath. This family, which by many authors is regarded only as a subfamily of the Scorpionidæ, is found only in the Americas. Only one genus occurs in our fauna.

Genus DIPLOCENTRUS (Dip-lo-cen′trus)

Three species belonging to this genus have been reported from within the limits of the United States.

Diplocentrus keyserlingi (D. keys-er-ling′i).—In this and the following species the lateral lobes of the tarsi are rounded, the ventral spines being in a curved row. There are ten to twenty teeth in the comb. It is a graceful species, light brown in color, and is not uncommon in Texas.

Fig. 32. STING OF DIPLOCENTRUS WHITEI

Diplocentrus whitei (D. whi′te-i).—This species, which doubtfully occurs within our limits, is nearly black in color. It is larger and rougher than *keyserlingi*. It has been reported from Texas and California.

Diplocentrus lesueurii (D. le-su-eu′ri-i).—This species was described from Florida and is known to occur in the West Indies. It differs from the two preceding ones in having only six to eight teeth in the comb. The lateral lobes of the tarsi are subtruncate, the ventral spines being in a straight row.

Family SCORPIONIDÆ (Scor-pi-on′i-dæ)

This family, which is widespread throughout the tropics of the world and contains the largest members of the order, agrees closely in most respects with the preceding one. The principal difference is in the sting which never has a spine or tubercle beneath it.

Only one genus occurs in our fauna.

Genus OPISTHACANTHUS (O-pis-tha-can'thus)

Only one species of this genus occurs in our fauna.

Opisthacanthus elatus (O. e-la'tus).— The cephalothorax is deeply emarginate on the anterior margin; the postabdomen is small; the chelæ large. Comb with from four to fourteen teeth. The adult is sometimes three and one half inches in length. It is a West Indian species, which is found in southern Florida.

Family CHACTIDÆ (Chac'ti-dæ)

These scorpions can be distinguished from all others found in the United States by the presence of only two lateral eyes on each side. The family is represented in this country by a single genus.

Genus BROTEAS (Bro'te-as)

In this genus, the last segment of the tarsus bears two rows of bristles on the under side (Fig. 33). Only one species has been found in the United States.

Broteas alleni (B. al'le-ni).— This is a small species, the length of the body and tail together of the female being but little more than one inch and of the male, about one and one half inch. The dorsum is beautifully polished and not at all tuberculate. The palpi are of medium size. The tail is short; in the female, it is not so long as the body. This species is found in southern California.

Fig. 33. LAST SEGMENT OF TARSUS OF BROTEAS

Family VEJOVIDÆ (Ve-jov'i-dæ)

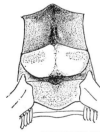

Fig. 34. STERNUM OF VEJOVIS

There is one spur on each side at the base of the last tarsal segment of the last pair of legs. There are three lateral eyes on each side. The sternum is usually broader than long, with a deep median furrow and with the lateral sides nearly parallel (Fig. 34). There is no spine under the sting. There are four genera of this family found in the United States:

A. Middle area of the comb either indistinct or composed of not more than six pieces.

Spiders and Their Near Relatives

    B. Lower margin of the movable finger of the cheliceræ with about five teeth. The sting normal. P. 30.
                                                                                    Uroctonus
  BB. Movable finger of the chelicerae with not more than three teeth. Sting of male swollen at base. (Fig. 36). P. 30.
                                                                                     Anuroctonus
AA. Middle area of the comb divided into at least eight small pieces.
    B. No tooth on the lower margin of the movable finger of the chelicerae. Penultimate tarsal segment of the three front pairs of legs not strikingly clothed with bristles above. P. 31.   Vejovis
  BB. A strong, brown tooth on the lower margin near the tip of the movable finger of the chelicerae. Penultimate tarsal segment of the three front pairs of legs with a comb of long bristles. P. 32.   Hadrurus

Genus UROCTONUS (U-roc-to′nus)

The lower margin of the movable finger of the chelicerae is armed with about five teeth. The edge of the fingers of the chelae bear one long row of tubercles flanked by isolated tubercles (Fig. 35). The sting of both sexes is of the usual form. The lung-slits are nearly oval.

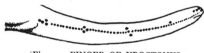

[Fig. 35 FINGER OF UROCTONUS

Uroctonus mordax.—This is the only known species of this genus. It is found on the Pacific Coast. It attains a length of nearly two and one half inches.

Genus ANUROCTONUS (A-nu-roc-to′nus)

The movable finger of the chelicerae is furnished with not more than three teeth. The lung-slits are elongate. The sting of the male is swollen at its base (Fig. 36).

Anuroctonus phæodactylus (A. phæ-o-dac′ty-lus).—This is the only known species of the genus. It has been found in Utah and in California. It attains a length of nearly two and one half inches.

Fig. 36. STING OF ANUROCTONUS

Genus VEJOVIS (Ve'jo-vis)

There are no teeth on the lower edge of the claw of the cheliceræ. The edge of the fingers of the chelæ bears one long row of tubercles flanked by isolated tubercles. The last segment of the tarsus of the last pair of legs is furnished with a distinct empodium and with a row of short tubercles on the lower side. The middle area of the comb is composed of from eight to many rounded pieces. The fulcra are also rounded. The next to the last segment of the tarsus of the first three pairs of legs is not furnished with a comb of bristles.

Most of our common species of this genus can be separated by the following chart, which was taken from Banks ('oo).

- A. Hand of the chelæ with distinct ridges or keels and more or less granulate.
  - B. On the under side of the first caudal segment the median keels are distinct and sharp, although fine, the sting is very slender and long. Occurs in the Far West.
    Vejovis punctipalpi
  - BB. There are no median keels on the under side of the first caudal segment or they are extremely indistinct, the sting is of ordinary length.
    - C. The hand is strongly keeled, there are no keels on the hind tibiæ, the colour is yellowish or greenish. Occurs from Nebraska to Idaho and in Utah and Nevada. *Vejovis boreus*
    - CC. Hand less sharply keeled, hind tibiæ with very plain keels, colour uniform reddish brown, legs paler. Occurs in Texas. *Vejovis mexicanus*
- AA. Hand smaller, without keels, the corners rounded and smooth.
  - B. On the under side of the first caudal segment the median keels are plain, but *not* indicated by black lines, hand very slender, the fingers longer than in *Vejovis spinigerus*, colour uniform yellowish. Occurs in New Mexico.
    Vejovis flavus
  - BB. No median keels on the under side of the first caudal segment or at most only indicated by black lines.
    - C. Under side of cauda not very dark, the keels all indicated

by black lines, palpi usually yellowish. Occurs in Texas, Arizona, and California. *Vejovis spinigerus*

CC. Under side of cauda, with the whole of the dorsum and the palpi dark reddish brown; no black lines indicating the keels on the cauda, smaller than the preceding species. Occurs from South Carolina to Texas.
Vejovis carolinus

Genus HADRURUS (Had-ru'rus)

There is a strong brown tooth on the lower margin of the movable finger of the cheliceræ. The last segment of the tarsus with a large empodium. The penultimate tarsal segment of the three front pairs of legs with a comb of long bristles.

Hadrurus hirsutus (H. hir-su'tus).— This is a very large and hairy species found in the Southwest. The penultimate tarsal segment of the first three pairs of legs is furnished with long hairs on the back.

Order SOLPUGIDA*

The Solpugids

The order Solpugida (Sol-pu'gi-da) includes a moderately large group of curious arachnids which are primitive in appearance. Although they cannot be considered rare within the regions they inhabit, they are not often encountered by the ordinary collector. They are chiefly nocturnal and hide away during the daytime, which may account in part for their unfamiliarity, but some are diurnal.

Figure 37 will serve to illustrate the form of these strange arachnids. Their most striking features are the enormous size of the chelicerae, and the segmented condition of the thorax. In the latter respect they resemble the Microthelyphonida and the Shizonotidæ already described; the segment bearing the second pair of legs (the fourth pair of appendages) being more or less distinct, and the segments bearing the last two pairs of legs

---

* The name *Solpugides* was proposed by Leach in 1815 as a family name; it is now spelled Solpugida for the sake of uniformity with other ordinal names of Arachnida. The name Solifugæ, which is used by some writers was proposed by Sundervall in 1833; there seems to be no good reason for substituting this for the older name. The name Galeodea, proposed for this order by Kirby and Spence in 1826, is used by some writers, evidently because the generic name Galeodes is older than the generic name Solpuga; but as the name Solpuga is still retained for one of the genera of this order, there appears no good reason for adopting the latter ordinal name.

resembling the abdominal segments in the degree of their distinctness. This is doubtless a very generalized feature.

The abdomen consists of ten segments, all of which are free. The beak, which bears the mouth at its tip, projects forward from between the basal segments of the pedipalps. There is a pair of eyes near the middle line on the front part of the head, and one or two vestigial eyes on each side of the head (Bernard '96). The chelicerae are two-jointed and chelate; compared to the size of the body, they are larger than in any other arachnids; and they are the only appendages fitted for crushing the prey. "Observers relate that, in order to bring the beak up to the wound in its prey, the animals work the chelicerae with a sawing motion, holding tight with one to drive the other deeper in." A peculiarity of these chelicerae is that the second segment is articulated to the lower side of the first segment, so that the pincers open and shut dorsoventrally. In some forms each chelicera bears on its upper side a remarkable sensory appendage, which is called the *flagellum;* this varies in shape greatly in the different genera. The function of the flagellum is not clearly understood, but recent studies indicate that it probably plays a role during mating. It is lacking in the Eremobatidae. The pedipalps are leg-like in form, but without claws. They bear no masticatory ridges; and it is evident that their chief function is that of feelers. At the tip of each pedipalp there is an invaginated sense organ (Bernard '96); these were formerly supposed to be suckers for clinging to objects. The first pair of legs have lost their locomotor function and doubtless resemble the pedipalps in function; their

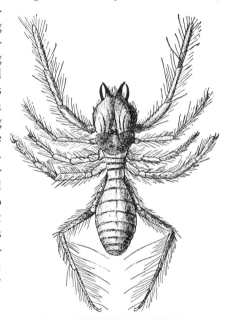

Fig. 37. EREMOBATES CINEREA
(after Putnam)

claws are vestigial. It might be said, therefore, that the solpugids have two pairs of pedipalps and only three pairs of legs. No other arachnids have two pairs of feelers. It will be remembered that in the Pedipalpida, where the first pair of legs are developed into feelers, the pedipalps are fitted for crushing the prey. The remaining three pairs of legs are fitted for locomotion, and bear tarsal claws. The trochanters of the last pair of legs consist each of two segments.

One of the most characteristic features of the solpugids are the racquet-organs, which are borne by the last pair of legs. These are T-shaped or racquet-shaped sense organs (Fig. 38), of which there are five on each hind leg, two on the coxa, two on the trochanter, and one on the femur. They have been described in detail by Bernard ('96).

Fig. 38.
RACQUET-ORGANS

The respiratory organs are tracheæ. There is a pair of spiracles behind the coxæ of the first pair of ambulatory legs; and the second and the third abdominal segments each bear a pair of spiracles on the ventral side; these are covered by opercula. Sometimes the fourth abdominal segment bears a single, median spiracle.

The opening of the reproductive organs is on the first abdominal segment, and is covered by a pair of opercula, which are regarded as vestiges of abdominal limbs.

Most solpugids spend the day under stones or other rubbish or in holes in the ground, and come forth at night to seek their prey; but some species are diurnal. They occur chiefly in desert regions, but sometimes they enter houses. They feed on insects, and it is said that they will attack and "devour" small vertebrates such as lizards (Hutton '43). Captain Hutton states distinctly that the Galeodes observed by him consumed an entire lizard except the jaws and part of the skin. Other instances in which solpugids are supposed to have eaten their prey are given by the Rev. J. J. Wood, in his "Natural History Illustrated," and quoted by Murray ('77). Still it is believed that the solpugids take only liquid food, which they suck from the bodies of their victims. (Bernard '96, p. 357.) Professor Cook reports that in southern California they enter hives and capture honey-bees, both workers and drones.

Pocock ('98) states that the females excavate subterranean burrows for the protection of themselves and their young. The process has been observed in the case of an Indian species. Choosing a suitable spot, the female proceeded to cut away the earth in a circle with her chelicerae, then kicked away the loosened fragments with her legs, or scraping them together into a heap with the pedipalps, pushed the pile by main force from the entrance of the burrow. At its opposite end, the eggs, about fifty in number and resembling a mustard seed in size and shape, were laid; they hatched about a fortnight afterward. For three weeks the young showed no sign of movement. They then moulted for the first time and started to crawl about on their own account, little copies in miniature of their mother, who mounted guard at the entrance and resolutely repelled all intruders, snapping without hesitation at every object thrust into the burrow.

The solpugids are exceedingly agile; on this account they have been called wind-scorpions, a name translated from the Arabic. One observer compares them to a piece of thistledown driven before the wind. "Often when going at full speed, in search for food, they will stop abruptly and begin hunting and feeling around a small spot, irresistibly calling to mind the behaviour of a dog checked in mid-course by the scent of game." (Pocock, '98.)

The Indian species referred to above is an expert climber, and has been seen to ascend trees to some height above the ground in search of prey. And it is also stated that in Egypt it is no uncommon thing to see *Galeodes arabs* climbing on to a table to get at the flies. To capture such quick and wary insects the solpugid adopts the tactics of the hunting-spider; instead of making a furious dart, as it would if the prey in sight were a beetle, it proceeds to stalk the flies in the most wary fashion, creeping toward them with such slowness and stealth that the movements of the legs are almost imperceptible, yet all the while drawing gradually nearer and nearer; then like a flash of light the intervening space is traversed, and the insect struck down and captured. (Mr. A. Carter quoted by Mr. Pocock '98.)

The solpugids are commonly believed to be venomous; but those who have studied them most carefully do not think that this is so. No poison glands have been found, and observers have allowed themselves to be bitten by solpugids without suffering anything worse than a passing pain from the wound.

Among the more important works on the structure of these animals are those of Dufour ('62) and Bernard ('96); and a general work on their classification was published by Simon ('79). The species of the United States were studied by Putnam ('83), and an analytical synopsis of them is given by Banks ('00). Of more recent date a comprehensive report has been published by Roewer ('34) in which the order is treated from a biological and systematic viewpoint for the whole world. Although the species from the United States are still imperfectly known, more than twenty-five names have been given to our solpugids, some of which are probably synonymous. Other species probably remain to be discovered in the southwestern part of our country.

Solpugids are most often encountered in the southern and western states and abound in the semi-arid expanses of our southwest. A few species, however, occur farther north, in the states of Montana, Idaho, and Washington, and even in the western provinces of Canada. Roewer recognizes ten families for the world of which only two are found in, and are peculiar to the Americas. They can be separated by the following chart.

A. Second and third tarsi with a small, dorsal, terminal spine in front of the paired claws; anterior margin of the cephalothorax subtruncate. P. 36.   EREMOBATIDÆ
AA. Second and third tarsi lacking such a spine; anterior margin of the cephalothorax subconical. P. 38. AMMOTRECHIDÆ

Family EREMOBATIDÆ (Er-e-mo-ba'ti-dæ)

The front margin of the cephalothorax is subtruncate, the principal tergite (propeltidium) being proportionately much broader than in the ammotrechids. The legs are relatively long and slender, the first one resembling the palpus in form but smaller and more slender, armed at the end with two very small claws which are often difficult to see. The pedipalp has the tarsus immovably joined to the metatarsus and is often provided with cylindrical, filiform, setiform, or otherwise differentiated setæ. The legs are normal, none of them specialized for digging into the earth. The first three legs have the tarsi consisting of a single segment and the fourth tarsus is generally three-jointed. The only exception is *Eremobax magnus* in which all the tarsi are one-jointed and the first one apparently lacks terminal claws. On the dorsal surface of the second and third tarsi just in front of the claws is present a stout spur. The tarsi are set with ventral spines which are distinctive in number and arrangement for each genus.

The cheliceræ are large and provided above with stout setæ, particularly in the males. Both fingers of the chelicera in the female are set with stout teeth. The fixed finger usually bears three anterior teeth, the most distal one commonly the largest. In the male the fixed finger is a slender keel or spur which completely lacks teeth and which is somewhat variable in form among the species. No flagellum is present, but various types of setæ are situated on the inner margin of the fixed finger and are collectively referred to as the "Flagellum complex." The setæ are usually simple, but special types are sometimes present.

The anal segment of the abdomen is normal, flattened, the longitudinal anal opening placed vertically. On the caudal margin of the fifth ventral sternite of the males are specialized setæ (ctenidia) which are distinctive for the species. They are rarely absent in the males and quite as rarely present in the females. The genital sternite of the female is sclerotized and provides good specific characters.

Most of our solpugids belong to this family which has its headquarters in the southwestern United States and in Mexico. The majority of the species are average in size, about three fourths of an inch in length, but some are much larger, attaining a total length of nearly two inches. Roewer divided this family into two subfamilies on the basis of the segmentation of the fourth tarsus.

A. All the tarsi one-jointed. P. 37.   EREMOHAXINÆ
AA. The first three tarsi one-jointed, but the fourth tarsus three-jointed. P. 38.   EREMOBATINÆ

Subfamily EREMOHAXINÆ (Er-e-mo-hax-i'næ)

Only one species of this subfamily is known, *Eremobax magnus*, from southern Texas. Adult females are one and one half inches long; the males somewhat smaller.

Subfamily EREMOBATINÆ (Er-e-mo-ba-ti′næ)

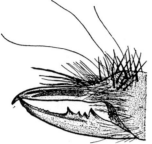

Fig. 39. CHELICERA OF AN EREMOBATID

About twenty species of this subfamily have been described from the United States. Representatives of eight of the ten known genera, as recognized by Roewer ('34), are found in our fauna. The genera are based on the spinal armature of the metatarsi and tarsi of the walking legs. Students are referred to that author for keys to the species and descriptive data.

Family AMMOTRECHIDÆ (Am-mo-trech′i-dæ)

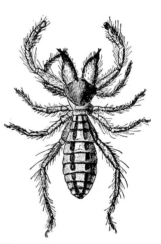

Fig. 40. AMMOTRECHONA CUBÆ (after Putnam)

The carapace is subconical or broadly rounded in front. The moderately long legs are normal, not modified for digging. The first metatarsus is unarmed beneath. The first tarsus is one-jointed, the others with one or more than one joint. The first tarsus is armed with a single spur representing the modified claws. The basal joint of the fourth tarsus is longest, slightly exceeding in length the two terminal joints taken together. The fixed finger of the chelicera is not differentiated in the males, being provided with a row of teeth in both sexes, essentially like the movable finger. The flagellum of the males is an awn-shaped process, immovably attached to the inner side of the chelicera at the base. The anal tubercle of the abdomen is flattened, and the longitudinal anal opening is vertical or terminal in position.

This family, which is found only in the New World, is distributed from the southern limits of the United States to southern South America. Only four species have been reported from north of Mexico. They all belong to the typical subfamily Ammotrechinæ and are characterized by having the first three tarsi one-jointed and the fourth tarsus three-jointed. *Ammotrecha californica*, described from California, and *Ammotrecha peninsulana*, recorded from Arizona, presumably belong in the genus *Ammotrechula*. The two eastern species, belonging in distinct genera, may be separated by the following chart.

A. Distal joint of fourth tarsus with one pair of ventral spines; carapace pale yellow or brown; dorsum of abdomen with a pair of narrow longitudinal dark bands. Found in Cuba and Florida. *Ammotrechona cubæ*

AA. Distal joint of fourth tarsus with two pairs of ventral spines; carapace dark brown; abdomen with a single broad brown stripe. Found in Texas. *Ammotrechula texana*

Order PSEUDOSCORPIONIDA

The Pseudoscorpions

The pseudoscorpions (Fig. 41) are small arachnids which resemble scorpions in the form of their pedipalps and of their body, except that the hind part of the abdomen is not narrow, as is the postabdomen of scorpions, and they have no caudal sting.

The body is flattened, which enables these creatures to live in narrow spaces, as beneath the bark of trees, between the leaves of books, and between boards in buildings. The cephalothorax is often smooth but

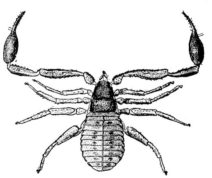

Fig. 41. PSEUDOSCORPION

may be crossed by one or two transverse furrows. The portion of the carapace lying in front of the eyes is referred to as the "cucullus." In some forms (Cheridiidæ and Garypidæ) it is narrowed and elongated; in others it is not especially conspicuous. This structure should not be confused with the cucullus or hood of the Order Ricinulei in which group it is a hinged, movable plate. There is no evidence that median eyes were ever present in the pseudoscorpions. When eyes are present, they are situated near the front of the carapace on each lateral margin where one or two eyes are found on each side. Four eyes are apparently the primitive number.

On the ventral side of the cephalothorax, the coxæ of the appendages ordinarily meet on the middle line so that there are almost no traces of a true sternum. In the genus *Sternophora* the coxæ are separated at their bases and the enclosed sternite is termed the pseudosternum.

The beak is well developed and consists of a labrum and labium (lophognath).

The cheliceræ are complexly developed structures which serve as grasping, spinning, cleansing, and sensory organs. They are two-jointed and chelate and are essentially horizontal in position, the movable finger forming the lateral member of the pincers (Fig. 43). The chelicerae vary considerably in size among the families. "The chelicerae are specialized for the performance of at least three, and probably four, functions and are correspondingly complexly developed. In the first place, they are grasping or holding organs and are commonly employed in holding the prey while transporting or feeding upon it. They are also utilized in transporting the grains of sand or bits of other material used in nest building. In correlation with this function, they are chelate or pincer-like and are provided with a more or less specialized armature of teeth. Secondly, they function as spinning organs and have generally a more or less complex spinneret (galea) developed on the tip of the movable finger. Thirdly, they are cleaning organs which are continually employed in life in grooming the chelæ of the pedipalps, which are the seat of the most important sensory structures possessed by these animals. Peculiar and highly characteristic combs or 'serrulæ' are developed in this last connection. Finally, the chelicerae are apparently

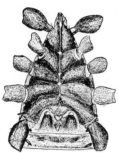

Fig. 42. VENTRAL ASPECT OF CEPHALOTHORAX AND OF TWO ABDOMINAL SEGMENTS

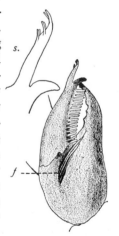

Fig. 43. CHELICERA OF CHELIFER
s, spinneret more enlarged   *f*, flagellum

Spiders and Their Near Relatives

the seat of certain important sensory functions, as witnessed by the development of the peculiar flagellum and the well-developed and ordinally constant lyrifissures." (Chamberlin, '31, page 62.)

The spinneret, which only in rare cases is seemingly absent, is situated on the outer side of the movable finger near the tip. Two main types are present. In some forms (Chthoniidæ, etc.) the spinneret is simply a slight eminence or tubercle (Fig. 44); in other groups it is prolonged into a slender, more or less branched appendage (galea) (Fig. 43) which is subject to considerable variation in form.

The fingers of the chelicerae bear outgrowths termed serrulæ which are keel-like structures set with fine teeth or blades. The *serrula exterior* is attached to the movable finger for varying degrees of its length, the distal two-thirds being free (as in *Chthonius*, Fig. 44), the distal one-third or fourth free in some species, or the whole keel being attached throughout its length (Fig. 43). The *serrula interior*, attached to the base of the fixed finger is subject to much more variation in form. The fixed finger is invariably provided with marginal teeth. The movable finger is typically dentate only in the Chthoniodea and Neobisiodea. In all other forms this finger is armed with a terminal lobe or process.

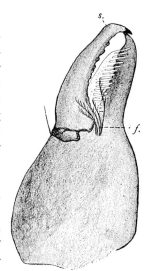

Fig. 44.
CHELICERA OF CHTHONIUS
s, spinneret   *f*, flagellum

The first segment or fixed finger bears on its inner side an appendage termed the flagellum (Figs. 43 and 44) which is probably sensory in function. It may consist of a tuft of eight to twelve pinnate setæ or a smaller number of blades which may be toothed, branched, or simple.

The six-segmented pedipalps are enormously developed and are terminally chelate, resembling those of scorpions. They serve as prehensile organs to capture and kill their prey and are at the same time the seat of important sensory organs. The coxæ are extended forward so as to form masticatory plates or they bear distinct endites; these serve to hold the prey in front of the mouth and perhaps also function as jaws. The other joints of the pedipalp are the trochanter, femur, tibia, and chela. The latter is furnished with tactile hairs and is the seat of the venom apparatus when it is present.

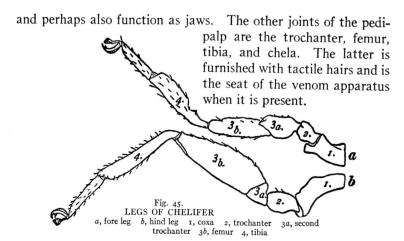

Fig. 45.
LEGS OF CHELIFER
a, fore leg   *b*, hind leg   1, coxa   2, trochanter   3*a*, second trochanter   3*b*, femur   4, tibia

All of the four pairs of legs are fitted for walking (Fig. 41). The segmentation of the legs in this order is somewhat different from the usual type in the Arachnida. The tibia is not divided, in other words a patella is wanting. On the other hand, with many pseudoscorpions, the femur consists of two, more or less distinct segments; when this division exists, the proximal segment may be termed the *pars basalis* or *pars tibialis*. The pars basalis is often referred to as the *second trochanter*. It is sometimes a very distinct segment (Fig. 45 *a*); but in other cases it is a triangular segment, closely joined to the femur (Fig. 45 *b*), and resembling the trochanter of many beetles.

Fig. 46. TARSUS OF CHELIFER

The number of tarsal joints is of great systematic importance in the order. In the Heterosphyronida the forelegs possess a miotarsus, which is equivalent to the metatarsus and telotarsus fused, and the hind legs have these elements distinct. In the Diplosphyronida the tarsi of all the legs are two-jointed (having both metatarsus and telotarsus); in the Monosphyronida all tarsi have these joints fused (miotarsus). The praetarsus is usually distinct; and there are two lateral claws. Below the claws there is a membranous empodium, as is illustrated in Fig. 46.

The respiratory organs are tubular tracheæ, which open by

two pairs of long slit-like spiracles on the second and third abdominal segments, on the ventral side near the lateral margins. In the form of the respiratory organs the pseudoscorpions differ greatly from the scorpions, near which they are commonly placed.

The opening of the reproductive organs is on the middle line between the second and third abdominal sterna. The female carries the eggs about, attached to the abdomen; they are on the outer surface of a sac which is attached to the wall of the opening of the reproductive organs by a slender and broad pedicel, the whole forming a small raspberry-shaped mass. The sac is filled with food material; and the young when hatched are furnished with a long sucking beak by means of which they obtain the food thus supplied. They attain their definite form in this position. The transformations have been described by Metchnikoff ('71), Bouvier ('96), and Barrois ('96).

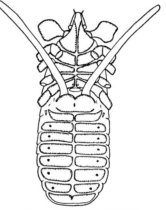

Fig. 47. VENTRAL ASPECT OF CHELIFER SHOWING THE RAM'S-HORN ORGANS EXPANDED

In the males lateral genital sacs are present under the genital operculum, close to the genital aperture. In some species these are capable of being extruded. "In the subfamily Cheliferinæ of the Cheliferidæ, they reach their highest and most peculiar development. Here they are completely evaginated and when retracted are thrown into a dense series of peculiar, reticulated folds that permit an enormous extension. In this group they constitute the 'ram's-horn organs' of many authors, the term going back to Menge. When fully expanded, a process that is doubtless accomplished by blood pressure, they are everted completely from the genital cavity and project forward as far as the cheliceræ in graceful curves. In these comparatively few species, these organs have the definite function of sexual display as previously shown." (Chamberlin, '31, p. 189.)

The true silk glands were discovered by Croneberg ('87). They are situated in the cephalothorax and open through a group of minute apertures near the tip of the movable finger of the chelicera, in the same position as the opening of the poison gland of spiders.

Spiders and Their Near Relatives

In those forms in which the spinneret is long and branched the openings of the silk ducts are on the tips of the branches; in *Obisium* they open on a blunt prominence, which may be regarded as a rudiment of a spinneret; in this case according to Bernard ('93a) there are about seven of these ducts in each chelicera. In others the number of ducts varies from one to ten.

The silk spun by the cheliceræ is used for the formation of a web or a cocoon in which the pseudoscorpion retreats during the moulting period and during the winter. Sometimes earth is mixed with the silk in forming the cocoon. The silk glands appear to be subject to periodic variations in their development; this is doubtless explained by the fact that they are needed only at certain and rather remote periods. It is now known that the function of the serrula is the manipulation of this silk.

The pseudoscorpions live under stones, beneath the bark of trees, in moss, under leaves on the ground, in the nests of bees, of ants, and of termites, and in the dwellings of man, where they are often found between the leaves of books. In the South they have been beaten from leaves of palmetto. It is believed that they feed chiefly on mites, psocids, and other minute insects. They are often found attached to insects, especially to flies and to beetles. Some writers think that they do not feed on these large insects, but merely use them as means of rapid locomotion; but according to the observations of Berg ('93) the pseudoscorpions kill the flies and eat them afterward. In many cases it seems probable that the body of a large insect serves as a good hiding place. Large beetles, especially those brought in from the tropics, are often found to harbor a score or more large pseudoscorpions, closely packed together under their elytra. It is doubtful that they could kill such large insects, even by concerted action. On the other hand it is possible that the venom is powerful enough to kill some of the smaller, soft bodied insects on which they are sometimes found.

An excellent account of the order with descriptions of the species occurring in France was published by Simon ('79) and later Balzan ('91) proposed an excellent classification. The most important, older work on the American pseudoscorpions is that of Banks ('95b) who gave a synopsis of the species known at that time. Many of our species have been described by that author. More recently there has appeared a monographic revision of "The Arachnid Order Chelonethida" by Dr. J. C. Chamberlin ('31) who, basing his system on an intensive analytical study of the morphology and biology of these creatures, has produced the classification which is now generally accepted. Although the systematic portion of this monumental paper deals only with the higher categories, going only to the genera, students will find it indispensable to any serious study of the Pseudoscorpionida. Most of our species are listed by Beier ('32) in his comprehensive treatment of the world fauna. The order is a difficult one for the beginning student because of the requirements in preparation and technique which must be fulfilled before any critical work can be attempted. Even many of the higher categories are based on structures which are not visible except after especial preparation. Because of these difficulties the remainder of the discussion here devoted to the order must necessarily be very brief.

The order Pseudoscorpionida is a fairly large one, about eight hundred species having been described from the world. Approximately one hundred and sixty genera are known. Inasmuch as the order has been neglected for the most part by arachnologists, many hundreds of species and numerous genera will undoubtedly be added when the group has been more thoroughly studied on the basis of sufficient material. The pseudoscorpions are fairly well represented in our fauna. At the present time more than a hundred species have been described, and they are placed in nearly forty genera. Several of the species are cosmopolitan. Chamberlin has divided the order into three suborders, all of which are found in the United States. They are based on the segmentation of the legs and may be separated as follows:

TABLE TO THE SUBORDERS

A. Legs with a dissimilar number of tarsal joints (heterotarsate). P. 45. Suborder HETEROSPHYRONIDA
AA. Legs with the same number of tarsal joints (homotarsate).
B. All legs with two-segmented tarsi. P. 46.
Suborder DIPLOSPHYRONIDA
BB. All tarsi comprising a single segment. P. 47.
Suborder MONOSPHYRONIDA

Suborder HETEROSPHYRONIDA

This suborder, which is also known as the Chthoniinea, is characterized by the dissimilar number of joints in the tarsi of the legs. The forelegs are seven-segmented, due to the fusion of the metatarsus and tarsus to form a single segment, termed a miotarsus; whereas the hind legs are eight-segmented, with both tarsal joints distinct. The femora of all legs are always distinctly divided into a basal and tibial portion. In both the known families the fingers of the pedipalp are without a venom apparatus. This condition, due to the loss of the apparatus in both fingers, is not unique for the suborder, for these glands and ducts have been lost sporadically in one or both fingers in the families of the other suborders. The primitive state is seemingly the presence of a well developed venom apparatus in both the fixed and the movable fingers.

The chelicerae are strongly developed, often very large, and especially well adapted for a prehensile function. Both fixed and movable fingers are provided with marginal teeth. The serrula exterior is free for a considerable portion of its length (Fig. 44). The spinneret is simplified, consisting ordinarily of an opaque, rounded tubercle (Fig. 44) which is often difficult to distinguish and which is sometimes completely absent.

Two families of this suborder are distinguished by Chamberlin. In the Dithidae, represented in our fauna by two species from the southeastern portion of our country, the femur of the fourth legs is only slightly shorter than the second femur. In the other family, the Chthoniidae, the fourth femur is decidedly shorter than the second femur. The Chthoniidae is poorly represented in this country as compared with Europe where there are numerous species. The thirteen that have been described up to the present time are placed in several genera. Most of the species of this series are quite slender forms with rather long legs and elongated pedipalps. They may have four, two or no eyes. In the latter case they are often cave species or live in comparable situations outside.

Some of the familiar species of the Chthoniidae found in the eastern part of our country are *Chthonius packardi*, which occurs in caves in Kentucky and Indiana; *Chthonius caecus*, known from caves in Virginia; and *Chthonius crosbyi*, which has been taken in New York State.

Suborder DIPLOSPHYRONIDA

This suborder is also known as the Neobisiinea.

The tarsi of all the legs are uniformly segmented, all of them having the metatarsus and tarsus distinct. The femora are in all cases divided into two distinct segments. The venom apparatus is always present in the fixed finger of the pedipalp, and is often present in both fingers.

The cheliceræ are either large with both fingers dentate (Neobisiodea) or they may be small, in which case the movable finger is not provided with teeth (Garypoidea). In the first case the serrula exterior is usually free for at least the distal fifth of its length; whereas in the latter case it is fused to the movable finger for its full length or nearly so. In this series the spinneret is usually a slender, translucent appendage, which is often ornamented with characteristic branches. When thus developed, it is referred to as a galea (Fig. 43, *s*).

Chamberlin has divided this suborder into seven families. All but one of them are found in our fauna. Two others are known only from North America, the Syarinidæ which is found only in the Rocky Mountain and Pacific Coast States, and the Menthidæ which occurs only in the subtropical deserts of western Mexico and the southwestern United States. The various families are differentiated chiefly by the dentition of the fingers of the pedipalp, and the presence and location of the venom apparatus and its accessories. Approximately fifty species pertaining to these families have been described from the United States.

The cephalothorax is variable in shape. In some families the sides are parallel or nearly so, and the cucullus is relatively narrow. In the Garypidæ the cephalothorax is subtriangular, greatly narrowed in front of the eyes, the broad cucullus narrowed and emarginated in front and marked with a median longitudinal suture. The number of eyes present in this section may be four, two, or none, as in the Heterosphyronida.

Representative species of this suborder found in the east are *Microcreagris rufulus*, common in the Atlantic States; *Microcreagris cavicola*, a blind species from caves in Virginia; *Neobisium carolinense*, a large species from North Carolina; and *Neobisium brunneum*, a small pseudoscorpion from the Northern States.

Suborder MONOSPHYRONIDA

This suborder is also known as the Cheliferinea.

The tarsi of all the legs are uniformly segmented, all of them comprising a single segment, a fused metatarsus and tarsus. The femora are divided into a basal and tibial segment in most of the families but in one group, the Cheridiodea, the femora comprise a single segment. The venom apparatus of the pedipalp is variously developed. It is completely lacking in the Pseudogarypidæ and present in the movable finger only in the Chernetidæ. In the other families it is present in the fixed finger or in both.

The cheliceræ are usually quite small, and the serrula exterior is attached throughout its length (Fig. 43). The spinneret is of the galeate type, being a slender appendage which is often elaborated by one or several branches.

Chamberlin recognizes nine families in this suborder, of which only five have been found in our fauna. One of them, the Pseudogarypidæ is peculiar to the western United States. Another rare family, the Sternophoridæ, is unique among the pseudoscorpions in having the coxæ separated, leaving a median longitudinal sternal element which is termed a pseudosternum. In all other families of the order the coxæ are contiguous on the midline and there is no indication of a sternum. A single species is known from the United States, but others occur in Mexico and Australia.

In this series of families the cephalothorax is moderately narrowed in front, and the sides are subparallel. The space in front of the eyes, the cucullus, is narrow. An exception is provided in the Pseudogarypidæ in which group the cephalothorax is subtriangular and the cucullus is broader, strongly produced in front into a transverse row of three protuberances. The American species of this series are always two-eyed or blind.

The most familiar member of this suborder is *Chelifer cancroides*, a domestic species, which is cosmopolitan.

Order RICINULEI (Ric-i-nu′le-i)

The Ricinulids

The curious, enigmatic arachnids of the order Ricinulei are regarded as the rarest of all arthropods. Of such infrequent occurrence are these creatures that during the past one hundred years, since the description of the first one by Guerin-Meneville in 1838, only thirty or forty examples have been collected and recorded in literature. The scanty material, however, has been rich in species, for at the present time sixteen are known from the world. They are placed in two genera, *Ricinoides* of tropical Africa and *Cryptocellus* of Central and South America.

These arachnids resemble ticks superficially in general appearance and further simulate them in their sluggish, deliberate movements. As a group they possess various peculiarities which have isolated them from the other orders and made their true position uncertain.

The integument is strongly chitinized, is of great thickness, and is covered rather thinly with various types of simple hairs. The arrangement of the tubercles and pits on the integument provides characters which are used to differentiate the species. The cephalothorax is compact, shows no signs of segmentation, and is usually marked by a shallow groove. Appended to the frontal edge is the cucullus, or hood, a structure which is considered to be homologous with a lobe present in the tailless whip-scorpions which bears the median eyes. No eyes are present in the Ricinulei. The cucullus is movable and, when bent downward, completely covers the mouth and the chelicerae. The cephalothorax seems to join broadly to the abdomen but the actual juncture is a narrow pedicel which is ordinarily hidden by the base of the abdomen when in close articulation with the caudal end of the cephalothorax. The chitinized dorsal ridges and the depressions into which fit the processes of the posterior coxae make up the "coupling device." The living animal is able to disengage the carapace from the abdomen so that the genital opening is exposed, seemingly a necessity during mating and egg laying. The abdomen is composed of nine segments of which the tergites of only four are easily distinguished, the third, fourth, fifth, and sixth. The latter three are each divided into a large median and two smaller lateral plates which are set off by thinner chitin of a paler color.

The sternum is a narrow longitudinal plate, completely hidden by the coxae which are contiguous on the middle line. The chelicerae, which are normally covered by the cucullus when the animal is not feeding, are two-jointed and chelate. Both movable and fixed fingers are provided with teeth.

The pedipalps are curious appendages capable of being rotated which are used presumably during the feeding process. Their coxae are fused together on the midline and are not provided with endites. The remaining joints of the appendage are the two trochanters, femur, tibia, and tarsus, the latter ending in a small chela.

Spiders and Their Near Relatives

The walking legs are relatively stout and of moderate length, the first one being the shortest and the second one the longest. The coxæ of the first three pairs are immovable and fit closely together on the midline. The fourth coxa, however, is movable and forms an integral part of the coupling device. The number of joints in the legs is quite variable. The first has the usual trochanter, femur, patella, metatarsus, and a one-jointed tarsus. The second has the same number of segments but the tarsus is five-jointed. The third leg is provided with two trochanters and a four-jointed tarsus; the fifth with two trochanters and a five-jointed tarsus. All the tarsi are provided with two smooth claws which are located in a depression at the end of the tarsi. The third leg in the males is modified into a remarkable structure which, by analogy, is presumed to be concerned with copulation. The form of this organ is shown in Fig. 48.

The respiratory organs are tracheal tubes which are situated at the caudal end of the cephalothorax, the inconspicuous apertures opening above the third coxæ.

The reproductive organs open at the base of the abdomen as in all arachnids. The aperture, which is apparently situated between the first and second abdominal somites, lies in a space which is completely covered over when the abdomen is coupled to the carapace. The males have a complicated copulatory apparatus on the third legs which is presumed to aid in the transfer of a spermatophore to the female during copulation. However, the exact use of this unique structure and the other details of the mating process have never been observed. The copulatory apparatus resembles in a superficial way the palpal organs of spiders. It is made up of two heavily chitinized, cylindrical elements which lie in a shallow cavity and various accessory apophyses on the proximal segments of the leg. The principal elements are apparently not hollowed out to act as a reservoir for or to conduct the sperm. Inasmuch as in many of the arachnids a spermatophore is transferred to the female during mating, it is reasonable to suppose that this structure acts as the agent in the transfer. The character of the "coupling" apparatus and the unusual location of the female genital opening lend support to this interpretation.

Practically nothing is known of the life span, the egg laying, the various stages of the immature animals, and the ecological preferences of these rare arachnids.

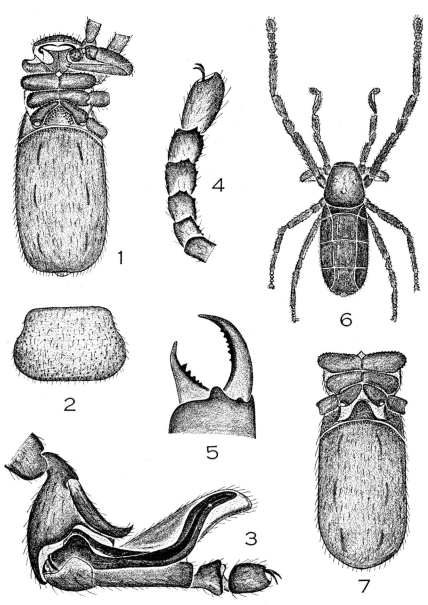

Fig. 48. CRYPTOCELLUS DOROTHEÆ

1. Ventral view of female, appendages omitted. 2. Cucullus of female, frontal view. 3. Basitarsus and tarsus of third left leg of male, lateral view, showing details of the copulatory device. 4. Tarsus of second left leg of female, lateral view. 5. Right chelicera of female, frontal view. 6. Dorsal view of female. 7. Ventral view of abdomen and base of carapace of male, showing the "coupling" apparatus.

Spiders and Their Near Relatives

The principal works on the Ricinulei are those of Hansen and Sorensen ('04) and Ewing ('29). In the first paper will be found a comprehensive treatment of the group from the viewpoint of history, structure, and taxonomy. The second paper deals for the most part with the American species.

Family RICINODIDÆ (Ri-ci-nod'i-dæ)

All the recent and fossil species are placed in a single family, the Ricinodidæ, a name which has supplanted the older, more familiar Cryptostemmatidæ of most authors. The fossil species are rather closely related to the living forms but they differ in having the second coxæ triangular in shape and not meeting on the midline. In *Curculioides* the dorsum of the abdomen is not divided up into the characteristic tergites present in the other genera. All the recent species are very similar in general appearance and structure. Ricinulids have been found only on the west coast of Africa and in the American tropics. The genus *Cryptocellus* is limited to the Americas, and ten species have been described on the basis of about two dozen specimens, most of them collected during the past few years. A single species is known from Texas.

Genus CRYPTOCELLUS (Cryp-to-cel'lus)

The first American representative of the order was collected by H. W. Bates on his trip to the Amazon in 1861. Westwood placed it in a genus apart from the African species, *Cryptocellus*. The penultimate segment of the second tarsus is much shorter than the terminal segment, whereas in *Ricinodes* the opposite is true. The claw of the chelicera is opposed by a single process on the basal segment. The two genera are otherwise very similar.

Ten species of *Cryptocellus* have been described from the Americas, most of them based on females. The majority of the species are found in Central America and in the Amazon basin. A single species from within the borders of the United States has recently been described.

Cryptocellus dorotheæ (C. dor-o-the'æ).—The color in both sexes is a bright rusty orange or red. This species is probably the smallest one yet to be described, measuring about one fifth of an inch in length. The whole animal is evenly covered with small, pale, round pits, and set with small tubercles on the margins of the sclerites. The body is relatively slender, much more so than the other species. In the female the abdomen is armed with large median tergites which are longer than or at most as broad as long; in the male these tergites are distinctly longer than broad. The structure of this species is illustrated in Fig. 48.

This interesting arachnid was taken at Edinburg, Texas, in the Rio Grande River Valley. The specimens of *Cryptocellus dorotheæ* so far collected seemed to prefer a sandy soil and were found shortly after rains which left the soil rather moist. They were found under a permanent cover such as afforded by slabs of concrete, heavy sheet iron, roofing material, etc., which had probably not been disturbed for several years. The creatures are very sluggish and move with considerable deliberation while they seem to feel their way along. Their movements simulate closely the average tick such as represented by males of the Ixodidæ while crawling over the ground.

Order PHALANGIDA*

The Harvestmen

The harvestmen are very common in most parts of the United States. They are well-known to children in this country under the name daddy-long-legs, but as this term is also sometimes applied to crane-flies, the name harvestmen is preferable. In some sections of the country the harvestmen are known as grandfather-graybeards. It was probably a misunderstanding of this name that led Wood ('68) to state that in northern New York they are called "Grab for Gray-bears." The writer spent his boyhood in northern New York and during that period heard only the name grandfather-graybeard applied to these creatures.

The name harvestmen was probably suggested by the fact that they are most often seen at harvest time. A similar term, *faucheurs* or hay-makers, is applied to them by the French. The Germans call them *Afterspinnen* or pseudo-spiders. Other English names are harvest-spiders and shepherd-spiders.

Most harvestmen can be recognized by their very long and

---

* The original form of the name used above was *Phalangita*, a family name proposed by Latreille in 1802. It was changed to Phalangides by Leach in 1815. It is now spelled *Phalangida* for the sake of uniformity with the other ordinal names of Arachnida. The name *Opiliones*, which is used by some writers, was proposed by Sundevall in 1833; there is no good reason for substituting this for the older name.

slender legs (Fig. 49), although some species have comparatively short ones. The carapace is indistinctly if at all segmented. The abdomen is not constricted off from the cephalothorax and is short and broad.

The sternites of the cephalothorax consist of a labium, which is well-developed in some forms and greatly reduced in others,

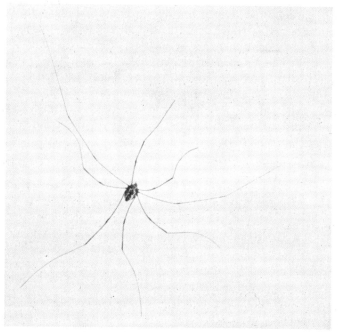

(Photograph by M. V. Slingerland)
Fig. 49. A HARVESTMAN

and a sternum, which is more or less reduced as a result of either the approximation of the coxæ or of a forward thrust of abdominal sternites.

The abdomen consists of nine segments, the anal tergite being the ninth. But the sternites of the abdomen are more or less reduced in number, sometimes to five, either by the consolidation of two or more or by the suppression of some of them, or by both methods. The first abdominal sternite is termed the *genital;* the second, the *tracheal;* and the eighth the *anal.* " In many genera of the suborder Mecostethi the eighth abdominal sternite shows indications of being composed of two sternites, which would bring the total number of sternites up to nine, thus equalling the tergites in number." (Pocock '02.)

The various forms of the labium, the sternum, and the sternites of the first two abdominal segments afford characters for limiting the larger divisions of the order.

The eyes are only two in number in our species and are situated on a prominent tubercle, like a lookout tower, near the middle of the cephalothorax, an eye looking out on each side.

Near the anterior margin of the cephalothorax, on each side opposite the attachment of the first pair of legs, there is a small opening; these are sometimes called the *lateral pores* and sometimes the *openings of Krohn's glands*. They are the openings of scent glands which have been described by Krohn ('67) and by Rossler ('82).

The respiratory organs are tubular tracheæ, which open by a pair of spiracles situated on the ventral side at the juncture of the lateral margins of the cephalothorax and the second abdominal sternite. Secondary spiracles have been found on the legs of some members of this order by Hansen (See Loman '96).

The reproductive organs open on the boundary between the cephalothorax and the abdomen on the middle line of the ventral aspect of the body; the opening is covered by a prolongation of the sternite of the first abdominal segment, which is called for this reason the genital sternite. The female is provided with a large ovipositor, and the male with a large penis; each of these structures is usually retracted within a sac. The form of these organs varies in different species and has been used by systematists for distinguishing species. It is well therefore to preserve specimens with these organs exposed (Fig. 50). This is "a simple operation, requiring only that the abdomen of the living specimen be compressed between the thumb and finger, when these organs will be extruded, and if the specimen is immediately dropped into alcohol will ordinarily remain exposed" (Weed '89a). As the first abdominal sternum is thrust far forward the opening of the reproductive organs in some forms is not far removed from the mouth.

Fig. 50. CADDO AGILIS WITH THE OVIPOSITOR EXPOSED

Spiders and Their Near Relatives

It is not as easy to distinguish the sexes of harvestmen as it is those of spiders, where a glance at the palpi is sufficient. With some harvestmen there are no obvious secondary sexual characters; but with most of them the sexes differ in size, colouration, and the proportions of the appendages. Usually the body of the male is smaller and shorter and more brilliantly coloured than that of the female; but the markings of the female are often replaced by a uniform tint in the male. The granulations and spines are usually more marked in the male; and the legs are usually longer. The form of the chelicerae and of the pedipalps often differ in the two sexes, but no general statement can be made of these differences.

The chelicerae are three-jointed and chelate, the third joint forming the finger of a chela. They are comparatively long and in some forms they are quite stout. The pedipalps are leg-like in form; but are much shorter than the legs; a tarsal claw is usually present; the coxae of the pedipalps bear endites, which are more or less membranous. A similar pair of more or less membranous endites are borne by the coxae of the first pair of legs. These two pairs of endites have been described as two pairs of jaws but in the forms examined by the writer they are hardly fitted for chewing. The legs are very long and slender except that the coxae are stout; but the coxae on account of their fixed condition appear to belong to the body instead of to appendages; the first two pairs of legs bear movable endites in some forms. The tarsi and sometimes other segments of the legs are divided into smaller portions by what are termed false articulations. The tarsal claws vary in number and in form.

Although the harvestmen have stilt-like legs, they do not raise the body much above the ground when they walk, but carry it low down, with the middle part of their legs high in the air. When disturbed they stand on six legs and move the second pair about in the air; this suggests that perhaps the tips of the second pair of legs are furnished with especially sensitive tactile organs.

In the North most harvestmen die in the autumn; a single species, *Leiobunum formosum*, is known to live over the winter as an adult instead of depositing eggs and dying in the autumn, as do the other species (Weed' 89 a). In the South more of them hibernate; they hide under rubbish during the winter.

The eggs are laid in the ground, under stones, or in crevices of wood. They are placed in position by means of the ovipositor which can be extruded to a great length (Fig. 50). Henking ('88) figures a female in the act of depositing her eggs in the ground, and represents the ovipositor three times as long as the entire body of the animal.

Unlike the eggs of spiders, those of the harvestmen which are laid in the autumn do not hatch till the following spring. The species observed by Henking remained six months in the egg. The young moulted as soon as they reached the surface of the ground; they were white as snow except the eyes, which were coal-black; the abdomen was pointed behind showing a one-jointed vestige of a postabdomen. In a short time the body becomes of the same colour as the adult. The newly hatched young use the second pair of legs as feelers in the same way as does the adult; but they are apt to hide under stones or other objects and rarely attract attention till midsummer, when they are more conspicuous.

The adults also ordinarily hide during the day, but at twilight wander about in search of food. Some species apparently often migrate from the fields where they were hatched to the vicinity of houses, barns, and out-buildings where they congregate in large numbers; occasionally they congregate in a similar manner on the trunks of trees.

It was said long ago by a German writer that "they spring and pounce upon their victim as the cat upon the mouse, and seize it with their palpi as if with hands." This is perhaps the foundation for the statement made in many books that the harvestmen feed on living insects. But Henking ('88), who kept large numbers of them under as nearly as possible natural conditions, found that they shunned living insects and fed only on those that they found dead. They also fed on various kinds of soft vegetables and fruits, from which they pressed the fluid by means of their cheliceræ. On the other hand, Weed ('93), states that *Liobunum politum* when in confinement eagerly devours plant lice; and Banks ('01) states that they feed mostly on living insects.

The chief works on the harvestmen of the United States are those of Wood ('68), Weed ('92*b* and '93), and Banks ('93*a*, '93*b*, '94*b*, and '01). A synopsis of most of our species is given in this last cited paper by Banks. All the American species are treated in a monumental paper by Roewer ('23) on the world fauna.

The order Phalangida, or Opiliones, is divided into three suborders, all of which are represented in our fauna.

  A. Genitalia exposed; stink glands opening at ends of tubercles at the sides of the carapace. P. 58.   CYPHOPHTHALMI
AA. Genitalia covered by a movable plate; stink glands not opening on tubercles.
    B. Palpi stout, the tarsi with a strong, reflexed claw; anterior legs with one tarsal claw, the posterior legs with two claws or a compound claw. P. 58.   MECOSTETHI
  BB. Palpi weak, the tarsi with a small claw or none; all legs with a single tarsal claw. P. 64.   PLAGIOSTETHI

Suborder CYPHOPHTHALMI (Cyph-oph-thal'mi)

These arachnids resemble some of the mites to such a marked degree that they are known as mite-like phalangids. The palpi are weak, not developed for grasping, and all the tarsi are armed with a single claw. The tarsi are composed of a single segment, not elongated and divided as in most other phalangids. Eyes are usually absent. The openings of the stink glands are situated on elevated tubercles. The abdomen shows nine tergites above, but in *Holosiro* their limits are obscure.

Genus HOLOSIRO (Ho-lo-si'ro)

The single species on which this genus was based (*Holosiro acaroides*) represents one of three species as yet to be described from the New World. The species was found in the Coast Range Mountains in western Oregon, and is remarkable for the complete lack of superficial indication of dorsal segmentation. Ewing ('32) gives a full description and figures of the species.

Suborder MECOSTETHI (Mec-os-te'thi)

In this suborder the sternum of the first abdominal segment extends but little if any in front of the hind coxæ and the sternum of the cephalothorax is narrow and long, lying between the coxæ of the first three pairs of legs (Fig. 51). It is to this fact that the subordinal name Mecostethi refers, being derived from the Greek words *mekos*, length, and *stethos*, breast. The hind tarsi are either furnished with two claws or with a compound claw; the tibia and tarsus of the pedipalps are more or less depressed; and the hind legs are usually the longest. This suborder is called the Laniatores (Lan-i-a-to'res) by some writers.

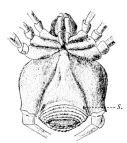

Fig. 51. VENTRAL ASPECT OF LIBITIODES SAYI *s*, spiracle

The suborder Mecostethi is represented in our fauna by three families which can be separated as follows:

A. Posterior tarsi with a dorsal lobe, a pseudonychium, which projects between the paired claws. P. 59. COSMETIDÆ
AA. Posterior tarsi lacking a pseudonychium.
    B. Posterior tarsi with two, separate claws. P. 62.
        PHALANGODIDÆ
    BB. Posterior tarsi with a compound claw: a large median claw with a lateral branch on each side. P. 61.
        TRIAENONYCHIDÆ

Family COSMETIDÆ (Cos-met'i-dæ)

The second pair of legs are without endites. The pedipalps are shorter than the body, with the tibia and tarsus depressed, and with the tarsal claw about half as long as the tarsus. The pedipalps are usually appressed to the face so as to fit over the cheliceræ. The eye tubercle is low, without coniform tubercles. The spiracles are exposed.

These curious phalangids are found for the most part in tropical regions where there are numerous genera and species. Three species belonging to this family, representing three genera, have been found in the warmer portions of the United States. All of them were previously referred to *Cynorta*. The genera and species can be separated as follows:

A. Fourth tarsus six-jointed.   *Libitiodes sayi*
AA. Fourth tarsus with more than six joints.
    B. Third and fourth legs of about the same thickness as first and second.   *Metacynorta ornata*

BB. Third and especially fourth legs thicker than the first and second. *Eucynortella bimaculata*

Libitiodes sayi (L. say′i).—This is our most widely distributed species of the group, occurring in the Gulf States and extending as far north as Missouri and Kansas. The body is nearly one fourth inch in length, of a rusty colour mottled with a darker shade. In well-marked individuals there is a Y-shaped yellow stripe back of the eyes (Fig. 52) and a transverse band near the hind edge of the carapace. The stem of the Y is sometimes wanting and sometimes it extends to the transverse band. Between the arms of the Y and the transverse band there are in some individuals indications of two other transverse bands. Sometimes all of the white marks are wanting. Near the middle of the carapace there is a pair of small tubercles, and there is a second, somewhat larger pair near the hind margin just in front of the transverse yellow band.

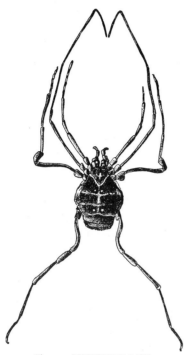

Fig. 52. LIBITIODES SAYI

Metacynorta ornata (M. orna′ta).—This species resembles the preceding in size and markings, but can be easily distinguished by the presence of a pair of prominent acute spines borne on a pair of lobes in the position occupied by the second pair of tubercles in the preceding species (Fig. 53). This species is found in the southeastern part of the United States.

Eucynortella bimaculata (E. bi-mac-u-la′ta).—This is a smaller species, measuring one sixth inch in length, found in southern California. The dorsum is brownish red, with two long, somewhat lunate white spots near the end of the carapace; the venter is red. The body is finely granulate, without large tubercles or spines.

Family TRIAENONYCHIDÆ (Tri-aen-on-ych'i-dæ)

The members of this family resemble the phalangodids very closely in appearance. The posterior tarsi are armed with a single claw which has a lateral branch on each side. However, in *Cyptobunus*, which was based presumably on a young individual in which the adult characters were not completely developed, the posterior tarsi are said to be armed with a single claw. The pedipalps are robust. Only two genera have been described from our fauna.

Genus SCLEROBUNUS (Scler-o-bu'nus)

Fig. 53.
METACYNORTA ORNATA

The posterior tarsi are armed with a single claw but it is a compound one, there being a spur or branch on each side toward the base. The palpi are not as long as the body. Eyes are present. Two species have been found in the west. These can be separated by difference in colour.

Sclerobunus robustus (S. ro-bus'tus).—The colour of the body is deep reddish, with the hinder segments finely bordered with brown. Body pyriform, two thirds as broad as long. Length of body one seventh inch. This species is common in Colorado.

Sclerobunus brunneus (S. brun'ne-us).—The colour of the body is brown, paler on the cephalothorax; legs brown, a little darker than the body, their tips yellow. The dorsum and legs have many little tubercles, each tipped with a stiff hair or bristle. The legs are short. This was described from Washington State.

Genus CYPTOBUNUS (Cyp-to-bu'nus)

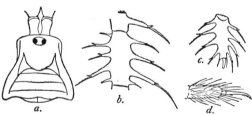

Fig. 54. CYPTOBUNUS CAVICOLUS
a, body and chelicerae *b*, tibia of pedipalp *c*, tarsus of pedipalp
d, tarsus of leg (after Banks)

The only known representative of this genus was recently discovered by Professor Cooley in a cave in Montana and was described by Mr. Banks. It differs from all other members of this suborder in having the posterior tarsi armed with a single simple claw (Fig. 54d). The legs are very slender; the second pair are plainly longer than the fourth. The eye-tubercle is rather large, not very high, rounded, with two large black eye-spots above. The tibiæ and tarsi of the palpi are depressed and armed laterally with long, slender spines, each spine consisting of a basal part, truncate at tip and a long terminal bristle arising beneath a short spur or apophysis (Fig. 54b).

Cyptobunus cavicolus (C. ca-vic'o-lus).— This is a pale whitish nyaline species, with black eye-spots. Figure 54a represents the outline of the body. The length of the body of the only known specimen is less than one twelfth inch.

Family PHALANGODIDÆ (Phal-an-god'i-dæ)

In this family the hind coxæ are united to the first abdominal segment at the base, but are free at the apex. The second pair of legs have distinct endites. The pedipalps are robust and as long or longer than the body. The spiracles are indistinct. The hind tarsi are armed with two, separated claws.

Our four genera can be separated as follows:

A. Eye-tubercle prolonged into a sharp spine.
  B. Eye-tubercle arising from the front margin of the cephalothorax. P. 63.   PACHYLISCUS
  BB. Eye-tubercle arising back from the front margin of the cephalothorax. P. 62.   NEOSCOTOLEMON
AA. Eye-tubercle high, but rounded at the end.
  B. Third and fourth tarsi four-jointed. P. 63. PHALANGODES
  BB. Third and fourth tarsi five-jointed. P. 63. EREBOMASTER

Genus NEOSCOTOLEMON (Ne-o-sco-to-le'mon)

The eye-tubercle arises some distance back from the anterior margin of the carapace and is prolonged into a long, sharp spine. Eyes are present, and they are widely separated. The posterior tarsi are five-jointed.

Two species are found in the United States.

Neoscotolemon spinifer (N. spi-ni'fer).—This species was found at Key West, Florida. The body is spiny. The femur of the palpus has two broad processes terminating in long spines and three short spurs near the middle of the joint on each side.

Neoscotolemon pictipes (N. pic-ti′pes).—This Cuban species has been taken by Mr. Banks at Falls Church, Virginia. There is only one short spur at the middle of the femur on each side.

Genus PACHYLISCUS (Pach-y-lis′cus)

The eye-tubercle arises from the anterior margin of the carapace and ends in a sharp spine. The pedipalps are shorter than the body. The following is our only species.

Pachyliscus californicus (P. cal-i-for′ni-cus).—The colour of the body is yellow, venter paler, legs whitish toward the tips. Body oval, truncate in front, rounded behind. Eye-tubercle large, arising from the anterior margin of the carapace, and about one half as long as it, much roughened and finely granulated. The eyes are near the base of the tubercle. The legs are short. Length of body one twelfth inch. Occurs in southern California.

Genus EREBOMASTER (Er-e-bo-mas′ter)

The eye-tubercle is high but it is rounded at the end. The posterior tarsi are five-jointed. A single species is known.

Erebomaster flavescens (E. flav′es-cens).—Body broad and stout, uniformly straw-yellow, including the appendages. The eye-tubercle is large and high; the eyes are distinct, black, and situated near the base of the conical tubercle. Pedipalps less than twice as long as the body, but nearly twice as long as the carapace. Length of body, one eighth inch. This is a cave species found in Kentucky, Indiana, and Virginia.

Genus PHALANGODES (Phal-an-go′des)

The eye-tubercle arises some distance back from the anterior margin of the carapace and is rounded at the distal end. Eyes may be present or absent. The pedipalps are long, sometimes longer than the body. The tarsus of the second pair of legs is about twice as long as the body in some species.

Our three species may be separated as follows:

Spiders and Their Near Relatives

- A. Eyes lacking; patella of palpus with two lateral spines on each side.   *Phalangodes armata*
- AA. Eyes present; patella of palpus with one lateral spine on each side, or completely unarmed.
  - B. Tarsus of palpus with two ventral spines; patella with one lateral spine on each side.   *Phalangodes californica*
  - BB. Tarsus of palpus with three ventral spines; patella unarmed.   *Phalangodes brunnea*

Phalangodes armata (P. ar-ma′ta).—This species is found in the Mammoth Cave and in other caves. The body is from one eighth to one sixth inch in length. All the legs are remarkably long and slender. The colour of the adult is whitish straw-yellow, of the young, white.

Phalangodes californica (P. cal-i-for′ni-ca).—The colour is pale yellowish with the dorsum of the abdomen rather darker. The eye-tubercle is low, rounded, close to the front margin. Length of body less than one twelfth inch. The species was described from specimens taken in Alabaster Cave, California.

Phalangodes brunnea (P. brun′ne-a).—Body yellowish brown, the abdomen mottled with irregular blackish markings, the posterior margins of ventral and last few dorsal segments margined with black. Eye-tubercle large and blunt. Length of the body one twelfth inch. This species occurs in North Carolina and Tennessee. It is found in leaves and debris in damp situations in the woods.

Suborder PLAGIOSTETHI
(Pla-gi-os-te′thi)

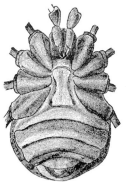

Fig. 55. VENTRAL ASPECT OF LEIOBUNUM VENTRICOSUM

In the members of this suborder the sternum of the first abdominal segment extends much in front of the hind coxæ (Fig. 55). A result of this forward thrust of this sternite is that the sternum of the cephalothorax is short and transverse; this suggested the name Plagiostethi, which is from the Greek words *plagios*, transverse, and *stethos*, breast. The tarsi of the legs are furnished each with but one simple claw; the tibia and tarsus of the pedipalps are cylindrical; and the second legs are the

Spiders and Their Near Relatives

longest. This suborder is called the Palpatores (Pal-pa-to'res) by some writers.

The Plagiostethi includes five families, four of which are represented in the United States; these can be separated by the following table:

   A. Last segment of the pedipalp with a claw at the end, this segment longer than the preceding one except in the male of *Protolophus*. P. 65.   PHALANGIIDÆ
AA. Last segment of palpus with no claw at the tip, this segment much shorter than the preceding one.
     B. Ocular tubercle prolonged into a plate which extends beyond the front margin of the cephalothorax. P. 79.   TROGULIDÆ
   BB. Ocular tubercle not prolonged into a plate.
      C. First and fourth coxæ lacking marginal spinules. P. 76.   ISCHYROPSALIDÆ
     CC. First and fourth coxæ with an anterior and posterior marginal row of spinules. P. 78.   NEMASTOMATIDÆ

Family PHALANGIIDÆ (Phal-an-gi'i-dæ)

In the Phalangiidæ the last segment of the pedipalps is much longer than the penultimate one and is armed with a small claw; the coxa of the fourth leg is united near its base on the posterior side to the tracheal sternite of the abdomen; the endite of the second leg is directed horizontally inward; and tibial spiracles are present.

The genera of the Phalangiidæ occurring in our fauna are separated by Banks ('01) as follows:

   A. Male with the palpus enlarged, last joint shorter than the penultimate; female with the patella of the palpus provided with a long branch; palpal claw smooth; a row of teeth on sides of coxæ. P. 67.   PROTOLOPUS
AA. Last joint of palpus longer than the penultimate; patella of palpus without a branch in the adult.
     B. Eye-tubercle of enormous size; three long spines on the femur of the palpus. P. 68.   CADDO
   BB. Eye-tubercle of normal size.

C. A group of spinules on the anterior margin of the cephalothorax and the eye-tubercle spinose; palpal claw smooth; rarely if ever with lateral rows of teeth on coxæ; frequently with spines on the femur of the palpus.
  D. Femur of palpus provided with prominent spines.
    E. No false articulations in the metatarsus of the first legs; eye-tubercle more remote from the anterior margin. P. 70.   Lacinius
    EE. At least one false articulation in the metatarsus of the first legs; eye-tubercle nearer to the margin. P. 70.   Odiellus
  DD. Femur of palpus not furnished with prominent spines.
    E. Femora and tibiæ of the first and third pairs of legs thickened; coxa of the second legs completely shut out by those of the first and third legs. P. 71.   Globipes
    EE. Femora and tibiæ of the first and third legs normal; coxa of the second legs not shut out by those the first and third legs.
      F Two prominent suprachliceral teeth; dorsum punctate. P. 71.   Homolophus
      FF. No such teeth.
        G. The femur of the first legs longer than the width of the body. P. 69.   Phalangium
        GG. The femur of the first legs shorter than the width of the body. P. 69.   Mitopus
CC. Either with the eye-tubercle smooth or, if spinose, then no group of spinules on the anterior margin of the cephalothorax.
  D. Palpal claw denticulate; adult with a row of teeth on the sides of the coxæ, distinct at least on the anterior side of the coxæ of the first pair of legs; legs usually long.
    E. Body very hard and granulate above and below; legs very short; fourth pair of legs nearly as long as the second pair. P. 76.   Mesosoma
    EE. Body softer, although often subcoriaceous; rarely granulate below; legs longer, fourth pair much shorter than the second pair.

> F. Femur of the first pair of legs longer than the body or in some females a little shorter, but longer than the width of the body. P. 73. LEIOBUNUM
> FF. Femur of the first legs shorter than the body, in females not so long as the width of the body. P. 75. HADROBUNUS
DD. Palpal claw smooth; no such rows of teeth on coxæ; legs usually shorter.
> E. Inner margin of patella of pedipalps extended; body very hard and rough; eye-tubercle spinose; coxæ of the third and fourth pairs of legs enlarged in the males. P. 72. TRACHYRHINUS
> EE. Inner margin of patella of pedipalps not extended, or at least with the body quite soft.
>> F Legs slender and long as in *Leiobunum;* coxæ of the second legs not shut out by the coxæ of the first and third pair of legs. P. 73. LEURONYCHUS
>> FF. Legs much shorter; femora of the first pair of legs thickened or not as long as the width of the body; eye-tubercle smooth.
>>> G. Femora and tibiæ of the first and third pairs of legs thickened; coxa of the second legs shorter than that of the first. P. 72. EURYBUNUS
>>> GG. Femora and tibiæ of the first and third pairs of legs not thickened; coxa of the second legs about as large as that of the first. P. 71. LEPTOBUNUS

Genus PROTOLOPHUS (Pro-tol'o-phus)

This genus differs from all other members of the Phalangiidæ occurring in our fauna in having the last segment of the pedipalps of the male shorter than the penultimate, and in that the patella of the pedipalps of the female is provided with a long branch. There is a median pair of large tubercles on the dorsal aspect of each of the first five abdominal segments. Two species are found in the Far West.

Spiders and Their Near Relatives

Protolophus tuberculatus (P. tu-ber-cu-la′tus).— The cephalothorax is gray, the dorsum of the abdomen darker, somewhat reddish brown; there is a broad, darker, median strip, extending from the anterior margin of the cephalothorax to the sixth abdominal segment. The abdominal tubercles are unarmed. Occurs in Texas and California.

Protolophus singularis (P. sin-gu-la′ris).— This species is similar to the preceding but differs in that the abdominal tubercles bear from two to four spines. Occurs in California.

Genus CADDO (Cad′do)

The eye-tubercle is of enormous size, covering the greater part of the cephalothorax (Fig. 56); it is wider than long, smooth, with a broad median furrow. The eyes are very large and situated at each side of the ocular tubercle. Only two species are known.

Fig. 56. CADDO AGILIS

Caddo agilis (C. ag′i-lis).— This is the larger of the two species, attaining a length of one sixth inch. The colour is brownish with a silvery white longitudinal median band. The trochanter of the pedipalps is armed below with only scattered stiff hairs. The femur of the pedipalps is armed with three large spine-tipped tubercles which are evenly spaced on the basal half (Fig. 57a).

This species has been found in New York, District of Columbia, and Ontario.

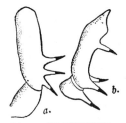

Fig. 57. TROCHANTER AND FEMUR OF THE PEDIPALP OF
a, Caddo agilis b, Caddo boopis

Caddo boopis (C. bo-o′pis).—The length of the body is only about one twenty-fifth of an inch. The color is dark brownish without the white median band. The trochanter of the pedipalps is armed beneath with a large tubercle bearing two spines, one large and one small. The femur of the pedipalps bears three large spines, two close together near the base and one near the middle of the segment (Fig. 57b).

As yet this species has been found only at Ithaca, N. Y. It was taken by Professor Crosby in sifting leaves on a heavily wooded bank.

Genus PHALANGIUM (Pha-lan'gi-um)

The last segment of the pedipalps is longer than the penultimate, and the femur of the pedipalps is not furnished with prominent spines; there is a group of spinules on the anterior margin of the cephalothorax and the eye-tubercle is spinose; and the femur of the first legs is longer than the width of the body. Two species occur in the United States.

Phalangium opilio (P. o-pil'i-o).—The pedipalpi are very long, in the male they are longer than the body, in the female they are longer than the width of the body; the second segment of the chelicera of the male is prolonged above in a spur, that of the female is normal. The length of the body is a little over one fourth inch. The species was described from Arkansas. It occurs abundantly in New York.

Phalangium parietinus (P. pa-ri-e-ti'nus).—The palpi are much shorter than in the preceding species, in the female they are scarcely as long as the width of the body; the cheliceræ of the male are normal. The length of the body varies from one fifth to a little more than one fourth inch.

This species is widely distributed in the Northern States. Weed in writing of it states as follows: "This species is pre-eminently what may be called an indoor form. It abounds especially in sheds, outhouses, and neglected board-piles, being rarely found in the open field. Its colour especially fits it for crawling over weather-beaten boards, making it inconspicuous against such a background. During the day it is usually quiet, but at dusk and on cloudy days it moves about quite rapidly."

Genus MITOPUS (Mit'o-pus)

The harvestmen of this genus agree with Phalangium in the characteristics given above except that the femur of the first legs is shorter than the width of the body. Three species occur in our fauna.

Mitopus californicus (M. cal-i-for'ni-cus).— The tibia of the second legs is much longer than the metatarsus of these legs; there is one false articulation in the metatarsus of the first legs; and the eye-tubercle is about its diameter from the anterior margin of the carapace. The species occurs in California.

Mitopus dorsalis (M. dor-sa′lis).— In the second legs the tibia is subequal to or shorter than the metatarsus; there is no false articulation in the metatarsus of the first legs; the colour of the body is gray and white. This is an Alaskan species.

Mitopus montanus (M. mon-ta′nus.)— This species agrees with the preceding in having the tibia of the second legs subequal to or shorter than the metatarsus; but differs in having two false articulations in the metatarsus of the first legs and in being brown or black and white. It occurs in New Hampshire.

Genus LACINIUS (La-cin′i-us)

In this and in the following genus there is a group of spinules on the anterior margin of the carapace, the eye-tubercle is spinose, the palpal claw is smooth, and the femur of the palpus is provided with prominent spines. In this genus there are no false articulations in the metatarsus of the first legs, and the eye-tubercle is more remote from the anterior margin than in the following genus.

Lacinius texanus (L. tex-a′nus).— The metatarsi of the first and third legs are banded in the middle, and there are very prominent spines at the tips of the femora and patellæ. Occurs in Texas.

Lacinius ohioensis (L. o-hi-o-en′sis).— The metatarsi of the first and third legs are not banded; the legs are longer than in the preceding species and the spines at the tips of the femora and patellæ are less prominent. Occurs in the Eastern and Middle States.

Genus ODIELLUS (O-di-el′lus)

There is a group of spinules on the anterior margin of the cephalothorax and the eye-tubercle is spinose; the palpal claw is smooth; the femur of the palpus is armed with prominent spines; and there is at least one false articulation in the metatarsus of the first leg.

Odiellus pictus.—The colour of the dorsum is mottled ash-gray with a dark central band extending the whole length of the body; this band is suddenly contracted near the posterior margin of the cephalothorax, then gradually expanded until it reaches the end of the anterior third of the abdomen, where it is suddenly contracted; behind this point the band is again widened and contracted. The body measures about one fifth inch in length. This species is found in New England; and is abundant in New York.

Genus HOMOLOPHUS (Ho-mol'o-phus)

There are quite prominent spines upon the cephalothorax and eye-tubercle and transverse rows of them on the abdomen. The legs are thicker than usual, the anterior femora being much thicker than the eye-tubercle is wide. There are two prominent suprachelicheral teeth.

Homolophus biceps (H. bi'ceps).— The dorsum is pale brownish with a great many small white spots and darker brown punctures; eye-tubercle with a broad white stripe above; and there is a bifid white stripe from the eye-tubercle to the anterior margin; venter, sternum, and coxæ whitish; spiracles and some spots on the venter brown. The male measures one sixth inch in length; the female, one fifth. Occurs from Colorado to Washington State.

Genus GLOBIPES (Glob'i-pes)

The principal character of this genus is the enlarged femora and tibiæ of the first and third pairs of legs; the eye-tubercle is low and with a few spines; the legs are short; but the femur of the second legs is longer than the body and more than twice as long as the femur of the first legs. There are no false articulations in the metatarsus of the first legs. The palpi are normal. A single species has been described.

Globipes spinulatus (G. spin-u-la'tus).— The colour of the body is brown or reddish brown, tip of abdomen more gray, dorsum somewhat mottled with brown. The female has two median white spots near the tip of the abdomen. Length of body one seventh inch. Occurs in California.

Genus LEPTOBUNUS (Lep-to-bu'nus)

The legs are short, with the segments but little thickened; the femur of the first legs is much shorter than the body; frequently the femur of the second legs is not as long as the body. The eye-tubercle is narrow and is usually smooth. Two western species have been described.

Leptobunus borealis (L. bo-re-a′lis).— The apices of the coxæ are dark; the tibia of the first legs is marked with one dark band; the palpus is lineate with brown. Occurs in Alaska.

Leptobunus californicus (L. cal-i-for′ni-cus).— The apices of the coxæ are pale; the tibia of the first legs is marked with two dark bands; and the palpus is not lineate. Occurs in California.

Genus EURYBUNUS (Eu-ry-bu′nus)

The eye-tubercle is very low and smooth; the cephalothorax with an elevation on the anterior margin bearing a few small spines. The segments of the dorsal shield of the abdomen are so closely united that their sutures are hardly discernible. The femora, patellæ, and tibiæ of the first and third legs are enlarged; the femur of the second legs is barely twice as long as that of the first legs, and a little longer than the body; the metatarsus of the first legs is without false articulations.

Eurybunus brunneus (E. brun′ne-us).— The body is very smooth; the fourth leg is nearly as long as the second. Occurs in California.

Eurybunus spinosus (E. spi-no′sus).— The body is armed with transverse rows of spinules; the fourth leg is much shorter than the second. Occurs in California.

Eurybunus formosus (E. for-mo′sus).— This is a beautiful species, which occurs in Texas. The dorsum is a rich dark brown, with a medium elongate white spot near the tip; the sides just behind the legs are also white, and the legs are ringed with white. The length of the body is nearly one third inch.

Genus TRACHYRHINUS (Trach-y-rhi′nus)

The inner margin of the patella of the pedipalps is extended at the tip into a spur-like prolongation; the body is very hard and rough; the eye-tubercle is spinose; and the coxæ of the third and fourth legs are enlarged in the males.

Trachyrhinus favosus (T. fa-vo′sus).— The body is very hard; the dorsum is nearly square, quite level, has projecting angles on the anterior lateral corners and is coarsely punctate. The colour is grayish, spotted with black, and with a central vase-like marking faintly indicated; the coxæ are unicolorous, and the legs mostly black. Occurs in Nebraska and Colorado.

Trachyrhinus marmoratus (T. mar-mo-ra'tus).— This species was described from New Mexico. It differs from the preceding in having the coxæ pale spotted with brown and in having the legs mostly pale.

Genus LEURONYCHUS (Leu-ron'y-chus)

Similar in most respects to *Leiobunum*, but with the palpal claw smooth, and without the lateral rows of teeth on the coxæ. Two species have been found in the Far West.

Leuronychus pacificus (L. pa-cif'i-cus).— The body is marked with a brown dorsal stripe; the patellæ of the legs are brown lined with white. This species occurs in California and in Washington.

Leuronychus parvulus (L. par'vu-lus).— This species differs from the preceding in lacking the dorsal stripe and in that the patellæ are not lineate. It occurs in Washington.

Genus LEIOBUNUM (Lei-o-bu'num)

In this genus the cuticula is soft or subcoriaceous. The anterior and lateral borders of the carapace are smooth. The eye-tubercle is relatively small, smooth, or rarely provided with small, slightly distinct tubercles; it is widely separated from the anterior border. The legs are very long and slender; the fourth pair are much shorter than the second pair. The palpal claw is denticulate.

The genus *Leiobunum* is the largest of this order in our fauna, including sixteen species. The males are separated by Banks ('01) as follows:

A   A distinct spur on the femur of the palpus; dorsum yellowish, without stripe; trochanters concolorous with the coxæ. Occurs in the Northern States.   *Leiobunum calcar*
AA. Without a spur on the femur of the palpus.
  B. Palpus, except the tarsus, mostly black.
    C. Dorsum dark, often with two large pale spots behind; trochanters usually pale. Occurs on the West coast.
                                                    Leiobunum exilipes
    CC. Dorsum pale yellowish, no stripe; trochanters and bases of legs black. Occurs in the Eastern States.
                                                 Leiobunum nigropalpi

BB. Palpus yellowish or brownish.
  C. Femur, patella, and tibia of the palpus plainly thickened, dorsum with a black stripe, trochanters dark brown. Found in the District of Columbia.
Leiobunum crassipalpis
  CC. Palpal segments not thickened.
    D. Femur of the palpus very long, curved, extending much above the surface of the cephalothorax, dorsum with a distinct black stripe, trochanters black. A widely distributed species.
Leiobunum vittatum
    DD. Femur of the palpus shorter, dorsum without a distinct black stripe.
      E. Dorsum dark, with two large yellowish spots on the union of the cephalothorax and abdomen. Occurs in California. *Leiobunum bimaculatum*
      EE. Not so marked.
        F. Apex of the tibia of the second pair of legs white, trochanters black, body short. Occurs in the Eastern States. *Leiobunum longipes*
        FF. Apex of the second tibia not white.
          G. Apex of the femur of the first pair of legs white preceded by a black band, coxæ pale outside. Occurs in New Mexico and in Texas.
Leiobunum townsendi
          GG. First femur not so marked.
            H. Trochanters dark, contrasted in colour with the coxæ.
              I. Legs black, dorsum dark brown, eye-tubercle slightly spinulate. Occurs in Ohio. *Leiobunum nigripes*
              II. Legs pale, sometimes marked with black.
                J. First femur barely as long as the body, tips of femora and tibiæ dark brown or black, eye-tubercle nearly smooth. Occurs in the Eastern States
Leiobunum formosum
                JJ. First femur plainly longer than the body, eye-tubercle spinulate, legs not so distinctly marked with black.

> K. Abdomen tapering behind, dorsum golden, base of femora not black, trochanters dark brown, species of moderate size. Occurs in New York. *Leiobunum verrucosum*
> KK. Abdomen rather short and broad, legs very long, small species.
>> L. Dorsum and trochanters light brown, bases of femora not black. Occurs in the Southern States. *Leiobunum bicolor*
>> LL. Dorsum yellowish, trochanters and bases of femora black. Occurs in Alabama. *Leiobunum speciosum*

HH. Trochanters pale, concolorous with the coxæ.
> I. Body short and broad, femur of the last pair of legs often black at base, small species. Occurs in the Eastern States. *Leiobunum politum*
> II. Body tapering behind, femur of the fourth legs not black at base, larger species.
>> J. Extremely spinose beneath, an impressed line on the last ventral segment, large species. Occurs in the Southern States. *Leiobunum flavum*
>> JJ. Moderately spinose, rather granulate, no impressed line, species of moderate size. Occurs in the Eastern States. *Leiobunum ventricosum*

Genus HADROBUNUS (Had-ro-bu′nus)

The body is large; the legs are moderately slender, the femur of the first legs is much shorter than the body, in the female shorter than the width of the body, there are several false articulations in the metatarsus of the first legs; the eyes are of normal size, the eye-tubercle is of moderate size and with a few denticles above; there are no spines on the anterior margin of the cephalothorax; the pedipalps are without spines, and with the last segment longer than the penultimate, the palpal claw is dentate. Two species have been described.

Hadrobunus grandis (H. gran'dis).— In this species the dorsum is finely spinulate; the legs are longer than in the following species, and in the female are not much marked with brown at the tips of the segments. It occurs in the Eastern States.

Hadrobunus maculosus (H. mac-u-lo'sus).— This species occurs in the Southern States and differs from the preceding in that the dorsum is more smooth and is marked with many small, round, pale spots; the legs are shorter and are more marked with brown.

Genus MESOSOMA (Mes-o-so'ma)

This genus differs from *Leiobunum* in that the body is very hard and granulate both above and below; the legs are very short; and the fourth pair of legs are nearly as long as the second pair.

Mesosoma nigrum (M. ni'grum).— The body is ovate; it is black above, light beneath, the legs are black except at the base; the ocular tubercle is destitute of spines but is armed with obtuse granules. Length of body about one fifth inch. This species occurs in the Southern and Western States.

Family ISCHYROPSALIDÆ (Isch-y-rop-sal'i-dæ)

This family differs from the Phalangiidæ, to which it is allied, in the following characteristics: the last segment of the pedipalps is shorter than the penultimate one and is clawless; the coxa of the fourth leg is not fused with the adjacent sternite of the abdomen; the endite of the second leg is directed vertically downward; and there are no tibial spiracles.

Representatives of three genera of this interesting family have been found in our fauna.

Genus TARACUS (Tar'a-cus)

The members of this genus can be separated at a glance from all other harvestmen occurring in our fauna by the great length of the chelicerae, which are longer than the body and project forward (Fig. 58). Three species have been found in the Far West.

Taracus spinosus (T. spi-no'sus).— This species differs from the other two in that the dorsum of the abdomen is spinose, and the cheliceræ are pale in colour. It measures about one twelfth inch in length. The colour is pale yellowish with the claws of the cheliceræ reddish brown. It was found in California.

Taracus pallipes (T. pal'li-pes).— In this and in the following species the dorsum of the abdomen is smooth and the cheliceræ are dark. In this species the cheliceræ are smooth. The cephalothorax is pale, with a broad, black, median stripe, as wide as the base of the cheliceræ; the abdomen is dark gray above with black spots; the venter is pale, darker near the tip. The length of the body is one fourth inch; the cheliceræ are one half longer. This species occurs on the Pacific Coast; it was described from the State of Washington.

Fig. 58. TARACUS PACKARDI

Taracus packardi (T. pac-kar'di).— The dorsum of the abdomen is smooth and the chelicerae are dark, as in the preceding species. But the chelicerae are rough and armed with very distinct spines on the basal segment. The length of the body is one fifth inch. This species has been found in Colorado and in New Mexico (Fig. 58).

Genus SABACON (Sa'ba-con)

This genus differs markedly from *Taracus* in the chelicerae which are of about average size and length. The fourth segment of the palpus is much thickened. In this and the preceding genus the coxae are not set with marginal denticles. A single species, which also occurs in Siberia, is known from the United States.

Sabacon crassipalpe (S. cras-si-pal'pe).—There are no false articulations in the femora of the third and fourth pairs of legs. The basal segment of the abdomen bears a row of spines, the median pair of these is much the largest; the next four segments have each a pair of humps crowned with stiff hairs. The colour of the body is pale, with a large brown spot on the cephalothorax; the eye-tubercle is black; and there is also a large brown spot on the front part of the dorsum of the abdomen. Length of body one tenth inch. The male has a prominent process on the basal joint of the chelicerae. This species, previously known as *Phlegmacera occidentale* and as *P. cavicolens*, occurs in the Pacific northwest and is also widely distributed in the East. It is often found in caves.

Genus TOMICOMERUS (To-mi-com'er-us)

This genus, which is related to *Sabacon*, contains a single known species, *Tomicomerus bryanti*, and has been found only on the snows of Mount St. Elias in Alaska. It differs from *Sabacon* in having distinct false articulations in the femora of the third and fourth pairs of legs.

Family NEMASTOMATIDÆ (Ne-mas-to-mat'i-dæ)

This family is distinguished from the Ischyropsalidæ by the presence of an anterior and posterior row of small spines or teeth on the first and fourth coxæ. The fourth segment of the palpus is scarcely thickened.

Two genera have been found in the United States.

Genus NEMASTOMA (Ne-mas'to-ma)

In this genus the carapace and dorsal sclerite of the abdomen are fused together and immovable.

Nemastoma modestum (N. mo-des'tum).—The eyes are well developed. The dorsum is granulated and armed with tubercles. Near the union of the cephalothorax and abdomen there are two diverging rows of four curved spines. The species has been found in California and in Washington State.

Nemastoma packardi (N. pack'ar-di).—Eyes are present. This is a cave-inhabiting species found in Utah. It differs from the preceding in that the dorsum is devoid of spines.

Nemastoma inops (N. i'nops).—This differs from the two preceding species in that the fourth segment of the palpus is less than twice as long as the fifth. It is a blind, cave-inhabiting species found in Kentucky.

Genus CROSBYCUS (Cros'by-cus)

This genus differs from *Nemastoma* in having the carapace and the dorsal sclerite of the abdomen distinctly separated by a suture.

Crosbycus dasycnemon (C. das-y-cne'mon).—This is a minute species, the body measuring less than one twenty-fifth of an inch in length. The legs are armed with short, acute, perpendicular spines and with long, slender hairs, which in their basal half stand at right angles to the leg, and in their distal half are bent forward and are somewhat matted. It has been found in Missouri and in North Carolina.

Family TROGULIDÆ (Tro-gu'li-dæ)

The Trogulidæ differs from the Nemastomidæ, with which it is closely allied, in that the sternites of the abdomen, except the genital and anal, are fused, do not overlap, and are marked by a median longitudinal sulcus; and in that the first and second abdominal sternites are widely rounded anteriorly and overlap considerably the proximal extremities of the two posterior pairs of coxæ; and also in having the eighth or penultimate tergite small and narrow, and not expanded laterally. The legs are shorter and stouter than in the preceding family; and the carapace bears a frontal process which conceals the cheliceræ and pedipalps.

This family is represented in our fauna by only two genera, both of which are restricted to the Pacific Slope. They are separated as follows by Mr. Banks:

A. Eye-tubercle projecting in the form of a spoon, two spines at each side on the anterior margin. P. 79.
ORTHOLASMA

AA. Eye-tubercle branched, a single club at each side on the anterior margin. P. 80. DENDROLASMA

Genus ORTHOLASMA (Or-tho-las'ma)

The cephalothorax is armed with a pair of spines at each side on the anterior margin. The eye-tubercle projects in front in the form of an almost flat, gradually widening plate; the tip of the eye-tubercle is rounded, with a more dense central rib and some lateral ribs connected by a membrane. The eyes are but partly visible from above.

Only two species of these harvestmen are known; both of these are from the Pacific Coast, and were described by Mr. Banks.

Ortholasma pictipes (O. pic'ti-pes).— The form of the body of this remarkable harvestman is represented by Fig. 59; the legs are long and slender, and are not represented in the figure in order to save space. The forward projecting process of the eye-tubercle has only four or five openings on each side; and the ribs of this process project beyond its edge.

This species was discovered in California by Dr. J. C. Bradley.

Ortholasma rugosa (O. ru-go'sa).— This species differs from the preceding in having six or more openings on each side of the

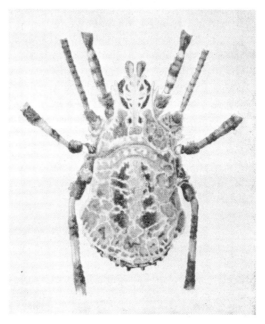

Fig. 59. ORTHOLASMA PICTIPES

forward projecting process of the eye-tubercle; and in that the femora and tibiæ of the legs are not banded. The colour of mature individuals is black, of young ones brownish. The length of the body is about one seventh inch.

This species is common in southern California.

Genus DENDROLASMA (Den-dro-las'ma)

The cephalothorax is armed with a club at each side on the anterior margin. The eye-tubercle projects forward in the form of a central support, with lateral branches somewhat connected at the tips, the whole forming an oval figure. The eyes are visible from above.

Dendrolasma mirabilis (D. mi-rab'i-lis). — The colour of the body is brown or black with the venter paler; the projecting part of the eye-tubercle and the club at each side are whitish. The sides and the hind margin of the cephalothorax are armed with a row of tubercles which are more or less connected, and there is a square of similar tubercles just behind the eye-tubercle. The dorsum of the abdomen has also many series of tubercles. The length of the body is one eighth inch. This species was discovered in Washington State and is the only known one of the genus.

Order ACARINA (Ac-a-ri'na)

The Mites

In this order the abdomen is unsegmented and is not constricted at the base, but is broadly joined to the cephalothorax,

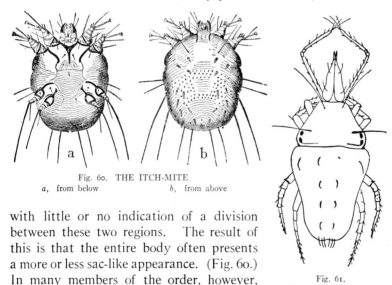

Fig. 60. THE ITCH-MITE
a, from below   *b*, from above

Fig. 61.
BDELLA PEREGRINA
(after Banks)

with little or no indication of a division between these two regions. The result of this is that the entire body often presents a more or less sac-like appearance. (Fig. 60.) In many members of the order, however, the body is divided into two regions which are commonly termed the cephalothorax and the abdomen (Fig. 61). But these regions do not correspond to those bearing the same names in the other orders of the

Spiders and Their Near Relatives

Arachnida. The so-called abdomen of mites includes the true abdomen and the last two thoracic segments; this is indicated by the fact that it bears the third and fourth pairs of legs.

In many mites the body is marked by numerous transverse, fine lines, which are so impressed as to appear like the divisions between minute segments, but the number of these divisions on the ventral aspect of the body may differ greatly from that of the dorsal.

Normally there are six pairs of appendages, as with other arachnids; these are the cheliceræ, the pedipalps, and four pairs of legs.

As a rule the cheliceræ consist of two segments and are often chelate. In many forms, however, the cheliceræ are slender,

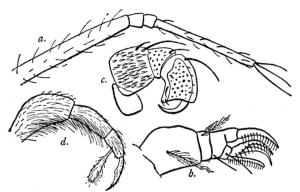

Fig. 62. SEVERAL TYPES OF PEDIPALPS
a, Bdella   b, Cheyletus   c, Arrenurus   d, Trombidium   (adapted from Banks)

needle-like, and fitted for piercing. This difference in form of the cheliceræ is not of great taxonomic value, as in several cases the two types exist within the same family.

The pedipalps consist of not more than five segments; they may be prominent or greatly reduced in size. They vary greatly in form and in function. In some mites, they are simple, filiform, and have a tactile function; in others they are specialized for predatory purposes, being armed with spines, hooks, or claws; and in still others they are chelate, the chela being used for clinging to some object. In some forms, more or less distinct endites are present. In the harvest-mites, and in some other families, the next to the last segment of the pedipalp is armed with one or two claws, and the last segment is often clavate and appears like an appendage of the preceding segment, in such cases it is termed the *thumb*. Several types of pedipalps are represented in Fig. 62.

The head segments, those bearing the cheliceræ and the pedipalps, are often more or less distinct from the following thoracic segment, and form what has been termed the *beak*, *rostrum*, or *capitulum* (ca-pit'u-lum); this division, however, is indistinct in some forms; and sometimes the beak is partly or completely retracted into the following segments. When the beak is thus retracted, the opening of the body from which the mouth-parts project is known as the *camerostoma* (cam-e-ros'to-ma). In *Uropoda* the base of the first pair of legs is also retracted into the camerostoma (Fig. 63).

Sometimes the basal segments of the pedipalps are united and form a *lip*, or *labium;* and above the cheliceræ there is in many forms a thin corneous plate, termed the *epistoma* (e-pis'to-ma). The sides of the epistoma may be united to the labium and thus form what is known as the *oral tube*, for it is through it that the cheliceræ are protruded.

In several families there is an organ termed the tongue or *hypostoma* (hy-pos'to-ma), which arises from the inner base of the beak, and may be divided or simple. In the ticks it is large and roughened with sharp teeth.

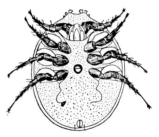

Fig. 63.
UROPODA, VENTRAL ASPECT
(after Banks)

Sometimes the hypostoma has a groove above, called the *vomer*.

As in other arachnids, the normal number of legs is four pairs; but almost invariably the newly hatched young has only three pairs, and in the family Eriophyidæ even the adults have only two pairs.

In the remarkable genus *Pteroptus*, which is parasitic on bats, the larval stage is passed within the body of the mother, and the young mite when born has four pairs of legs.

That the possession of only three pairs of legs by larval mites is an adaptive characteristic is shown by the fact that the embryo of certain forms, as *Gamasus* and *Ixodes*, has four pairs of legs, one pair of which is aborted before the birth of the larva, and is again developed when the larva transforms to a nymph.

Spiders and Their Near Relatives

The legs consist of from five to seven segments. When only five segments are present they are the coxa, trochanter, femur, tibia, and tarsus; the tibia may be divided forming a patella, and sometimes the femur is also divided. In some forms the terminal segments of the legs are divided by more or less distinct false articulations. The tarsus ends in one or two claws; and in many cases an empodium is present; this may be either a cup-shaped sucker or may be claw-like.

One or more pairs of lateral eyes are usually present; as a rule these are sessile, but sometimes they are elevated on pedicels. Rarely the eyes are situated near the middle line of the body.

There is great difference in the position of the openings of the abdomen. Normally the genital aperture is on the ventral surface of the abdomen near the base; but in some forms, as the ticks, the genital segment is pushed forward between the legs so far that it is close to the mouth; and in other forms the genital opening is at the extreme tip of the body, and the anus is upon the dorsal aspect of the abdomen.

Mites breathe either by tubular tracheæ or by the general surface of the body. In those forms that possess tracheæ there are great differences as to the position of the spiracles; these variations have been used by some writers in limiting the principal divisions of the order.

In the course of their postembryonic development, mites undergo a metamorphosis. Most species are oviparous, some are ovoviparous and a few are viviparous. From the egg there appears a form which normally has only three pairs of legs; this is known as the larva. The use of the term larva in this connection is appropriate; for this form resembles the larva of an insect with a complete metamorphoses in being adaptive, as is shown by the temporary reduction of the number of legs. The larva feeds for a time and then, after a resting stage, the skin is shed and the eight-legged nymph appears. Usually at least, the added pair of legs is the fourth. There may be one or more nymphal moults, and the successive instars may differ in appearance, but all lack a completely developed genital orifice. At the close of the nymphal life there is a resting stage during which the nymph transforms to the adult, there being a histolysis and a rebuilding of some of the internal organs as in the pupæ of insects.

Variations from this usual course of transformation occur.

Thus in the genus *Pteroptus* the larval stadium is passed within the body of the mother, and in the genus *Pediculoides* the young are retained within the body of the mother till they reach the adult stage. And in many there has been developed a peculiar nymphal instar, known as the *hypopus* (hyp'o-pus), the function of which is the distribution of the species. The hypopus differs greatly in structure from the preceding and the succeeding nymphal instars, both of which have normal legs and mouth-parts. In the hypopus there is no mouth opening and no mouth-parts; and the legs are short and not fitted for walking. On the ventral surface near the tip of the abdomen there is an area provided with several sucking disks by means of which the hypopus attaches itself to an insect or other animal and is transported to some other locality, where it may find a suitable breeding place. When this is reached the hypopus transforms into a normal nymph.

The order Acarina is one of great biologic interest and of equally great economic importance; but the minute size of most of the species makes its study somewhat difficult, and comparatively few students are attracted to it. In fact few people realize the immense number of individuals and of species of mites that are about us. In a recent catalogue, Banks ('07) enumerates 450 species, representing 133 genera, that are found within the United States. And it is probable that not more than one third of our species have been described.

It is impractical to treat the Acarina in this work as fully as are treated the other orders of Arachnida, without greatly exceeding the desirable limits of space. The student who wishes to study this group seriously must make use of more special works. Fortunately a most excellent treatise on the mites of this country has been published by Banks ('04) who has given much attention to the order and who has described many of our known species. In the preparation of the account given here, I have freely used this work by Banks. Among other important American papers is a *Review of the Genera of the Water-mites*, by Wolcott ('05); a *Revision of the Ixodoidea, or Ticks of the United States*, by Banks ('08), papers on the Oribatoidea by H. E. Ewing, and studies on the gall-mites are being published by Parrott, Hodgkiss, and Schoene at the Geneva, N. Y., Experiment Station. There are many European works on this order; the most important is Canestrini's *Prospetto dell' Acarofauna Italiana*, published at

Padua, in parts, which appeared in the years 1885 to 1899. An older standard work is that of Megnin, *Les Parasites et les Maladies Parasitaires*, Paris, 1880.

Twenty-two of the families of mites are known to be represented in our fauna.* These twenty-two families are grouped into eight superfamilies. We have space here for only a brief discussion of the superfamilies. The eight superfamilies are separated by Banks as follows:

  A. Abdomen annulate, prolonged behind; very minute forms; many with but four legs. P. 93. DEMODICOIDEA
AA. Abdomen not annulate nor prolonged behind; all with eight legs in the adult state.
    B. With a distinct spiracle upon a stigmal plate on each side of the body, usually below, above the third or fourth coxæ or a little behind; palpi free; skin often coriaceous or leathery; tarsi often with a sucker.
      C. Hypostome large, furnished below with many recurved teeth; venter with furrows; skin leathery; large forms, usually parasitic. P. 89. IXODOIDEA
      CC. Hypostome small, without teeth; venter without furrows; body often with coriaceous shields, posterior margin never crenulate; no eyes. P. 90
                                                        GAMASOIDEA
    BB. No such distinct spiracle in a stigmal plate on this part of the body.
      C. Body usually coriaceous, with few hairs; with a specialized seta arising from a pore near each posterior corner of the cephalothorax; no eyes; mouth-parts and palpi very small; ventral openings of the abdomen large; never parasitic; tarsi never with a sucker. P. 91.
                                                        ORIBATOIDEA
      CC. Body softer, without such specialized seta.
        D. Living in water. P. 88. HYDRACHNOIDEA
        DD. Not living in water.
          E. Palpi small, three-jointed, adhering for some distance to the lip; ventral suckers at genital opening or near anal opening usually present; no eyes; tarsi often end in suckers; beneath the

---
* In Banks' Treatise twenty-five families are described; but in his later catalogue three of these families are united with other families.

Spiders and Their Near Relatives

skin on the venter are seen rod-like epimera that support the legs; body often entire; adult frequently parasitic. P. 92   SARCOPTOIDEA

EE. Palpi usually of four or five joints, free; rarely with ventral suckers near genital or anal openings; eyes often present; tarsi never end in suckers; body usually divided into cephalothorax and abdomen; rod-like epimera rarely visible; adults rarely parasitic.

    F. Last segment of palpi never forms a "thumb" to the preceding joint; palpi simple, or rarely formed to hold prey; body with but few hairs. P. 87.   EUPODOIDEA

    FF. Last segment of palpi forms a "thumb" to the preceding, which ends in a claw (a few exceptions); body often with many hairs. P 87.
                                              TROMBIDOIDEA

Superfamily EUPODOIDEA (Eu-po-doi'de-a)

The Eupodoidea includes two families, the Eupodidæ and the Bdellidæ, each of which contains only a small number of known American species. These are predaceous mites which feed on small insects or insect eggs. They live as a rule in moist places, in moss, among fallen leaves, on rotten bark, and in other similar situations; but some are found on the leaves of trees. Many of the species are red or marked with this colour. But our most conspicuous red mites do not belong here; these are the harvest-mites described later.

Superfamily TROMBIDOIDEA (Trom-bi-doi'da-e)

The Red-spiders and the Harvest-mites

The Trombidoidea includes six families, two of which are of general interest. These are the Tetranychidæ or "red-spiders" and the Trombidiidæ or harvest-mites.

The Tetranychidæ (Tet-ra-nych'i-dæ) are of considerable economic importance on account of their injuries to cultivated plants. They are well-known under the common name "red-spider." In the colder portions of the country they are common pests in greenhouses and in the warmer and drier regions they often infest fruit trees in the open air to a serious extent. The common species of the greenhouse is *Tetranychus telarius* (T. te-la'ri-us). To this family belongs also the clover mite, *Bryobia pratensis* (Bry-o'bi-a pra-ten'sis), which infests clover and other annual plants as well as fruit trees.

The species that infest greenhouses can be kept in check by keeping the plants moist, spraying them every day. But the best remedy is flowers of sulphur applied either as a dry powder or as a spray. A spray of kerosene emulsion is also effective, killing the eggs as well as the mites.

The family Trombidiidæ (Trom-bi-di'-i-dæ) includes the well-known harvest-mites. These are always red in colour, but some are much darker than others. Most of the species are of moderate or large size. A common species in the Eastern States, *Trombidium sericeum* (Trom-bid'i-um se-ric'e-um—) is found in moist woodlands and often attracts attention by its bright red colour and silky vestiture.

In the harvest-mites the body is divided into two regions, the cephalothorax and abdomen; but the so-called abdomen bears the third and fourth pairs of legs. The body and legs are covered either with bristles or with feathered hairs. There are two pairs of lateral eyes; in many cases these are borne on pedicels.

The six-legged larvæ of some species are the "red-bugs" of the Southern States that attach themselves to man, and cause serious annoyance. They burrow beneath the skin and produce inflamed spots. They can be killed by the use of a sulphur ointment. The mature mite is not parasitic, but wanders about feeding on small insects.

Superfamily HYDRACHNOIDEA (Hy-drach-noi'de-a)

The Water-mites

The Hydrachnoidea includes two families, the Hydrachnidæ and the Halacaridæ. The Halacaridæ (Hal-a-car'i-dæ) is a small family of marine mites. The species that occur along the coasts of North America have not been studied, only a single species having been described; but of the Hydrachnidæ about one hundred North American species are known. These represent twenty-five genera.

In the Hydrachnidæ (Hy-drach'ni-dæ) the body is short, usually high, and sometimes nearly spherical. There is no division between the cephalothorax and abdomen. They have one or two pairs of eyes, which in some cases are situated near the median line.

In most species the adult mite lives free in the water, but some are parasitic in the gills of mollusks. Nearly all of the species live in fresh water; but some are found in brackish water, and a few live in the littoral zone of the sea. The free species feed on small crustacea, infusoria, and minute insect larvæ.

The larval mites are often found attached to aquatic insects; they appear like club-shaped eggs attached by the small end.

Superfamily IXODOIDEA
(Ix-o-doi'de-a)

The Ticks

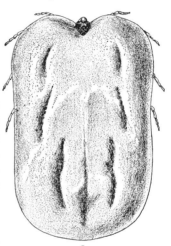

Fig. 64. THE CATTLE TICK
a, female *b*, male

Of all mites, the ticks are the ones that most often attract the attention of those who are not making a study of this order. This is due to the fact that they attach themselves to man and to domestic animals and are exceedingly annoying pests. This is especially true in the warmer portions of the country. Among the ticks are also the most important members of the order from an economic standpoint as certain species transmit serious diseases.

Ticks are parasitic on mammals, birds, and reptiles. In most cases they do not seriously injure their host, but in others they cause serious inflammation and swelling of the infested part. Among those that transmit diseases the Southern cattle tick is the most important one that occurs in this country.

It is the Southern cattle-tick, *Boophilus annulatus* (Bo-oph'ı-lus an-nu-la'tus) (Fig. 64) that transmits the tick-fever, or Texas-fever as it was formerly called, from one animal to another. This disease has caused the death of thousands of cattle in this country; and makes it extremely difficult to introduce cattle from the North into the South. This tick is found only in the South and the Federal Government has established a quarantine line across which cattle are not allowed to be shipped or driven without being freed from the ticks.

The Department of Agriculture at Washington and several of the Southern Experiment Stations have issued publications giving the life history of this tick in great detail and suggestions as to methods of preventing the spread of the disease; these can be obtained on request by any one interested in the subject. In some of these publications this tick is designated as *Margaropus annulatus* (Mar-gar'o-pus an-nu-la'tus).

The cause of the disease is a protozoan parasite *Babesia bovis* (Ba-be'si-a bo'vis), which lives in the blood of cattle, destroying the red blood corpuscles and which is transferred from one animal to another by the tick. Even the young of ticks that have infested diseased cattle may transmit the disease.

Superfamily GAMASOIDEA (Gam-a-soi'de-a)

The Gamasid Mites

The Gamasid mites are often observed attached to beetles and other terrestrial insects upon which they spend a part of their life. It was formerly believed that these mites were parasites of the insects, but it has been found that they are nymphs which secure transportation in this way, a method of distribution of the species.

A few species are parasitic on birds, bats, and small mammals; some are found in ant-nests; a great many occur among fallen leaves, and some live in decaying substances, either animal or vegetable. Most species prey on small insects or on other mites. Among those found on insects, some are attached by an anal pedicel formed of excretions.

Among the species that are parasitic on birds there is one *Dermanyssus gallinæ* (Der-ma-nys'sus gal-li'næ), which is a serious pest of poultry. The mites of this species hide in cracks and crevices by day, but at night crowd upon the fowls and suck their blood. This pest can be destroyed by spraying the poultry houses with kerosene.

All of our species are now included in the family Gamasidæ of which nearly fifty species have been described representing eighteen genera.

Superfamily ORIBATOIDEA (Or-i-ba-toi'de-a)

The Oribatid Mites

The oribatid mites are now classed as a single family, the Oribatidæ; this is a large one, nearly one hundred and fifty species being known from North America alone. Nearly all of these were described by either Mr. Banks or Mr. Ewing; the number of species will doubtless be greatly increased when our fauna is more thoroughly explored.

In most cases these mites are distinguished by the presence of a pair of large club-shaped setæ arising from near the posterior corners of the cephalothorax (Fig. 65). These setæ are known as the pseudo-stigmatic organs, from the fact that the trichopores from which they arise were formerly believed to be spiracles.

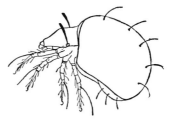

Fig. 65. AN ORIBATID MITE (*Hoploderma sphærula*) (after Banks)

With most of the Oribatidæ the cuticula is coriaceous, giving the mites some resemblance in appearance to small beetles. For this reason they are sometimes termed "beetle mites"; but as this name is often applied to those gamasid mites that are found on beetles it is not distinctive.

The resemblance of the oribatid mites to beetles is increased by the fact that the suture separating the cephalothorax and abdomen is usually quite distinct. Many of the common forms are shining black in colour.

These mites are of little or no economic importance; some of them are supposed to injure grass to a slight extent; but most of them are found among moss, on the bark of trees, and on the ground under wood, bark, stones, and fallen leaves. Most of the species feed on vegetable matter, a few are found on decaying animal matter, and some bore into decaying wood.

Spiders and Their Near Relatives

Superfamily SARCOPTOIDEA (Sar-cop-toi'de-a)

The Itch-mites and Others

The Sarcoptoidea include seven families of mites; but several of these families are small and include only forms that are not likely to attract the attention of the ordinary observer. Some of these forms, however, are of great scientific interest, as for example *Pediculoides,* which lives parasitically upon insects, and which gives birth to sexually mature young. It is in one of the families of this group, the Tyroglyphidæ, that those mites that have the remarkable migratory stage, the hypopus, described on an earlier page, are found. It is to this family also that belong certain species which attack food products as cheese, sugar, flour and dried meats; these are often of considerable economic importance. The disease known as "grocers' itch" is caused by these mites which sometimes spread from infested materials to the hands of those handling them.

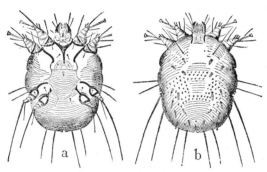

Fig. 66. SCARCOPTES SCABEI

To the family Sarcoptidæ belong the itch-mites, which are so-called because they often burrow within the skin of man, causing the disease known as the itch. The diseases of domestic animals known as scabies or mange are also due to members of this family.

The species attacking man is *Sarcoptes scabei* (Sar-cop'tes sca'be-i) (Fig. 66). It is apt to infest the soft skin between the fingers and may spread to other parts of the body. The mites make burrows within the skin and the eggs are deposited within the burrows. The spread of the disease is due to the spreading of the mites from one person to another; this is often brought about by hand-shaking. The remedy most often used is a sulphur ointment, by means of which the mites can be killed.

The disease known as sheep-scab is caused by the mite

Sarcoptes ovis. This disease is combatted by dipping the sheep in some poisonous solution. A "sheep-dip" approved by the Federal Government is made of extract of tobacco or nicotine solution and flowers of sulphur. It contains five hundredths of one per cent. of nicotine and two per cent. flowers of sulphur; or, for the first dipping in lieu of the sulphur, more nicotine is used so that the solution contains not less than seven hundredths of one per cent. of nicotine. A second dipping of the infested animal should be made six or eight days after the first, in order to kill the mites that have hatched in the interval.

Superfamily DEMODICOIDEA (De-mod-i-coi'-de-a)

The Gall-mites and Others

In this superfamily are grouped certain very minute mites that agree in having the abdomen annulate and prolonged behind; it includes two families.

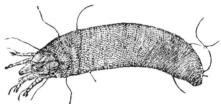

Fig. 67. ERIOPHYES PYRI

The more important of these two families is the Eriophyidæ or gall-mites. The members of this family have only four legs and these are five-jointed. The mites live on plants and often within galls.

Some species produce galls which resemble those made by plant lice or aphids in having an open mouth. These galls may be trumpet-shaped or they may be but slightly elevated resembling a blister.

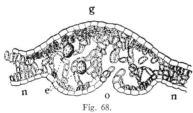

Fig. 68.
DIAGRAM OF GALL OF ERIOPHYES PYRI
g, gall   *n,n*, normal structure of leaf
o, opening of gall   *e*, eggs   (after Soraur)

A common disease of the pear and apple, known as the pear-leaf blister, is produced by *Eriophyes pyri* (Er-i-o-phy'es py'ri) (Fig. 67). The blisters characteristic of the disease are swellings of the leaf, within which there is a cavity affording a residence for the mites. Figure 68 represents a section of a leaf through one of these galls. Here the leaf is seen to be greatly thickened at the diseased part. On the lower side there is an opening through

Spiders and Their Near Relatives

which the mite that started the gall entered, and from which young mites developed in the gall can escape, in order to start new galls. In addition to the swelling of both surfaces of the leaf its internal structure is seen to be modified. In some parts there is a great multiplication of the cells, and in others a large part

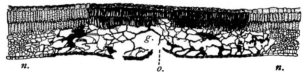

Fig. 69. GALL OF ERIOPHYES PYRI
g, gall *n,n*, uninjured part of leaf *o*, opening of gall

of the cells have been destroyed. Two eggs of mites are represented in this gall. As the season advances, and the galls become dry and brownish or black, the thickening of the leaf becomes less marked. In fact, in some cases there is a shrinkage of the parts affected. Figure 69 represents a section through a leaf collected and studied in October.

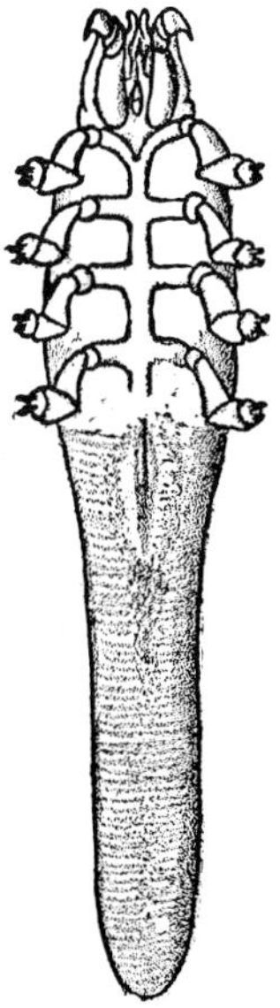

The most conspicuous of the abnormal growths on leaves caused by mites are bright coloured patches, often mistaken for fungoid growths; these occur on the leaves of various trees, and are due to an abnormal development of the epidermal plant cells or a deformation of the plant hairs, caused by the mites. This type of abnormal growth is termed an *erineum* (e-rin′e-um).

The members of the family Demodecidæ resemble the gall-mites in having the body annulate and prolonged behind; but differ in having eight legs which are three-jointed. It includes a single genus, *Demodex*, the species of which are found in the sebaceous glands and hair-follicles of various mammals. The species that infests man is *Demodex folliculorum* (Dem′o-dex fol-lic-u-lo′rum) (Fig. 70). It was formerly supposed to be the cause of "blackheads" or comedones (com-e-do′nes) on the face; but it has been found that the mites occur in healthy as well as diseased follicles.

Fig. 70. DEMODEX FOLLICULORUM (after Megnin)

CHAPTER II: THE EXTERNAL ANATOMY OF SPIDERS

SPIDERS resemble the allied animals described in the preceding pages in having the segments of the body grouped in two regions, the *cephalothorax* (ceph-a-lo-tho'rax) and the *abdomen* (Fig. 71), in having four pairs of legs fitted for walking, and in having the antennæ modified into organs of prehension, the *chelicerœ*.

The spiders differ from other Arachnida in having the abdomen unsegmented and joined to the cephalothorax by a narrow stalk. There is a single small family of spiders, *Liphistiidæ*, in which the abdomen is segmented; but representatives of this family have been found only in the East Indies. In the genus *Tetrablemma*, found in Ceylon, the abdomen bears a series of plates, which are evidently vestiges of a segmented condition. But in all American spiders the abdomen is sac-like.

THE CEPHALOTHORAX

In spiders the cephalothorax, like the abdomen, is unsegmented; although frequently the head and the thorax are slightly separated by a furrow, the *cervical groove* (Fig. 72). In such cases, most writers refer to the head as the cephalic part, or the *pars cephalica*, and to the thorax as the thoracic part, or the *pars thoracica;* but the simple terms *head* and *thorax* are sufficient for all purposes, and will be used in this book when it is necessary to refer to the principal divisions of the cephalothorax. The hard integument forming the dorsal wall of the cephalothorax is termed the *carapace* (car'a-pace).

THE HEAD

The head is that portion of the cephalothorax which bears the eyes and the so-called mouth-parts, the appendages that are used for seizing and chewing their prey. It is either slightly or not at all separated from the thorax; but it is almost always

The External Anatomy of Spiders

easily distinguished. It is usually wedge-shaped behind; the lateral portions of the thorax extending a considerable distance on each side of it (Fig. 72).

THE EYES.— The eyes are all simple, resembling in appearance the ocelli or simple eyes of insects; in none of them is the

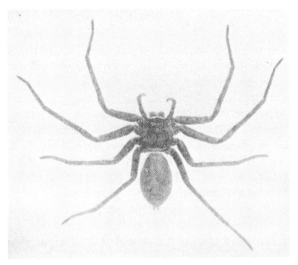

Fig. 71. A SPIDER, SHOWING THE DIVISION OF THE BODY INTO CEPHALOTHORAX AND ABDOMEN

outer layer divided into facets as in the compound eyes of insects. They are usually situated near the front end of the head; in some cases they are grouped upon a tubercle (Fig. 72); in others, they are separated so as to occupy nearly the whole width of the head (Fig. 73).* The normal number of the eyes is eight; but two, four or six of them may be wanting and certain cave spiders are blind. The number and the arrangement of the eyes furnish characters which are much used in classification, as is shown in later portions of this book.

In works on the classification of spiders, two types of eyes are distinguished, the *nocturnal eyes* and the *diurnal eyes*. The so-called nocturnal eyes are found in spiders that live in the dark or that frequent shady places; they are distinguished by being pearly white in colour. The so-called diurnal eyes lack the pearly lustre and are variously coloured. This distinction

---
* See page 10 on the primitive position of the eyes.

The External Anatomy of Spiders

is not a good one; it is discussed at length in the next chapter, where the structure of nocturnal eyes is explained.

The anterior median eyes differ from the other six eyes in a remarkable manner as regards their intimate structure and mode of development. These eyes have been termed the *postbacillar* (post-bac'il-lar) *eyes;* and the other six eyes, the *prebacillar eyes*. The differences in structure between these two types of eyes are discussed in the next chapter.

In many spiders the eyes are arranged in two transverse rows each containing four eyes (Fig. 73); and this is regarded as the normal arrangement. Special names are applied to the different eyes, the names being suggested by their relative positions when thus arranged. The names are *anterior median*, applied to the two intermediate eyes of the first row; *posterior median*, the two intermediate eyes of the second row; *anterior*

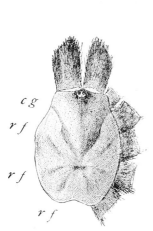

Fig. 72.
CARAPACE OF A TARANTULA
c g, cervical groove
r f, radial furrows

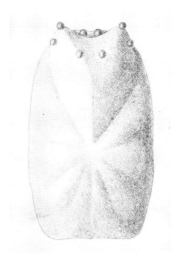

Fig. 73. MODEL OF THE CARAPACE OF A SPIDER VIEWED FROM ABOVE. THE ANTERIOR ROW OF EYES APPEARS TO BE PROCURVED

lateral, one at each end of the first row; and *posterior lateral*, one at each end of the second row. Some writers refer to the lateral eyes as the *side eyes*.

The rows of eyes are frequently curved. When the lateral eyes of a row are farther forward than the median eyes the row

The External Anatomy of Spiders

is said to be *procurved;* when the lateral are farther back than the median eyes the row is *recurved.* In determining whether a row of eyes is procurved or recurved, a line passing through the centre of the eyes is considered, and not one tangent to either anterior or posterior border of them, and in case of the anterior row they should be viewed from in front. In this case if the anterior lateral eyes are higher than the anterior median — i. e., farther from the edge of the clypeus — the row is recurved, although when seen from above the lateral eyes may be farther forward than the median eyes and consequently the row is apparently procurved. This is illustrated by the accompanying two views of a model (Fig. 73 and Fig. 74).

Fig. 74.
THE MODEL REPRESENTED IN FIG. 73 VIEWED FROM IN FRONT; THE ANTERIOR ROW OF EYES IS RECURVED

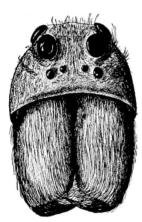

Fig. 75.
EYES IN THREE ROWS, LYCOSA CAROLINENSIS

Sometimes the curvature is very great, and the posterior median eyes are widely separated; then the eyes are said to be arranged in three rows (Fig. 75); and in some cases the eyes are in four rows, each pair of median eyes and each pair of lateral eyes constituting a row (Fig. 76).

Fig. 76.
EYES IN FOUR ROWS, LYSSOMANES

THE AREAS OF THE HEAD.— Special names are applied to different areas of the head; but the areas thus designated are not limited by sutures and consequently the names applied to them do not have the definite morphological significance that similar names have in descriptions of insects. The areas most commonly recognized are the following:

The eye-space.— That part of the head which is between the rows of eyes is termed the *eye-space.*

The median ocular area.—The space limited by the four median eyes and including that occupied by these eyes is termed the *median ocular area.*

It is also termed by some writers the *ocular quadrangle* or the *ocular quad*. In descriptions of the jumping spiders (Attidæ), the term *ocular quadrangle* refers to the space occupied by all of the eyes; and in this book this term is used only in this sense.

In the Lycosidæ the four posterior eyes outline a trapeziform area, which is described as the *quadrangle of the posterior eyes*.

The ocular tubercle.— In many spiders the eyes are situated on an elevated portion of the head; this portion is termed the *ocular tubercle*.

The clypeus (clyp′e-us).— The space between the eyes and the first pair of appendages, the chelicera, is termed the *clypeus*. Frequently, in the tables, reference is made to the width of the clypeus. This is the distance from the front edge of the clypeus to the eyes nearest that edge.

The face.— That part of the head which can be seen when the spider is observed from directly in front is termed the *face* (Fig. 77). This term is used in descriptions of certain spiders belonging to the second suborder. It includes the clypeus and a part or the whole of the eye-space.

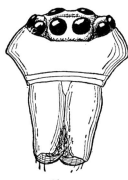

Fig. 7.
FACE AND CHELICERÆ OF THERIDION

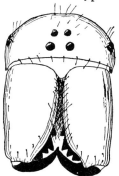

Fig. 78.
FACE AND CHELICERÆ OF ARANEA

The front.— The anterior portion of that part of the head lying immediately back of the clypeus is termed the *front*; the front includes the eye-space, but has no definite limit behind.

THE CHELICERÆ (che-lic′e-ræ).— The first pair of appendages of the head are the *chelicera*. They are situated in front of and above the mouth, from which they are separated by the rostrum, and consist each of two segments, a large basal one, and a terminal, claw-like one (Fig. 78). These are the appendages with which the spider seizes and kills its prey. Near the tip of the claw there is the opening of a poison gland (Fig. 79).

The chelicera are modified antennæ; they are homologous

with the second antennæ of Crustacea, but not with the antennæ of insects.*

It has been supposed by many writers that the cheliceræ correspond to the mandibles of insects, and they have, therefore, been called *mandibles*, a term that cannot properly be applied to them. The term *falces*, proposed by Blackwall in 1852, is also used by many writers, but there is no good reason for discarding the much older term cheliceræ.

The lateral condyle of the chelicera.— In many spiders there is at the base of each chelicera, on the lateral face, a smooth prominence, an articulating condyle (Fig. 80, *l c*); this may be termed the *lateral condyle of the chelicera*. This condyle is wanting in many families; and its presence or absence is a useful character in classification. It is sometimes called the *basal spot of the chelicera*.

When the condyle is present, the margin of the clypeus is usually marked with a dark spot at the point of articulation with the condyle.

The furrow of the chelicera.— Usually there is a furrow in the basal segment of the chelicera for the reception of the claw when it is closed (Fig. 81); and often there is on one or both sides of this furrow a row of teeth (Fig. 83).

In the tarantulas, where the claw moves vertically, the gins of this furrow may be designated as the inner and the outer respectively; in the true spiders, as the upper and the lower.

The scopula (scop'u-la).— Frequently there is on the upper side of the furrow of the chelicera, a brush of hairs (Fig. 82, *s*); this is the *scopula of the chelicera*. Scopulæ are found also on other appendages.

The rake of the chelicera.— In certain tarantulas that burrow in the ground the extremity of the basal segment of the chelicera is armed with several rows of strong teeth, which are used by the spider in excavating its burrow; these constitute the *rake of the chelicera* (Fig. 83).

The claw of the chelicera.— The claw of the chelicera is very hard, curved, and pointed. On its concave face, there are usually

---

* In the Crustacea (lobsters, crabs, and allies) there are two pairs of antennæ, the *first antennæ* or antennules and the *second antennæ*, commonly called the antennæ. In spiders the first antennæ are lost, appearing only in the embryo; and the second antennæ are modified into prehensile organs, the cheliceræ. In insects the first antennæ are retained and function as feelers, while the second antennæ are almost invariably lost: they are retained, however, in a vestigial condition in some Thysanura.

Fig. 79. TIP OF CLAW OF CHELICERÆ

Fig. 80. HEAD AND CHELICERÆ OF A SPIDER
l c, lateral condyle

Fig. 81. CLAW AND PART OF THE BASAL SEGMENT OF A CHELICERA
a, articular sclerite *k*, toothed keel
o, opening of the poison gland

Fig. 82. CHELICERA OF AGELENA NÆVIA WITH SCOPULA
s, scopula

Fig. 83. CHELICERA OF A TARANTULA SHOWING THE RAKE OF THE CHELICERA

Fig. 84. A CHELATE CHELICERA, PHOLCUS

Fig. 85. ROSTRUM AND ENDITES OF THE PEDIPALPS OF ARGIOPE, DORSAL VIEW
r, rostrum

two delicate keels, of which the lower is usually finely and regularly toothed (Fig. 81, *k*). The claw is traversed by the duct of the poison gland, which opens near the tip of the convex side (Fig. 81, *o*). The position of this opening is such that it is not closed by the pressure of the claw against the victim, and it allows the venom to flow into the wound made by the claw.

The claw of the chelicera is freely movable, but only in one plane. In the tarantulas, the claws of the two cheliceræ move parallel to each other in vertical planes; while in the true spiders they move obliquely inward and backward toward each other.

The articular sclerite.— At the base of the claw and on the side toward the furrow, there is a small sclerite (Fig. 81, *a*); this is the *articular sclerite of the chelicera*. This sclerite may be a vestige of an intermediate segment of the chelicera (See Bernard, '96, p. 322).

Chelate chelicerœ.— In most cases the chelicera of spiders are of the uncate type (See p. 12); but in a few spiders the terminal portion of the inner margin of the basal segment is prolonged so as to oppose the tip of the claw and thus form a chela (Fig. 84).

THE ROSTRUM.— The *rostrum* or upper lip is a single, median, appendage-like part of the head, which resembles in form and position the labrum of an insect; and it is quite probable that the two are homologous. The rostrum is situated below or behind the cheliceræ and between the second pair of appendages, the pedipalps.

By removing the cheliceræ, the rostrum can be seen lying between and upon the endites of the pedipalps (Fig. 85, *r*). On the dorsal surface of the rostrum there is a longitudinal, median keel, which is clothed with a band of hairs.

THE EPIPHARYNX.— On the ventral surface of the rostrum there is a plate which may be termed by analogy the *epipharynx*. This plate is strongly chitinized, and is marked by many transverse striæ, which lead to a central longitudinal slit, opening into a longitudinal tube within the rostrum (Fig. 86,).

As the tube in the rostrum extends back to the beginning of the œsophagus, the function of these striæ in the epipharynx is, evidently, to collect the fluid pressed from the spider's victim and to conduct this fluid to the tube, from which it can flow to the œsophagus.

THE LABIUM (la'bi-um).— The ventral wall of the head is

Fig. 86.
THE EPIPHARYNX
OF ARGIOPE
a, tip *b*, basal portion

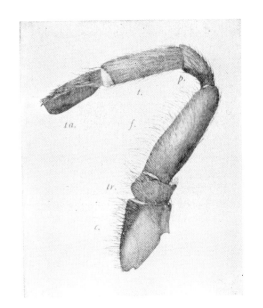

Fig. 88. PEDIPALP OF A TARANTULA
EURYPELMA, VENTRAL SURFACE
c, coxa *f*, femur *p*, patella *t*, tibia *ta*, tarsus
tr, trochanter

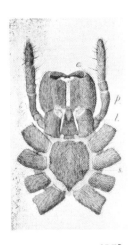

Fig. 87. MOUTH-PARTS
AND STERNUM OF
AMAUROBIUS
c, chelicera *l*, labium
p, pedipalp *s*, sternum

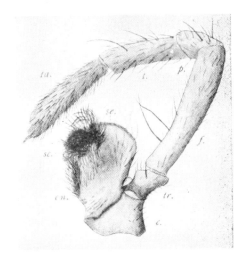

Fig. 89. PEDIPALP OF A TRUE SPIDER,
TRACHELAS
c, coxa *en*, endite *f*, femur *p*, patella *sc*, scopula
se, serrula *t*, tibia *ta*, tarsus *tr*, trochanter

103

formed of a single sclerite, which is usually more or less movable, and which on account of its position has been termed the lower lip or *labium*. Like the rostrum, the labium is situated between the second pair of appendages, the pedipalps (Fig. 87), the rostrum occupying a more dorsal position, and the labium, a more ventral one.

The labium of a spider is not homologous with the labium of an insect, which is formed of a pair of united appendages.

THE MOUTH.— The mouth cavity is situated between the base of the pedipalps, which form the sides of this cavity. The roof of the mouth is formed by the epipharynx; and the floor by a sclerite, the tip of which lies upon the labium.

The mouth is fitted for the reception of only liquid food. The spider cuts and presses its victim; and when it is sucked dry, the hard parts are thrown away.

THE PEDIPALPS (ped'i-palps).— The second of the two pairs of appendages of the head are the *pedipalps*. They are situated one on each side of the mouth; and are more or less leg-like in form, especially in females. By some writers these appendages are termed the *maxillæ;* but as they are not homologous with the maxillæ of insects, this is an undesirable use of the term.

Each pedipalp consists of a series of six segments which are named, beginning with the one next to the body: *coxa, trochanter, femur, patella, tibia,* and *tarsus* (Fig. 88). These terms are the same as those applied to the corresponding segments of the leg. The pedipalps differ from the legs in that the tarsus consists of a single segment, there being no metatarsus; and there is never more than one tarsal claw, which is wanting in all males and in the females in many genera.

A special set of terms for the segments of the pedipalps has been much used and is included here for reference. According to this system the basal part is termed the maxilla and the following segments, beginning with the trochanter, are designated as the *axillary, humeral, cubital, radial* and *digital.*

The endites.— In the tarantulas, except *Atypus,* the coxa of the pedipalps closely resembles the coxa of a leg (Fig. 88); but in the true spiders the coxa bears a plate, which is the crushing part of the organ, this plate is the *endite* (Fig. 89, *en*). On the dorsal surface of the organ there is a suture separating the endite from the basal part of the coxa. This suture is present

The External Anatomy of Spiders

even in the tarantulas where the endite is but little developed (Fig. 90). The endite is also known as the *maxillary plate;* but this is an undesirable term, as the pedipalps are not homologous with maxillæ.

The scopula (scop'u-la). — The internal border of the endite in many cases bears a brush of hairs, this is the *scopula of the pedipalp* (Fig. 89, *sc*).

The serrula (ser'ru-la). — In addition to the scopula the endite usually bears near its distal margin a keel which is finely toothed, this is the *serrula* (Fig. 89, *se*; and Fig. 91). The serrula doubtless plays an important part in lacerating the prey so as to set free the juices upon which the spiders feed.

Fig. 90. PEDIPALP OF TARANTULA, EURYPELMA; DORSAL SURFACE

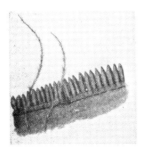

Fig. 91. THE SERRULA OF THE ENDITE OF TRACHELAS

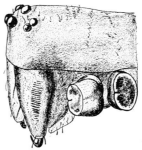

Fig. 92. PART OF LEPHTHY-PHANTES NEBULOSUS, SHOWING THE FILE ON THE CHELICERA

The palpus. — The coxa and the endite taken together are regarded as the trunk of the pedipalp and the remaining segments as an appendage, which is termed the *palpus*.

The External Anatomy of Spiders

In females the tarsus of the palpus resembles the tarsus of a leg except that it consists of a single segment and bears only a single claw or none. But in the males the tarsus of the palpus is more or less enlarged and is very complicated in structure. As the details of this structure vary greatly in different groups of spiders much use is made of them in classification; for this reason a special discussion of this part is given later. The characteristic features of the palpi of males are not fully developed until the spider reaches maturity; in young males the palpus appears merely as a simple club-like organ.

THE STRIDULATING ORGANS OF THE MOUTH-PARTS.— In the sheet-web weavers (Linyphiidæ), the external face of the cheliceræ is furnished with a file-like series of ridges (Fig. 92) against which the inner face of the femur of the pedipalps is rubbed to produce a sound. And in several genera of tarantulas also the mouth-parts are furnished with stridulating organs.

THE PALPI OF MALE SPIDERS

INTRODUCTION

The remarkable modification of the palpi of the males of spiders into organs for the transference of the seminal fluid to the female at the time of pairing of the sexes attracted the attention of naturalists at a very early date; and the great variety of forms presented by these organs has led systematists to make much use of them in taxonomic work. In practically all of the more important works on the classification of spiders there are figures and descriptions of the palpi of males.

Notwithstanding the general recognition of the value of these organs for taxonomic purposes our knowledge of their structure is very inadequate. Several important contributions to this subject have been published and are well-known, notably those of Westring ('61), Menge ('66), Bertkau ('75 and '78), Wagner ('87), Van Hasselt ('89), and Chamberlin ('04 and '08). Still we find, even in the more recent publications, figures of palpi given with almost no effort to identify their parts; and even when some of the parts are named we find different terms applied to homologous parts in the descriptions of different genera.

The necessity of selecting, from the many terms that have been proposed for parts of the palpi, a set to be used in this volume and the need of terms for parts that had not been described led me to make a special study of the subject. The results of this study were published recently (Comstock '10); but the more important of them are repeated here so that they may be available for use in this book.

THE MORE GENERALIZED TYPES OF PALPI

In all spiders the external opening of the reproductive organs of the male is on the lower side of the abdomen near its base, in the epigastric furrow. Some time before pairing the seminal fluid is emitted from this opening and is stored in a tubular cavity in an appendage of the last segment of the palpus, where it is retained until the pairing of the sexes, and from which it then passes to the spermathecæ of the female.

The transference of the seminal fluid from the opening of the reproductive organs to the receptaculum seminis of the palpus has been observed by several

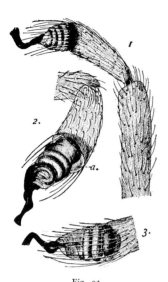

Fig. 93.
TARSUS OF FILISTATA
HIBERNALIS
1, lateral aspect   2, oblique view
3, mesal aspect

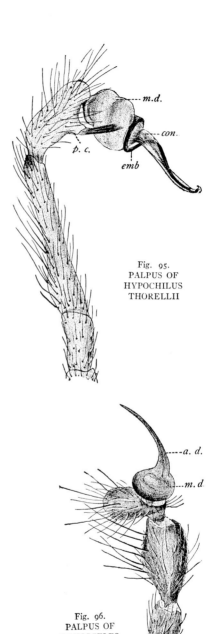

Fig. 95.
PALPUS OF
HYPOCHILUS
THORELLII

Fig. 94.
DIAGRAM OF THE
RECEPTACULUM
SEMINIS

Fig. 96.
PALPUS OF
LOXOSCELES
RUFESCENS

writers. From the published observations it appears that the male spins a delicate web upon which he emits the seminal fluid; after which the fluid is taken up by the palpus.

The genital appendage of the palpus of the male is exceedingly complicated in structure in the more specialized spiders, as in the Argiopidæ; but it is comparatively simple in some of the more generalized families. A few illustrations of the simpler forms will be given here.

THE FILISTATA TYPE OF PALPUS.— In *Filistata hibernalis*, which is a very common house spider in the South, is found the most simple type of male palpus that I have seen among spiders. In the males of this species, the distal end of the last segment of the palpus, the tarsus, contains a coiled tube (Fig. 93); this is the *receptaculum seminis*. The proximal portion of this tube is slightly enlarged and ends blindly; the distal part is slender and extends through a slender, twisted prolongation of the tarsus ending at its tip by an open mouth. The modified terminal portion of the tarsus, which contains the receptaculum seminis, is the *genital bulb*. By looking directly at the tip of the palpus, instead of at one side of it, it can be seen that the base of the bulb is situated in a cavity in the end of the main part of the palpus (Fig. 93, *a*);* this cavity is the *alveolus* (Menge '66). The slender prolongation of the bulb, which contains the terminal portion of the receptaculum seminis is the *embolus;* the embolus is often termed the *style*.

A study of the palpus of *Filistata* gives a clue to the probable course of the evolution of the genital bulb. It is evident that the bulb is a specialization of the tip of the tarsus, and its most striking feature is the presence within it of the coiled receptaculum seminis. Regarding the origin of the receptaculum seminis, the fact that it is furnished with a transversely striated intima, like the intima of a trachea, indicates that it is merely an invagination of the body-wall. In its primitive form, it was probably a cup-like depression in the tip of the tarsus.

In its most perfect form, as seen in the more specialized spiders, the receptaculum seminis consists of three quite distinct parts: first, the proximal end of it, the *fundus*, is enlarged so as to form a pouch, the wall of which is more delicate than that of the other parts (Fig. 94, *fu.*); I have not been able to see tænidia in the intima of this part, and infer that it serves as a compressible bulb; second, the intermediate portion, the *reservoir*, is a large coiled tube occupying the middle division of the genital bulb (Fig. 94, *res.*), in this part the tænidia of the intima are well-developed and are sometimes very prominent; third, the terminal portion constitutes the *ejaculatory duct*, this is a slender tube traversing the apical division of the bulb (Fig. 94, *ej. d.*) the wall of this duct is often dark in colour, which renders it easy to trace the course of the duct in an expanded bulb.

The tracing of the course of the ejaculatory duct is often the only method by which the embolus can be recognized in a complicated palpus; for when the embolus is small or when it is lamelliform a slender apophysis may be mistaken for it.

After the stage represented by *Filistata* had been reached, a shifting in the position of the bulb occurred in most spiders. Instead of occupying a terminal position, at the tip of the tarsus, it has moved to one side of the tarsus in all spiders known to me except *Filistata*. In the tarantulas and in *Hypochilus thorellii*, the most generalized in many respects of the true spiders, the genital bulb is nearly terminal, but is, nevertheless, distinctly on one side of the tarsus (Fig. 95). In other spiders it has moved to a greater or less extent toward the base of the tarsus, which it has nearly reached in many, as for example in *Loxosceles rufescens* (Fig. 96).

In *Hypochilus* (Fig. 95) and in *Loxosceles* (Fig. 96), the alveolus is comparatively small; but in many spiders it is large, resulting in the tarsus being more or less cup-like in form; this is shown in some of the figures of the more specialized palpi given later. This cup-like form of the tarsus, as distinguished from its appendage, the genital bulb, suggested for it the name *cymbium*, which is the classical name of a small drinking vessel.

---

*In the figures of palpi of males given in this chapter, uniform abbreviations are used for the names of the parts. A list of these is given at the close of the account on page 121.

The External Anatomy of Spiders

THE TARANTULA TYPE OF PALPUS.— In those spiders that are commonly known in this country as tarantulas, and which represent the more generalized of the two principal divisions of the order Araneida, there exists a comparatively simple type of palpus; but in none of them that I have seen, or of which I have seen figures, is it as generalized as is the palpus of *Filistata*.

In the palpi of the tarantulas, the genital bulb has migrated to one side of the tarsus; but it is still near the tip of this segment of the palpus (Fig. 97). A striking feature of the bulb is that it is divided into two distinct segments. The smaller basal segment may be termed *the basal division of the bulb* (Fig. 97, *b. d.*). The larger segment consists of two parts: a large stout part, which may be termed *the middle division of the bulb* (Fig. 97, *m. d.*), and a slender terminal portion, which may be termed *the apical division of the bulb* (Fig. 97, *a. d.*); there is, however, no distinct line between the middle and the apical divisions, the one gradually merges into the other; but in the more specialized palpi these two divisions are distinctly separated.

In the articulating membrane which joins the

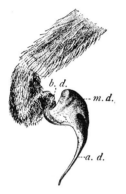

Fig. 97. TARSUS OF EURYPELMA

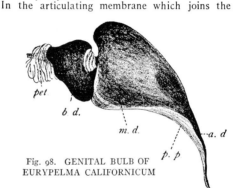

Fig. 98. GENITAL BULB OF EURYPELMA CALIFORNICUM

bulb to the tarsus, there is on one side a distinct sclerite, which can be seen by removing the bulb from the alveolus (Fig. 98, *pet.*); this is doubtless homologous with what has been termed the *petiole* in more specialized palpi.

The greater part of the wall of the bulb in the tarantula type of palpus is very densely chitinized but there is a longitudinal area on the concave side of the middle and apical divisions which is comparatively soft (Fig. 98, *p.p.*); it may be that this part is distended by blood pressure at the time of pairing as is the hæmatodocha in the more specialized palpi; but upon this point I have no data. This soft strip may correspond to that portion of the spiral type of embolus, described later, that I have designated the *pars pendula*.

THE PALPUS OF LOXOSCELES.— In certain genera of the true spiders, the palpi are as simple as in the tarantula type. In *Loxosceles* of the family Scytodidæ, for example (Fig. 96), although the bulb has migrated nearly to the base of the tarsus, the bulb itself is very simple in structure. The basal division of the bulb is inconspicuous; the middle division is nearly spherical, and the apical division is long and slender. Here the receptaculum seminis is differentiated into the three parts described above; the reservoir is large, while the ejaculatory duct is very slender.

THE PALPUS OF DYSDERA.— In the family Dysderidæ two quite distinct types of palpi occur. In *Ariadna* the palpus resembles very closely that of *Loxosceles;* but in *Dysdera* it is of a very different form (Fig. 99); this is due to the fact that the apical division of the bulb is not slender, and is sharply differentiated from the middle division, its wall being much less densely chitinized. But there is on each margin a distinct sclerite; and this part of the bulb bears distinct apophyses. At the tip of the apical division there appears to be the beginning of a separation into embolus and conductor.

The External Anatomy of Spiders

A summary of the parts of the tarsus in the more generalized types of palpi of males is shown by the following table:

Body of tarsus or cymbium, containing the alveolus
Genital bulb
    Internal parts
        Receptaculum seminis
        Fundus
        Reservoir
        Ejaculatory duct
    External parts
        Petiole
        Basal division
        Middle division
        Apical division or embolus

THE INTERMEDIATE TYPES OF PALPI

There are palpi which hold an intermediate position as regards complexity of structure between the comparatively simple tarantula type and the exceedingly complex forms to be described later. These intermediate types occur in widely separated portions of the araneid series; but agree in their more essential characteristics; for sake of brevity, I will discuss only a few examples of the intermediate types; and will then pass to a description of forms in which the maximum number of parts are found.

Fig. 99. PALPUS OF DYSDERA INTERRITA

The most important characteristic of these intermediate types is that the apical division of the bulb is separated into two, more or less nearly, parallel parts. One of these parts contains the ejaculatory duct of the receptaculum seminis, this is the *embolus;* the other is intimately associated with the embolus and is known as the *conductor of the embolus,* or the *conductor of the style,* or, simply, as the *conductor.*

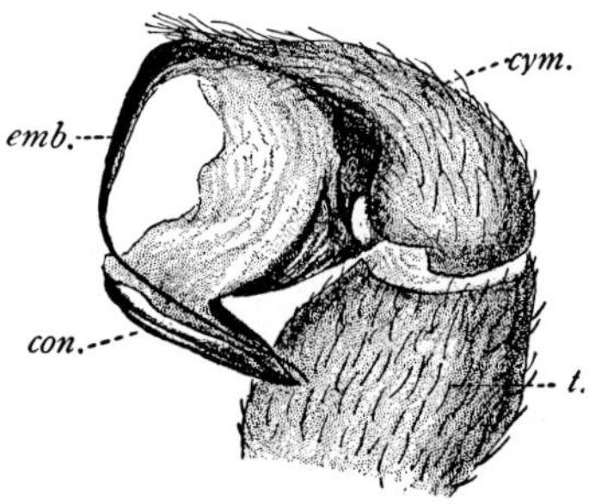

A comparatively simple example of this group of palpi is that of *Atypus bicolor.* Here the terminal part of the conductor is a broad concave plate (Fig. 100), in which the terminal portion of the embolus rests.

A more complicated form of the apical division of the bulb exists in *Hypochilus thorellii* (Fig. 95). Here the embolus is coiled about the conductor, the terminal part of which is concave so as to support the terminal portion of the embolus; the tip of the conductor bears a delicate membranous flap.

Fig. 100.
TARSUS OF ATYPUS BICOLOR

In *Hypochilus* the tarsus bears a branch which supports a prominent bunch of bristles (Fig. 95, *p. c.*); this may be a rudimentary form of paracymbium, a part that is well-developed in *Pachygnatha.*

A somewhat similar condition exists in *Pachygnatha* (Fig. 101). Here the proximal part of the embolus is coiled about the conductor, which is a broad twisted plate; and the terminal portion of the embolus is supported by the corresponding part of the conductor. When at rest the apical division of the bulb

The External Anatomy of Spiders

rests in the concave tip of the cymbium; but in the specimen figured the bulb has been extended so as to show the parts better; and the embolus and conductor have been separated at the tip.

In *Pachygnatha* the tarsus is divided into two distinct parts, which are joined by a movable articulation at the base. The larger part is the cymbium (Fig. 101, *cym.*) the smaller part, the *paracymbium:* this is termed by some writers *the accessory branch of the tarsus* (Fig. 101, *p. c.*). The cymbium and the paracymbium resemble the other segments of the palpus in the nature of their cuticula and in the fact that they are clothed with hairs.

THE MORE SPECIALIZED TYPES OF PALPI

In the development of the bulb of the male palpus in the more specialized families of spiders there has been evolved an exceedingly complicated organ, which is difficult to understand, on account of its small size and the fact that when

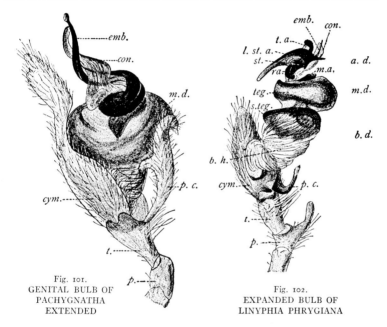

Fig. 101.
GENITAL BULB OF
PACHYGNATHA
EXTENDED

Fig. 102.
EXPANDED BULB OF
LINYPHIA PHRYGIANA

at rest it is compactly folded. Fortunately when such a palpus is boiled in a solution of caustic potash (10%) the bulb expands so that its parts can be seen; and if preserved in glycerine, it remains flexible, so that it can be easily manipulated. The expanded bulbs figured below were prepared in this way. Even with the best of preparations, it is sometimes difficult to make out the relation of parts; this can be most easily accomplished by the use of a stereoscopic binocular microscope.

The extreme specialization of the palpi of males is marked chiefly by the development of hæmatodocha, to be described later, and by an increase in the number of distinct parts and appendages of the bulb. The maximum degree of specialization is to be found in the Araneinæ, of which the palpi of several species of *Aranea* are described later. The understanding of the relation of the parts of the bulb in this genus will be facilitated by a study first of a more simple form, such as is found in the Linyphiidæ.

The External Anatomy of Spiders

THE LINYPHIA TYPE OF PALPUS.— The very common *Linyphia phrygiana* will serve as an example of the Linyphiidæ.

As in *Pachygnatha*, just described, the body of the tarsus of *Linyphia* consists of two parts; the *cymbium* (Fig. 102, *cym.*), and the *paracymbium* (Fig. 102, *p. c.*). The *alveolus* is a circular cavity near the base of the cymbium.

When the bulb is expanded, the three divisions of it are distinctly separated; there being a slender neck between the basal division (Fig. 102, *b. d.*) and the middle division (Fig. 102, *m. d.*) and also a similar slender neck between the middle division and the apical division (Fig. 102, *a. d.*).

The wall of the basal division of the bulb consists of two parts; the basal hæmatodocha, and the subtegulum.

The basal hæmatodocha.— The genital bulb is attached to the cymbium, within the alveolus, by means of a sac-like structure, which, ordinarily, is inconspicuous or completely concealed by other parts of the bulb, but which is very conspicuous in the expanded bulb (Fig. 102, *b. h.*). This has been named the *hæmatodocha* from the fact that at the time of pairing it is distended with blood. The wall of the hæmatodocha appears to consist of elastic connective tissue; hence the name *spiral muscle* applied to it by Menge is inappropriate. In fact no muscle tissue has been found within the genital bulb. As similar extensible blood sacs are present in more distal parts of the bulb of many spiders, I suggest that this one be termed the *basal hæmatodocha*.

The subtegulum.— The proximal end of the basal hæmatodocha is attached to the cymbium, the distal end, to a ring-like sclerite, for which I propose the term *subtegulum* (Fig. 102, *s. teg.*) .

The middle division of the bulb.— The middle division of the bulb (Fig. 102, *m. d.*) is that part which contains the chief portion of the receptaculum seminis, the *reservoir;* its wall is the tegulum, and it bears an appendage, the median apophysis.

The tegulum.— The term tegulum was applied by Wagner to all of the more densely chitinized parts of the wall of the genital bulb; but as it is desirable that the different sclerites should bear distinctive names, I propose that this term be restricted to the sclerite that forms the wall of the middle division of the bulb. In *Linyphia*, the tegulum, in this restricted sense is a ring-like sclerite (Fig. 102, *teg.*).

The median apophysis.— Arising within the distal margin of tegulum there is an appendage, only the tip of which is shown in the view of the bulb figured here (Fig. 102, *m. a.*); this is the *median apophysis*. In many spiders this appendage is very conspicuous; and to it have been applied several names. In fact in several instances a writer has applied different names to this part in his descriptions of different genera. Among the names that have been applied to it are *clavis*, *unca*, and *scopus*. The term median apophysis occurs frequently in descriptions, and is the older name for this part.

The median apophysis is articulated to the middle division of the bulb near the point from which the apical division arises; and in some cases, as in *Aranea*, it appears to be more closely articulated with a basal segment of the apical division, the radix, than it is with the tegulum.

The apical division of the bulb.— This division includes that portion of the bulb which lies distad of the middle division; it consists of two subdivisions: the conductor and the embolic subdivision. The embolic subdivision is traversed by the ejaculatory duct and is composed of several distinct parts. In fact the multiplication of parts of the embolic subdivision is the most characteristic feature of the more specialized types of palpi as contrasted with the intermediate type described above.

The conductor.— The conductor (Fig. 102, *con.*) is easily recognized by its relation to the embolus, which rests upon it, and by its membranous texture. Its attachment to the middle division of the bulb is by means of an exceedingly delicate membrane.

In *Linyphia* the embolus rests upon the conductor throughout its length; but in many genera the palpi of some of which are described later, the function of the conductor is evidently to protect the tip of the embolus in the unexpanded bulb. In many cases the embolus is very long while the conductor is short; but in every case the embolus in the unexpanded bulb occupies such a position that its tip is protected by the conductor.

In most cases the conductor can be recognized at a glance by its peculiar texture; sometimes it is chitinized to a considerable extent, but even then it usually has a membranous margin; and in any case it can be recognized by its relation to the tip of the embolus in the unexpanded bulb.

The embolic subdivision.— Closely connected with the membranous base of the conductor is the base of a separate subdivision of the apical division of the bulb; as this portion bears the embolus it may be termed the *embolic subdivision*.

The radix and the stipes.— Immediately following the membranous neck that connects the middle and the apical divisions of the bulb and parallel with the membranous base of the conductor, there are two segments of the embolic subdivision; to the basal one of these I apply the term *radix* (Fig. 102, *ra.*); and to the second, the term *stipes* (Fig. 102, *st.*). For a more distinctly segmented condition of the base of the embolic subdivision see the figures

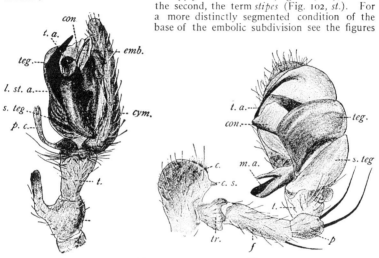

Fig. 103. PALPUS OF LINYPHIA PHRYGIANA

Fig. 104. PALPUS OF ARANEA FRONDOSA

of *Eriophora circulata* given later (Fig. 109 and 110), where the corresponding parts bear the same letters.

The embolus.— The organ through which the ejaculatory duct opens, the *embolus*, is comparatively simple in *Linyphia*, being a short spine-like part (Fig. 102, *emb.*).

The lateral subterminal apophysis.— In *Linyphia phrygiana* there is developed a remarkable plate-like apophysis, which serves to protect the exposed face of the unexpanded bulb. In Fig. 102 (*l. st. a.*), only the edge of this apophysis is shown; but in Fig. 103, the broader face of it is represented. I designate this the *lateral subterminal apophysis* as it occurs on the lateral aspect of the unexpanded bulb, and also to distinguish it from an apophysis developed on the opposite face of the bulb in a subterminal position, which occurs in certain other genera.

The terminal apophysis.—The embolic subdivision ends in a strongly chitinized lobe, which may be designated the *terminal apophysis* (Fig. 102, *t. a.*). To apophyses of this kind Menge applied the term *retinacula;* but as this term predicates their function, which in some cases is obviously not that implied by the name, I prefer apophysis with a modifying term indicating the position of the particular apophysis described.

The External Anatomy of Spiders

The Aranea Type of Palpus.— I have selected the palpus of *Aranea frondosa* as an example of an extremely specialized palpus. In Fig. 104 the entire palpus, with the bulb unexpanded, is represented slightly twisted so as to show the ventral aspect of the proximal segments and the lateral aspect of the bulb.

The proximal segments of the palpus.— This account of the palpi of male spiders is devoted almost entirely to a discussion of the parts of the tarsus, the proximal segments being well understood; there are, however, a few features of these segments in the aranea type that merit attention here.

Upon the coxa there is a prominent spur, *the coxal spur* (Fig. 104, *c. s.*); and upon the inner side of the femur near its base, there is a groove, the *femoral groove*, into which the coxal spur fits when the palpus is extended forward. The presence or absence of this spur and groove is an important generic characteristic in the Araneinæ.

The patella bears two prominent spines at its apex (Fig. 104, *p.*). This is also true in the males of several other

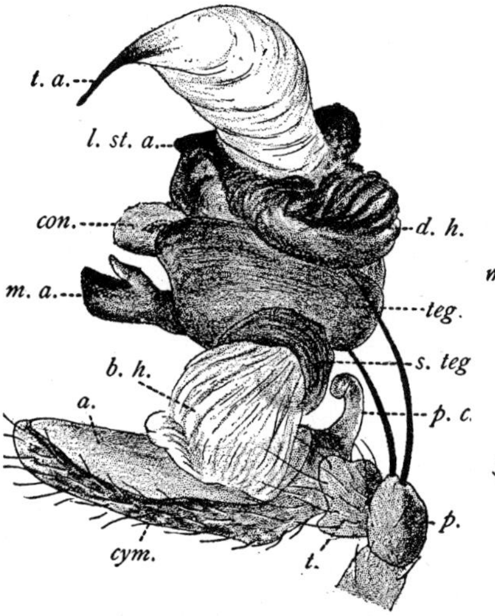

Fig. 105. LATERAL ASPECT OF AN EXPANDED BULB OF ARANEA FRONDOSA

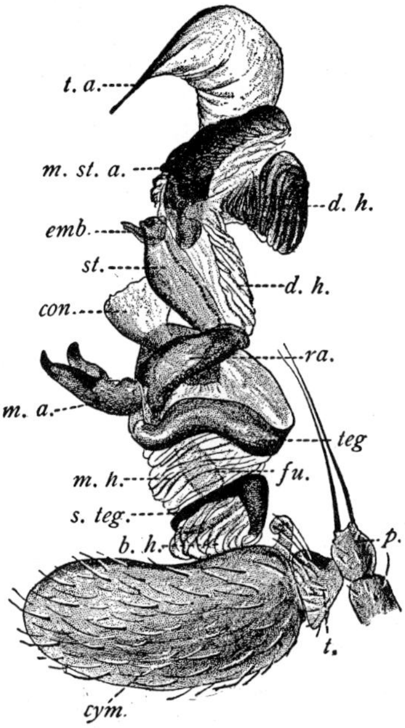

Fig. 106. MESAL ASPECT OF AN EXPANDED BULB OF ARANEA FRONDOSA

genera; but in the greater number of genera of the Araneinæ there is only a single spine in this position.

The most striking feature of the tibia is its shortness, it being of about the same length as the patella.

The tarsus.— As in *Linyphia*, the tarsus of *Aranea* consists of two parts: the cymbium and the paracymbium. But in *Aranea* the paracymbium (Fig. 105, *p. c.*) is merely a prominent apophysis arising from the base of the cymbium and is not articulated with the cymbium by a movable joint as in *Linyphia* and in *Pachygnatha*. The alveolus is much more extended than it is in *Linyphia*; here it occupies nearly the whole length of the cymbium (Fig. 105, *a*).

The unexpanded bulb.— In the unexpanded bulb of *Aranea frondosa*, the subtegulum (Fig. 104, *s. teg.*), tegulum (Fig. 104, *teg.*), and a terminal lobe of the apical division of the bulb, bearing a long and slender terminal apophysis (Fig.

The External Anatomy of Spiders

104, *t. a.*) are visible. Two prominent appendages can also be seen; the median apophysis (Fig. 104, *m. a.*) and the conductor (Fig. 104, *con.*)

The expanded bulb.— Two figures of the expanded bulb are given here; Fig. 105 represents the lateral aspect of the bulb, the aspect that is exposed when the bulb is not expanded; and Fig. 106, the mesal aspect, the one that is next the cymbium in the unexpanded bulb.

The basal hæmatodocha is essentially the same as in *Linyphia* (Figs. 105 and 106, *b. h.*).

The subtegulum is a ring-like sclerite but its form is like that of a seal-ring being narrow on the mesal aspect of the bulb and wide on the lateral aspect (Fig. 106, *s. teg.*). This wider part of the subtegulum is all of it that is commonly observed and has been termed the *lunate plate*.

The specimen represented in Fig. 106 was more fully expanded than that used for Fig. 105. In the more expanded specimen there is evident a large hæmatodocha between the subtegulum and the tegulum; this I designate the *middle hæmatodocha* (Fig. 106, *m. h.*) The dark axial object seen through the wall of the middle hæmatodocha is the fundus of the receptaculum seminis (Fig. 106, *fu.*).

The tegulum is also a ring-like sclerite, which is broad on the lateral aspect of the bulb (Fig. 105, *teg.*), and is narrow on the mesal aspect (Fig. 106, *teg.*).

The median apophysis (Fig. 104 and 105, *m. a.*) is a conspicuous appendage, which projects from the ventral side of the bulb. Although the position of this appendage in *Linyphia*, in which the middle and apical divisions of the bulb are distinctly separated, shows that the median apophysis is an appendage of the middle division, in *Aranea* it appears to be articulated with the base of a proximal segment of the apical division, the radix.

The conductor (Fig. 106, *con.*) arises at the base of the apical division and is closely connected with the tegulum.

The radix (Fig. 106, *ra.*) is much larger than in *Linyphia*. Here it forms the wall of one side of the basal segment of the embolic division of the apical division. That this is the case is more clearly shown in the bulb of *Eriophora circulata* (Figs. 110 and 111, *ra.*), where the segmentation of the embolic subdivision is much more marked.

The stipes (Fig. 106, *st.*) is also much larger than in *Linyphia;* it is articulated with the distal end of the radix. Like the radix, the stipes forms the wall of one side of a segment of the embolic subdivision of the bulb, a fact which is also well shown in the bulb of *Eriophora circulata* (Figs. 110 and 111, *st.*).

The embolus is borne by the embolic subdivision distad of the stipes; it projects ventrad between the distal end of the stipes, which is mesad of it, and the conductor, which is laterad of it in the unexpanded bulb. In the specimen represented in Fig. 106, the distal end of the stipes and the embolus have been pushed away from the conductor in the expanding of the bulb.

The distal hæmatodocha.— The most striking feature of the embolic subdivision in the aranea type is the presence of a large hæmatodocha, which when expanded overshadows all other parts. This hæmatodocha I designate the *distal hæmatodocha* (Figs. 105 and 106, *d. h.*). It is doubtless due to the development of this hæmatodocha that the radix and the stipes are restricted to one face of their respective segments of the apical division in *Aranea frondosa*, the remaining parts of the wall of these segments forming a part of the distal hæmatodocha.

The mesal subterminal apophysis.— On the mesal aspect of the bulb, there arises from the distal hæmatodocha a prominent apophysis (Fig. 106, *m. st. a.*); this may be termed the *mesal subterminal apophysis*.

The lateral subterminal apophysis.— On the lateral aspect there is also an apophysis borne by the distal hæmatodocha (Fig. 105, *l. st. a.*); this may be termed the *lateral subterminal apophysis*.

The terminal apophysis.— In *Aranea frondosa*, the tip of the embolic subdivision of the bulb ends in a spear-shaped apophysis (Fig. 105, *t. a.*); this may be termed the *terminal apophysis*.

THE PALPUS OF ARANEA OCELLATA.— A glance at the palpus of *Aranea ocellata* will show that it is of essentially the same type as that of *Aranea frondosa*

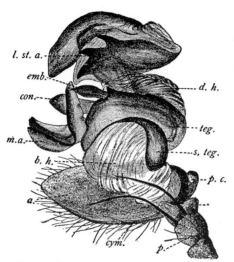

Fig. 107. LATERAL ASPECT OF AN EXPANDED BULB OF ARANEA OCELLATA

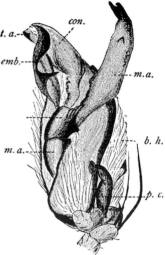

Fig. 108. UNEXPANDED BULB OF ERIOPHORA CIRCULATA

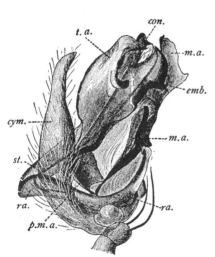

Fig. 109. UNEXPANDED BULB OF ERIOPHORA CIRCULATA

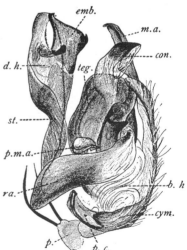

Fig. 110. EXPANDED BULB OF ERIOPHORA CIRCULATA

The External Anatomy of Spiders

but is different in some details. It is figured here to illustrate the kind of variations in form that serve to distinguish closely allied species (Fig. 107). The median apophysis differs markedly in form from that of *A. frondosa;* the tegulum bears a small but distinct apophysis; the lateral subterminal apophysis bears two prominent teeth; and the terminal apophysis is lacking, the embolic subdivision ending in a blunt lobe.

THE PALPUS OF ERIOPHORA CIRCULATA.— The most striking modification of the aranea type of palpus, taking the palpus of *Aranea frondosa* as typical, is that of *Eriophora circulata*, which is the most complex palpus that I have studied. In the unexpanded bulb, there appears to be no resemblance to the bulb of *Aranea frondosa.* In *Eriophora circulata* (Figs. 108 and 109), the bulb is very large and the cymbium comparatively small and narrow (Fig. 109, *cym.*). The basal hæmatodocha (Fig. 108, *b. h.*) is conspicuous, which is a result of the other parts of the bulb being twisted into unusual positions. The median apophysis is large and projects beyond the tip of the bulb (Fig. 108, *m. a.*). But the most remarkable feature is an elbowed structure on the mesal aspect at the base of the bulb (Fig. 109). The fact that the ejaculatory duct can be traced throughout the length of this elbowed structure gave the first definite clue to the relations of the parts of the bulb. The part containing the ejaculatory duct evidently

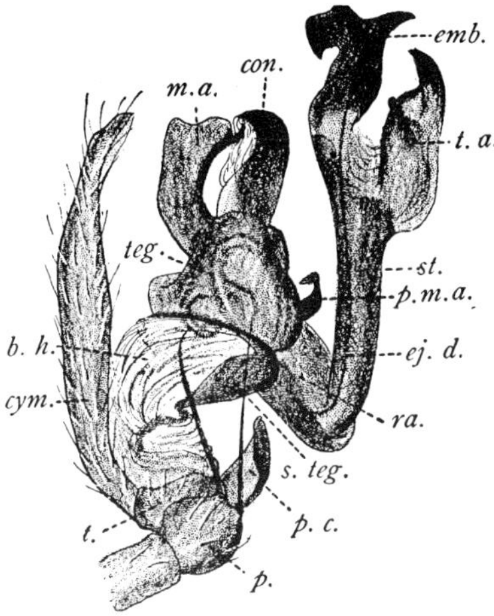

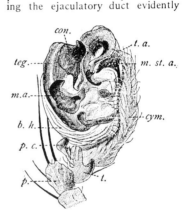

Fig. 111. EXPANDED BULB OF ERIOPHORA CIRCULATA

Fig. 112. UNEXPANDED BULB OF ARANEA GIGAS

pertains to the apical division of the bulb, although it appears to arise from the base of the bulb.

When the bulb of *Eriophora circulata* is expanded and untwisted, as occurs in the process of expansion, the relations of the parts are more easily seen. Figures 110 and 111 represent two views of a preparation of this kind. If Fig. 111 be studied it will be seen that the relations of parts are essentially the same as in *Aranea frondosa* (Fig. 106); the basal hæmatodocha, subtegulum, and tegulum follow in the same sequence; the median apophysis and the conductor project from beneath the tegulum in the corresponding positions, and the elbowed structure, which in the unexpanded bulb appears to arise at the base of the bulb is here clearly seen to be the embolic subdivision of the bulb. The most remarkable differences are the lack of a prominent distal hæmatodocha and the fact that the radix (Fig. 111, *ra.*) and stipes (Fig. 111, *st.*) are each a complete cylinder, instead of merely forming one face of the wall of a segment of the apical division, as in *Aranea frondosa.* At the distal end of the stipes, between this part and the embolus and the terminal apophysis, there is a vestigial distal hæmatodocha (Fig. 110, *d. h.*).

The External Anatomy of Spiders

In this species there is an apophysis, which like the median apophysis, is joined by a flexible articulation to the tegulum within the cup-like cavity formed by the distal margin of the tegulum (Figs. 109 and 111, *p. m. a.*); this may be termed the *paramedian apophysis*. As I have not found this apophysis in other palpi, I do not consider it a fundamental part.

THE PALPUS OF ARANEA GIGAS.— The preceding species, *Eriophora circulata*, and several others have been separated from *Aranea* by Pickard-Cambridge and placed in the resurrected genus *Eriophora* of Simon. The peculiar form of the genital bulb in *Aranea circulata* appears to sustain this separation. But in the palpus of *Aranea gigas* (Figs. 112 and 113) we find a form intermediate between the aranea type and what may be termed the eriophora type.

In the unexpanded bulb of *Aranea gigas* (Fig. 112) the parts are twisted so as to render the basal hæmatodocha conspicuous as in *Eriophora circulata*; but otherwise there is little similarity in appearance to either this species or to *Aranea frondosa*.

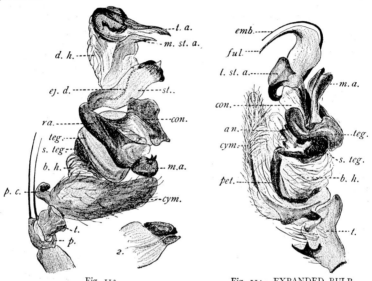

Fig. 113.
EXPANDED BULB OF ARANEA GIGAS

Fig. 114. EXPANDED BULB OF DOLOMEDES FONTANUS

In the expanded bulb (Fig. 113) it can be seen that the embolic subdivision, is intermediate in form between the two types, resembling the aranea type in having a large distal hæmatodocha and a well-developed median subterminal apophysis; and resembling the eriophora type in the form of the embolus, which is lamelliform (Fig. 113, 2).

THE PISAURID TYPE OF PALPUS.— In the family Pisauridæ there is a type of palpus which, while it resembles the aranea type in its more general features, differs from that type in several important particulars. The palpus of *Dolomedes fontanus* (Figs. 114 and 115) may be taken as an example of this type.

A study of an expanded bulb of this species (Fig. 114) reveals the following characteristics: There is a well developed *petiole* of the bulb (Fig. 114, *pet.*), which in this species consists of two nodes with an unchitinized internode. The *sub-tegulum* bears very prominent *anelli* (Fig. 114, *an*), which are described in a later paragraph. The *median apophysis* is prominent (Fig. 114, *m. a.*). The *conductor* (Fig. 114, *con.*) is extremely membranous. The *radix* and the *stipes* are not developed as distinct segments. The *embolus* is of the spiral type (Fig. 114, *emb.*)

The *terminal apophysis* is modified into an organ for the support of the embolus (Fig. 114, *ful.*), which may be termed the *fulcrum* of the embolus. This type of terminal apophysis has been termed, incorrectly, the conductor. The true conductor in this species, as in all others studied, is an organ whose function is to protect the tip of the embolus in the unexpanded bulb. At the base of the terminal apophysis, at the point where the embolus arises, there is a lamelliform *lateral subterminal apophysis* (Fig. 114, *st. a.*).

In the unexpanded bulb (Fig. 115), the long *embolus* makes a curve in the distal end of the alveolus beyond the end of the bulb. The *fulcrum* is applied against the embolus on its concave side, and has a furrow on its distal face within which the embolus rests. The distal part of the *conductor* is wrapped about the tip of the combined embolus and fulcrum, serving, as in all other cases observed, as a protection to the tip of the embolus.

THE THOMISID TYPE OF PALPUS.— In the family Thomisidæ, there occurs a striking modification of the palpus, which consists of the absence of the conductor of the embolus and, in certain members of the family, of a specialization of one edge of the cymbium for the protection of the tip of the embolus in the unexpanded bulb. To this specialized part of the cymbium I have applied the term *tutaculum* (tu-tac'u-lum); it is described and figured in the general account of the Thomisidæ on a later page. It is best developed in the genus *Xysticus*.

THE ANELLI OF THE SUBTEGULUM.— In *Aranea*, the chitinized part of the wall of the basal division of the bulb, the subtegulum, is reduced to a ring-like sclerite (Fig. 116, 1.); but in certain other genera, the subtegulum is cup-shaped or basket-like. In *Agelena nævia* (Fig. 116, 2.), one side of the subtegulum is greatly thickened; at the proximal end of this thickening, which corresponds to the lunate plate, there is a condyle, which articulates with the petiole; and at the distal end, there is a condyle,

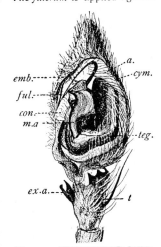

Fig. 115. UNEXPANDED BULB OF DOLOMEDES FONTANUS

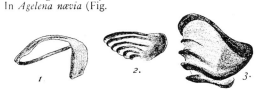

Fig. 116. THREE KINDS OF SUBTEGULUM

which articulates with the tegulum. The other side of the cup-like subtegulum contains in its wall several parallel, incompletely ring-like sclerites; these may be termed the *anelli* of the subtegulum. In *Dolomedes fontanus* (Fig. 116, 3), the anelli of the subtegulum are greatly thickened and form prominent, projecting ridges.

It is probable that the presence or absence of the anelli of the subtegulum, and their nature when present will afford characters of use for taxonomic purposes.

THE DIFFERENT TYPES OF EMBOLUS.— The form of the embolus varies greatly in different species of spiders. Two principal types can be recognized, the connate and the free; and the free type includes three subtypes.

The connate type of embolus.— In the connate type, the embolus is not separate from the middle division of the bulb, but is merely a more slender continuation of it, as in the tarantulas, *Loxosceles* (Fig. 96), and *Ariadna*.

The free type of embolus.— In the free type of embolus, there are one or more movable articulations between the embolus and the middle division of the bulb.

The External Anatomy of Spiders

In the free type, the embolus varies greatly in form; but the different forms can be grouped under three heads; coniform, lamelliform, and spiral.

A coniform embolus.— In this type, there may be a broadly expanded base; but the projecting part of the embolus is a straight or slightly curved cone. The embolus of *Aranea frondosa* (Fig. 117, 1.) is an example of this type.

A lamelliform embolus.— In this type the embolus is flattened, and may bear a greater or less number of apophyses; an example of this type is found in *Lepthyphantes minuta* (Fig. 117, 2.).

A spiral embolus.— In the spiral type, as seen in *Agelena*, for example, the embolus is long, slender and coiled; and, in a well-expanded specimen, it is seen to be composed of three distinct parts; first, the wall of the convex side is densely chitinized, forming a gutter-like sclerite, which may be termed the *trunk of the embolus (truncus)* (Fig. 117, 3, *t. e.*); second, the greater part of the wall of the embolus is membranous, and forms a loose flap along the concave side of the organ, which contains the ejaculatory duct; this flap (Fig. 117, 3, *p. p.*) may be designated the *pars pendula* of the embolus; third, at the distal end of the pars pendula, there is a triangular, chitinized area, through which the ejaculatory duct opens (Fig. 117, 3, *a. s.*), this may be termed the *apical sclerite* of the embolus.

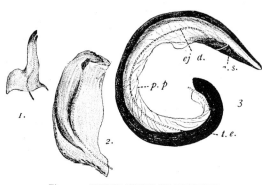

Fig. 117. THREE TYPES OF EMBOLUS

The pars pendula and the apical sclerite may be completely withdrawn into the trunk of the embolus, so that only the latter is visible; the embolus then appears to be merely a strongly chitinized style; it is in this condition that it is usually seen and described.

CONCLUSION

In the preparation of this account many palpi other than those figured here have been studied; and it is believed that the series examined has been sufficiently large to warrant the conclusions given regarding the fundamental parts of the genital bulb. There remains to be determined the manner in which the different types of palpi have been specialized in other families of the order, and the details of the modifications characteristic of genera. This, however, is too great an undertaking to be attempted at this time; and must be left for those who monograph the different families.

I wish, however, to urge the importance of describing palpi from expanded specimens. A large proportion of the figures of palpi that have been published being of unexpanded examples, show comparatively little of the structure of this organ. The labour involved in expanding the bulb of a palpus is very little; a preparation can be made in five minutes; and in no other way can so much be done to make possible a description that will describe.

The following tabular statement shows the relations of the fundamental parts of the tarsus in the more specialized types of palpi; not all of these parts are invariably present, and frequently subordinate apophyses are developed.

Body of the Tarsus
    Cymbium, containing the alveolus
    Tutaculum (in the Thomisidæ)
    Paracymbium

Genital Bulb
  Internal parts
    Receptaculum seminis
      Fundus
      Reservoir
      Ejaculatory duct
  External parts
    Basal division of the bulb
      Basal hæmatodocha
      Petiole
      Subtegulum
        Lunate plate
        Anelli of the subtegulum
    Middle division of the bulb
      Middle hæmatodocha
      Tegulum
      Median apophysis
      Paramedian apophysis
    Apical division of the bulb
      Conductor
      Embolic subdivision
        Radix
        Stipes
        Embolus
          Body of embolus
          Pars pendula
          Apical sclerite of the embolus
      Distal hæmatodocha
      Lateral subterminal apophysis
      Mesal subterminal apophysis
      Terminal apophysis, sometimes developed into a fulcrum

NAMES OF THE PARTS OF THE PALPUS AND ABBREVIATIONS USED FOR THEM IN THE ILLUSTRATIONS

Accessory branch = paracymbium.
Alveolus, *a*.
Anelli of the subtegulum, *an*.
Apical division of the bulb, *a. d.*
Apical sclerite of the embolus, *a. s*
Basal division of the bulb, *b. d.*
Basal hæmatodocha, *b. h*.
Clavis = median apophysis
Conductor of the embolus, *con.*
Coxa, *c*.
Coxal spur, *c. s*.
Cymbium, *cym*.
Distal hæmatodocha, *d. h*.
Ejaculatory duct, *ej. d*.
Embolic subdivision of the bulb, *e. s*.
Embolus, *emb*.
Femur, *f*.
Fulcrum, *ful*.
Fundus of the receptaculum seminis, *fu*.
Lateral subterminal apophysis, *l. st. a*.
Lunate plate = subtegulum in part.
Median apophysis, *m. a*.
Mesal subterminal apophysis, *m. st. a*.
Middle division of the bulb, *m. d*.
Middle hæmatodocha, *m. h*.
Paracymbium, *p. c*.
Paramedian apophysis, *p. m. a*.
Pars pendula of the embolus, *p. p*.
Patella, *p*.
Petiole of the bulb, *pet*.
Radix, *ra*.
Receptaculum seminis, *r. s*.
Reservoir, *res*.
Scopus = median apophysis.
Spiral muscle = hæmatodocha.
Stipes, *st*.
Style = embolus.
Subtegulum, *s. teg*.
Tegulum, *teg*.
Terminal apophysis, *t. a*.
Tibia, *t*.
Trochanter, tr.
Trunk of the embolus, *t. e*.
Tutaculum, *tu*.

THE THORAX

The thorax is that part of the cephalothorax which bears the four pairs of legs. It is either slightly separated from the head by a furrow, or completely coalesced with it.

THE TERGUM OF THE THORAX.— The dorsal aspect of the thorax, or the *tergum*, is by far the most prominent part of this

The External Anatomy of Spiders

region of the body. It extends forward on each side of the wedge-shaped hind end of the head, and covers to a greater or less extent the sides of the thorax. It is often marked by a *median furrow* and several *radial furrows* (Fig. 118); these are lines along which muscles are attached to the inner surface of the body-wall. The median furrow is sometimes called the *dorsal groove*.

THE STERNUM.— The plate forming the ventral wall of the thorax is the *sternum* (Fig. 119, s.); it occupies the entire space between the two rows of legs; and usually each lateral margin bears four notches for the reception of the coxæ of the legs.

The sigilla (sig'il-la).— In many tarantulas the sternum is marked by circular or oval, impressed, bare spots, normally four on each side; these are termed the *sigilla*. It should be noted that sigilla is a plural noun; the term sigillæ, sometimes used, is, therefore, incorrect.

THE EPIMERA (ep-i-me'ra).— Correlated with the great development of the tergum of the thorax there is a marked reduction of the lateral portions, each of which usually consists of a narrow band between the legs and the tergum. In some cases this band consists of a series of four sclerites, one above each leg. These sclerites are termed the *epimera;* each epimeron represents the pleural portion of a segment.

THE LEGS.— There are always four pairs of legs. Each leg consists of seven segments which are named, beginning with the one next the body as follows: *coxa, trochanter, femur, patella, tibia, metatarsus* and *tarsus* (Fig. 120). The tarsus usually bears at its tip two or three *claws*.

In certain spiders, which have very long and slender legs, the tarsus is rendered flexible by a large number of secondary joints or "false articulations."

The prætarsus (præ-tar'sus).— The claws are outgrowths of a small terminal portion of the leg which is not ordinarily distinguished from the tarsus. This terminal part with its appendages has received the name *prætarsus* (de Meijere '01). As a rule the prætarsus in spiders is withdrawn into the projecting ventral part of the tarsus; but sometimes it is a distinct segment (Fig. 121). This segment is termed by Simon ('92, p. 52) the *onychium*, an unfortunate use of the term, as ordinarily it is used as a synonym of empodium. (See p. 123.)

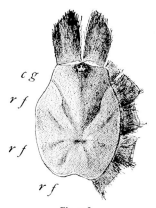

Fig. 118.
CARAPACE OF A TARANTULA

Fig. 119.
VENTRAL ASPECT OF
CEPHALOTHORAX
OF ULOBORUS
s, sternum

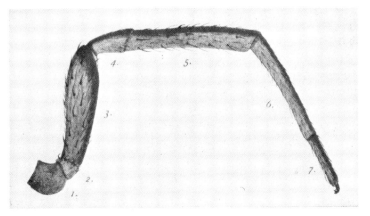

Fig. 120. LEG OF A SPIDER
1, coxa   2, trochanter   3, femur   4, patella   5, tibia   6, metatarsus   7, tarsus

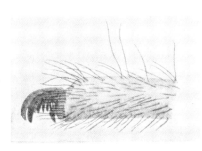

Fig. 121. TIP OF THE LEG OF A SPIDER
SHOWING THE PRÆTARSUS

Fig. 122. CLAWS AND ACCESSORY
CLAWS OF A SPIDER

The External Anatomy of Spiders

The claws.— Each tarsus of the legs is armed with two or three claws. When three claws are present, they can be designated as the *paired claws* and the *third claw* respectively.

The *paired claws* are placed side by side at the tip of the upper surface of the tarsus, or of the prætarsus if it is distinct; they are usually armed with a series of teeth (Fig. 121).

The *third claw* is a modified empodium (see below); when present, it is situated below the paired claws. It is smaller than the paired claws, and is sometimes armed with a small number of fine teeth. Usually the terminal portion is bent down rather abruptly (Fig. 121).

Fig. 123. A TENENT HAIR FROM THE EMPODIUM OF CLUBIONA

The empodium (em-po'di-um). — When the terminal portion on the middle line of the prætarsus projects between or below the paired claws it is termed an *empodium*. Sometimes the empodium is claw-like; it then constitutes the third claw described above; sometimes it is a cushion-like pad or adhesive lobe; and sometimes it is not developed as a distinct part. Many terms have been applied to this part; among them are *arolium, onychium, palmula, plantula* and *pulvillus*.

The accessory claws.— In many spiders, especially those that spin webs, the tip of the tarsus is armed with several claw-like spines; these have been termed the *accessory claws* (Fig. 122).

Fig. 124. TIP OF TARSUS OF CLUBIONA WITH TERMINAL TENENT HAIRS

The accessory claws are very different morphologically from the true claws, being modified hairs each produced by a trichogen; while the true claws are spine-like projections of the body-wall formed by many hypodermal cells.

The External Anatomy of Spiders

The terminal tenent hairs.— The tarsi of spiders are armed, in many species, with hairs of the type known as tenent hairs, i. e., hairs that are dilated at the extremity (Fig. 123), and which serve to aid the animal in clinging to smooth surfaces, probably as in insects by means of an adhesive fluid excreted through the cavity of the hair. A bundle of such hairs which exists in certain spiders at the tip of the tarsus just below the claws may be designated as the *terminal tenent hairs* (Fig. 124). This bundle is often divided into two by a smooth line; these bundles are the *fasciculi unguiculares* or claw-tufts of certain writers. A small sclerite upon which the bundle or bundles of terminal tenent hairs are borne is sometimes referred to as the *hypopodium;* this is sometimes covered by the empodium.

The scopulæ (scop'u-læ).— The lower surface of the tarsus and the metatarsus are often armed with tenent hairs. A brush of such hairs is termed a *scopula* (pl. *scopulæ*). The scopula

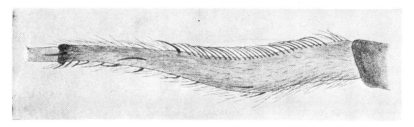

Fig. 125. CALAMISTRUM OF ULOBORUS GENICULATUS

and the terminal tenent hairs are frequently both present. A thick clothing of ordinary hairs in this position is not a scopula.

The calamistrum (cal-a-mis'trum).— Those spiders that possess the peculiar spinning organ known as the cribellum have also on the upper margin of the metatarsus of the hind legs one or two rows of curved spines (Fig. 125); these constitute the *calamistrum*. The calamistrum plays the part of a hackle in the formation of the hackled band, characteristic of the webs of these spiders.

Male spiders on reaching the adult stage either lose the cribellum and calamistrum or retain them in a vestigial condition; but such males can be recognized by the wide separation of the fore spinnerets.

The lyriform organs.— Near the distal extremity of each segment of the leg except the tarsus there are one or more minute

The External Anatomy of Spiders

organs, which have been termed the *lyriform organs* (Fig. 126). These are believed to be sense organs, and are discussed under that head in the next chapter.

Leg formulæ.— The relative lengths of the different pairs of legs often afford distinctive characteristics, which are much used in descriptive works. The four pairs of legs are numbered from before backward; and in descriptions their relative lengths are indicated by a formula consisting of the numerals 1, 2, 3, and 4 arranged in the order of the relative lengths of the legs of the spider described beginning with the numeral indicating the longest pair of legs. Thus the formula 1 4 3 2 indicates that the first pair of legs is the longest and the second pair the shortest. Where two or more pairs of legs are of equal length the fact is indicated by placing a bar or bracket above the numerals, thus the formula 2 $\overline{3\ 4}$ 1 indicates that the third and fourth pairs of legs are of the same length.

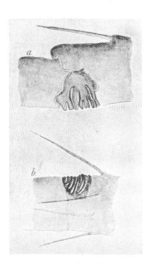

Fig. 126. LYRIFORM ORGANS OF ARGYRODES

Spine formulæ.— The number and arrangement of the spines on the segments of the legs are often indicated by formulæ. Thus if there are two rows of spines of three each under the tibia of the first legs and two rows of two each under the metatarsus of these legs, the fact is indicated as follows: spines, tib.I., 3–3, met. I, 2–2.

THE ABDOMEN

The abdomen of spiders is more or less sac-like and almost always without traces of segmentation except in a very small area at its caudal end. In most spiders the abdomen is of an elongated, rounded form; but an exceedingly great variation in the form of this region exists.

THE PARTS OF THE ABDOMEN.— Owing to the almost total lack of segmentation comparatively few distinct parts can be recognized in the abdomen; spiders differ greatly in this respect from insects, in which the abdomen is composed of a

The External Anatomy of Spiders

series of well-marked segments. The following parts can be recognized in the abdomen of spiders.

The pedicel.— In all spiders the abdomen is joined to the thorax by a slender stalk, the *pedicel*, which is usually concealed from above by the convexity of the part immediately following it. In certain spiders, which are ant-like in form, the pedicel of the abdomen is conspicuous; but in many of these a prolongation of the thorax enters into the composition of the pedicel. In the dorsal wall of the pedicel there is a slender, longitudinal sclerite, which is termed the *lorum of the pedicel;* in some genera the lorum is divided by a transverse

Fig. 127. LORUM OF THE PEDICEL OF PISAURINA

Fig. 128. ARANEA FRONDOSA, SHOWING FOLIUM

suture (Fig. 127). In some forms there is also a sclerite in the ventral wall, the *plagula* (plag'u-la). The plagula is well-developed in the Dysderidæ.

The muscle-impressions.— On the dorsal aspect of the abdomen, there are several pairs of muscle-impressions; these are small, hardened, depressed points, which, like the radial furrows of the thorax, indicate the points of attachment of muscles to the body-wall. There is also near the base of the abdomen a median, unpaired muscle-impression, which is more or less prominent. In some spiders there are muscle impressions on the ventral aspect of the abdomen.

The folium.— The dorsal aspect of the abdomen in many spiders is marked by an area with scalloped margins (Fig. 128); this area on account of its leaf-like outline is termed the *folium* (fo'li-um). The folium is frequently marked with spots; and its margin is often a double stripe, the inner part of which is dark and the outer part light.

The postabdomen.—At the caudal end of the abdomen there is a small, conical or semicircular portion, which, when seen from above, appears to consist of two or three segments; this is the greatly reduced *postabdomen*. In the embryo, the postabdomen is much more distinct, resembling in a striking degree the postabdomen of scorpions (see Korschelt and Heider '99, Vol. III., p. 50).

The epigastrium.— On the ventral aspect of the abdomen, the basal portion is usually more convex than the remainder of this aspect of the abdomen and is sometimes more densely chitinized; this area is termed the *epigastrium* (Fig. 129, *e.*).

The epigastric furrow.— A furrow separating the epigastrium from the more caudal portion of the abdomen is known as the *epigastric furrow* (Fig. 129, *ef.*). In this furrow there is, on the middle line of the body, the opening of the reproductive organs, and at each end, the opening of a lung.

The furrow of the posterior spiracle.— This is a transverse furrow on the ventral aspect of the abdomen, a short distance in front of the spinnerets, in which is situated the posterior spiracle, in most of those spiders that have only three spiracles. In a few cases, as in *Anyphæna* and *Glenognatha*, the third spiracle is situated near the middle of the ventral aspect of the abdomen.

The abdominal sclerites.— Owing to the lack of segmentation in this region of the body comparatively few sclerites can be recognized on the abdomen, still in some spiders the abdomen bears one or more distinct sclerites. There may be a *dorsal sclerite*, one on the dorsal aspect of the abdomen; an *epigastric sclerite*, one on the epigastrium; a *ventral sclerite*, one in the region behind the epigastric furrow; and an *infra-mammilliary sclerite*, a semicircular one situated in front of the spinnerets, i. e., below the spinnerets when the spider is hanging suspended by a thread.

The spiracles.— The spiracles or openings of the respiratory organs, are of two quite distinct kinds; one leading to lung-like organs, the *lung-slits*, and one leading to tracheæ of the ordinary type, the *tracheal spiracles*.

The External Anatomy of Spiders

The lung-slits.— Of these there are either one or two pairs. The first pair, which is always present except in one small, exotic family (Caponiidæ) is situated one at each end of the epigastric furrow. In the tarantulas and in the family Hypochilidæ there is a second pair of lung-slits behind the first (Fig. 129).

The tracheal spiracles.— In the true spiders except the family Hypochilidæ there are ordinary tracheæ in addition to the book-lungs. In most cases the ordinary tracheæ open by a single spiracle, which is usually situated on the middle line, a short distance in front of the spinnerets; but sometimes this single spiracle is near the middle of the ventral aspect of the abdomen. In a few spiders there

Fig. 129. VENTRAL ASPECT OF ABDOMEN OF A TARANTULA
e, epigastrium
ef, epigastric furrow

Fig. 130. DYSDERA CROCATA A TRUE SPIDER WITH A PAIR OF TRACHEAL SPIRACLES BEHIND THE LUNG-SLITS

is a pair of tracheal spiracles situated just behind the lung-slits; these can be distinguished from lung-slits by the absence of external indications of book-lungs. Compare Figs. 129 and 130.

The epigynum (e-pig′y-num).— The internal reproductive organs open on the middle line of the abdomen in the epigastric furrow. In the male the opening is simple, without any intromittant organ, the palpi taking the place of such an organ. In the female the two ovaries open by a common opening, which in some spiders is not accompanied by any specialized chitinous structure; this condition exists in *Tetragnatha* and other genera. But in most of the true spiders (Argiopoidea) there is connected with this outlet a more or less complicated apparatus, which is termed the *epigynum.*

The External Anatomy of Spiders

As the epigynum varies greatly in form in different species, even in closely allied ones, it often affords the most distinctive characteristics for recognizing species; but these features are not fully developed until the spider reaches maturity.

In its more simple form the epigynum is merely a lid-like plate covering or accompanying the opening of the oviduct; it is a chitinized portion of the cuticula of the segment in front of the epigastric furrow; and it is situated on the middle line of the body at or very near the hind margin of this segment.

As explained in the following chapter, there are connected with the internal reproductive organs of the female one or more pouches for receiving and retaining the spermatozoa, the *sperma-*

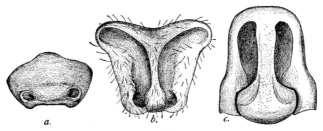

Fig. 131. DIFFERENT TYPES OF EPIGYNA
a, Pirata montanus   b, Trabea aurantiaca   c, Lycosa pikei

thecæ. When there is a single spermatheca, as in *Segestria*, it opens on the middle line, behind the opening of the oviduct. But in most spiders there are two spermathecæ which open separately, one on each side of the opening of the oviduct; and when there is a specialized epigynum, these openings are connected with it. In fact it seems probable that the primary function of the epigynum is to receive and direct the palpal organ of the male, and that the various specific forms of epigyna are correlated, in each case, with corresponding specific differences in the palpus of the male; these peculiarities tending to prevent the union of individuals of different species. Frequently, when there is a well-developed epigynum, the openings of the spermathecæ are on the outer face of the epigynum and can be easily seen (Fig. 131, *a b*); as their position varies in different species it is often indicated in specific descriptions. Some writers refer to them as the *openings of the epigynum.*

Even when the openings of the spermathecæ are covered by

The External Anatomy of Spiders

a part of the epigynum, they are sometimes visible in mounted specimens in which the epigynum has been rendered transparent (Fig. 132).

Care should be taken not to mistake the glands of the spermathecæ, which are often dark and show through the body-wall, for the openings of the spermathecæ. (See account of these glands in the following chapter.) A secondary function of the epigynum, as an ovipositor, is discussed later.

An example of a comparatively simple epigynum is that of *Pirata montanus* (Fig. 131, *a*); this is a nearly plain plate, with the two openings of the spermathecæ near the posterior lateral corners. A somewhat more complicated form is illustrated by the epigynum of *Trabea aurantiaca* (Fig. 131, *b*); in this the plate is depressed or furrowed longitudinally, and the depressed area is divided by a ridge-like elevation, which divides the depression into two furrows or channels, each of which leads to the opening of the spermatheca of the corresponding side; this ridge-like elevation has been termed the *guide* by Chamberlin ('04), as its function "seems clearly to be that of a guide to the male embolus, controlling the course of the latter and facilitating its entrance to the spermatheca." In many cases the guide extends laterally on each side at its posterior end; this is true to a slight extent in the epigynum of *Trabea*, but more markedly so in that of many species of *Lycosa*, where the lateral expansions often conceal the openings of the spermathecæ, as in the epigynum of *Lycosa pikei* (Fig. 131, *c*.). In some epigyna the posterior portion of the median part of the guide and the anterior edge of the lateral extensions of it are extended horizontally in plate-like expansions, these are termed, by Chamberlin ('08) the lateral plates or *alæ* of the guide.

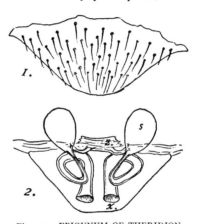

Fig. 132. EPIGYNUM OF THERIDION DIFFERENS
1, surface view 2, view when made transparent *x*, openings of the spermathecæ *s*. spermathecæ *z*, tube leading from the spermatheca to the vagina (after Emerton)

The External Anatomy of Spiders

A more complicated form of epigynum is found, for example in spiders of the genus *Aranea*, where there is developed an appendage which is usually soft and flexible, and which is termed the *scape* or *ovipositor*, as it is believed to have a function analogous to that of the ovipositor of insects (Fig. 133). In some cases, however, the form of the scape is such that it is difficult to see how it can function as an ovipositor; and it may be that in these cases it functions during copulation instead of during oviposition. The scape of *Aranea gemma* (Fig. 134) will serve to illustrate this point. Here the scape is immovable and the spoon-shaped tip seems fitted to receive some part of the palpus of the male rather than to be of use in placing the eggs. When there is a well-developed scape the tip of it is usually more or less spoon-shaped, this part of the scape is termed the *cochlear*. The basal plate of the epigynum which bears the scape, and which forms a vaulted, porch or hood that covers the opening of the oviduct was named by McCook ('89–'93) the *atriolum* (a-tri′o-lum).

A still more complicated form of epigynum is found in some of the Linyphiidæ and Argiopidæ, where the ovipositor consists of two finger-like projections: first, the more common one, the scape, which arises from the atriolum, and consequently in front of the openings of the oviduct; and second, one which arises behind the opening of the oviduct; this is termed the *parmula* (par′mu-la). Each of these projections may be grooved on the face next its fellow, the two grooves forming a tube (Fig. 135).

The entrance to the reproductive organs of the female is termed the *vulva;* it is an open space covered by the epigynum. Some writers apply the term *vulva* to the epigynum.

In some spiders there is a well-defined sclerite on each side of the epigynum; these sclerites are termed the *lateral sclerites*.

The epigastric plates.— There is on each side in front of the lung-slit a hard plate; these are termed the epigastric plates.

THE SPINNING ORGANS.— The spinning organs are situated near the caudal end of the abdomen, on the ventral aspect, and consist usually of three pairs of spinnerets, to which is added in certain spiders another organ, the *cribellum*.

The spinnerets.— The spinnerets are finger-like in form and usually six in number (Fig. 136), although sometimes there are only four, and the number may be reduced to a single pair.

The pairs of spinnerets are usually designated as the *upper,*

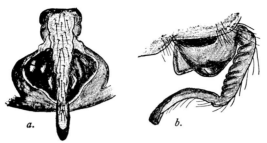

Fig. 133. EPIGYNUM OF ARANEA ANGULATA VAR. SILVATICA
a, ventral view   *b*, lateral view

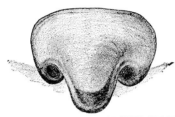

Fig. 134. EPIGYNUM OF ARANEA GEMMA

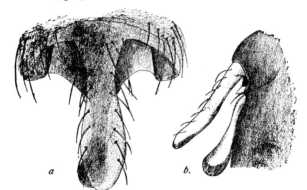

Fig. 135. EPIGYNUM OF BATHYPHANTES NIGRINUS
a, ventral view   *b*, lateral view

Fig. 136. LATERAL ASPECT OF BODY OF AGELENA SHOWING
THE SPINNERETS OF ONE SIDE

The External Anatomy of Spiders

the *lower*, and the *median*, these terms referring to their relative positions when the spider is hanging suspended by a thread; but as very many spiders never assume this position, it is better to designate the three pairs of spinnerets as the *fore, middle,* and *hind* respectively.

The hind spinnerets usually consist each of two segments, but sometimes of three or even four segments; the middle spinnerets are not segmented; the fore spinnerets consist almost always of two segments each.

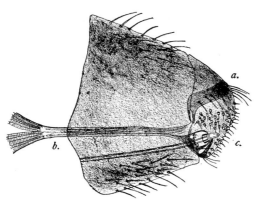

Fig. 137. FORE SPINNERET OF ARANEA
a, chitinous ring around the spinning-field   *b*, tendon of flexor muscle   *c*, outlet of cylindrical gland (after Bucholz and Landois)

The sides of the spinnerets are comparatively firm, but the terminal portion is membranous; this membranous terminal portion constitutes the *spinning field*. The spinning field is always surrounded by hairs, some of which are simple and some are barbed. These are movable and appear to have some function in the spinning; it is probable that some of them at least are tactile.

Sometimes, at least, the spinning field is surrounded by a chitinous ring, to one side of which is attached the tendon of a flexor muscle (Fig. 137), which probably moves the spinneret as a whole.*

---

*The spinnerets represent the fourth and fifth pairs of abdominal legs (the other abdominal legs are lost in the course of the development of the embryo). Each of these legs acquires, at an early embryonic period, a biramose form, like the primitive appendages of the Crustacea, consisting of an axis to which is attached an inner endopodite and an outer exopodite. Spinning glands may develop on both rays, and it is thus suggested that the primitive number of spinnerets was eight. No existing spiders are known in which these eight spinnerets are fully developed as functional organs; in *Liphistius*, however, (which also shows the generalized character of a segmented abdomen) the full number are present, but only the four external ones (exopodites) possess functional spinning glands. In the true spiders, we find present six functional spinnerets, the four large ones being derived from the exopodites of the fourth and fifth pairs of abdominal appendages, while the middle spinnerets are developed from the endopodites of the fifth pair, the endopodites of the fourth pair being altogether wanting or else concerned in the formation of the cribellum or colulus. (Korschelt and Heider, III., p. 79.)

The External Anatomy of Spiders

The spinning tubes.— There are many small tubes distributed over the surface of the spinning field; these are the *spinning tubes;* it is through them that the silk is expelled from the body. Usually each spinning tube consists of two segments, a stouter basal one and a more slender terminal one (Fig. 138). Some spiders have one hundred or more of these spinning tubes on each spinneret. The number of spinning tubes varies greatly in different species, and

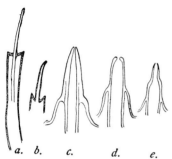

Fig. 138. THE SPINNING TUBES OF ARANEA SEEN IN OPTICAL SECTION
a, of aciniform glands *b,* of pyriform glands *c,* of ampullate glands *d,* of cylindrical glands *e,* of aggregate glands (after Apstein)

Fig. 140. THE SPINNERETS AND THE CRIBELLUM OF AMAUROBIUS

Fig. 139. HIND SPINNERET OF STEATODA SHOWING TWO SPIGOTS WITHOUT A TIP (after Apstein)

Fig. 141. THE CRIBELLUM OF AMAUROBIUS MORE ENLARGED

also in different individuals of the same species. It may be that the number increases with age; but we lack observations on this point.

Although the spinning tubes are almost invariably borne by the tip of the spinneret, in a few cases there is one or more on the basal segment of a two-jointed spinneret.

There are four different types of spinning tubes: the long and cylindrical (Fig. 138, *a*); the short and conical (Fig. 138, *b*); a form much larger than the preceding, which is termed a *spigot* (Fig. 138, *c, d, e*); and a spigot without a tip (Fig. 139).

Different kinds of silk are spun from the different kinds of spinning tubes.

The External Anatomy of Spiders

The cribellum (cri-bel'lum).— In certain families there is, in front of the spinnerets, an additional spinning organ, which on account of its sieve-like appearance when slightly magnified, is named the *cribellum*. The cribellum consists of a transverse plate, which is usually divided by a delicate keel on the middle line of the abdomen into two equal parts (Fig. 140 and Fig. 141). On each of these areas, which are often a little convex, there are very many spinning tubes, the number greatly exceeding the number of spinning tubes borne by the spinnerets. The number of spinning tubes varies greatly, however, in different species; thus Bertkau ('82) found 500 in the undivided cribellum of *Diotima* and about 4800 in each half of the cribellum of *Stegodyphus lineatus*, making 9600 in all. Or, to select observations based upon genera represented in the United States, Bertkau found over 300 spinning tubes in each half of the cribellum of a *Dictyna*, and about 3600 in the undivided cribellum of an *Uloborus*. It is evident that the threads issuing from the spinning tubes of the cribellum are exceedingly small. These threads are used, doubtless, in the formation of a part of the hackled band which is a characteristic feature of the webs of spiders possessing a cribellum and a calamistrum.

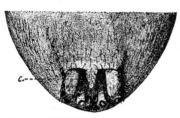

Fig. 142. THE TIP OF THE ABDOMEN OF LOXOSCELES *c*, colulus

THE COLULUS (col'u-lus).— Immediately in front of spinnerets there is in certain spiders a slender or pointed appendage, the *colulus* (Fig. 142, *c.*). The function of this organ is unknown. It is wanting in the tarantulas, the Drassidæ, and those families in which the cribellum exists. It has been suggested by Menge that the colulus is the homologue of the cribellum.

THE ANUS.— The posterior opening of the alimentary canal is situated just behind the group of spinnerets, on the lower side of the last segment of the postabdomen.

THE STRIDULATING ORGANS.— In some of the comb-footed spiders, Theridiidæ, the males possess a stridulating organ, consisting of a scraper on the abdomen and a file on the thorax; this organ is described in the account of that family. Stridulating organs borne by the mouth-parts are described on page 2†

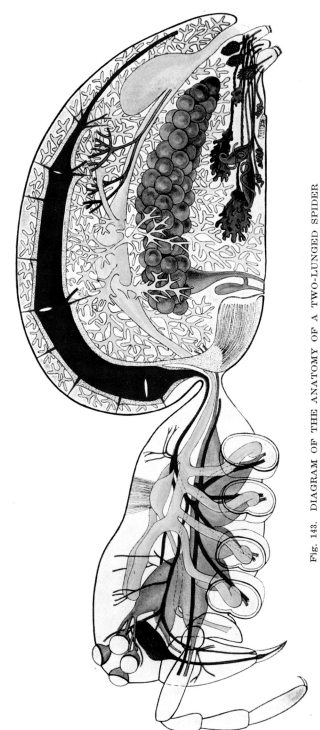

Fig. 143. DIAGRAM OF THE ANATOMY OF A TWO-LUNGED SPIDER

CHAPTER III: THE INTERNAL ANATOMY OF SPIDERS

THE more general features of the internal anatomy of spiders are illustrated by the accompanying plate (Fig. 143) upon which the different systems of organs are indicated by different colours, except that the muscles and the endosternites are omitted, in order that the diagram should not be too complicated. This plate represents the organs projected upon a vertical, median section of the body.

The black outline of the figure represents the body-wall; the respiratory organs which are, morphologically, infoldings of the body-wall, are indicated by the same colour; the alimentary canal is coloured yellow; the blood vascular system, red; the nervous system, blue; the poison gland, the silk glands, and the Malpighian vessels, green; and the reproductive organs, purple.

THE BODY-WALL

a. THE THREE LAYERS OF THE BODY-WALL

In that division of the Animal Kingdom to which spiders belong, the phylum Arthropoda, the outer covering of the body serves as a skeleton as well as a protecting shield. It is more or less firm in texture, and to its inner surface and to inward projections of it are attached the muscles that serve to move the body and its appendages; and within the cavity bounded by it are located the viscera. In other words, the body-wall is a firm tube containing the softer structures. The appendages of the body, that is, the legs, the mouth-parts, and the spinnerets, are also tubular, and the cavity of each communicates with the general body cavity.

Three, more or less distinct, layers can be recognized in the body-wall of a spider; first, the outer protecting layer, the *cuticula;* second, an intermediate cellular layer, the *hypodermis;* and third, an inner, delicate, membranous layer, the *basement membrane.*

The Internal Anatomy of Spiders

As the outer and inner layers are derived from the hypodermis, it will be described first.

The hypodermis.— The hypodermis is the active, living part of the body-wall. It consists of a single layer of cells (Fig. 144, *hy*).

Certain of the hypodermal cells become highly specialized and produce the hollow hairs with which the body is clothed; such a hair-forming cell is termed a *trichogen* (Fig. 144, *tr*), each trichogen remains connected through a pore in the cuticula, the trichopore, with the cavity of the hair it produces.

Other hypodermal cells are modified so as to form glands, which open through pores in the cuticula, either directly or through the tubular hairs. These glands may be either unicellular or multicellular.

The cuticula.—Outside of the hypodermis, there is a firm layer, which pro tects the body and serves as a support for the inter nal organs; this is the cuticula. The cuticula con sists of two layers: first, an outer, usually thinner, layer, which contains the pigments that produce the colours of the outer sur-

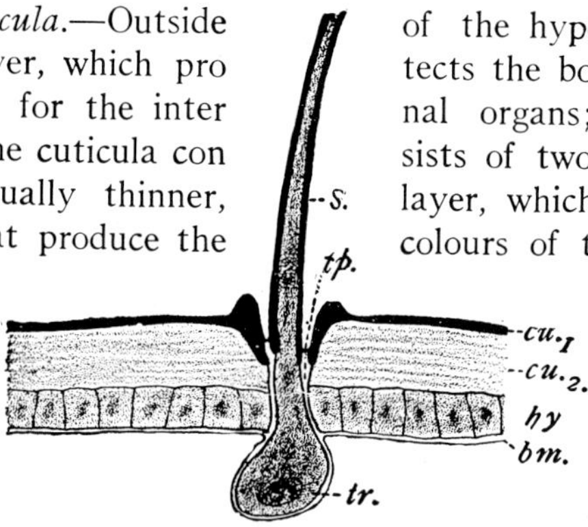

Fig. 144. DIAGRAM OF A SECTION OF THE BODY-WALL
cu.1, primary cuticula *cu.2*, secondary cuticula *hy*, hypodermis
bm, basement membrane *tr*, trichogen *s*, seta *tp*, trichopore

face, the primary cuticula (Fig. 144, *cu. 1*); and second, an inner, usually thicker, layer, devoid of pigment, the *secondary cuticula* (Fig. 144, *cu. 2*). The secondary cuticula is laminate; the laminæ remind one, when the cuticula is seen in section, of the lines of growth of an exogenous tree.

The well-known firmness of certain parts of the body-wall is due to the presence in the cuticula of a substance which has been termed *chitin*. This substance bears some resemblance in its physical properties to horn; but it is very different from horn in chemical composition.

The Internal Anatomy of Spiders

When freshly formed, as is done before each moult, the cuticula is flexible and elastic, and certain portions of it, as at the nodes of the body and of the appendages, and throughout the greater part of the wall of the abdomen in spiders, remain so. But the greater part of the cuticula of the cephalothorax and of the appendages becomes firm and inelastic; this is probably due to a chemical change resulting in the production of chitin. What the nature of this change is or how it is produced is not yet known; but it is evident that a change occurs; we may speak, therefore, of *chitinized cuticula* and of *non-chitinized cuticula*.

The hinge-like movements of the body and of the appendages are made possible by the non-chitinized condition of the cuticula at the joints.

A portion of the body-wall that is sharply distinguished from surrounding portions by being chitinized is called a *sclerite;* and a narrow line between two sclerites is termed a *suture*.

The basement membrane.— The inner ends of the hypodermal cells are bounded by a very delicate, but more or less distinct, membrane, the basement membrane (Fig. 144, *bm*).

b. THE CUTICULAR APPENDAGES

Under the head of cuticular appendages are included those outward prolongations of the body-wall that do not form an integral part of it, being separated from it, in each case, by a flexible joint; such appendages are the hairs or setæ and the movable spines.

The fixed spines, as for example those borne by the abdomen of *Micrathena* (Fig. 145), are not regarded as appendages. These form an integral part of the body-wall and differ only in size from mere nodules on the surface of the body.

The hairs or setæ.— These vary greatly in their external form; but no distinction, except in degree of stiffness, can be made between the extremely slender and very flexible type commonly called hairs and the stiffer one commonly

Fig. 145. MICRATHENA

The Internal Anatomy of Spiders

termed setæ; the two are morphologically the same, and they grade by insensible degrees into each other.

As regards their internal structure, the hairs of spiders, like those of insects and other arthropods, differ markedly from the hairs of mammals, being hollow.

Each hair of a spider arises from a more or less cup-like cavity in the cuticula of the body-wall, which may be made more pronounced by a ring-like elevation of the cuticula surrounding it (Fig. 144); but the latter feature is not an essential one. The sheath of the hair is continuous with the cuticula of the body-wall, but at its base it is infolded and more or less flexible, thus forming the joint. The cup-like cavity in the cuticula from which the hair arises is situated at the end of a pore in the cuticula, the *trichopore*. The trichopore and the base of the hair is filled with a prolongation of the hypodermal cell that formed the hair, the *trichogen* (Fig. 144, *tr*).

Fig. 146. SECTION OF A SETA
n, nerve (after Hilton)

Fig. 147. GLANDULAR HAIR OF A CATERPILLAR
(after Hilton)

The above may be regarded as the essential features of an arthropod hair. Certain writers refer to *protecting hairs*, which are supposed to present only these features, and to serve merely to protect the body. My own observations, based, it is true, chiefly on the hairs of insects, lead me to doubt the existence of hairs that are merely protective in function.

In most cases, if not in all, each hair is supplied with a minute nerve, connecting the hair with a subhypodermal nerve plexus (Fig. 146); this feature has been carefully worked out with insects, and it is not probable that it is radically different in spiders. Hairs furnished with a nerve are regarded as organs of special sense; in most cases they are organs of touch, the *tactile hairs*; these are discussed later.

A very important modification of the ordinary type of a hair is that known as the glandular hair. These are hairs that serve as outlets of hypodermal glands (Fig. 147); several kinds of these, that are found in insects, have been described. The type of glandular hair that interests most the student of spiders is that known as the tenent hairs. These are found on the terminal segments of the legs and serve to aid in walking. For figures of tenent hairs see page 124.

c. THE ENDOSKELETON

In the Arthropoda, as already stated, the outer covering of the body serves as a skeleton as well as a protecting shield; but in addition to the firm outer layer of the body-wall, there are found, within the body-cavity certain hard parts which serve for the attachment of muscles; these constitute the *endoskeleton*.

The presence of these parts that constitute the endoskeleton forms only an apparent exception to the statement that the body-wall[1] constitutes the skeleton; for these parts are merely infolded portions of the body-wall.

The apodemes.— The infoldings of the body-wall that serve for the attachment of muscles vary greatly in form; they may be merely a slight ridge or they may be a prominent projection into the body cavity. In the latter case such a projection is termed an *apodeme*.

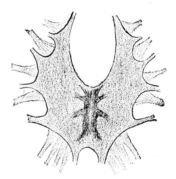

Fig. 148. THE ENDOSTERNITE OF THE CEPHALOTHORAX OF ARANEA (after Schimkewitsch)

The endosternites.— In the cephalothorax of spiders there is a horizontal plate above the ventral ganglion and below the alimentary canal, to which many muscles are attached. This is known as the *endosternite;* but as there are endosternites in the abdomen also, it may be specifically designated as the *endosternite of the cephalothorax* (Fig. 148).

The question of the origin of this endosternite has been much discussed; but it has been shown by Bernard ('96) that it is formed by the fusion and expansion of the tips of four pairs of apodemes, a pair extending into the body cavity in front of each pair of legs. The position occupied by these apodemes is that of the original intersegmental spaces, vestiges of which remain as lines across the sternum in very young spiders. The endosternite is, therefore, a part of the body-wall, like the more simple apodemes

The Internal Anatomy of Spiders

The endosternites are termed *aponeurotic plates* by some writers, who believe that they are formed by the coalescence of the tendons of muscles.

A so-called aponeurotic plate, which occurs in the pedicel of the abdomen, has been described. This, however, occupies a very different position from that of the endosternites, being near the dorsal body-wall above the aorta. The source of this plate has not been determined.

In the sac-like portion of the abdomen, there are vestiges of an endoskeleton. These vestiges are three centres of attach-

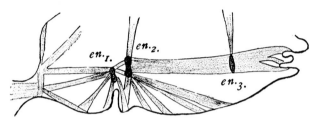

Fig. 149. THE THREE ABDOMINAL ENDOSTERNITES AND THE PRINCIPAL MUSCLES ATTACHED TO THEM
en 1, *en* 2, *en* 3, first, second, and third abdominal endosternites (after Schimkewitsch)

ment of muscles, one opposite the epigastric furrow, one a short distance back of this, and one near the furrow of the posterior spiracle (Fig. 149, *en 1, en 2, en 3*); these may be designated as the *first, second,* and *third abdominal endosternites* respectively.

The endosternites of the sac-like part of the abdomen are described by authors as intermediary tendons; but it seems clear to me that they are serially homologous with the endosternite of the cephalothorax. Correlated with the loss of the abdominal appendages there has been a reduction of the connection of the endosternites with the body-wall.

THE MUSCULAR SYSTEM

A striking feature of the muscles of spiders, and also of other arthropods, is the distinctness of the muscular fibres; each muscle being composed of a number of distinct, more or less isolated, straight fibres; and the fibres constituting a muscle are not enclosed in a common tendinous sheath, as is the case with vertebrates.

In appearance the muscles are either colourless and transparent, or yellowish white; and are of a soft, almost gelatinous consistence. When properly treated with histological reagents, and examined with a microscope of moderately high power, they present numerous transverse striations, like the voluntary muscles of vertebrates. Unstriped muscles are rare with arthropods; certain muscles, which have been supposed to be unstriped, show striations when properly treated.

The chief muscles are attached to the inner surface of the body-wall, to infoldings of it, the apodemes, and to the endo-sternites described in the preceding section. There are also prominent muscles attached to the alimentary canal, especially to the sucking stomach (Fig. 150).

It is not within the province of this book to enter into a description of the separate muscles; there is space for only a few general statements.

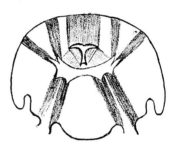

Fig. 150. DIAGRAM OF A TRANSVERSE SECTION OF THE CEPHALOTHORAX SHOWING THE PRINCIPAL MUSCLES
(after Schimkewitsch)

The muscles of the cephalothorax are extremely well developed; while those of the abdomen are greatly reduced. This is correlated, to a considerable extent, with the greater number and size of the appendages of the cephalothorax, the appendages of the abdomen, excepting the spinnerets, having been lost, and to the presence in the cephalothorax of a powerful sucking stomach.

The muscles of the body-wall.— The muscles of the body-wall, which form the most prominent feature of the muscular system of the larvæ of insects that are commonly studied, are very greatly reduced in spiders. This is doubtless due to the fact that the segments of the cephalothorax are fused in spiders so that there can be no movements between them, and that the abdomen is practically unsegmented. There are present, however, vestiges of both circular and longitudinal "skin-muscles."

The muscles of the appendages.— A very large proportion of the muscles of the cephalothorax serve to move the appendages. Some of these have their origin on the inner surface of the body-wall and others, on the endosternite; in each case they extend to the basal segment of the appendage. There are also muscles within the appendages that produce the movement of the separate segments of the appendages. Figure 151, which represents the musculature of a leg of Thelyphonus, one of the Pedipalpida, will serve as an illustration of the arrangement of the muscles in the leg of an arachnid.

The muscles of the sucking stomach.— The muscles of the sucking stomach are also very prominent; these are discussed in the section of this chapter treating of the alimentary canal.

The muscles of the abdomen.— It is much more difficult to trace out the muscles of the abdomen than it is those of the cephalothorax. In the abdomen the muscular system is greatly reduced and other viscera are so greatly developed that the muscles are overshadowed by them. Schimkewitsch ('84) gives a diagram showing the arrangement of the principal muscles of the abdomen of Aranea (Fig. 149). In this figure are represented the three vestiges of the endoskeleton of the sac-like part of the abdomen (Fig. 149, *en 1, en 2, en 3,*); these I have designated as the first, second, and third abdominal endosternites respectively.

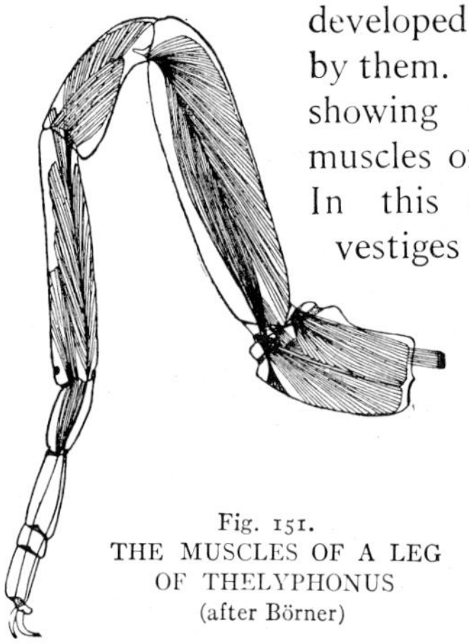

Fig. 151.
THE MUSCLES OF A LEG OF THELYPHONUS
(after Börner)

The most prominent of the muscles of the abdomen is a series extending from the pedicel of the abdomen to the three endosternites successively and from the third endosternite to the spinnerets. These have been named the longitudinal ventral muscles.

From the second and third endosternites there extend two pairs, one from each, of dorsoventral muscles. These are the muscles that are attached to the two pairs of more prominent muscle impressions, that appear externally as depressed points in the dorsal wall of the abdomen.

To the first abdominal endosternite are attached in addition to the longitudinal muscles, muscles from the book-lungs, and a muscle from the opening of the reproductive organs.

To the second abdominal endosternite are attached several muscles extending to the body-wall.

The Internal Anatomy of Spiders

The muscles extending backward from the third abdominal endosternite are those that move the spinnerets.

THE RESPIRATORY ORGANS

In the order Araneida, two types of respiratory organs are found; first, the book-lungs; and second, the tubular tracheæ. Some members of the order possess only one of these types; but the greater number of spiders possess both.

The number of respiratory organs is reduced in spiders to two pairs; and in most true spiders the second pair open by a single spiracle on the middle line of the body.

The more generalized spiders preserve two pairs of book-lungs. This condition is retained by *Liphistius*, all of the tarantulas and by a single family of the true spiders, the Hypochilidæ. Excepting the Hypochilidæ, most true spiders have a single pair of book-lungs; but in the Caponiidæ, which is not represented in the United States, both pairs of respiratory organs have been transformed into tubular tracheæ.

The accompanying figures (Fig. 152, 1-4) show in a diagrammatic way the distribution of the two types of respiratory organs among the families of spiders; in these figures the tracheal spiracles can be distinguished from lung-slits by the absence of external indications of book-lungs. The distribution is as follows:

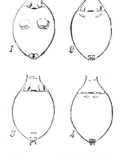

Fig. 152. DIAGRAMS SHOWING THE DISTRIBUTION OF THE TWO TYPES OF RESPIRATORY ORGANS
1, Liphistius, the tarantulas, and Hypochilidæ
2, Filistatidæ, Oonopidæ, and Dysderidæ
3, all true spiders except the five families mentioned here
4, Caponiidæ

1. With two pairs of book-lungs; *Liphistius*, the tarantulas (Avicularioidea), and the Hypochilidæ.

2. With one pair of book-lungs and one pair of tracheal spiracles; the Filistatidæ, the Oonopidæ, and the Dysderidæ.

3. With one pair of book-lungs and a single tracheal spiracle; all true spiders (Argiopoidea) except the Hypochilidæ, Filistatidæ, Oonopidæ, Dysderidæ, and Caponiidæ.

4. With two pairs of tracheal spiracles; the Caponiidæ.

The book-lungs.— These are sacs filled with air each of which open by a slit-like spiracle on the ventral aspect of the abdomen

The Internal Anatomy of Spiders

near its base. Their position is indicated externally on the living animal as pale trapezoidal spots with rounded angles (Fig. 153). The first pair open, one on each side, in the epigastric furrow; when there is a second pair they open a short distance back of the first pair.

From the anterior wall of each sac there project a series of horizontal, leaf-like folds, which has suggested the name of book-

Fig. 153.
VENTRAL ASPECT
OF THE ABDOMEN OF
A TARANTULA

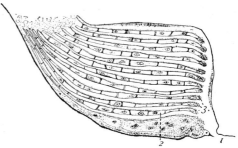

Fig. 154. DIAGRAM OF A BOOK-LUNG
1, lung-slit
2, space filled with blood
3, leaves of the book-lung

lungs (Fig. 143). But as these leaves are also attached to the lateral walls of the sac, the lung has been more aptly compared to that form of a letter-file that consists of an envelope divided by many partitions into numerous compartments (Lang '91).

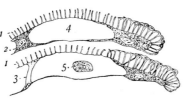

Fig. 155. THE TIPS OF TWO LEAVES
OF A BOOK-LUNG
1,1, palisade of spines on dorsal lamella
2, vertical support between lamellæ contracted
3, vertical support stretched
4, blood-cavity
5, globule of blood (after Berteaux)

Each leaf of the book-lung consists of a flattened, sac-like fold of the body-wall, and hence of two lamellæ. These lamellæ are connected at frequent intervals by vertical supports; and the leaves are kept apart, so that the air can circulate between them, by a palisade of vertical, knobbed spines on the upper surface of each (Fig. 154). At and near the free edge of the leaves, these spines are longer, and, instead of being knobbed, divide at the tip into several branches, which anastomose with branches of adjacent spines (Fig. 155).

The Internal Anatomy of Spiders

The blood passes from the body-cavity into the lumen of each leaf, and the respiratory process takes place through the walls of the leaves.

A hint regarding the mode of origin of this highly specialized respiratory apparatus is afforded by the comparatively simple lung-sacs of the Microthelyphonida, referred to on page 15.

The tubular tracheæ.— The distribution of tubular tracheæ among the different families of spiders, and the number of tracheal spiracles are indicated by Fig. 152. In most cases the tracheæ open by a single spiracle, situated on the middle line of the ventral aspect of the abdomen; sometimes this spiracle is near the middle

Fig. 156.
TUBULAR TRACHEÆ
OF ANYPHÆNA
(after Bertkau)

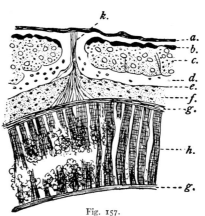

Fig. 157.
SECTION OF HEART
AND PARTS LYING ABOVE IT
(after Vogt et Yung)

of the length of the abdomen, but usually it is only a short distance in front of the spinnerets. In a few spiders there is a pair of tracheal spiracles situated just behind the lung-slits. These can be distinguished from lung-slits by the absence of external indications of book-lungs.

The tracheæ are paired organs even when they open by a single spiracle; this is well-shown by the figure of the main trunks of the tracheæ of *Anyphæna* (Fig. 156). As to the histological structure of the tracheæ, they, like the tracheæ of insects described in many text-books, present the same layers as does the body-wall, namely, a chitinous intima lining the cavity and continuous with the cuticula of the body-wall; an epithelial layer surrounding this and continuous with the hypodermis; and, bounding the epithelium next to the body-cavity, a basement membrane, continuous with the basement membrane of the body-wall. As with insects the chitinous intima is thickened spirally. The ramifications of the smaller tracheal branches appear to be much less extended than are the tracheæ of insects; this is probably due to the book-lungs being the chief organs of respiration.

THE CIRCULATORY SYSTEM

The circulatory system of spiders is an incomplete one; that is, for only a portion of its course does the blood flow in blood-vessels; during a part of its course it enters the body-cavity, as is the case with insects, where it fills the space not occupied by the internal organs. The extent of the vascular system, however, is much greater with spiders than it is with insects.

The heart.— The heart resembles the heart of an insect in position, being situated in the abdomen a short distance within the middle line of the dorsal body-wall, and above the intestine. In many spiders it is sunken to a greater or less depth in the mass of alimentary tubules; it is then difficult to expose it by dissection. Its position is represented in Fig. 143; and in Fig. 157 are represented some details not shown on the coloured plate. In this figure, *a*, represents the body-wall; *b* and *c*, a lobe of the alimentary tubules, of which *b* is the chalky layer; *d*, the dorsal longitudinal venous sinus; *e*, the pericradium, to be described later; *f*, the pericardial cavity; *g*, the longitudinal muscular layer of the heart; *h*, the annular muscles of the heart; and *k*, a cardiac ligament.

The wall of the heart is composed of three distinct layers: an inner, very delicate, structureless intima; an intermediate muscular layer; and an outer, connective tissue layer. The intima is so delicate that it is usually difficult to demonstrate its presence. The muscular layer consists chiefly of annular muscles; but there are also longitudinal fibres. The connective tissue layer is composed of slender fibres between which elongated nuclei can be seen.

The heart in spiders is a simple tube not divided into chambers by valves, as is the heart of insects.

The Internal Anatomy of Spiders

The more general features of the heart of different types of spiders are represented diagrammatically by the drawings in Fig. 158, which are based on a series of figures given by Causard ('96). The first drawing (1) represents the heart of a four-lunged spider; the second (2), the usual form of the heart in the two-lunged spiders; and the third (3), the heart of *Dysdera*. The

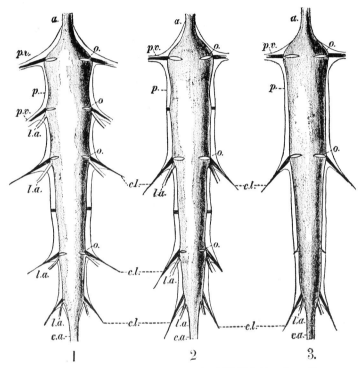

Fig. 158. DIAGRAMS OF HEARTS OF SPIDERS (after Causard)

heart is somewhat enlarged toward the anterior end and tapers toward the posterior end. From the anterior end extends the *aorta* (*a*); from the posterior end, the *caudal artery* (*c a*); and from its sides, the *lateral abdominal arteries* (*l a*). The cavity of the heart is more or less expanded at regular intervals, forming paired *diverticula*, from which the arteries arise. The wall of the heart is pierced by a series of paired openings, the *ostia* (*o*). From the heart extend a considerable number of strands, the *cardiac ligaments* (*cl*). At some distance from the heart, but completely surrounding it, is a thin-walled sac, the *pericardium* (*p*). Between the pericardium and the heart is the *pericardial cavity* (*pc*). The first one or two pairs of cardiac ligaments are larger than the others and function as *pulmonary veins* (*pv*). Each of the above-mentioned parts is described more in detail below.

The ostia of the heart.— The wall of the heart is pierced by several pairs of openings by which the blood is received into the heart; these are the *ostia* (Fig. 158, *o*). In the four-lunged spiders there are four pairs of ostia (Fig. 158, 1); in most of the two-lunged spiders there are three pairs (Fig. 158, 2); but in some of the two-lunged spiders, as in *Dysdera* (Fig. 158, 3), the number is reduced to two.

The diverticula of the heart.— As indicated above, the cavity of the heart is more or less expanded at regular intervals; and these expansions form paired diverticula. In the more generalized spiders, six pairs of these diverticula can be recognized. The four pairs of ostia of the heart in these spiders are in the walls of the first, second, third, and fifth pairs of diverticula respectively (Fig. 158, 1); the fourth pair of diverticula are midway between the third and fourth pairs of ostia, and bear only vestigial ligaments; the sixth pair is near the caudal end of the heart. In more specialized spiders the number of diverticula is more or less reduced; those which are most constantly prominent are the first, third, and fifth pairs, in the walls of which are the three pairs of ostia characteristic of most two-lunged spiders; the sixth pair are also usually well-marked, as from them arise a pair of arteries.

The aorta.— The aorta (Fig. 158, *a*) is the forward prolongation of the heart. It extends through the pedicel of the abdomen into the cephalothorax (Fig. 143). As it enters the cephalothorax, it gives off some small branches which ramify among the muscles of the posterior part of the cephalothorax. It then extends forward to a point near the sucking stomach, where it divides into two trunks, one passing forward on each side between the sucking stomach and the cæcal ring. Each of these trunks sends branches to the appendages, to the eyes, to the muscles, and to other organs of its side. This account of the course of the branches of the aorta is based on the figures and descriptions given by Causard ('96) and those by Petrunkevitch

The Internal Anatomy of Spiders

('10 b). It differs markedly from that commonly given in the text-books of zoölogy.

The lateral abdominal arteries.— In the four-lunged spiders, there are four pairs of lateral abdominal arteries; these extend from the second, third, fifth, and sixth pairs of diverticula (Fig. 158, 1, *la*). In the two-lunged spiders, the first of these four pairs is lacking (Fig. 158, 2); and in *Dysdera*, the first and second pairs are lacking and the third is greatly reduced in size (Fig. 158, 3).

The caudal artery.— The caudal artery is the backward prolongation of the heart (Fig. 158, *ca*). It divides into many small branches, which supply blood to the spinnerets and to the abdominal viscera.

The pericardium.— At some distance from the heart, but completely surrounding it is a thin-walled sac, which is termed the *pericardium* (Fig. 158, *p*). The space between the pericardium and the heart is the *pericardial cavity* (Fig. 158, *pc*).

The cardiac ligaments.— The heart is supported by a considerable number of ligaments. These extend from the heart to the body-wall; some of them extend dorsally, some ventrally, and some laterally. In Fig. 158, only the laterally extending ligaments are represented. In Fig. 157 a dorsally extending ligament (*k*) is shown; and several of them are represented in Fig. 143. The laterally extending cardiac ligaments are termed, by some writers, the *alary muscles*, by analogy with the "wings of the heart" of insects. But Causard ('96) has shown that the cardiac ligaments are not muscles, but are formed of connective tissue.

Each cardiac ligament is formed by a bundle of connective tissue fibres, which is inserted in the heart, and is enclosed in a tubular prolongation of the pericardium (Fig. 158, *cl*). The cardiac ligaments that extend laterally have been designated the *exocardiac ligaments*; those that extend dorsally, the *epicardiac ligaments*; and those that extend ventrally, the *hypocardiac ligaments*.

The pulmonary veins.— In the basal part of the abdomen of the four-lunged spiders, there are two pairs of large vessels which convey the blood from the two pairs of book-lungs to the pericardial cavity; and in the two-lunged spiders, there is a single pair of these vessels; these are the *pulmonary veins* (Fig. 158, *pv*).

The Internal Anatomy of Spiders

According to Causard, the pulmonary veins are large hollow exocardiac ligaments, which serve as veins in addition to their function of holding the heart in place. In fact he regards the other exocardiac ligaments as reduced pulmonary veins, which have lost their venous function, and become merely ligaments.

The venous circulation.— From the arteries the blood passes into the spaces of the body-cavity between the viscera, bathing the various organs. It then passes into two longitudinal sinuses, one dorsal and one ventral, which lead to the base of the abdomen. Here it is purified by the book-lungs; and then it passes through the pulmonary veins to the pericardial cavity, from which it enters the heart through the ostia.

The blood is also purified to a greater or less extent by tubular tracheæ as it is in insects; but the extent of the ramifications of these tracheæ varies greatly in different spiders.

THE ALIMENTARY CANAL

The alimentary canal of arthropods is a tube extending from one end of the body to the other. With some arthropods, as the larvæ of insects, it is a nearly straight tube; with others, as most adult insects, it is greatly lengthened, being much longer than the body and, consequently, more or less coiled within it. This increase in the length of the alimentary canal gives an increased surface for the performance of its functions.

In spiders, and other arachnids an extended area of the digestive tract is obtained in a different and peculiar manner; instead of a lengthening of the intestine, there have been developed extensive diverticula, extending in different directions from it (Fig. 143).

Three chief regions of the alimentary canal are recognized; namely, the fore-intestine, the mid-intestine, and the hind-intestine. These regions differ in their histological structure, a difference due to different embryological origin. The fore-intestine and the hind-intestine are developed from an infolding in each case of the ectoderm, the germ layer from which the body-wall is formed. And these regions are lined with a chitinous layer, the *intima*, which is a continuation of the cuticula of the body-wall. On the other hand, the mid-intestine is developed from the entoderm and is lined by an epithelium.

The Internal Anatomy of Spiders

The fore-intestine is that portion of the alimentary canal that is developed from an infolding of the ectoderm at the anterior end of the embryo, the *stomodæum*. It consists of the pharynx, the œsophagus, and the sucking stomach.

The histological structure of the fore-intestine is the same as that of the body-wall; the lumen being lined with a chitinous intima, a continuation of the cuticula; and, surrounding this, there is an epithelial layer, continuous with the hypodermis; and this in turn is bounded on the side toward the body-cavity by a basement membrane. The muscular coat of the fore-intestine, which is well-marked in some arachnids, is absent in the true spiders. Figure 159 represents a small portion of the wall of the sucking stomach, in this the layers are shown and also the attachment of some muscles.

Fig. 159. SECTION OF WALL OF SUCKING STOMACH

The pharynx.— The pharynx or mouth-cavity is situated between the rostrum and the sternum. It extends in a more or less nearly vertical direction (Fig. 143). It is bounded in front by a plate borne by the rostrum, which may be termed, by an analogy drawn from the anatomy of insects, the *epipharynx*; and is bounded behind by a plate borne by the sternum, which by a similar analogy may be termed the *hypopharynx*.

By bringing together the lateral margins of the epipharynx and the hypopharynx, a tube-like cavity is formed into which the spider sucks the juices pressed from its prey by the mouth-parts, the tufts of hairs borne by the mouth-parts preventing the escape of the liquid food.

Fig. 160. THE EPIPHARYNX

The epipharynx is marked by many transverse striæ (Fig. 160) which lead to a longitudinal slit in its middle line, which in turn leads to the opening of the œsophagus. Thus if the mouth-cavity be closed as tightly as possible by the pressing together of the epipharynx and the hypopharynx there will still remain an open way for the passage of the last drop of fluid pressed from the prey.

At the upper end of the pharynx near the entrance to the œsophagus there is a straining apparatus consisting of hairs which prevent the entrance of solid food into the œsophagus.

Several pairs of muscles extend from the walls of the pharynx to the body-wall; one of these, the retractor of the pharynx, is attached to the upper end of the pharynx and is sometimes very large.

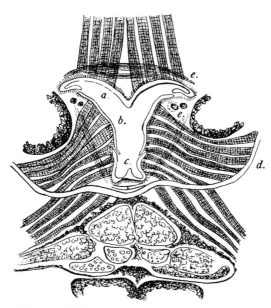

Fig. 161. CROSS-SECTION OF THE SUCKING STOMACH AND ADJACENT PARTS
a, dorsal plate *b*, lateral plate *c*, ventral plate *d*, endosternite *ee*, sphincter muscles (after Vogt et Yung)

Within the rostrum there is an unpaired gland, the *pharyngeal gland*, which opens into the pharynx near the beginning of the œsophagus.

The Œsophagus.— This is that part of the fore-intestine that connects the pharynx with the sucking stomach; it extends in a more or less nearly horizontal direction and consequently nearly at right angles to the pharynx. It is tubular in form; and the intima of its dorsal and lateral walls is thick and strong, forming an inverted gutter-like sclerite, which is transversely striated. The epithelium of the œsophagus is greatly reduced. The hind portion of the œsophagus is surrounded by the central nervous system; the course of this part of it is indicated in Fig. 143 by dotted lines.

The sucking stomach.— The sucking stomach is an enlarged portion of the fore-intestine, whose function is indicated by its name. It is situated just behind the point where the fore-intestine emerges from the nervous collar and rests upon the endosternite (Fig. 143). The chitinous intima of the fore-intestine is greatly thickened in this region, affording a firm support for powerful muscles. In a cross-section of the sucking stomach (Fig. 161), it can be seen that the thickened intima consists of four longitudinal plates; a dorsal and a ventral one, and one on each side. From the dorsal plate very strong muscles extend to the ental surface of the median furrow of the cephalothorax; and from the ventral and the lateral plates muscles extend to the endosternite. These act to enlarge the cavity of the organ. Opposed to the muscles just described, the sucking stomach is supplied with sphincter muscles (Fig. 161). These it will be observed are outside of the wall of the intestine and do not, therefore, strictly speaking, form a muscular coat of the intestine.

b. THE MID-INTESTINE

The mid-intestine begins a short distance back of the sucking stomach; it differs greatly from the fore-intestine both in histological structure and in general form.

The distinctively characteristic feature in the structure of the mid-intestine is the absence of a chitinous intima; this portion of the alimentary canal being lined with a layer of cells, the essential digestive element of the organ (Fig. 162, ep). This digestive epithelium rests on a basement membrane, outside of which there is a peritoneal layer composed of fibres, presumably of connective tissue (Fig. 162, p). No muscular coat has been observed; but outside of the peritoneal layer there is a cellular layer, which has been termed by several writers, but I believe incorrectly, a peritoneal layer (Fig. 162, f). This so-called peritoneal layer, I believe to be the fat-body, which is described later.

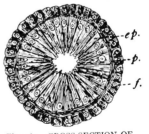

Fig. 162. CROSS-SECTION OF MID-INTESTINE (after Bertkau)

The most striking feature of the mid-intestine of spiders is the presence of an extensive system of diverticula by means of which the extent of the digestive epithelium is greatly augmented. In the cephalothorax, these diverticula are in the form of large

The Internal Anatomy of Spiders

simple cæca; but in the abdomen they are much-branched **tubules,** forming a great mass, which has the general appearance of a gland. These diverticula may be designated as follows:

The anterior cæca.— At the beginning of the mid-intestine a cæcum extends forward on each side of the dorsal muscle of the sucking stomach (Fig. 163 *ac*). These two cæca often unite in front of this muscle, thus forming what has been termed the ring-stomach, or the *cæcal ring*.

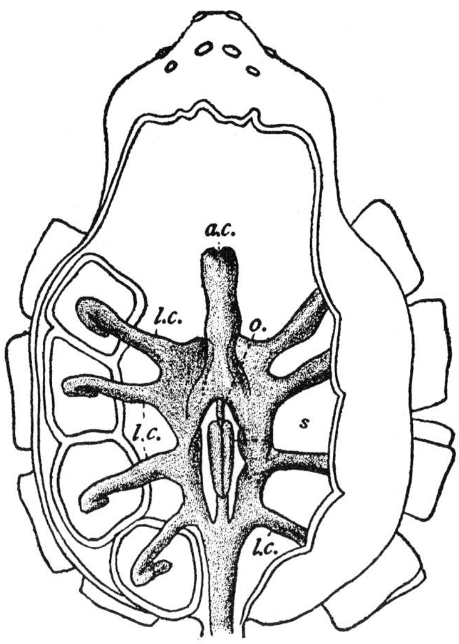

The lateral cæca. — From each anterior cæcum, four lateral cæca arise. These extend laterally and ventrally and each ends in or near a coxa of a leg (Fig. 163, *lc*).

The alimentary tubules. — In the anterior part of the abdomen, there arise, from a somewhat expanded portion of the mid-intestine, several much-branched dorsal diverticula; these constitute the alimentary tubules. These are developed to such an extent that they form the most prominent part of the viscera of the abdomen.

Fig. 163. DORSAL VIEW OF THE CEPHALOTHORACIC PORTION OF THE ALIMENTARY CANAL
o, œsophagus *s*, sucking-stomach *ac*, anterior cæca, united in front *lc*, lateral cæca (after Leuckart)

The mass of alimentary tubules was described by the early writers as the liver and by later ones as a digestive gland; but it has been shown that the epithelium of these tubules does not differ from that of the main portion of the mid-intestine; and that the food penetrates to the tips of the tubules and is absorbed by the tubules.

c. THE HIND-INTESTINE

The hind-intestine is the terminal portion of the alimentary canal; it is that part that is developed from an infolding of the ectoderm at the posterior end of the embryo, the *proctodæum*. It, consequently, like the fore-intestine, resembles the body-wall in its histological structure, there being a chitinous intima, an epithelial layer and a basement membrane. It differs from the fore-intestine in possessing a more or less well-developed muscular coat.

The stercoral pocket.— The most striking feature of the hind-intestine is the presence of a large, bladder-like diverticulum of its dorsal wall, which is situated near the place of union of the mid-intestine and hind-intestine (Fig. 143). It serves as a reservoir of fecal matter and is termed, therefore, the *stercoral pocket* (ster'co-ral).

The Malpighian vessels.— There is a pair of tubes which open into the alimentary canal near or at the place of union of the mid-intestine and hind-intestine; these are the Malpighian vessels (Mal-pe'ghi-an or Mal-pig'hi-an). Each of these tubes is branched and the branches ramify among the alimentary tubules. They are supposed to correspond to the kidneys in function.

Some writers maintain that Malpighian vessels open into the mid-intestine. If this should prove to be true, they cannot be homologous with the Malpighian vessels of insects, which are developed from the proctodæum.

The rectum.— The rectum is a simple tube extending from the stercoral pocket to the anus.

THE ADIPOSE TISSUE

In spiders the greater part of the adipose tissue or fat-body is intimately associated with the mid-intestine and its diverticula. In the case of the simple cæca of the mid-intestine, situated in the cephalothorax, the fat cells form a continuous investing sheath, which has been termed, incorrectly, the peritoneal coat. In the abdomen, the fat-body completely fills the spaces between the alimentary tubules, and surrounds them as a whole.

In many spiders the fat-body presents a chalky appearance due to the presence in it of fine particles, which are believed to be crystal-like remains of the food. These "fecal crystals" are sometimes very abundant in the dorsal superficial portion of the mass; and, showing through unpigmented portions of the cuticula, produce conspicuous white markings.

The Internal Anatomy of Spiders

In the cephalothorax, between the nervous system and the body-wall, there is a layer of fat cells which appear to perform a function similar to that of the urate cells of insects.

THE REPRODUCTIVE ORGANS

The two sexes are distinct in all Arachnida; and in each sex, the reproductive organs lie in the abdomen, and open near its base.

a. THE REPRODUCTIVE ORGANS OF THE MALE

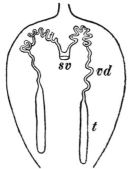

Fig. 164. DIAGRAM OF THE REPRODUCTIVE ORGANS OF THE MALE
t, testes *v d*, vasa deferentia
s v, seminal vesicle (after Bertkau)

The testes and their ducts. — The *testes* are situated in the anterior part of the abdomen, between the ventral body-wall and the longitudinal ventral muscles. They are two long tubes (Fig. 164, *t*); which are closed behind, and are continued in front as two, long, thin, and often much coiled sperm ducts, the *vasa deferentia* (Fig. 164, *vd*). The two vasa deferentia open into a common pouch, the *seminal vesicle* (Fig. 164, *sv*); and the seminal vesicle opens in turn through a single opening on the middle line of the body in the epigastric furrow.

The palpal organ. — There is no copulatory organ directly connected with the outlet of the reproductive glands; but the seminal fluid is transferred to the female at the time of the pairing of the sexes by means of a highly specialized appendage of the palpus of the male. This organ is described in detail in the preceding chapter.

b. THE REPRODUCTIVE ORGANS OF THE FEMALE

The internal reproductive organs of the female consist of the *ovaries*, the *oviducts*, the *uterus*, the *vagina*, and one or more pouches for the reception of the seminal fluid at the time of pairing, the *spermathecæ*. The vagina and the spermathecæ are invaginations of the body-wall and, like other invaginations of the body-wall, are lined with a chitinous intima, which is a continuation of the cuticula; while the uterus is lined with an epithelium similar to that lining the oviducts.

The Internal Anatomy of Spiders

In certain generalized spiders, as *Atypus, Segestria, Dysdera, Tetragnatha*, and others, the spermathecæ are diverticula of the vagina. In these cases there is only a single external opening of the reproductive organs. But in most spiders, the spermathecæ are more or less detached and have separate external openings. In these cases there are three external openings of the reproductive organs, the opening of the vagina, and a pair of openings of the afferent ducts leading to the spermathecæ. In this type of reproductive organs, a tube connects the spermathecæ of each side with the vagina. Figure 165 is a diagrammatic representation of this type.

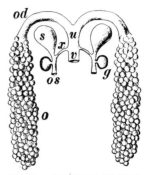

Fig. 165. DIAGRAM OF THE INTERNAL REPRODUCTIVE ORGANS OF THE FEMALE OF ARANEA
o, ovary *od*, oviduct *u*, uterus *v*, vagina *s*, spermatheca *os*, opening of the spermatheca *x*, tube connecting the spermatheca with the vagina *g*, gland of the spermatheca

The ovaries.— The ovaries are situated in the abdomen below the intestine and above the longitudinal ventral muscles (Fig. 143). They are surrounded by silk glands and the alimentary tubules, and are two broad tubes bearing numerous ovarian follicles and appearing like a cluster of grapes. The free ends of the ovaries are sometimes grown together so as to produce a ring-like structure (Fig. 166).

Fig. 166. RING-LIKE OVARIES OF A SPIDER
o, ovaries
od, oviduct
s, spermathec
(after Lang)

The size of the ovaries varies greatly, depending on the state of the development of the eggs. Just before the egg-laying period, they become greatly distended and occupy a considerable part of the abdominal cavity.

The oviducts.— Each ovary opens through a short *oviduct;* and these in turn open into a common pouch, the *uterus*, which leads to the *vagina;* and the *vagina* opens on the middle line of the body in the epigastric furrow.

The spermathecæ.— The spermathecæ vary in number. In those spiders where the spermathecæ are diverticula of the vagina, as described above, there are, according to Engelhardt ('10), two in *Pachygnatha*, three in *Tetragnatha*, and five in *Segestria;* and according to Bertkau ('75) twenty-eight in *Atypus*. In those spiders where the spermathecæ are detached there may be either one, two, or three connected with the afferent duct of each side.

The duct leading from the external opening of a spermathecæ to the large reservoir of this organ frequently bears a prominent enlargement with densely chitinized walls (Fig. 165, *g*). This organ was described by Schimkewitsch ('84), who states that the chitinous layer of its wall is pierced by numerous pores and is surrounded by an epithelium composed of very high cells, which are probably glandular. I, therefore, propose the term *glands of the spermathecæ* for these organs. As the chitinous wall of these glands is often dark in colour, they may be seen through the body-wall and appear as a pair of dark spots, which may be mistaken for the openings of the spermathecæ.

The external openings of the afferent ducts of the spermathecæ are in the epigynum, which is described in the preceding chapter.

THE NERVOUS SYSTEM

a. THE CENTRAL NERVOUS SYSTEM

In fully developed true spiders, the central nervous system is entirely concentrated into the cephalothorax. In early embryonic stages there is a pair of ganglia in each segment of the body; but in the course of the embryonic development a cephalization of the ganglia takes place which results in their being consolidated into a single mass surrounding the œsophagus (Fig. 143). In the tarantulas, however, there is a small ganglion in the pedicel of the abdomen.

The brain.— That part of the central nervous system that lies above the œsophagus is termed the brain; from it arise the optic nerves and the nerves of the chelicerae.

The subœsophageal ganglion.— That portion of the central nervous system lying below the œsophagus is termed the *subœsophageal ganglion*. From this ganglion arise the nerves extending to the pedipalps and those extending to the legs; and from the posterior end extend two large abdominal nerves. The brain and the subœsophageal ganglion are not distinct, as they are in insects, but are consolidated into a single mass surrounding the œsophagus.

The Internal Anatomy of Spiders

b. THE VISCERAL NERVOUS SYSTEM

A visceral nervous system has been proven to exist in spiders and in other Arachnida. It consists of an unpaired nerve connected with the brain by paired nerves and running along the œsophagus and stomach. Ganglia connected with the ventral chord have also been described as belonging to the visceral nervous system (Lang).

c. THE EYES

The number and the position of the eyes have been discussed in the opening pages of the preceding chapter; here we have to do with their internal structure.*

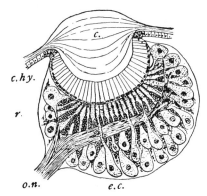

Fig. 167.
DIAGRAM OF A POSTBACILLAR EYE
c, cornea chy, corneal hypodermis r, retina
ec, eye-capsule on, optic nerve (after Widman)

The eyes of spiders are o the type known as the several layered ocellus, each eye consisting of two or more layers of cells behind a single lens (Fig. 167). The lens (Fig. 167, c) is termed the *cornea;* the outer of the two more prominent layers (Fig. 167, c. hy), the *corneal hypodermis;* and the inner of these two layers (Fig. 167, r), the *retina.* The eye is enclosed in a sheath (Fig. 167, ec), the *eye-capsule;* in some eyes there is a layer of reflecting cells behind the retina, the *tapetum;* and in some eyes there are pigment cells, which vary in position in different eyes.

The cornea.— The cornea or lens of the eye (Fig. 167, c), is merely a modified portion of the cuticula of the body-wall, with which it is continuous. It is a lens-shaped thickening of the cuticula, which is free from hairs and, consequently, is not traversed by trichopores, and in which the primary cuticula is devoid of pigment, permitting the passage of light to the nervous portion of the eye.

---

*Much has been published regarding the eyes of spiders. The works that I have consulted most in the preparation of the account given here are the following: Grenacher '79; Graber '80; Schimkewitsch '84; Bertkau '86; Widman '07; and one, as yet, unpublished paper, by Mr. George D. Shafer.

The Internal Anatomy of Spiders

The cornea is moulted and renewed in the same manner as other portions of the cuticula of the body-wall, and at the same time.

The corneal hypodermis.— Immediately beneath the cornea is a layer of cells (Fig. 167, *c. hy*) which is continuous with the hypodermis of the body-wall, and which is merely a more or less modified portion of the hypodermis.

Several different names have been applied to this part. It has been termed the *vitreous layer*, or the *vitreum*, a term suggested by its analogy with the vitreous humour of the eye of vertebrates; the *lentigen*, because it produces the lens; and the *corneal hypodermis*, a self-explanatory term. I prefer the last name notwithstanding its greater length.

The retina.— Lying immediately behind the corneal hypodermis, is a layer composed of highly specialized visual cells; this is the retina (Fig. 167, *r*).

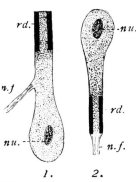

Fig. 168. DIAGRAMS OF TWO VISUAL CELLS
1, postbacillar   2, prebacillar
nu, nucleus
nf, nerve fibre   *rd*, optic rod

Each *visual cell* consists of several parts; there can be distinguished a cell-body, in which there is a nucleus (Fig. 168, *nu*); the cell-body is connected with the central nervous system by means of a nerve-fibre (Fig. 168, *nf*); and there is also a hard structure, which is known as the *optic rod* (Fig. 168, *rd*).

The optic rod is the distinctively characteristic feature of a visual cell. The form of this part varies in different eyes, and, sometimes in different parts of the same eye; it is commonly double, occupying two faces of one end of the cell with a thin layer of the cell-body extending between the two elements, but sometimes the two parts are united.

As the optic rods of adjacent visual cells are placed side by side, there is a well-marked layer of the retina occupied by the rods; but the position of this layer is not always the same. Two types of eyes are recognized which are distinguished by the position of the nuclei of the retina with reference to the position of the layer of optic rods. In one type the nuclei lie in front of the optic rods or *bacilli*, as the rods are often termed (Fig. 169);

the eyes of this type are termed *prebacillar* (pre-bac'il-lar) *eyes*. In the other type the nuclei lie behind the optic rods; such eyes are termed *postbacillar eyes*.

The anterior median eyes are postbacillar; all other eyes of spiders are prebacillar.

In the prebacillar eyes (all eyes except the anterior median) the nerve fibres join the visual cells at the end of the cell bearing the rod (Fig. 169); while in the postbacillar eyes (anterior median) the nerve fibres join the visual cells between the nucleus and the rod (Fig. 167).

The two types of eyes differ in the method of their development, but this is a phase of the subject into which we cannot enter here; it is sufficient to state that on account of their peculiar

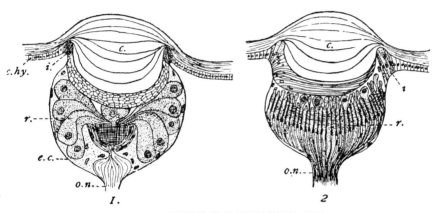

Fig. 169. DIAGRAMS OF A PREBACILLAR EYE
c, cornea *chy*, corneal hypodermis *r*, retina *e c*, eye-capsule *o n*, optic nerve
i, iris pigment (after Widman)

method of development the anterior median eyes are termed by some writers the *inverted eyes*, while all the other eyes are termed *erect eyes*.

The eye-capsule.— The eye-capsule is a sheath enclosing the eye. It appears to be continuous with the sheath of the optic nerve; it is sometimes termed the *post-retinal membrane*.

The tapetum.— In the eyes of many animals there is a structure that reflects back the light that has entered the eye causing the well-known shining of the eyes in the dark. This is often observed in the eyes of cats and in the eyes of moths that are attracted to our lights at night. The part of the eye that causes this reflection is termed a *tapetum* (ta-pe'tum). The supposed function of a tapetum is to increase the effect of a faint light on the visual organs, the light being caused to pass through the retina a second time when it is reflected from the tapetum, which is behind the retina.

The structure of the tapetum varies greatly in different animals; in the cat and other carnivores it is a thick layer of wavy fibrous tissue; in insects it is a mass of fine tracheæ; in the crustacea it is formed of a light-coloured pigment in the accessory pigment cells of the eye; and in spiders it consists of a layer of cells behind the retina containing small crystals that reflect the light.

The form of the tapetum layer varies in the eyes of different spiders; three types have been described. In the first type it lies immediately within the eye-capsule and encloses the whole of the retina. As the inner end of the eye-bulb is more or less compressed, the tapetum, as Mark has aptly described it in *Agelena*, is a canoe-shaped structure, with a fissure along the keel, through which the nerves pass to the retina. This type is termed the *funnel-form* type.

The second-type of tapetum is also canoe-shaped, with a slit at the bottom, and is also termed *funnel-form;* but it differs from the first type in that the funnel is more shallow, enclosing for the most part, only the proximal rod-bearing ends of the visual cells; and it does not lie immediately within the eye-capsule, owing to the fact that the nuclear ends of many of the visual cells are bent back under the sides of the tapetum cup and consequently lie between it and the eye-capsule. This form of a tapetum has been figured by Widman, whose figures I copy. Figure 169 1, represents a section of a prebacillar eye of *Amaurobius* cut at right angles to the longer axis of the tapetum; and (Fig. 169, 2,) represents a section lengthwise of the tapetum slit.

These two figures show a peculiar condition of the corneal hypodermis which exists in some eyes of spiders. The nuclear ends of the cells are crowded to one side; the result is that in one section these cells are cut lengthwise; in the other, transversely.

The third type of tapetum is in the form of a shallow cup whose sides are divided into many parallel strips by a series of cross slits running out each way at right angles to the long slit at the bottom of the cup; this type is termed the *grate-form* type.

The Internal Anatomy of Spiders

The nerve fibres enter the retina through the cross slits as well as through the long one. Figure 170 is a diagram representing one end of a grate-form tapetum, the figure is a composite one, the upper half representing the tapetum, as observed by Shafer, of *Lycosa helluo*, and the lower half, that of *Pardosa lapidicina*.

In this figure, *l. sl.* represents the longitudinal slit and *c. sl.*, the cross slits; both of these are filled with pigment excepting the spaces occupied by the nerve fibres, sections of which are represented as white spots. The white spaces, *t.*, alternating with the cross slits, represent the tapetum strips. Extending in from the rim of the cup, which is heavily pigmented, there is a strip of pigment, *pg. st.*, upon each tapetum strip. In each half of the figure, the rod end, *r.*, of several visual cells is represented. These cells are surrounded by pigment except where they rest upon the tapetum.

The pigment cells.— Cells containing a dark pigment occur in various situations in and about the eyes of spiders. These pigment cells, limiting as they do the transparent tracts that may be followed by the light, doubtless play an important part in the operation of the eye.

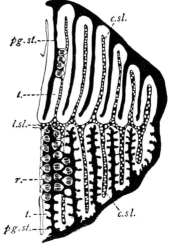

Fig. 170. DIAGRAM OF ONE END OF A GRATE-FORM TAPETUM
(after Shafer)

A ring of hypodermal cells about the rim of the eye and immediately under the edge of the lens that contains pigment is termed the *iris* (Fig. 169, *i*). The iris pigment occurs both in the undifferentiated hypodermis and in the marginal cells of the corneal hypodermis.

In eyes of the postbacillar type (the anterior median eyes) there are interstitial pigment cells between the visual cells (Fig. 167,). In some species the pigment cells extend to the corneal hypodermis; in others they reach only the proximal end of the optic rods.

In eyes of the prebacillar type (all eyes except the anterior median) differing conditions occur in different species. In eyes with a funnel-form tapetum there is no pigment in front of the tapetum; but in the second type of funnel-form tapetum-eyes described above, there may be interstitial pigment-cells between the nuclear ends of those visual cells that bend around behind the tapetum. And in both the first and the second types of funnel-form tapetum-eyes there are pigment cells between the nerve fibres behind the tapetum.

In the grate-form tapetum-eyes, not only are there pigment cells between the nerve fibres behind the tapetum, but they extend above the tapetum as well, surrounding the optic rods.

In the prebacillar eyes of many spiders there is found a pigmented layer similar to the funnel-form tapetum in form and position and, doubtless, homologous with it. This layer has been termed, unhappily, a "pigmented tapetum." But as it cannot reflect light, owing to its being pigmented, it cannot properly be called a tapetum.

Finally, in some cases, a prebacillar eye is divided by a thick sheet of pigment into two portions, a larger median and a smaller lateral. In these *divided eyes* the visual cells differ in form in the two parts of the retina; and the form of the tapetum may differ in the two parts.

Nocturnal eyes and diurnal eyes.— In the case of many spiders that live in the dark or that frequent shady places some of the eyes are pearly white. It is believed that such eyes are fitted for seeing in a faint light; they have been termed, therefore, *nocturnal eyes*. On the other hand, those eyes that are dark in colour are termed *diurnal eyes*. Considerable use is made of this distinction in the classification of spiders. It is of interest, therefore, to determine whether the pearly white eyes are fitted for nocturnal vision or not; and if so, to determine in what respect they differ from the so-called diurnal eyes.*

Strictly speaking, any eye furnished with a tapetum, that is with a reflecting structure so placed behind the retina that the light entering the eye is made to traverse the retina a second time, must be considered a nocturnal eye.

It does not follow, however, that every eye provided with a

---

*At my suggestion, the working out of this problem was undertaken by Mr. George D. Shafer in the entomological laboratory of Cornell University. The morphological data given below as well as some of that given above are drawn from a thesis by Mr. Shafer, which, at the present writing, is unpublished.

tapetum can be recognized as a nocturnal eye by its colour when viewed externally. In order that the eye may give forth the pearly lustre it must be comparatively free from pigment between the tapetum and the cornea; for if this is not the case, the reflected light will be absorbed to such an extent that the eye will not glisten.

As no postbacillar eyes (anterior median) have been found in which there is a tapetum, it is probable that these eyes are never nocturnal. It should be noted, however, that in a few cases, as in *Meta* for example, the anterior median eyes have the pearly white lustre. A careful study of these eyes shows that there is no tapetum present, the retina immediately behind the optic rods being densely packed with interstitial pigment cells. The pearly lustre is due, therefore, to some other cause than the presence of a tapetum, and these eyes are only apparently nocturnal.

The pearly white prebacillar eyes of several genera (*Pholcus, Theridion, Agelena, Clubiona, Trachelas, Meta,* and *Amaurobius*) were studied by Mr. Shafer. It was found that in each there is a funnel-form tapetum, and that there is no pigment between the tapetum and the cornea. These eyes are, therefore, truly nocturnal; and the pearly lustre is doubtless due to the presence of the tapetum.

In the prebacillar eyes of *Pardosa, Lycosa,* and *Misumena*, for example, a grate-form tapetum was found. These eyes are, therefore, strictly speaking, nocturnal. But as in the grate-form tapetum-eyes the pigment cells extend in front of the retina, surrounding the optic rods, but little light can be reflected out of the eye. On this account these eyes are apparently diurnal.

To sum up the results it can be said that certain diurnal eyes (the anterior median eyes of *Meta*, for example) are apparently nocturnal; and that many nocturnal eyes (those having a grate-form tapetum) are apparently diurnal. In other words the presence or absence of a pearly lustre is not sufficient to determine whether an eye is nocturnal or diurnal.

The eye-muscle.— Each anterior median eye is provided with a muscle extending from the body-wall to the eye-capsule. No eye-muscles have been observed connected with any of the prebacillar eyes in true spiders.

The Internal Anatomy of Spiders

d. THE LYRIFORM ORGANS

There are found in several of the orders of the Arachnida organs of a peculiar type, the function of which has not been determined, but which have been supposed to be organs of hearing. They occur in spiders, in the Phalangida, the Pedipalpida, and in the Pseudoscorpionida, and are known as the *lyriform organs*.

These organs vary greatly in form but agree in their essential features. The distinctly characteristic external feature is a slit in the cuticula in which there is a wider, oval, central portion, and, on each side of this, a long, narrow portion, which may or may not taper to a point. At the bottom of this slit there is a thin layer of the cuticula, so that the slit does not reach the body-cavity. These slits occur in the sternum and in the cuticula of most of the segments of the legs, except the tarsi, and of the mouth-parts. They are extremely small; but are very constant in their form and in their position.

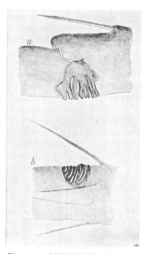

Fig. 171. LYRIFORM ORGANS

In the sternum the slits may occur singly or in groups. But the more characteristic form of these organs, and the one that suggested the name lyriform, is found on the appendages, where they usually occur near the distal end of a segment; but sometimes they are found near the middle of a segment. In this type there are several slits grouped side by side; and the cuticula surrounding the group of slits, as well as that of the spaces between them is greatly thickened (Fig. 171). The spaces between the slits suggest the cords, and the outer rim of the organ, the frame of a lyre, hence the name lyriform. In Fig. 171, *a* represents the lyriform organ in the metatarsus of the fourth leg of *Argyrodes trigonus;* and *b* that found in the same position in *Theridion frondeum*.

As to the internal structure of the lyriform organs, it has been found that there is beneath each slit a nerve-end-cell. This cell is elongate, with a nucleus at its proximal end, near its connection with a nerve, and with the distal end long and slender, and extending to the thin layer of cuticula limiting the slit within. According to Bertkau ('78), who discovered these organs, the nerve-end-cell terminates in a small chitinous rod which rests against the thin layer of cuticula; but this rod is not described by later writers, and is not now believed to exist.

The observations of Bertkau led to the belief that the lyriform organs are auditory in function. But the failure of other observers to discover scolopalæ or auditory pins in the nerve-end-cells connected with them has made this conclusion doubtful.

A detailed description of the lyriform organs has been published by Mr. N. E. McIndoo since the preceding account was put in type. (Proc. Acad. Nat. Sci. Phil. 1911, pp. 375–418). The results of Mr. McIndoo's experiments indicate that the lyriform organs are olfactory in function.

e. THE ORGANS OF TOUCH

The organs of touch of spiders are distributed generally over the surface of the body. They are hollow hairs or setæ each furnished with a nerve. The structure of such hairs has been described on page 140, and is illustrated by Fig. 146.

It is probable that the nerves of the hairs are connected with a subhypodermal nerve-plexus as is well-known to be the case with insects. Traces of such a nerve-plexus have been observed by Bernard ('96) in the Solpugida.

f. THE ORGANS OF TASTE AND OF SMELL

Very little has been written regarding the organs of taste and of smell of spiders; and it is still a question whether organs of either of these senses exist among spiders or are wanting. It may be, as indicated above, that the lyriform organs are olfactory.

THE GLANDS OF SPIDERS

With the exception of the poison glands and the silk glands, the glands of spiders have received comparatively little attention. The moulting of spiders has been described in great detail by Wagner ('88); and reference is made by him to the moulting fluid

The Internal Anatomy of Spiders

which facilitates the shedding of the old cuticula; but I find no reference to the moulting fluid glands. Doubtless there are in spiders, as there are in insects, modified hypodermal cells that produce the moulting fluid.

Several glands have been described connected with the digestive organs that may function as salivary glands. These occur in the endites and in the rostrum. But the most prominent of the glands connected with the mouth-parts are those that open through the cheliceræ and produce the venom.

Glands which open in the coxal segment of the third pair of legs in tarantulas and in some true spiders have been observed; these are termed *coxal glands*.

a. THE POISON GLANDS

The poison-glands are two in number and are situated in the true spiders in the anterior part of the cephalothorax (Fig. 143). Each gland discharges its product through a long slender duct which opens near the tip of the claw of the chelicera of the corresponding side of the body. In the tarantulas each poison gland is situated in the basal segment of a chelicera.

The glands are sac-like in form; the lumen of the sac serves as a reservoir of venom; the wall is composed of excreting cells, supported by a layer of connective tissue, and there is a layer of muscle fibres surrounding the sac. The fibres of the muscle layer are arranged in a spiral manner.

b. THE SILK-GLANDS

The term silk is ordinarily used to designate the thread spun by the silkworm, *Bombyx mori*, from which nearly all of our silken fabrics are made; but by entomologists the term is also applied to other similar products. In this sense, silk is produced by various animals and is used by them for many different purposes,

Among insects there is considerable variation in the structure and position of the silk organs; thus while with caterpillars the silk is produced by modified salivary glands, and is spun from a spinneret situated near the mouth; with ant-lions it is produced by modified Malpighian tubes and is spun from the caudal end of the alimentary canal. It is obvious, therefore, that silk-pro-

The Internal Anatomy of Spiders

ducing organs have arisen independently in different races of animals.

The silk organs of spiders are the most complicated silk organs known. This condition might be expected from the fact that a single species of spider spins several distinct kinds of silk; thus, for example, an orb-weaving spider spins five different kinds of silk.

A detailed account of the different kinds of silk spun by spiders is given in the following chapter; and in the preceding chapter the external spinning organs are described. In this we have to consider the internal silk organs, the glands that produce the silk.

As spiders spin several kinds of silk it is obvious that there are several kinds of silk glands differing in function. Seven different kinds of silk glands have been recognized. These differ in form, in number, in colour, in the structure of their ducts, and in the nature of their products.

No spider has been found to possess all of the seven kinds of silk glands; but three of the kinds have been found in all species studied from this point of view; and a fourth is wanting in only two families. The three other kinds are each characteristic of a particular group of spiders, and no two of them are found together. Each of the three groups of spiders that possesses a characteristic kind possesses also the first four kinds. Hence the presence of five kinds in a single spider is common.

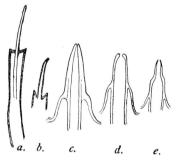

Fig. 172. SPINNING TUBES

Much has been written on this subject; but the paper that has been of the most service in the preparation of this account is one by Apstein ('89). I have followed Apstein quite closely in my account of the structure of the glands; but in several cases my own observations do not confirm his conclusions regarding their function.

The different kinds of silk are spun from different types of spinning tubes (Fig. 172), which are described on page 135.

The aciniform or berry-shaped glands.— There are four clusters of these glands, one for each of the *hind* and *middle* spinnerets

The Internal Anatomy of Spiders

(Fig. 143). They have been found in all species studied. In *Aranea* there are about one hundred in each cluster or four hundred in all; but in some spiders there are only a few of them. They open through the first type of spinning tube, those with a long, more or less curved base, and a slender straight tip (Fig. 172, *a*).

The name applied to these glands was suggested by the botanical term *acinus*, which is applied to one of the kernels of a fruit like a raspberry. A single gland resembles an acinus in form; each of the four clusters of these glands resembles a berry.

The aciniform glands are nearly spherical and consist of an epithelium invested by a peritoneal membrane (Fig. 173). The epithelium is not continued into the duct. All parts of the gland give the same reaction to stains.

I have determined by direct observation that the swathing band is formed of silk from these glands. This fact is easily determined by feeding a spider confined in a glass tube.

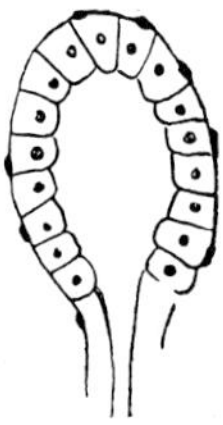

The pyriform or pear-shaped glands.— There are two clusters of these glands, one for each of the *fore* spinnerets (Fig. 143). They have been found in all species studied. In *Aranea* there are about one hundred in each cluster, or two hundred in all, but some spiders have only a few of them. They open through spinning tubes of the second type (Fig. 172, *b*), in which the basal part is short, the tip twice as long as the base and curved.

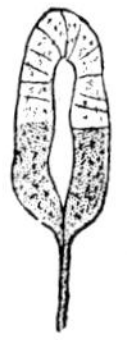

Fig. 173. AN ACINIFORM GLAND

Fig. 174. A PYRIFORM GLAND

These glands are elongate and consist of an epithelium invested by a peritoneal membrane (Fig. 174). The epithelium is not continued into the duct. The part of the gland next its mouth gives a darker colour reaction to stains. Owing to this these glands are easily distinguished in stained preparations.

The function of the pyriform glands is the formation of the attachment disks. The making of these disks by a spider confined in a vial can be easily observed.

The ampullate or bellied glands.— These are referred to by certain writers as the cylindrical glands; but this name has been applied to the type of glands described next. There are usually four ampullate glands; but frequently there are six or eight, and

The Internal Anatomy of Spiders

in some spiders there are as many as twelve. They are large, long glands, more or less cylindrical in outline (Fig. 143, and Fig. 175). The glands consist of an epithelium invested by a peritoneal membrane. The epithelium is continued into the duct, which also has a chitinous intima. Each gland opens through a spigot (Fig. 172 c). In the usual condition, when there are four of these glands, the spigots are located one on the inner side of each of the *fore* and *middle* spinnerets. The ampullate glands occur in all species examined. They are termed ampullate because each is furnished with an ampulla or sac-like dilation.

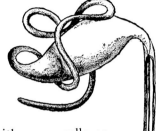

Fig. 175. AN AMPULLATE GLAND

The drag-line and the dry threads of webs are made of silk from the ampullate glands. The fact that the drag-line is composed of only two or four strands and that it comes from these glands was discovered by Warburton ('90).

The source of the elastic element of the viscid thread of the orb-weaving spiders has not been determined. It appears to come from the centre of the group of spinnerets; as it consists of two strands of considerable size, it is evidently spun from two spigots. As the drag-line of orb-weavers usually consists of two strands, I infer that in these spiders two of the ampullate glands have been so modified that they produce elastic silk. Wilder ('66) demonstrated that the yellow silk of *Nephila*, which is the viscid silk, is spun from the fore spinnerets.

The cylindrical glands.— These glands are commonly present in female spiders; but are wanting in the Dysderidæ and in the Attidæ. There are usually six of them; but sometimes there are many. In males there is a smaller number or they are wanting. In the usual case, where there are six cylindrical glands, one opens on the outside of each middle spinneret, and two open on the inside of each hind spinneret. Each opens by a spigot similar to the spigots of the ampullate glands, but with a wider opening (Fig. 172, d).

These glands lie directly on the ventral side of the body cavity; they are long, cylindrical, of uniform diameter, and convoluted (Fig. 176). They resemble in structure the ampullate glands.

The function of these glands is the production of the silk of which the egg-sac is made. Spiders of the two families in which they are wanting do not spin true egg-sacs. This is the only case of a sexual difference in the silk glands of spiders.

The aggregate or treeform glands.— These are found only in the orb-weaving spiders and in the Linyphiidæ and Theridiidæ. They are a symmetrical in form, being irregularly branched and lobed. (Fig. 177.) The duct bears on its middle part many knobs which are filled with cells. There are six of these glands, four large and two small. Three open near together on the inner surface of each *hind* spinneret, through spigots the tip of which is pointed (Fig. 172 *e*). The term aggregate was suggested by the lobed form of the glands, which gives them a compound appearance.

It is believed that the function of these glands is to secrete the viscid drops of the viscid and elastic thread. As these glands are found in the Theridiidæ, as well as in the orb-weavers, it is important to determine whether viscid threads occur in the webs of these spiders. It is certain that they are uncommon or at least are inconspicuous, but they have been observed in the webs of *Linyphia, Steatoda,* and *Theridion.*

The lobed glands.—These occur only in the Theridiidæ and are either two or four in number. They open on the *hind* spinnerets through spigots without a tip (Fig. 178). The lobed glands are very irregular in form, consisting of a mass of irregular lobes (Fig. 179).

The silk of which the swathing film of the Theridiidæ is made is secreted by the lobed glands. Correlated with their presence is the development of only a small number of aciniform glands.

The cribellum glands.— These are found only in those families in which a cribellum and calamistrum are present. They open on the cribellum through exceedingly small pores (Fig. 141, p. 135). The glands are exceedingly numerous; they are spherical and are often grouped; that is several many be enclosed in a common sheath of peritoneal membrane.

It is probable that the function of the cribellum glands is to secrete the woof of the hackled band.

Figure 180 represents the relative position of the silk glands in a spider.

Fig. 176.
A CYLINDRICAL GLAND

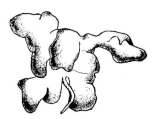

Fig. 179.
A LOBED GLAND

Fig. 177.
AN AGGREGATE GLAND

Fig. 178.
HIND SPINNERET OF STEATODA,
SHOWING TWO SPINNERETS
WITHOUT A TIP (after Apstein)

Fig. 180.
SECTION OF ABDOMEN
SHOWING SILK GLANDS
(after Apstein)

The Internal Anatomy of Spiders

THE SILK GLANDS AND THE SPINNING TUBES OF SPIDERS

THE GLANDS			THE SPINNING TUBES				
Name	Number	Fore Sp.†	Middle Sp.†	Hind Sp.†	Type of Tube	Families	Function
Aciniform or berry-shaped	4 clusters*		100 more or less or only a few	100 more or less or only a few	First	All species studied	Swathing band
Pyriform or pear-shaped	2 clusters*	100 more or less or only a few			Second	All species studied	Attachment disks
Ampullate or bellied	4, usually, often 6 or 8, maybe 12	1 on inner side	1 on inner side		Spigot	All species studied	The dragline and the dry thread of webs, perhaps also elastic silk
Cylindrical	6 in female, sometimes many; less or wanting in male		1 on outer side	2 on inner side	Spigot	Wanting in the Attidæ and the Dysderidæ	The egg-sac
Agreggate or treeform	6 or 4; 4 large and 2 small			3 on inner side	Spigot with pointed tip	Argiopidæ, Linyphiidæ and Theridiidæ	Probably the viscid drops
Lobed	2 or 4			1 or 2, near the tip	Spigot without tip	Theridiidæ	Swathing film
Cribellum	Very many				Very small	Cribellatæ	Probably the woof of the hackled band

*There may be 100 more or less in each cluster or only a few.
†The number of spinning tubes given in the table is for a single spinneret in each case.

176

CHAPTER IV: THE LIFE OF SPIDERS

THE careful observer of the ways of spiders soon learns that the members of the different families differ greatly in their modes of life; and that even within the limits of a single family striking variations exist. In fact each species, while agreeing with closely allied forms in the more general features of its life history, exhibits specific pecularities.

There is given in later pages of this book, so far as the limits of space will permit, what is known regarding the characteristic habits of each of the families of North American spiders, and of some of the more common species in each. But in order to avoid repetition those features that are common to all or to many species are discussed here.

1.— METHODS OF STUDY

The study of the life of spiders is essentially a field study; a few species live in our dwellings, and some others can be kept in confinement, but the great majority of them can be best observed in the open air; and so abundant are these creatures that one need never lack materials for study during the warmer parts of the year.

Spiders live in a great variety of situations. A few species seem to prefer the angles of buildings; many are usually found among shrubbery; some stretch their webs high in trees; others build among the low herbage of meadows and pastures; the members of several families run over the surface of the ground or lurk beneath stones; and certain kinds dig burrows in the earth. The student should therefore pursue his studies in as great a variety of situations as possible.

A collecting outfit.— For the collection and preservation of spiders only a small equipment of apparatus is necessary. A collecting outfit should include bottles of alcohol for the preservation of specimens, empty bottles for bringing home specimens alive, a pocket lens, a piece of black cloth, preferably velvet, to place behind webs when studying them, and, most important of all, material for taking notes.

Methods of collecting.— The most instructive method of collecting spiders is by carefully seeking for them and observing their habits before taking specimens; this is especially true of the web-building species. Many other species are to be found by overturning stones and other objects lying on the ground and by tearing off the bark from dead trees, logs, and stumps.

But if it is desired to make as complete a collection as possible of the species of a locality, other methods must also be used. Of these those known as sweeping and as sifting are the more important. In sweeping, an insect net is used and the foliage of shrubs and trees is beaten, and herbage is swept, as in the well-known method of collecting insects. In sifting, dead leaves are collected from the ground and shaken in a large sieve over a sheet of paper.

The apparatus devised by Professor Berlese of Florence, and described by Dr. L. O. Howard in *Entomological News* for February, 1906, is very useful for collecting the small spiders that live among dead leaves and other rubbish. This apparatus consists of a metal cylinder in which the rubbish is placed and to which heat is applied causing the spiders and other animals in it to move down into a receptacle placed to catch them

The preservation of specimens.— Owing to the softness of the body, spiders cannot be well preserved dry. The preservative fluid that is usually employed is alcohol; it should not be diluted, but used of the full strength of commercial alcohol. Care should be taken not to put too many specimens in a bottle, and if the specimens are large the alcohol should be changed after one or two days, as the alcohol first used will be diluted by the fluid of the body.

If specimens are to be preserved permanently, the bottles should be provided with rubber stoppers, as the alcohol is liable to evaporate from bottles closed with cork. In some of the larger collections the specimens are put in small corked bottles and these bottles are stored in larger bottles or fruit jars filled with alcohol and tightly closed.

The egg-sacs and many kinds of nests of spiders can be mounted and preserved dry; but in most cases it is impracticable to preserve specimens of the webs; here photographs must take the place of the actual specimens.

It is possible, however, to make beautiful mounts of the central and more characteristic portion of orb-webs. This is done by mounting the web between two plates of glass. Take two pieces of glass of the same size; prepare one of these by fastening a narrow strip of adhesive paper along the edges of one face; then carefully press the other piece of glass against the web to be preserved; the viscid silk will cause the web to adhere to the glass, and if the operation be carefully performed the relation of the different lines will be preserved; the lines that extend beyond the edge of the glass should then be wiped away, using care not to disturb the position of the lines upon the glass; then cover the web with the other piece of glass which will be held a short distance away from the web by the strip of paper on its edge; the two pieces of glass are then fastened together by pasting a strip of gummed paper over the edges of them. The kind used in mounting lantern slides is the most available form of adhesive paper for this purpose.

The laboratory equipment.— The laboratory should be furnished with a dissecting microscope, which is the most convenient type of microscope for use in the study of the classification of spiders, and with a compound microscope. I use also, with great satisfaction, the Pfeiffer dissecting microscope which gives an erect image and which is made to fit upon the stand of an ordinary dissecting microscope. For anatomical work the stereoscopic microscope, made by Zeiss, is very useful. There is also needed a supply of watch glasses, forceps, dissecting needles, a measuring rule, and a glass tube fitted with a rubber bulb, like a pen filler, for picking up and transferring minute specimens in alcohol.

Some method of storing the specimens in systematic order, so that allied forms shall be kept together and so that any desired specimen can be easily found, should be adopted. I use bent neck vials, stoppered with rubber, mounted on blocks, and stored in systematic sequence in the same kind of drawers as we use for the storing of insects. In some collections the bottles are stored in drawers fitted with series of holes for the reception of the bottles; and in other collections the small bottles are stored in larger ones, as already described, and these are arranged on the shelves of a cabinet.

The Life of Spiders

On photographing spiders.— It is exceedingly difficult to induce living spiders to pose for their photographs. Sometimes, in the case of large species, where it is not necessary to approach the camera too closely to the spider, a good photograph can be obtained of one in its natural attitude, and I have taken photographs of running spiders by enclosing them in a plate glass box and photographing through one side of the box. As a rule, however, it is necessary to kill a spider before a satisfactory photograph can be made of it.

The specimen should be killed in a cyanide bottle, and then either posed dry on a piece of cardboard or placed in a porcelain dish of alcohol and photographed with a vertical camera. This must be done promptly while the legs and palpi are flexible and before they become curled up. Sometimes, in the case of brightly marked species, better results are obtained if a colour screen is used.

On photographing spider webs.— If one wishes to take photographs of spider webs, special preparations should be made for it. Photographs of webs taken in the field are seldom satisfactory. Only rarely can a web be found in a suitable position as regards lighting and background, and in a satisfactory state of preservation.

When possible, the spider whose web it is desired to photograph, should be induced to build a web where it cannot be injured by wind or insects and where the nature of the lighting and of the background can be controlled. Many of the pictures in this volume are from photographs of "made to order" webs.

These webs were made in the glass house of our insectary, which resembles an ordinary greenhouse. Any room suitable for photographic work would serve for this purpose.

The most easily obtained webs were those of spiders that build upon shrubs or trees. Having found such a web in the field, the branch upon which it was built was carefully cut and carried, with the spider and web upon it, into the insectary and firmly fixed into the earth in a large flower pot. This permitted the moving of the branch without disturbing the web, when it was desired to pose it for a photograph.

If the spider was not unduly disturbed it remained on its web in the new situation, and in most cases repaired the web during the first or second night following. Usually the outer framework

The Life of Spiders

was left but the central and more characteristic portions of the web were entirely rebuilt. This new web being protected from wind and from insects often remained in a perfect condition for a considerable time, giving ample opportunity for the study and the photographing of it.

Owing to the extreme delicacy of the threads of which a web is made very careful lighting of it is necessary. By simply moving the flower pot containing the branch, the web can be put in the most favourable position.

As to the background, good results were obtained by hanging a piece of black velvet behind the web; this substance reflects very little light and makes a good contrast with the light lines of the web. But the most perfect results were obtained by the use

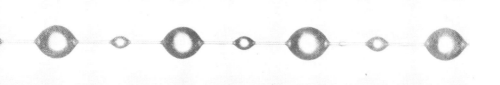

Fig. 181. VISCID SILK OF ARANEA

of a dark hole for a background. This was obtained by using an open box about five feet deep, three feet wide, and three feet long. This was lined with black velvet and placed upon its side. The web to be photographed was then placed before the opening of the box. The bottom of the box, which served as the background, being in the shadow of the sides reflected almost no light.

Frequently the getting of a photograph of a web was not so simple as in the case of those built on shrubs and trees; some of the special devices used for this purpose are described in the accounts of the webs given later.

On photographing threads of silk.— After making many unsuccessful attempts to make photomicrographs of silk, I devised a very simple method. Two small pins are inserted in the end of a narrow strip of cork; these pins are about one third inch apart; by pressing these pins against a thread which it is desired to photograph a section of the thread is obtained stretched between the pins; thus mounted, the thread can be kept from injury by contact with any object; the piece of cork bearing the pins and silk is placed on the stage of a microscope so that the silk is in front of the objective. It is now easy to study the silk; and by combining the microscope with a camera in the ordinary way, to make a photomicrograph of it. Figure 181 was made from a photograph obtained in this way.

II.—THE DEVELOPMENT OF SPIDERS

The egg-sac.— The eggs of spiders are laid in a mass and are usually protected to a greater or less extent by a covering of silk, the *egg-sac* or *cocoon;* but sometimes they are merely agglutinated and are without a silken covering; this is the case in some of the Pholcidæ.

Sometimes the egg-sac is so thin that it does not conceal the eggs, being merely a loose web of threads; in other cases it is very elaborate in structure. Several types of egg-sacs are discussed later in the discussion of the motherhood of spiders.

The life within the egg-sac.— As a rule the eggs hatch in a comparatively short time after they are laid. The egg-shell is broken, in the case of certain spiders at least, by a tooth on the base of the pedipalps. Although the eggs hatch soon after they are laid, the spiderlings may remain within the egg-sac for a long time. Very many of the spiders living in the North pass the long winter as spiderlings within the egg-sac. With these the eggs are laid in the autumn, the spiderlings soon hatch, but remain in the egg-sac till the following spring. Thus it may happen that a spider whose entire life is one year in duration spends nearly or quite one half of this period within the egg-sac. On the other hand, the young of many species leave the egg-sac soon after hatching.

According to the observations of Dr. B. G. Wilder ('73a) some of the young of *Miranda aurantia* feed upon their weaker brothers and sisters; so that from an egg-sac that in early winter contains a large number of spiderlings there emerge in the spring a much smaller number of partly grown spiders. How general this habit of cannibalism is has not been determined. McCook states that the young of the orb-weavers prey upon each other after they leave the egg-sac.

The moulting of the cuticula.— Spiders, like insects, shed, from time to time, an outer layer of the cuticula of the body. This process is termed moulting and the cast skin is called the exuviæ. The number of the moults and the intervals between them varies with different species, and has been determined in only a few cases. Blackwall states that *Epeira catophylla* moults five times and *Tegenaria civilis,* nine times. Wagner found that the young of *Trochosa singoriensis* moulted four times before beginning to lead an independent life.

The cuticula of the cephalothorax and of its appendages, the legs and mouth-parts, is firm and inelastic, being hardened by a horny substance known as chitin. Owing to this it will not stretch enough to allow for the growth of the spider. The result is, that from time to time, the cuticula becomes too small for the spider and must be shed.

Fig. 182. TARSUS OF A MOULTING SPIDER, DOLOMEDES SCRIPTUS

But before this is done a new cuticula is formed beneath the old one; then the old skin bursts open, and the spider casting it off is clothed in a soft skin, which stretches to accommodate the increased size of the body. Very soon, however, this new skin becomes hardened with chitin, and after a time it in turn must be shed.

Before the old cuticula is shed it is loosened by a fluid, the moulting fluid, which is excreted from glands that open through the new cuticula. Figure 182 represents the tarsus of a moulting spider; here the old skin can be seen separated from the new; this is especially clear in the region of the claws, there being apparently two pairs of claws; but the outer pair is merely the loosened cuticula of the claws.

After the old cuticula is loosened, it splits along the sides of the body and in front of the eyes; the slit being just above the base of the legs and mouth-parts and extending back on each side upon the abdomen. That portion of the old cuticula that had covered the dorsal aspect of the cephalothorax is lifted off like a lid and remains attached to the abdominal portion (Fig. 182 bis). Through the opening thus made the body is worked out, and the appendages one after the other are pulled from their old coverings.

Fig. 182 bis. MOULTED SKIN OF A SPIDER, DOLOMEDES URINATOR, MALE

In addition to the mechanical reason for the moult, the allowing of the increase in size of the body, there are doubtless physiological reasons for it also. In spiders the old cuticula of the abdomen is about as elastic as the new one, still the moult includes this part of the body as well as the firmer portions. At each moult the spider is clothed with a complete coat of new hairs. When the function of the hairs as sense organs is considered, it is seen that their renewal at intervals is very important. Other physiological reasons for the moult have been suggested, but as yet they are not well understood.

Transformations.— In the course of their development, spiders undergo comparatively slight changes in form. Their development is comparable to that of those insects that undergo an incomplete metamorphosis. They increase in size, and the proportions of the different parts of the body undergo a greater or less change; but there is no marked change in form as there is with those insects that undergo a complete metamorphosis.

The most marked change takes place at the last moult, when the development of the sexual organs is completed. Previous to this moult, the tarsus of the pedipalp of the male is merely a club-like segment in appearance; but when the cuticula is shed for the last time, the exceedingly complicated organ already described is disclosed. So also in the case of the female of those species that have an epigynum, this organ is not exposed till the last moult takes place. At this time too, the abdomen of the female may undergo a marked change in form, due to the growth of the eggs within it. The males of the hackled-thread weavers lose the cribellum and calamistrum at the last moult or have these organs greatly reduced.

The reproduction of lost organs.— The reproducing of legs that have been lost by immature spiders is frequently observed. If a leg be lost by a young spider the wound soon heals, and at the succeeding moult the bud of a new leg appears. This bud increases in size at each succeeding moult; and in time, if the process begins early enough in the life of the spider, a functional leg is obtained. Figure 183 represents a spider in which two legs, the left fore leg and the right hind leg, were being reproduced; this is a mechanical reproduction of a photograph of a spider in our collection.

Fig. 183.
A SPIDER REPRODUCING LOST LEGS

III.—THE FOOD OF SPIDERS

All spiders are carnivorous. Their prey consists chiefly of insects; but they will feed on other spiders that they can overcome, even on weaker members of their own species. This cannibalism is not confined to that of young spiders described in the preceding section, but is also true of adults. It is a common occurrence for the female to destroy the male of its own species, which is smaller and weaker.

Besides insects and spiders other small animals are occasionally destroyed by these voracious creatures. The most striking instance of this is the destruction of small birds by a large South American tarantula, the bird-spider (*Avicularia avicularia*).

The habits of this spider were first described by Madam Merian (1647-1717). Most writers have doubted the truth of her statement; but it has been confirmed by Bates ("A Naturalist on the Amazon," p. 80). Although Simon is among those that doubt Madam Merian's account, he states that the Avicularia that he caught in Venezuela ran on the trunks of trees with astonishing rapidity (Simon '92, p. 170). And as the so-called bird-spider is very large, the body measuring two inches in length and the legs expanding seven inches, it is certainly large enough and strong enough to overcome some of the smaller birds. In fact several apparently well-authenticated instances of the destruction of small vertebrate animals, including birds, a mouse, a fish, and a snake, by spiders that were much smaller than the Avicularia are given by McCook ('89. I. 235). But of course all such occurrences are very exceptional.

The mouth of a spider is fitted only for the taking of liquid food. This it presses from its victim by means of its cheliceræ and the endites of its pedipalps. Observations on spiders in confinement indicate that some of them at least require water frequently; but it is evident that others obtain moisture only by sucking the liquids of their victim.

Although spiders are extremely voracious, they are capable of enduring long fasts. Blackwall states that a female *Theridion quadripunctatus* was known to exist for eighteen months without nutriment in a vial closely corked.

IV.—MEANS BY WHICH SPIDERS OBTAIN THEIR PREY

Great differences exist among spiders as regards the ways in which they obtain their prey; some, the wandering spiders, stalk their prey; others lie in ambush for it; many trap it by means of snares; and a few live as commensals.

The wandering spiders.— Among the more familiar examples of wandering spiders are the wolf-spiders (Lycosidæ), the jumping spiders (Attidæ), and most of the crab-spiders, (Thomisidæ). These run about in search of their prey and pounce upon it when an opportunity offers.

The ambushing spiders.— Certain of the crab-spiders, as some species of the genus *Misumena*, hide in flowers and capture insects that visit them, resembling in this respect the ambush-bugs; and many of the mining spiders lie in ambush at the entrance of their burrows, and spring forth to seize insects that come near.

The net-building spiders.— The great majority of sedentary spiders, as those that do not wander in search of prey are termed, spin webs or snares for the trapping of insects, and wait either upon or near the webs where they can easily rush upon an entangled insect.

The commensal spiders.— Excellent examples of feeding at the same table with other species, or commensalism as it is termed, is exhibited by species of the genus *Argyrodes*. These small spiders live in the snares of larger web-building species and feed upon the smaller entrapped insects that are neglected by the owner of the web.

Means of destroying the prey.— Spiders destroy their prey by means of venom secreted by a pair of glands in the cephalothorax; the ducts from these glands open, one on each side, through a minute pore near the tip of the claw of the chelicera.

Most of the web-building species swathe their victims in a sheet of silk. The act of swathing can be easily observed by throwing a large insect into the web of a *Miranda*. The spider first rushes at the insect and pierces it with the claws of its chelicerae, and then darts back into a position of safety; this may be repeated several times; or, if the spider is not afraid of its victim, the biting may be omitted. Then the spider approaches the insect and pulling out a sheet of silk from its spinnerets with one hind leg thrusts the sheet against the insect. In doing this the spider uses first one hind leg and then the other. In the case of a large *Miranda* this sheet of silk is sometimes an inch in length, the body of the spider being held that far from the insect; under these conditions the sheet can be seen to be composed of a very large number of parallel threads, probably a thread from each of the small spinning tubes of all the spinnerets enters into the composition of the sheet. As soon as the sheet is fastened to the insect the spider rolls the insect over and over and thus wraps it in its shroud.

V.—THE SILK OF SPIDERS

The silk glands have been described in the chapter on internal anatomy; and the spinning organs, in the one on external anatomy;

The Life of Spiders

we have now to consider the nature and uses of the different kinds of silk.

The dragline.— Most spiders as they move from place to place spin a thread, which marks their course; this thread has been termed the *dragline*. Draglines are the most commonly observed threads of spiders; it is by them that spiders drop from an elevated position to a lower one; it is of them that the irregular nets are largely composed; and the foundation of an orb-web is made of draglines.

There is a commonly accepted error regarding the structure of a dragline. As the number of spinning tubes is large in the spiders most commonly studied, the orb-weavers, it has been inferred that in spinning a line the spider emits a delicate thread from each, and that all of these, sometimes several hundred in number, blend into one. This is not the case; the dragline is composed of a small number, usually only two, comparatively large threads. These, judging by their size are evidently spun from spigots.

Fig. 184.
AN ATTACHMENT DISK

The attachment disks.—If a spider that is running along some object and spinning a dragline as it goes be carefully observed, it will be seen to fasten this line at frequent intervals to the object on which it is. This is done by making what may be termed an *attachment disk;* such a disk is pictured by Fig. 184; it is composed of a large number of minute looped threads.

The method of making an attachment disk can be easily observed by enclosing a living spider in a bottle, and watching it with a lens as it fastens its dragline to the sides of the bottle. This I have done many times; and in all of the cases that I have observed, the spider applied the fore spinnerets to the surface of the glass and by quickly spreading the organs apart and bringing them together again two or three times made the disk. An examination of the disk with a microscope shows that it is composed of a large number of fine threads. It is evident therefore that the silk of which it is made issues from the numerous small spinning tubes of this pair of spinnerets and is therefore the product of the pyriform glands. The spiders that I have seen make attachment disks were *Aranea frondosa* and a species of *Amaurobius*.

The following interesting observation was reported to me by Prof. Cyrus R. Crosby: — "The other day when attempting to capture an *Aranea trifolia* she attached her thread but had no time to change to the dragline and so spun a *band* of silk a foot long before she was able to make a new attachment and start the dragline."

The swathing band.— The band of silk with which the orb-weaving spiders envelop their prey may be termed *the swathing band*. The method of making this band is described on an earlier page. I have watched the process many times, but owing to the timidity of the spiders observed, I have never been able to determine to my complete satisfaction whether the silk comes from all of the spinnerets or only from the hind and middle ones. It is easily seen that the band is composed of a great number of fine threads and it appears to come from all of the spinnerets. If this is true, it is the product of both the pyriform and the aciniform glands. I have observed the making of the swathing band only by orb-weaving spiders; probably other spiders have similar habits.

The swathing film.— Spiders of the family Theridiidæ use in swathing their prey a film of silk which differs from the swathing band of the orb-weavers. This silk is emitted from two or four spigots, one or two as the case may be, on each of the hind spinnerets. These spigots are the outlets of the lobed glands; which have been found only in this family.

The viscid thread.—In the webs of orb-weaving spiders there occurs a peculiar viscid thread; this is the spiral line forming the larger part of the orb. If this line be touched, ever so lightly, it will adhere to the object touching it, and when this object is removed the line will stretch greatly. This is the trapping portion of the web; the viscid nature of the silk causes it to adhere to an insect touching it; and the elasticity of the thread prevents the insect from breaking it at once and allows the insect to become entangled in other turns of the spiral; in this way the prevention of the escape of the insect is assured if the victim be not too powerful.

When highly magnified this viscid and elastic line is seen to be composed of two elements (Fig. 181): first, the axis of the thread consisting of two strands; and second, a series of globular drops borne upon this axis; the axis is the elastic element of the thread and the drops the viscid portion. In the webs of some of our larger spiders, the viscid drops on the spiral line can be seen with the unaided eye (Fig. 185).

It is evident that the elastic thread is spun from two spigots; but as yet I have been unable to locate these spigots and consequently have not traced the silk back to the glands producing it. The viscid thread appears to come from the centre of the group of spinnerets; this may indicate that one of its elements is produced by the ampullate glands that open on the middle spinnerets; if this is so it is probably the elastic portion that is so produced. I can see no foundation for the belief of Apstein ('89, p. 40) that the central part of the viscid thread is produced by the aciniform glands; these open through small spinning tubes and produce fine threads like those of the swathing band.

The silk of which the viscid drops are composed is believed to be secreted by the aggregate or treeform glands which open through spigots on the hind spinnerets (Apstein '89, p. 39). This silk is poured forth upon the elastic thread in a continuous sheet; but it breaks up into drops almost immediately, a result of the surface tension of the liquid. I have observed, when watching a large *Aranea gigas* spin its viscid spiral, that the thread as it was pulled from the spinnerets was smooth; but an instant later, the viscid drops appeared on that part of this section that had been put in place between two radii. It should be noted, however, in this connection, that the smooth portion of the thread, the section between the spinnerets and the radius to which it was fastened last, is greatly stretched; and that as soon as it is fastened to another radius, this tension is relaxed, which would result in a massing of its outer coat, and thus facilitate the formation of the drops.

The hackled bands.— By the term *hackled bands* may be designated the distinctively characteristic threads spun by spiders having a cribellum and a calamistrum. I suggest the term band for these threads because they are flat and more or less ribbon-like structures. This feature, however, is only apparent when they are greatly magnified; to the unaided eye they appear as threads.

Fig. 185. CENTRE OF WEB OF ARANEA GIGAS

The structure of the hackled bands differs in different families; and sometimes, in different genera of the same family. But in all that I have examined, the band consists of two elements: first, two or more longitudinal strands, the supporting part of the band, this may be termed the *warp;* and second, a viscid, sheet-like portion, supported by the warp, this may be termed the *woof*.

The warp in the hackled band of *Uloborus* and of *Hyptiotes* (Fig. 236) consists of two, straight, parallel threads; in that of Dictyna of a pair of greatly curled threads; and in the band of *Amaurobius* (Fig. 250) of two straight threads and two greatly curled ones. In *Filistata* there are a supporting double thread and two pairs of curled threads differing greatly in size and arrangement (Fig. 285).

In most cases the woof is an amorphous sheet; but in the bands of *Uloborus* and of *Hyptiotes* it consists of a series of regular, overlapping lobes.

Very little is known regarding the source of the two or more kinds of silk that are used in forming a hackled band. The size of the threads forming the warp indicates that they are spun from spigots; and it seems probable that the woof is derived from the cribellum; but I have been unable to verify these conjectures.

The method of making a hackled band is also not well understood. It is easy to see that the spider in doing it places the calamistrum of one hind leg beneath the spinnerets and makes a rapid combing motion. It is probable that the lobed nature of the warp of *Uloborus* and of *Hyptiotes* is produced in this way; but is the amorphous sheet that forms the woof of other genera thus formed? If so, how are the curled threads of the warp given their characteristic form?

The silk of the egg-sacs.— The silk of the egg-sac of many spiders presents an appreciably different appearance than does other silk spun by the same spiders. According to Apstein ('89) this silk is produced by the cylindrical glands of the female. These, in the forms studied by this writer, are six in number and open by spigots, one on each of the middle spinnerets and two on each of the hind ones. These glands are said to occur as a rule only in females and to be wanting in the Attidæ and in the Dysderidæ.

VI.— THE TYPES OF WEBS OF SPIDERS

The snares or webs of different spiders differ greatly in structure. An imperfect classification can be made which will indicate in a general way these differences; but the various types are connected by intermediate forms; and in some cases, a spider makes a composite web, one that includes two parts representing two types of webs. The following classification indicates the principal types of webs and will serve to define the terms applied to them in the later portions of this book.

Irregular nets.— The maze of threads extending in all directions that is built by the domestic-spider, *Theridion tepidariorum* (Fig. 322) is a good example of an irregular net. Most members of the Theridiidæ and some other spiders spin webs of this type.

Sheet-webs.— The most familiar example of a sheet-web is that of *Linyphia phrygiana* (Fig. 408). Here the principal part of the web consists of a more or less closely woven sheet extended in a single plane and consisting of threads extending in all directions in that plane with no apparent regularity of arrangement.

Funnel-webs.— The principal part of a funnel-web is sheet-like in structure; but webs of this type differ from the true sheet-webs in having a tube extending from one edge; this tube leads to the retreat of the spider or serves as a retreat. The web of the common grass-spider, *Agelena nævia* (Fig. 662) illustrates well this type. Frequently a very loose irregular net is spun above the sheet of a funnel-web; this impedes the flight of insects and causes them to fall upon the sheet where the spider, rushing from its retreat, can capture them.

Orb-webs.— The characteristic feature of an orb-web is that the central portion, the part lying within the supporting framework, consists of a series of radiating lines of dry and inelastic silk which support a thread of viscid and elastic silk. Different webs of this type vary greatly in structure; the principal variations are the following:

In the more symmetrical of the orb-webs the viscid line extends throughout the greater part of its length as a spiral line although near the outer edge of the orb it is looped back and forth a few times on the lower side of the web. The web of *Aranea frondosa* (Fig. 186) is an excellent illustration of this kind of a web. Such webs are termed *complete orbs*.

Fig. 186. WEB OF ARANEA FRONDOSA

The Life of Spiders

In the webs of some spiders the viscid thread extends in but few if any spiral turns; the great part of thread being looped back and forth on the radii. This is well-shown in the web of *Nephila* (Fig. 439) and in that of *Metepeira labyrinthea* (Fig. 187). Such a web is termed an incomplete orb.

In an ordinary incomplete orb there is a lack of regularity in the position of the turns of the viscid thread, different loops ending on different radii. But the species of *Zilla* and certain species of Aranea sometimes make an incomplete orb in which the viscid line is omitted from a definite sector of the orb (Fig. 470). As this type of web is more often made by the species of *Zilla* than by other spiders, I suggest that it be designated as the *zilla type of orb-web*.

In the zilla type of web there is a trapline extending from the hub to a retreat above the web; this trapline is opposite the vacant sector of the web.

The making of a zilla type of web is not even a specific characteristic. I have had a spider under observation in our insectary make a complete orb one day and replace it the next day by one of the zilla type.

Fig. 187.
WEB OF METEPEIRA LABYRINTHEA

A single known species, *Henztia basilica*, makes, according to the observations of Dr. McCook ('89 I. 164), a complete orb and then pulls it into a dome-shaped structure. Such a web is termed a *domed orb*.

The web of *Theridiosoma gemmosum* (Fig. 415) differs from other orb-webs in lacking a definite centre or hub, the radii being grouped into several irregular groups or rays. This web was named by Dr. McCook *a ray-formed orb-web*.

The web of *Uloborus* (Fig. 239) is a complete orb, but it

195

differs from the webs described above in the nature of the viscid silk; it is *an orb-web with a hackled band*.

Triangular web.— The remarkable web of *Hyptiotes* (Fig. 246) represents a distinct type, which has been named *the triangular web*.

Irregular webs with hackled bands.— Under this head may be classed all of the webs made by our common cribellate spiders excepting *Uloborus* and *Hyptiotes*. The webs of the different genera included here show striking differences, which are described in the accounts of the spiders given later.

Sheet and irregular net webs.— The webs of certain spiders, as *Linyphia marginata* (Fig. 405) and *Linyphia communis* (Fig. 400) are composite in nature being composed of a definite sheet combined with an irregular net.

Orb and irregular net.— The web of *Metepeira labyrinthea* (Fig. 187) represents also a composite type of web; here an orb-web is combined with an irregular net. There are species of Aranea that make this type of web.

VII.— THE BUILDING OF AN ORB-WEB

Few if any of the structures built by lower animals are more wonderful than the webs of an orb-weaver; it is of interest, therefore, to follow the steps by which one is made.

The bridge.— In making its web an orb-weaver first spins a number of lines extending irregularly in various directions about the place where its orb is to be. This is the outer supporting framework. Often the first line spun is a bridge between two quite distant points so situated with regard to each other that it would not be possible for the spider to carry a dragline from one to the other in the ordinary way. For example, webs are often built between two trees or shrubs above herbage which would make it impossible for the spider to carry a line over the intervening ground; and it is not an unusual occurrence to see a web stretched between shrubs separated by a running brook.

In these cases the first line was formed in this way: The spider, after selecting a point for beginning its operations, lifts up the abdomen away from the object on which it is standing and spins out a thread, which is carried off by a current in the air. After a time the thread strikes some object and adheres to it; the spider then pulls the line tight, and fastens it where it is standing. It then has a bridge along which it can easily run.

The foundation lines.— Having constructed this bridge, which is usually strengthened by passing back and forth over it and adding a dragline to it on each trip, it is easy for the spider to stretch the other lines that are to serve as the outer framework of the web. In doing this it fastens a thread to one point, and then walks along to some other point, spinning a dragline as it goes, and holding it clear of the object on which it is walking by means of one of its hind legs. When the second point is reached the thread is pulled tight and fastened in place by an attachment disk.

If it is desired to spin a second bridge below one already made, the spider has only to fasten a line to a point below one end of the first bridge and then walking up the supporting object, and across the bridge, and down the other support to the right point, spinning a dragline as it goes; this is then pulled tight and fastened; and a second bridge is formed; between these bridges the orb can be built. To complete the outer supporting framework of the web other lines are stretched in a similar manner between the bridges and from them to the supporting objects. These are placed in such a way as to leave an open space, more or less irregular in outline, in which the orb is to be built. The lines forming this framework have been named the *foundation lines.*

The radii.— After making the foundation lines, the radiating lines of the orb are formed. The first step in this operation is to stretch a line across the open space in the framework so as to pass through the point which is to be the centre of the orb. In doing this the spider may start on one side, and be forced to walk in a very roundabout way on the outer framework to the opposite side. It carefully holds the new line up behind it as it goes along, so that it shall not become entangled with the lines on which it walks; one or both hind feet serve as hands in these spinning operations. The spider then goes to the point where the centre of the orb is to be, and fastens another line there, it then walks back to the outer framework spinning a line as it goes and holding the new line clear from the one on which it is walking; the new line is fastened to one of the foundation lines. In this way all of the radiating lines are made, the spider returning to the centre of the web to begin each radius. In some cases at

Fig. 188. CENTRE OF WEB OF METARGIOPE TRIFASCIATA

least, if not usually, the spider returns to the centre along the radius just made and spins a line which it does not hold clear and which adheres to the radius, thus doubling it. In some observations that I have made it was easy to see that the radius was thicker behind the spider than in front of it. As the dragline consists of two lines, when a radius is doubled it consists of four; this was well shown in the photograph from which Fig. 192 was made.

The hub.— The centre of the web, where the radii converge, is strengthened by a mesh or net-work of lines termed *the hub*. A part of the hub is made while the radii are being stretched, the spider working a little on the hub on each return to the centre of the web; after all the radii are stretched the hub is completed.

The nature of the hub of the webs of different species of spiders differs greatly and forms one of the distinctively characteristic features of them. Three types of hubs were named by McCook ('89 1. 54); these are *the meshed hub, the sheeted hub,* and *the open hub*.

The meshed hub is formed of a series of irregularly shaped meshes, through which one can often trace the continuation of the radii as zigzag lines; the radii being pulled out of their direct course in the making of the hub. In some webs there is but little regularity in the meshes of the hub (Fig. 185); but in others the hub is a very beautiful structure.

The sheeted hub consists of a closely woven sheet of silk spun upon the net-work. This type is well-shown in the webs of *Metargiope* (Fig. 188).

The open hub resembles the hub of a wagon, in being a firm structure supporting the spokes or radii and having an open space in the centre (Fig. 189). This type of hub is characteristic of the webs of *Micrathena* and of those of the Tetragnathidæ.

The notched zone.— Immediately outside of the hub there is an area in which there are a few turns of a spiral line; the number of these turns varies from four or five or even less to ten or more. In spinning this line, which is done immediately after the hub is made, the spider attaches it to each radius lengthwise for a short distance instead of crossing the radius at right angles. The pulling of the line taut between the radii pulls them out of their direct course and gives to this area the appearance that suggested the

Fig. 189. CENTRE OF WEB OF MICRATHENA SAGITTATA

Fig. 190. HUB OF WEB OF MANGORA GIBBEROSA

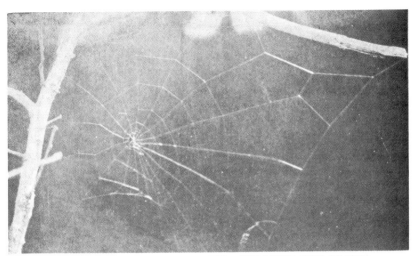

Fig. 190 bis. BEGINNING OF AN ORB-WEB. THE LOWER PART OF THE WEB PHOTOGRAPHED WAS BROKEN

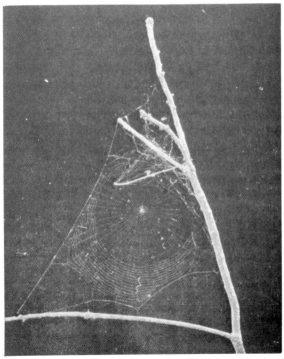

Fig. 191. A HALF-MADE WEB OF ARANEA LABYRINTHEA

name *the notched zone.* (Fig. 190.) This method of fastening the spiral line adds greatly to the strength of this region of the web.

The spiral guy-line.— After the completion of the centre of the web, the next step is to stay the radii by a spiral line, which is a continuation of the line of the notched zone, and which extends spirally over the entire area that is to be occupied by the orb. As the function of this line is merely to hold the radii in place during subsequent operations in the making of the web I have called it the spiral guy-line. The turns of this spiral are as far apart as the spider can conveniently reach and it crosses the radii at right angles (Fig. 190 *bis*). In Fig. 191 which represents a web at a somewhat later stage in its construction, four or five turns of the spiral guy-line can be seen between the hub and the outer completed portion of the web.

The viscid spiral.— All of the threads spun up to this stage in the construction of the web, excepting the attachment disks by which threads are anchored, are dry and inelastic. The spider now proceeds to stretch upon the radii a viscid and elastic line, which is the most important part of the web, the other lines being merely a framework to support it. In spinning the viscid line the spider begins at the outer margin of the orb, and passing around it fastens this line to each radius as it goes. At first the spider does not pass entirely around the web but makes a greater or less number of loops on the lower part of the web. In the web represented by Fig. 191, which was the web of a spider that makes an incomplete orb, there are many of these loops; but in most orb-webs there are only a few, the spider passing entirely around the web after making a few loops.

The turns of the viscid spiral are placed quite close together and the spiral guy-line is cut away turn by turn as the viscid spiral approaches it. The remnants of the spiral guy-line remain attached to the radii and form a series of minute specks which can be seen in large webs after their completion (Fig. 188).

In spinning the viscid line the spider fastens it to a radius by means of a small attachment disk; and then moves on pulling out the thread from the spinnerets; but before the thread is fastened to another radius the spider takes hold of it with either the claws or with a spine of one hind leg and straightening out this leg pulls out from the spinnerets more of this thread; the spinnerets are then applied to the next radius and the thread fastened in place; after which the spider takes away its hind leg and the thread contracts to the length of the space between the two radii. In Fig. 192 the fine threads of the attachment disk connecting the viscid thread to the radius can be seen.

The pulling out of the viscid line by the hind leg before it is fastened between two radii leaves it relaxed when put in place; so that it can be easily stretched by an entangled insect, and thus ensures the insect being caught in other turns of the viscid spiral.

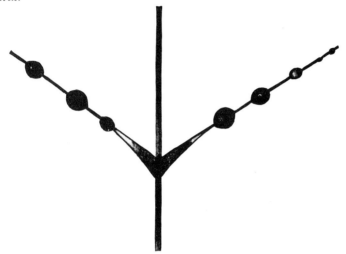

Fig. 192. FROM A PHOTOMICROGRAPH OF A RADIUS AND A VISCID LINE

The free zone.— The viscid spiral is not continued entirely to the notched zone, an area of greater or less width being left, free from any spiral line; this is termed the *free zone*. In some webs this feature is not conspicuous, but in others the free zone is of considerable width and furnishes a means by which the spider can pass freely from one side of the web to the other.

The stabilimentum.— Some of the orb-weavers strengthen their webs by spinning a zigzag ribbon across the centre or below the hub (Fig. 188). This ribbon has been termed the *stabilimentum*. It consists of a large number of minute threads resembling the swathing band and is doubtless spun from the small spinning tubes of either the pyriform or of the aciniform glands.

The Life of Spiders

The stabilimentum is sometimes an elaborate structure. This is especially true in the webs of the young of *Metargiope trifasciata* (Fig. 193). In fact I have observed that immature spiders are more apt to make a stabilimentum than old ones. I have noted this especially with *Metargiope* and with *Uloborus*. The spiders of the genus *Cyclosa* make a stabilimentum composed largely the dry skins of their victims.

The trapline. — When the orb is completed, many orb-weavers rest on the hub and wait there for their prey; if disturbed they either drop to the ground, spinning a dragline as they go, up which they ascend to the web later, or they rush off at one side to the support of the web. In many other species the spider has a tent above or at one side of the orb to which it retreats when disturbed and in which it waits for the ensnarement of its prey. This retreat is connected with the hub of the web by one or more threads which serve as a means of passage to and from the web. When waiting for its prey, the spider rests in the retreat with one or more of its feet upon the line leading to the hub of the orb; by this means it can feel any disturbance of the web. This use of the thread connecting the retreat with the orb has suggested the name *trapline* for it.

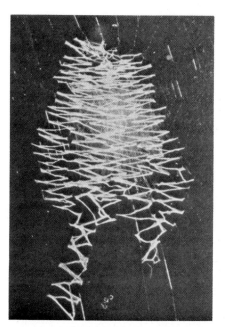

Fig. 193. STABILIMENTUM OF METARGIOPE

The tracks of orb-weavers. — In running about on their orbs, spiders ordinarily use the radii as means of support; and in passing to and from their retreat, they use the trapline, avoiding, if possible, contact with the viscid spiral line. But it sometimes happens, especially if a spider be frightened, that the orb is

Fig. 194. WEB OF ARANEA FRONDOSUS, REDUCED

The Life of Spiders

injured by the passage of the spider. This is apt to occur when the angle between the trapline and the plane of the orb is small; the spider, when rushing up the trapline, may touch and bring together several turns of the viscid line with each forward movement of one of its legs, a projecting knee reaching the orb. Such a trail is shown on the left side of the orb represented by Fig. 194. The trapline is shown indistinctly behind the orb; the retreat was in the curled leaf at the upper left corner of the orb.

VIII.—THE NESTS OF SPIDERS

The nest-building habit is found both among spiders that build webs and among those that do not; and the forms of the nests are even more numerous than are the types of webs.

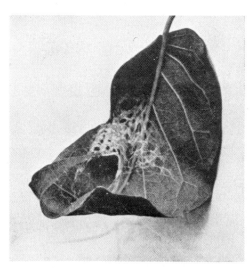

Fig. 195. RETREAT OF ARANEA THADDEUS

Among the spiders that live on the ground are some that make their nests in the earth. The most common nests of this type are the tunnels of the wolf-spiders (Lycosidæ). These vary from a simple vertical shaft with a delicate silken lining to one in which the entrance is surmounted by a turret or watch-tower (Fig. 722). Even more remarkable than these are the nests of the trap-door spiders in which the entrance is furnished with a hinged door (Fig. 216), and, frequently, with a second door at some distance from the entrance; and still more remarkable is the silken, tubular nest of the purse-web spider. (Fig. 228.)

Very many spiders, representing widely different families, make nests by folding or rolling leaves and lining the enclosed space with silk. This type of nest is made by the Clubionids, by some crab-spiders, by some of the jumping spiders, and by various others.

Some web-building species live upon their webs and have no definite retreat, merely running away from the web or dropping to the ground when disturbed. But very many of them build a nest near the web in which they watch for their prey and to which they retreat when frightened. In the funnel-web weavers this is a silken tube connected with the sheet which forms the principal part of the web. This funnel either serves as the retreat or leads into a more distant crack. Many of the orb-weavers roll or fold a leaf or tie several leaves together and line the space thus formed with silk. This nest is usually built above the web or at one side and there is a trapline leading from it to the centre of the orb, by which any disturbance of the web is communicated to the spider waiting in its nest, and down which it rushes to get its prey. A very elaborate nest of this kind is made by the lattice-spider, *Aranea thaddeus* (Fig. 195). The young of many species make their retreats entirely of silk while the adults of the same species use a folded leaf or a bunch of leaves.

IX.—THE PAIRING OF SPIDERS

There are many very interesting facts connected with the pairing of spiders; but the scope of this work admits of only a brief reference to them here. The anomalous specialization in the male of a pair of appendages of the head, the palpi, as secondary sexual organs is described in the chapter on external anatomy. In this place I wish merely to call attention to the behaviour of certain spiders at the mating season.

The sedentary spiders are the ones most easily observed at this time, owing to the fact that the females remain upon or near their webs and are therefore easily found, or are frequently accidently discovered, when mating.

With many species of these spiders, there is a striking disparity in the size of the two sexes, the female being many times larger than the male; but with others the difference in size is not so marked.

The males of the orb-weavers, for example, resemble the females in habits, making webs like them, until they reach maturity. They then wander in search of their mates. After which they are found in or near the webs of the females. The approach of the male to the female is made with great care; for if the female is not ready to receive his advances, she is apt to pounce upon him and destroy him.

In the case of the jumping spiders, Attidæ, the courting habits are very remarkable. The males, of some species at least, dance before the females, and display their bright colours in a way that is evidently intended to attract the attention of the female. This has been described in detail by Mr. and Mrs. Peckham ('89 and '90).

X.—THE MOTHERHOOD OF SPIDERS

Among the more interesting details in the life of spiders is the provision made by the females for the care of their young. Here as elsewhere in the economy of these creatures we find variety, the methods employed by one differing greatly from those employed by another. There is, however, within the limits of each of the families a considerable degree of uniformity in the plan adopted, although each species has its own way of carrying out this plan.

Fig. 196. EGG-SACS OF CYCLOSA BIFURCA

The first step is the care of the eggs. These are never laid singly, as is the case with many insects, but are invariably laid in one or more masses; and each mass is protected by a covering of silk, the egg-sac or cocoon.

Most spiders deposit all their eggs at one laying and enclose them in a single egg-sac; but in certain species the egg-laying is extended over a considerable period of time and a series of egg-sacs is formed. This is true of *Metepeira labyrinthea* and of the species of *Cyclosa* (Fig. 196).

The egg-sac is not merely a covering made in a haphazard way; but is a more or less elaborate structure, made in a definite manner characteristic of the species; it is frequently as easy to recognize the species from a study of the egg-sac as it is from a study of the spider herself.

The Life of Spiders

The simplest kind of an egg-sac is merely a mesh of threads holding the mass of eggs together; but so delicate that the eggs can be seen through it; such an egg-sac is made by *Pholcus* (Fig. 320). Other spiders, as some species of *Aranea,* enclose the eggs in a fluffy mass of silk which effectually conceals them, but which

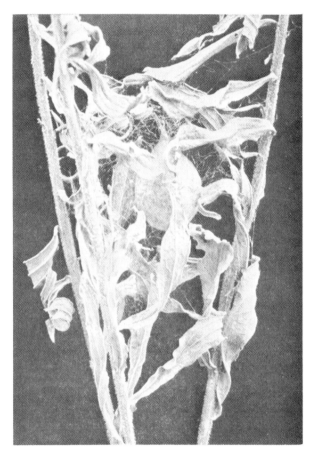

Fig. 197. EGG-SAC OF MIRANDA AURANTIA

has no definite outline. While still others make an egg-sac of a very definite form and consisting of several layers of silk, differing in texture. The egg-sac of *Miranda aurantia* (Fig. 197) is a good illustration of this type. If one of these be opened it will be found that the eggs are enclosed in a silken cup, which is surrounded by a thick layer of flossy silk, which in turn is enclosed in the firm, brown, closely woven outer covering that gives the characteristic pear-shaped form to this sac.

Many egg-sacs are more or less nearly spherical in outline. This is true of some of those that are suspended in the web as is the case with those of the common domestic spider, *Theridion tepidariorum;* many others are lenticular in form as for example, the egg-sacs of the Drassids (Fig. 198). While still others depart widely from either of these simple forms. The egg-sac of *Miranda aurantia,* mentioned above, is pear-shaped; that of the closely allied *Metargiope tr'faciata* is cup-shaped with a flat top; and the egg-sac of *Argyrodes trigonum* reminds one of a beautiful Grecian vase (Fig. 199).

The outer covering of most egg-sacs is opaque; but in a few cases it is translucent; the beautiful elliptical egg-sac of the cave spider, *Meta menardii,* (Fig. 200) does not conceal the enclosed mass of eggs; neither does that of the egg-sac of *Aranea trifolium.*

Some spiders after they have completed the silken portion of the egg-sac add a protecting layer of some foreign substance. The grass-spider, *Agelena nævia,* makes its egg-sacs beneath the loose bark of a dead tree, or in some other similar situation, and covers them with bits of rubbish (Fig. 201). A beautiful pear-shaped egg-sac (Fig. 202) which has a complete coating of mud is often found attached by a cord of silk to the lower surface of a stone or a piece of wood lying on the ground. The maker of this mud-coated egg-sac has not been determined as yet. The egg-sacs of *Tetragnatha* have the appearance of bearing a coating of foreign matter due to the presence of curiously twisted tufts of silk differing in colour from the rest of the egg-sac (Fig. 203).

Most spiders fasten their egg-sac in some secure position. It is often built against some object as the side of a stone or the branch of a tree or the side of a building. Many are attached to trees beneath the loose bark. The labyrinth-spider (p. 465) fastens its series of egg-sacs to a strong silken cord that holds it in place through the winter, and the species of *Miranda* anchor theirs by means of many strong silken threads (Fig. 197). Some species suspend the egg-sac, by a cord, in or near the web. In *Theridiosoma* (Fig. 417) the sac is fastened to twigs or to the sides of cliffs in the damp localities frequented by this spider; this, too, is the case with *Ero furcata* which envelops its egg-sac in a

Fig. 199.
EGG-SAC OF ARGYRODES TRIGONUM

Fig. 201.
EGG-SAC OF AGELENA NÆVIA

Fig. 200.
EGG-SAC OF META MENARDII

Fig. 202.
EGG-SAC, MAKER UNKNOWN

The Life of Spiders

silken net (Fig. 567) and suspends it by a cord to the sides of damp cliffs.

An egg-sac which has interested me greatly is one that I have found attached to the branches of shrubs and of trees. This egg-sac (Fig. 204) is made within a rolled leaf, the petiole of which is securely fastened to the twig by a band of silk, so that it is held in place through the winter. This egg-sac resembles in miniature the cocoon of the Promethea moth. I have been unable to determine what spider makes it.

The maternal duties of the spiders whose egg-sacs are referred to in the preceding paragraphs end with the making and fastening in place of the egg-sac. The spider dies soon after this labour is performed, and the spiderlings when they hatch must shift for themselves. But with certain other spiders this is not the case.

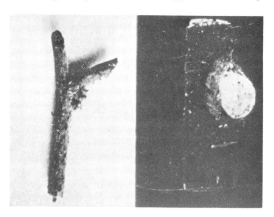

Fig. 203. EGG-SACS OF TETRAGNATHA

In some species of the genus *Dictyna*, the female makes her egg-sac within the web; the young soon hatch; and mother and young live together for a considerable period within the same web. Here the young are protected from attack by the presence of the mother; and are saved the necessity of making a web to catch their prey.

More remarkable than this are the evidences of maternal care exhibited by the nursery-web weavers, Pisauridæ. These spiders spin no web for catching insects but stalk their prey. After the egg-sac is made the female carries it with her, under her body wherever she goes (Fig. 683). When the spiderlings are about to emerge from the egg-sac, the mother takes it to the top of some herb or to the tip of a shrub, and fastens it in the centre of a nursery made by spinning a web over the leaves. She then posts herself as a guard on the outside of the nursery (Fig. 684).

The Life of Spiders

The wolf-spiders, Lycosidæ, also carry their egg-sac with them while they stalk their prey. The egg-sac in this case is dragged behind the spider, which appears to be a more convenient method than that employed by the nursery web-weavers. The wolf-spiders make no nursery; and when the spiderlings emerge from the egg-sac they pass to the back of the mother and are carried about pappoose-like for a considerable period. (Fig. 715.)

XI.— THE VENOM OF SPIDERS

There is a very general belief that spiders are to be feared on account of the venomous nature of their bites; and it is probable that the feelings of repugnance with which these creatures are commonly regarded are due to this belief; but there is really very little reason for it.

It is true that spiders secrete poison with which they kill

Fig. 204. EGG-SAC IN ROLLED LEAF

their prey; but it does not necessarily follow that the poison that would kill a fly would harm a man. And, too, spiders are exceedingly timid and think only of escape when approached by man. Even when they are caught in one's hand they merely try to get away and do not, like a bee or a wasp, endeavour to facilitate their escape by stinging their captor. During my study of spiders I have collected thousands of specimens and have taken very many in my hand but have never been bitten by one. Still the anxious reader may ask what would have happened had you been bitten?

As already stated, (p. 170), the venom apparatus consists of a pair of glands in the cephalothorax, or one in the basal segment of each chelicera, from each of which a duct leads to a small opening near the tip of the chelicera of the same side (Fig. 79). This opening is so placed that it is not closed by the pressure of the bite but allows the venom to flow into the wound.

The Life of Spiders

It is well-known what results follow the bite of an insect by a spider; let us now consider what follows when a spider injects its venom into man.

Several of the more prominent arachnologists, including Mr. Blackwall ('55) of England, and Baron Walckenaer ('37) and M. Duges ('36) of France, have made experiments to determine the effect on man of the bite of spiders. Each of these experimenters caused himself to be bitten by spiders; and all agree that the effects of the bites did not differ materially from those of pricks made at the same time with a needle.

I have given considerable attention to this question with the result that I firmly believe that in the North at least there is no spider that is to be feared by man. I have endeavoured to trace to their source some of the newspaper stories of terrible results following the bite of a spider; but have not found a bit of evidence that would connect a spider with the injury in any of the cases investigated. It often happens that a person suffering from blood poisoning produced in some unknown way infers that he has been bitten by a spider; or the inference may be made by some one else. When an enterprising reporter writes up the incident for a newspaper, the spider bite is not referred to as an inference, but as a fact.

The so-called Tarantula (*Heteropoda*) that is frequently brought to the North in bunches of bananas is often described as the cause of serious injury. This, however, although a large spider, is an inoffensive one. Mr. John T. Lloyd informs me that he has collected scores of specimens of this species with his hands in Samoa, where it is abundant and has never been bitten by it.

Although we have in the North no spider that is to be feared, it is quite possible that in the South it is different. I confess that I should not like to be bitten by one of the larger tarantulas of that region, although I know of no well-authenticated case of a person being bitten by one.

The spiders of the genus *Latrodectus*, of which we have a common representative in the South, are feared wherever they occur, and it is quite possible that they are more venomous than other spiders. See the account of *Latrodectus mactans* on a later page.

The conclusion of the matter is this: In the North there is no common spider that is to be feared; in the South, there is a single easily recognized species that is believed by some people to be dangerous. Surely one should not hesitate to study spiders on account of their venomous nature.

XII.—THE AERONAUTIC SPIDERS

Although spiders like man possess only legs as organs of locomotion, like man they are able to travel through the air by artificial means. Long before the invention of balloons or of aeroplanes, spiders had solved the problem of aerial navigation.

The method adopted by spiders differs greatly from any of those adopted by man in his efforts to navigate the air. The spider's method is more closely analogous to that by which many seeds, as those of the dandelion, are carried long distances; and the object is the same, the distribution of the species.

Attached to the dandelion seed there is a pappus consisting of a bundle of fine threads. These, when the seed is mature and fitted to start upon its journey, spread apart forming a very light, nearly spherical mass, which is easily carried away with the attached seed by the wind. Thus the seeds of a single plant are scattered over a wide area instead of too thickly seeding the place of their origin.

With spiders it is not the seeds or eggs that are distributed but the spiders themselves; and the object by which they are buoyed up when carried by the wind differs from the pappus of a seed in consisting largely of a single thread, or a bundle of parallel threads.

It is usually, but not invariably, very young spiders that exhibit the aeronautic habit; and exhibitions of it are most often observed in warm and comparatively still autumn days. At this time great numbers of young spiders, of many different species, climb each to the top of some object. This may be a fence post, the top of a twig, the upper part of some herb, or merely the summit of a clod of earth. Here the spider lifts up its abdomen and spins out a thread, which if there is a mild upward current of air is carried away by it. Occasionally the spider will attach a small flocculent mass to this thread which will increase the force of the current of air upon it. This spinning process is continued until the friction of the air upon the silk is

The Life of Spiders

sufficient to buoy up the spider. It then lets go its hold with its feet and is carried off by the wind.

That these ballooning spiders are carried long distances in this way is shown by the fact that they have been met by ships at sea hundreds of miles from land. And the showers of gossamer which are occasionally observed are produced by ballooning spiders.

It often happens that spiders attempt to fly when the wind is too strong and the threads they emit are not carried up but are merely blown against some nearby object. I have known in-

Fig. 205. A SEA OF GOSSAMER

stances in which large fields have been covered with a gauze of silk in this way. Members of a Country Club have reported to me that the grass of their links was so thickly covered with silk of spiders that the shoes and trousers of the players were greatly soiled by it.

Sometimes sheets of silk are formed by the massing together of the threads of myriads of spiders in such situations that the sheets are finally torn away by the wind and later are rained down far from the place of origin. This is the explanation of the showers of gossamer which are occasionally recorded.

An account of remarkable sheets of gossamer, based on the observations of Mr. J. B. Lembert of Yosemite, Cal., has been published by Dr. L. O. Howard (Proc. Ent. Soc. Wash. III, 191). These sheets are formed on the sides of the Yosemite Valley, are of yearly occurrence, and of great extent.

On one occasion I saw a ploughed field that was covered with a sheet of silk. It was evident that an immense number of small spiders had attempted to fly but that the wind had blown their thread merely from the crest of one furrow to another. Although the field was completely covered with the sheet of silk, so delicate was this fabric that it was invisible except where the light of the sun was reflected directly to the eye of the observer; the appearance being like that of the wake of the moon on slightly disturbed water (Fig. 205).

CHAPTER V:

Order ARANEIDA (Ar-a-ne'i-da)

THE SPIDERS

THE order Araneida includes only the spiders. These differ from the other Arachnida in having a sac-like abdomen which is joined to the cephalothorax by a slender pedicel. Spiders have been derived from an ancestral stock in which the abdomen was distinctly segmented. Most of the higher spiders retain no definite or superficial indications of this segmentation, although it is possible to find evidences of it in the bodies of very young spiders. In the more generalized four-lunged spiders external segmentation is retained to a greater or lesser degree. In many of the trap-door spiders and the funnel-web spiders there is present at the base of the abdomen a distinct hard plate which marks the position of the second abdominal somite and is homologous with the first dorsal tergite of *Liphistius*. In the folding-door tarantulas (Antrodiætinæ) three or four of these tergites are present, the greater number usually in the males. These tergites are set with curved setæ and resemble closely those found in the Liphistiidæ. However, it is in this last group of spiders that the superficial segmentation of the abdomen is most clearly shown. The dorsum of the abdomen of *Liphistius* and *Heptathela* is supplied with from six to twelve distinct tergites.

The segments of the cephalothorax are closely united so that none are distinct. The head portion may be separated from the thorax by a more or less profound groove, or all indications of such sutures may be completely obliterated. Eight eyes are present in the majority of spiders, a number apparently the primitive one for the order. They have been subject to important changes in position and to reduction in number. Six eyes are characteristic of some families, due to the early loss of the anterior median eyes. However, this number appears as an aberrancy in many families which are normally eight-eyed. In relatively few

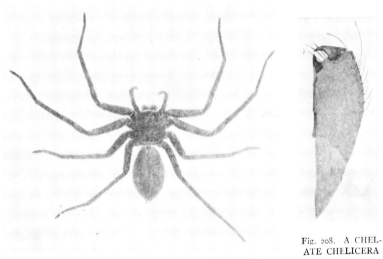

Fig. 206. A SPIDER, SHOWING THE PEDICEL OF THE ABDOMEN

Fig. 208. A CHELATE CHELICERA

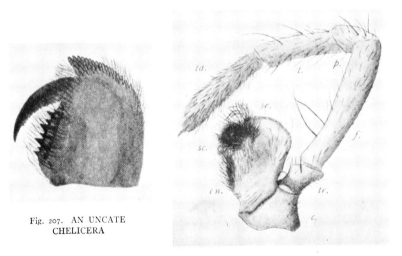

Fig. 207. AN UNCATE CHELICERA

Fig. 209. PEDIPALP OF A SPIDER ENDITE

Order Araneida

species the number has been even further reduced, to four and to two. An aberrant condition is the complete loss of eyes as found in some cave spiders. That in many cases this blindness is a recently acquired character is evidenced by the fact that eight-eyed, six-eyed, and totally blind species often belong to the same genus.

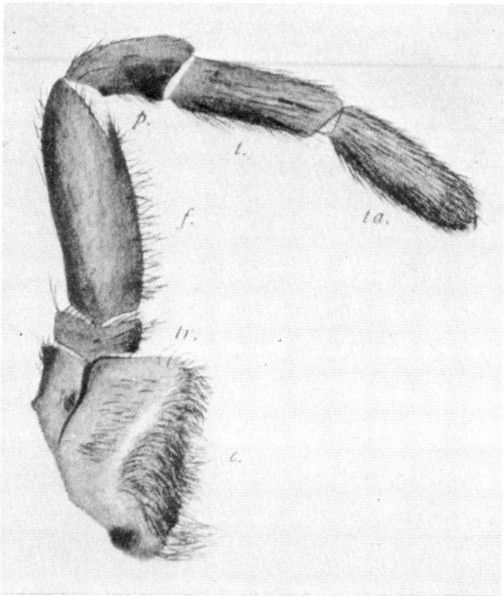

The chelicerae are two-segmentated and usually uncate, that is the claw is folded back into a groove in the basal segment, like the blade of a pocket-knife into the handle (Fig. 207). In rare cases they are chelate, there being a prolongation of the first segment which is opposed to the claw, thus forming a pincer-like organ (Fig. 208). The chelicerae are the instruments for transferring the venom to the body of the prey, and contain the poison ducts or even the glands themselves.

Fig. 210. PEDIPALP OF EURYPELMA

The pedipalps are more or less leg-like and are furnished in the larger number of families with more or less distinct endites (Fig. 209 *en*). They are rudimentary in many of our four-lunged spiders (Fig. 210). In all the true spiders and in some of our tarantulas (*Atypus, Hexura*, etc.) they are well developed. In the male spider the terminal portion of the pedipalp is modified into an organ for the transference of the seminal fluid. The degree of perfection or modification of this organ is almost in direct ratio with the position of the spider in the phylogenetic series. The palpi of the Avicularioidea are quite generalized. In a relatively few genera (*Atypus, Hexura, Antrodiætus*, etc.) the palpi are of an intermediate type, due to the development of a conductor of the embolus to embellish the otherwise simple organ. In the Argiopoidea specialization of the whole palpus as a primary and accessory copulatory apparatus

Order Araneida

has produced a highly complicated structure. A correlative evolution has occurred in the epigynum of the female.

The four pairs of legs are all fitted for walking. Their tarsi bear three or two claws. The presence of the median, unpaired claw generally indicates that silk is an indispensable material in the life of the spider. The loss of this claw ordinarily signifies that the spider is able to place much less reliance on silk. Compensating for its loss in many groups is the development of adhesive tarsal pads which allow the spider greater freedom of movement.

Two pairs of abdominal legs are preserved in a more or less two-branched form and serve to carry the outlets of the silk glands (see p. 134). These appendages are termed the spinnerets. In *Liphistius* four pairs are still functional, but in all other spiders three pairs, or even fewer, are present.

Fig. 211. A TARANTULA, EURYPELMA HENTZII

The order Araneida is divided conveniently into two suborders based on the articulation of the cheliceræ. These groups correspond to those adopted by Simon in his comprehensive treatment of the spiders of the world, his *Araneæ theraphosæ* and *Araneæ veræ*. More recent workers have divided the order into three or more divisions. Students are referred to the papers of Petrunkevitch listed in the bibliography for a detailed discussion of the modern system of classification. For the general purposes of this book the grouping proposed by Simon will be followed, but other names are applied to his divisions.

Those spiders in which the cheliceræ are paraxial (i.e., the plane of articulation vertical, as shown in Fig. 211) are discussed in Chapter VI: The Superfamily AVICULARIOIDEA.

Those spiders in which the cheliceræ are diaxial (i.e., the articulation horizontal, as shown in Fig. 229) are discussed in Chapter VII: The Superfamily ARGIOPOIDEA.

CHAPTER VI:

Superfamily AVICULARIOIDEA (A-vic-u-la-ri-oi′de-a)

THE TARANTULAS

THIS superfamily includes the large spiders that are well known in the warmer parts of our country as tarantulas, the trapdoor spiders, and certain other closely allied forms.

By many writers this group of families is termed the Mygalomorphæ; but as the old generic name *Mygale*, long used for these spiders, was first given to a genus of mammals, it and its derivatives cannot be properly applied to spiders.

The common name tarantula was first applied to an European spider which does not belong to the superfamily, *Lycosa tarentula*, the spider that was credited with causing tarantism by its bite. Tarantism was formerly supposed to be an hysterical disease which was common in Southern Europe in medieval times. It was characterized by an inordinate desire for dancing. It arose in individuals that believed they had been bitten by this spider and was then contagious. The sole cure was music of a specific kind which would incite the victim to dance and throw off the poison. It is now believed that the myth of the tarantula afforded a convenient pretext for the continuation of the heathen rites of the Bacchantes when such exhibitions were banned by the Christians. The spider itself is now believed to be no more poisonous than other species of its kind.

The name tarantula has been transferred in this country to the members of this superfamily and is so firmly established in the language of the people of the South and West that it would be impossible to change it. I have therefore adopted it. In this sense it is spelled tarantula, not tarentula. This matter is further complicated by the fact that the generic name *Tarantula* must be used for a genus of tailless whip-scorpions having been first used in this sense (see p. 20). But the confusion need not be great, as these whip-scorpions occur only in the extreme southern parts of our country, and are known by the latter name

Superfamily Avicularioidea

The spiders of this division are distinguished sharply from the so-called true spiders by the articulation of the cheliceræ. These stout appendages are joined to the cephalothorax in such a way that their long axis is parallel to the body and the claw moves vertically (Fig. 224). All the members of this division have two pairs of book lungs. The poison glands are contained entirely within the basal joint of the cheliceræ. The abdomen is segmented only in rare cases.

Fig. 212. PROFILE OF LIPHISTIUS (after Warburton)

The spinnerets are variable in number, two or three pairs being most commonly present. However, four pairs are found in *Liphistius* and only a single pair in *Diplothele*. The heart may have five, four, or even three pairs of ostia, the highest number found only in *Liphistius*.

The spermathecæ of the female genital system are completely hidden beneath the integument, and no true epigynum with separate external openings has been developed. The male palpus is correspondingly generalized, the principal elements remaining relatively simple, without conspicuous elaboration or adornment. The only exception is the intermediate type of palpus found in such genera as *Atypus*, *Antrodiætus*, and *Hexura*, in which there is a conductor of the embolus.

Twelve recent families of this superfamily are recognized but only four of them occur within the area considered by this book. The families Liphistiidæ and Heptathelidæ, which have been found so far only in southern Asia, are briefly discussed at this point because of their great scientific interest. They are usually given subordinal rank in the Araneida under the name Liphistiomorphæ, and were called the Mesothelæ by Pocock because of the advanced position of the spinnerets.

The genus *Liphistius*, the only known genus of the family Liphistiidæ, contains seven species which are doubtless the most generalized living members of the order Araneida. In general appearance (Fig. 212) they resemble the trap-door spiders of the family Ctenizidæ. The most striking characteristic of the genus is the presence on the dorsum of the abdomen of a series of twelve,

Superfamily Avicularioidea

well-marked tergal plates. The ventral surface of the first and second segments are furnished also with sternal plates covering the genital aperture and the two pairs of book lungs. The two pairs of abdominal legs that are preserved as spinnerets are each two-branched, thus forming eight spinnerets. These are situated near the middle of the ventral aspect of the abdomen far in front of the anal tubercle. The heart is provided with five pairs of ostia. In *Heptathela*, the only genus of the Heptathelidæ, the posterior median spinnerets are fused together to form a median colulus. Only four pairs of ostia are present in the heart. The two know species are from Japan and China. The liphistiids live in tunnels and close them with trap doors.

The members of the Avicularioidea are essentially of tropical and subtropical distribution. Within the limits of the United States they are found principally in the extreme southern portions. A very few species, notably the Atypidæ in the east and the Antrodiaetinæ in the west, occur in the northern states. Many tarantulas are of such great size that they attract attention wherever they occur, are captured as curiosities, and find their way into almost all collections of animals. The nests of certain other species, the trap-door spiders, are also commonly found in collections.

As regards their habits the American species whose ways have been observed fall into four groups.

In the first of these groups are placed those that either choose any kind of a retreat, living under stones or rubbish on the ground and in cracks of trees, or dig a very simple cell, which they line with a slight web of silk. They are for the most part active forms which often wander about in search of their prey. The legs are provided with scopular hairs and the two-clawed tarsi have terminal claw tufts. To this group belong the bird-spiders or tarantulas of the family Aviculariidæ.

The second group comprises the burrowing species which dig tunnels in the ground of variable form and usually close them with a trap door. The females are rarely found outside their burrows and the males wander only during the mating period. To this group belong the trap-door spiders, Ctenizinæ, and the folding-door tarantulas, Antrodiætinæ, both of the family Ctenizidæ.

The third group includes those tarantulas that line a crevice or simple burrow with silk and continue the silk above the ground

Superfamily Aviculariodea

in webs resembling those of the grass-spiders, the funnel-web tarantulas of the family Dipluridæ.

The fourth group comprises the purse-web spiders of the family Atypidæ. They make a tunnel in the earth, line it with silk, and prolong it above the surface as a closed tube. The atypids are known chiefly from temperate regions and have penetrated farther north than any other members of the superfamily.

The members of the last three groups have three tarsal claws and lack claw tufts.

As regards the position of the Avicularioidea in the phylogeny of the Araneida, it is clear that they represent the more generalized group. Their ancestors are known as fossils from the Carboniferous system in the Paleozoic rocks. The time of origin of the true spiders is still obscured in the incompleteness of the paleontological record. They are well differentiated as a group in the Mesozoic, and have also been reported as contemporary with the Avicularioidea in the Paleozoic rocks.

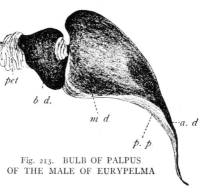

Fig. 213. BULB OF PALPUS OF THE MALE OF EURYPELMA

Nearly seventy names have been applied to spiders of the Avicularioidea from our fauna, and there is still considerable doubt regarding the distinctness of some of the forms described. But on the other hand, as these spiders live in regions that have been explored comparatively little, and as many of them are perfectly concealed in their nests during the daytime, doubtless, many other species remain to be discovered.

The four families of this superfamily found in America north of Mexico can be separated by the following chart:

TABLE OF FAMILIES OF THE AVICULARIOIDEA

A. Tarsi with three claws, sometimes scopulate beneath, but never with terminal claw tufts.
   B. Chelicerae with a well developed rastellum; terminal joint of hind spinnerets usually conical, shorter than the basal joint. P. 226.   CTENIZIDÆ

Superfamily Avicularioidea

    BB. Cheliceræ lacking rastellum, or at most with a weakly developed one consisting of a series of short spines; hind spinnerets usually long, widely separated, the terminal joint as long as the basal one or greatly exceeding it.
        C. Labium immovably joined to the sternum; endites strongly developed; sternum with eight sigilla. P. 247.   ATYPIDÆ
        CC. Labium free; endites rudimentary except in Hexura; sternum with four or six sigilla. P. 245.
                              DIPLURIDÆ
AA. Tarsi with ventral scopulæ, with two claws, and with terminal claw tufts. P. 239.   AVICULARIIDÆ

Family CTENIZIDÆ (Cten-iz′i-dæ)

The Trap-door Spiders

This family includes most of the species that burrow into the soil and cap the opening with a hinged trap door. Many of them are large, handsome spiders, but it is unfortunately true that their retiring habits make them difficult forms for study. Two of the three recognized subfamilies are found within the limits of the United States and may be distinguished as follows:

    A. Postabdomen situated immediately above the spinnerets; furrow of the chelicera well marked; median furrow of the carapace transverse or elliptic; palpus of male without a conductor of the embolus. P. 226.
                              Subfamily CTENIZINÆ
AA. Postabdomen situated a considerable distance above the spinnerets; furrow of the chelicera indistinct; median furrow of the carapace longitudinal; palpus of male with a conductor. P. 236.   Subfamily ANTRODIÆTINÆ

Subfamily CTENIZINÆ (Cten-i-zi′næ)

The Trap-door Spiders

The trap-door spiders resemble the typical tarantulas of the family Theraphosidæ in having the postabdomen situated immediately above the spinnerets. They are distinguished from the bird-spiders in lacking a bunch of terminal tenent hairs on the tarsi

Superfamily Avicularioidea

and in almost always having the third claw well-developed; and they differ from the web-making tarantulas in having the cheliceræ furnished with a rake (Fig. 214).

Fig. 214. CHELICERA OF PACHYLOMERUS SHOWNG THE RAKE

The lack of the bundle of terminal tenent hairs on the tarsi is correlated with the sedentary habits of these spiders; the possession of a third claw aids them in climbing up the sheet of silk with which they line their burrows; and the rake of the cheliceræ serves to dislocate the earth, when digging their burrows, and to make it into a ball, which they throw up with their hind legs; these are strong and armed with strong spines.

To this subfamily belong the larger number of the tarantulas that have attracted attention on account of their architectural skill. They dig tunnels in the ground, which, in most cases, are closed with a hinged lid (Fig. 215); this fact has suggested the popular name of trap-door spiders, which is applied to them.

After the tunnel is dug, the wall is usually coated with a coat formed of earth and saliva. Inside of this wall there is spun a layer of silk, which sometimes hangs free, but usually adheres to the earth. Sometimes the silken lining is restricted to the upper portion of the tunnel.

The wall of the tunnel is usually so firm, perhaps owing to the earth being cemented in some way, that the nest retains its form when the surrounding soil is dug away.

Different types of nest are made by the different species. In a few cases the mouth of the tunnel is *open;* but as a rule it is *closed* by a lid or door. The nests also differ in three other particulars: first, the tunnel may be either *simple* or *branched;* second, the door of a closed nest may be a simple flap of silk and dirt, *the wafer type* (Fig. 215), or it may be a

Fig. 215. ENTRANCE TO TUNNEL OF TRAP-DOOR SPIDER, THE WAFER TYPE

Superfamily Aviculariodea

thick stopple, with its edges accurately bevelled to fit the bevelled opening of the tunnel, *the cork type* (Fig. 216); and third, there may be either one or two doors to the nest, the *single door type* and the *double door type*. The second door, when there are two, is some distance below the first door.

In all cases the door is provided on one side with a hinge which is merely a continuation of the wall of the tube into the layer of silk that forms the foundation of the door.

The inner surface of the door presents the same appearance as the silken lining of the tube, being

Fig. 216. ENTRANCE TO TUNNEL OF TRAP-DOOR SPIDER, THE CORK TYPE

a firm layer of silk; but the outer surface of the door is covered with earth and made to simulate in a very perfect manner the surface of the surrounding soil, so that, when the door is closed, very careful observation is necessary to detect the presence of the nest. In those cases in which the nest is built in soil covered with moss, moss is planted by the spider upon the door of the nest. Many nests of this kind are figured by Moggridge ('73–'74); and I have some before me as I write, which were taken in California (Fig. 217); these are old specimens, and the moss is dry and withered so that it is comparatively inconspicuous.

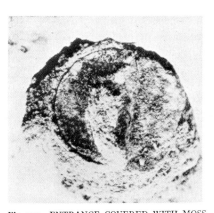

Fig. 217. ENTRANCE COVERED WITH MOSS

The wafer door and the cork door represent two quite distinct types. In the wafer type the door consists of a single layer of silk covered with soil; the edge of the door, which is thin, merely overlaps the edge of the tunnel; and there are no holes in the lining of the door for the reception of claws

Superfamily Aviculariodea

In the cork type, the door is thick, and its edge is bevelled so as to fit accurately the bevelled opening of the tunnel (Fig. 216). Near the edge of the inner surface of the door, at the point farthest from the hinge, there are two holes; these are to receive the claws of the chelicerae when the spider is holding the door closed; and, according to the observations of Moggridge, the door in the cork type consists of many layers of silk each furnished with a sloping rim of earth. He represents fourteen layers of silk and earth which went to make a single cork door examined by him. These layers were successively larger and larger beginning with the innermost, and he believes that the latter constituted the first door the spider ever made, and that the consecutive layers mark successive stages in the enlargement of the nest. His observations were made on the nest of a French

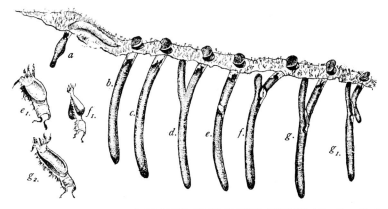

Fig. 218. DIFFERENT TYPES OF NESTS OF TRAP-DOOR SPIDERS (after Moggridge)

species. I have been unable to satisfy myself that the same thing is true of the door of our California species. I have, however, taken apart only the door of a single nest.

Figure 218, which is copied from Moggridge ('74), represents the different types of nests of the trap-door spiders of this subfamily; another type, described later, is made by a member of the Brachybothriinae. At *a* is represented the nest of a French species of *Atypus*, one of the Atypidae; this is included here merely for contrast; *b* represents the cork type of nest; Moggridge believed that in this type the tunnel is always simple, but Atkinson ('86) describes a branched nest with a door of the cork type.

Superfamily Aviculariodea

The remaining figures illustrate the different forms of nests that have the door of the wafer type as follows: *c*, a single-door, unbranched wafer nest; *d*, a single-door, branched wafer nest; *e*, a double-door, unbranched wafer nest, and *e1*, the lower door of the same; and *f*, *g*, and *g1*, three double-door, branched wafer nests, *f* and *g* differing in the form of the second door, shown at *f1*, and *g2*.

In the case of several of our genera, the habits are unknown; the species having been described from specimens that were collected without observations being made on their habits. Students living in the regions where these spiders occur have excellent opportunities for adding to our knowledge of the ways of these remarkable creatures.

The described species, some of them doubtfully attributed to our fauna, represent about a dozen genera. Of that number eight are fairly well known and can be separated as follows:

TABLE OF GENERA OF THE CTENIZINÆ

A. First tarsus lacking scopular hairs beneath, with simple setæ, and set with a series of stout lateral spines; teeth of claws in a single series.
    B. Abdomen hard, truncated and discoidal at the end, the margin with a fringe of bristles. P. 232.
                                                                          CYCLOCOSMIA
  BB. Abdomen relatively soft, rounded behind.
    C. Tibia of third legs with a deep depression at base above; first leg of male without spurs or other modification. P. 231.   PACHYLOMERIDES
    CC. Tibia of third legs normal; first metatarsus of male with a spur. P. 233.   BOTHRIOCYRTUM
AA. First tarsus with scopular hairs beneath (sometimes only simple hairs in the males) and not set with stout spines; teeth of claws in a single or double series.
    B. Posterior sigilla of sternum large, elongated, twice as long as broad, set close together near the midline much nearer each other than to the sternal margins; cheliceræ bearing within at apex a stout, toothed process; anterior claws unidentate at base.
        C. Only first leg of male modified; embolus of male palpus

serrate; anterior lateral eyes of female much larger than anterior median eyes. P. 234.
MYRMECIOPHILA

CC. First and second legs of male modified; embolus of male palpus a simple spine; eyes of first row subequal in size in females. P. 235. EUCTENIZA

BB. Posterior sigilla oval in shape, not decidedly larger than the pair in front, placed either near the margins of the sternum or about as far from each other as the margins; anterior claws pluridentate; cheliceræ rounded at apex.

C. Anterior tarsi and metatarsi subequal in length. P. 235. AMBLYOCARENUM

CC. Anterior metatarsi decidedly longer than the tarsi.

D. Coxa of pedipalp with spinules limited to the inner basal corner. P. 236. APTOSTICHUS

DD. Coxa of pedipalp with spinules scattered from base to apex. P. 235. ACTINOXIA

Genus PACHYLOMERIDES (Pach-y-lom'e-ri-des)

As indicated in the table above, this genus differs from the other genera of this subfamily by the presence of a deep depression in the basal part of the upper surface of the tibia of the third pair of legs (Fig. 219). This depression is smooth, and of a deeper colour than the remainder of the segment; on each side of this depression there is a narrow membranous line.

Fig. 219. TERMINAL PORTION OF THE THIRD LEG OF PACHYLOMERIDES

The figure represents the tibia of a female. In the only male that I have seen the depression in this segment is not so marked as in the female; and the membranous line is wanting.

The representatives of this genus are common in the warmer parts of the United States. Their burrows are furnished with doors of the wafer type. Several species have been described from the southern Atlantic States, one from Arizona, and one

Superfamily Aviculariodea

from California. As yet insufficient material has been studied to make it possible to indicate with any degree of certainty the distinctive characters of the species.

Pachylomerides audouini (P. au-dou-in'i).—The most common species of this genus in the southeastern part of our territory is believed to be the one described by Lucas in 1837 under this specific name. It is a large species, the body of the adult measuring one and one third inches in length. The male is shining black in color, has very long pedipalps. The first leg is not armed with spurs or patches of stout setæ.

Genus CYCLOCOSMIA (Cy-clo-cos'mi-a)

There are two genera of tarantulas, one, *Chorizops*, represented by a single species found in Mexico, and the other *Cyclocosmia*, represented by two species, one from China and one from the southern United States, that differ from all others by the remarkable form of the abdomen, the caudal end of which is truncate and discoidal (Fig. 221). In *Cyclocosmia* the tibia of the third pair of legs lacks the depressions characteristic of the two preceding genera, and the eyes are not widely separated.

Fig. 221. CYCLOCOSMIA TRUNCATA

Cyclocosmia truncata (C. trun-ca'ta).—This is a rare species which has been redescribed recently by Gertsch and Wallace ('36). Mature females are often very large, averaging somewhat more than an inch in length. The colour is dusky brown to black, the end of the abdomen being black. The first row of eyes is nearly straight, the second row moderately recurved. The abdomen is abruptly truncated behind and very hard, the distal portion being in the form of a round disk which is fringed with setæ. The disk is marked with six circular impressions which are shown in Hentz's figure (Fig. 221).

The form of the abdomen led Hentz, who described the species, to believe that this part of the body was used to plug up the entrance to its burrow. Hentz's figure (Fig. 221) shows the spider by a burrow without a lid. However, it is now known that *Cyclocosmia* caps the burrow with a trap door of the wafer type. The burrows are straight, cylindrical, and almost vertical in position, and are placed in the banks at the bottom of ravines. At the top they are of such a diameter that it would be impossible for the abdomen of the spider to plug the burrow. *Cyclocosmia* evidently derives some protection from its calloused abdomen for it proceeds head downward into the tunnel to a point where it is possible to completely close the opening. When in this position the spider is practically immune to the advances of predaceous or parasitic enemies that may invade the burrow for food or for a victim for its egg or larval parasite. The male is still unknown.

Specimens of this species have been taken so far only in Florida, Alabama, and Louisiana. Since the time of Hentz scarcely more than a dozen examples have been collected.

Genus BOTHRIOCYRTUM (Both-ri-o-cyr′tum)

The tibia of the third legs lacks the depression found in Pachylomerus; the tarsi of the first and second pairs of legs are not scopulate, but are furnished with strong spines; the tarsal claws are armed with but few teeth, in most cases with a single tooth, where there are more than one they are in a single series. The clypeus is nearly horizontal, and is twice as broad as the transverse diameter of the anterior lateral eyes.

Only two species of this genus have been described; one of them is the common California species, the other is a Mexican species which perhaps also occurs in our fauna.

The Common Californian Trap-door Spider, *Bothriocyrtum californicum* (B. cal-i-for′ni-cum).— The adult female measures one and one sixth inches in length. When alive the general colour of the whole spider is a dark blackish chocolate brown, the legs and cephalothorax being darker than the abdomen. In alcohol, the cephalothorax is deep reddish yellow brown; and the abdomen a dull yellowish brown colour. The sternum is not marked by sigilla. The anterior tarsi have only a single tooth at the base, those of the posterior tarsi have at the base four slender and recurved teeth of which the second and the fourth are longer than the others.

This is the species whose nest is most often seen in collections. The nest is an unbranched tube furnished with a single door of the cork type.

Superfamily Avicularioidea

Genus MYRMECIOPHILA (Myr-mec-i-oph′i-la)

In this and in the following genera of this subfamily the fore tarsi at least are furnished with a scopula, although sometimes in the case of males it is not dense. In this genus and in the following the posterior sigilla are large, near together, and widely distant from the margin of the sternum. In this genus the chelicerae are furnished on the inside, at the apex, with a blunt and toothed process. The claws, at least the anterior ones, are furnished with a single tooth at the base.

Of the four species described from the United States, the following is the best known and most widely distributed. It will serve as an example for the genus.

Myrmeciophila fluviatilis (M. flu-vi-at′i-lis).—The body of the female measures two thirds inch in length. The cephalothorax is dull olive with a rufous tinge; the abdomen dull yellowish; there is a broad longitudinal dorsal band of delicate brown, from which branch on each side seven bands of the same colour, extending down midway of the abdomen. In living specimens the femora are delicate light olive colour.

Fig. 222. MYRMECIOPHILA FLUVIATILIS, MALE

The male of this species is illustrated in Fig. 222. It is smaller and more slender than the female and has longer legs. The metatarsus of the first leg is bent, slender at the base and enlarged at the tip. The embolus of the palpus is provided with a series of teeth.

The burrow of this species is of great interest because it is usually provided with an inner side branch which is closed with a door. The outside door has been described as being of the cork type, but in *Myrmeciophila torreya* it is of the wafer type. The tunnels are often made within an ants' nest; in which case the spider probably feeds upon the ants.

This species is found in the southeastern United States from North Carolina to southern Texas.

Superfamily Avicularioidea

Genus EUCTENIZA (Eu-cten-i'za)

This genus is closely allied to *Myrmeciophila*. In the females the eyes of the first row are subequal in size, whereas in the males the lateral eyes are larger than the median eyes. The metatarsus of the first leg of the male bears a terminal process beneath from which originate several long spines. The second metatarsus is also modified below near the middle where there is a slight elevation which bears spines.

Species of this genus have been described from Mexico. Our only species, as yet undescribed, was taken in southern Texas.

Genus ACTINOXIA (Ac-ti-nox'i-a)

This genus differs from the preceding in that the interior apical angle of the chelicerae is rounded or slightly convex, and in that the claws of the tarsi are furnished with several teeth near the base.

Three species have been found within the limits of the United States. The following is probably the best known one.

Actinoxia versicolor (A. ver-sic'o-lor).—The adult female measures a little more than one inch in length; the male is one half as long. The endite of the pedipalp is armed with small spines which are scattered from the base to the apex.

This species has been recently studied by Mr. C. P. Smith ('08) in California. He states that it is the most common trap-door spider of the Santa Clara Valley, and of the foothills and cañons on either side. The burrow is long, narrow, branched, and furnished with a thin trap-door of the wafer type.

Genus AMBLYOCARENUM (Am-bly-o-ca-re'num)

The fore tarsi are scopulate; the posterior sigilla are small and much farther from each other than from the margin of the sternum; the teeth of the rake of the chelicerae are numerous and irregular; the anterior tarsi and metatarsi are short, robust, and nearly equal in length; the metatarsi are unarmed except by the apical spines.

Only one species has been found in our fauna.

Amblyocarenum talpa (A. tal'pa).— I have not seen this species. It was described from a female, which was three fourths inch in length. It occurs in California. This was first described

Superfamily Avicularioidea

as *Cyrtauchenius talpa;* but the genus *Cyrtauchenius* has been restricted by Simon to certain species found in Africa.

Genus APTOSTICHUS (Ap-tos'ti-chus)

This genus is closely allied to the preceding; but it differs in that the anterior metatarsi are longer than the tarsi and armed below with many spines.

Three species have been described; all are from California.

Aptostichus atomarius (A. at-o-ma'ri-us).— The body of the female is a little more than one half inch in length. The cephalothorax is dull reddish yellow, with a pale yellow pubescens; the cervical groove is nearly straight. The abdomen is dull yellow, with an ashy yellow pubescens.

Aptostichus clathratus (A. cla-thra'tus).— The body of the female is one half inch in length. The cephalothorax is a dull dark chestnut and nearly glabrous; the cervical groove is procurved. The abdomen is dark bluish in front and brick coloured behind.

Aptostichus stanfordianus.— The adult female measures from two fifths inch to a little more than one inch in length. The cephalothorax is tawny; the abdomen is yellowish brown, marked above with a median series of dark brown blotches, and lateral series of short linear spots. The male is unknown.

This species, recently described by Mr. C. P. Smith, is found in California. It makes a short burrow furnished with a trap-door.

Subfamily ANTRODIÆTINÆ (An-tro-di-æ-ti'næ)

The Folding-door Tarantulas

There are two small subfamilies of tarantulas, this and the Hexurinæ, that differ from the other typical tarantulas and agree with the family Atypidæ, in having the postabdomen situated a considerable distance above the spinnerets; and in that the furrow of the cheliceræ is indistinct. Their endites are moderately well developed, intermediate between the typical tarantulas where they are scarcely evident and the Atypidæ in which group they are strongly developed. Two or three distinct tergites are present at the base of the abdomen above, reminiscent of *Liphistius*.

These two subfamilies can be distinguished by the presence

or absence of the rake of the cheliceræ; it being present in the Antrodiætinæ and absent in the Hexurinæ.

In the possession of a rake of the cheliceræ the Antrodiætinæ spiders resemble the Ctenizinæ or trap-door spiders; and this resemblance is correlated with a striking resemblance in habits. Species of *Antrodiætus* excavate tunnels in the earth and like the Ctenizinæ close their tunnels with a door. But in the case of the single species whose habits have been carefully described this door consists of two semicircular parts which meet on the middle line of the opening, like a pair of folding doors. It may be that this habit is not shared by other members of the subfamily; but until more is known of their habits the Antrodiætinæ may be designated popularly as the folding-door tarantulas.

This subfamily includes few genera. Three of them are restricted to our fauna.

TABLE OF GENERA OF THE ANTRODIÆTINÆ

A. With only four spinnerets. P. 237.   ANTRODIÆTUS
AA. With six spinnerets.
   B. Median furrow of the carapace longitudinal; male without a conductor of the embolus; pedipalps half as long in male as first legs. P. 238.   ATYPOIDES
  BB. Median furrow of the carapace round; male with a conductor of the embolus and with pedipalps as long as first legs. P. 239.   ALIATYPUS

Genus ANTRODIÆTUS (An-tro-di-æ'tus)

The members of this genus can be easily recognized by the possession of the family characteristics given above and the fact that they have only four spinnerets. Twelve species have been described, but with our present lack of knowledge of the group it is impossible to state how many of these species are valid. The first one was described by Hentz as *Mygale unicolor* and this name was subsequently made the genotype of *Antrodiætus* by Ausserer. This name must supplant the more familiar *Brachybothrium* of Simon.

An account of the habits of a species belonging to this genus, is published by Atkinson ('86) under the name *Nidivalvata marxii*.

Superfamily Aviculariodea

Professor Atkinson's observations were made at Chapel Hill, N. C., and are in part as follows:

"This species begins the excavation of its tube by parting the earth from a central point with its anterior legs and palpi turning around at the same time so as to push the earth on all sides. It works with exceeding rapidity." "When beginning the nest in a patch of moss the spider will dive down into the moss and begin turning rapidly in all directions, at the same time spinning threads to fasten together the pieces of moss around and over it. I have watched four different ones make the nest, two beginning in moss which I had placed over the earth, and two beginning in soil. Two of these I had make a nest several times, and thus far every one has first entirely closed the entrance to the tube by building a sort of dome over it. Later, in one case a week, cutting through this and making the folding-door. Usually in making the dome, earth is placed on and about the edge of the tube, occasionally applying viscid liquid and spinning threads over it. Then the spider would, with its anterior legs and palpi pull the edge over the tube. This operation would be repeated until the dome was complete. When moss is convenient the door is made almost entirely of moss and silk; each door is a surface of a half circle, is hung by a semicircular hinge, and the two meet, when closed, in a straight line over the middle of the hole."

A specimen of Brachybothrium, which Mr. Banks determines as "apparently *B. accentuatum*," was presented to me by Mr. Paul Hayhurst, who collected it at Columbia, Mo. It was found in a burrow sixteen inches deep; this burrow was very much like that of a large tiger beetle larva, and was without a lid. The specimen is a female, and measures nine sixteenths inch in length.

Genus ATYPOIDES (At-y-poi'des)

In this genus there are three pairs of spinnerets; the median furrow of the thorax is longitudinal; the palpus of the male is without a conductor of the embolus; and the pedipalps of the male are only half as long as the first legs.

Only the following species is known:

Atypoides riversi (A. riv'er-si).— The adult male measures one half inch in length. The cephalothorax is greenish brown; the abdomen, dull purplish brown. The cheliceræ have an extraordinary appearance due to the presence of a long, projecting curved apophysis at the base. This is cylindrical, obtusely pointed, and densely clothed at and near its extremity, above and on the sides, with long coarse bristly black hairs.

The female resembles the male in colour. But instead of the long projection on the cheliceræ, there is on each a simple, strong, subconical prominence directed a little backward.

This species was described by Rev. O. P. Cambridge from specimens collected at Berkeley, Cal. by Mr. J. J. Rivers. It is stated that it tunnels in banks mostly by streams, forming a tubular projection above ground of any material at hand, woven up with silk, making no trap-door, but closing the aperture at times.

Genus ALIATYPUS (Al-i-at'y-pus)

As in the preceding genus, there are three pairs of spinnerets; but the median furrow of the thorax is a rounded pit; the palpus of the male is furnished with a conductor of the embolus, and the pedipalps of the male are as long as the first legs.

Aliatypus californicus (A. cal-i-for'ni-cus).— The body measures about one half inch in length. The cephalothorax and legs are pale; the abdomen brownish or grayish with the venter light.

This species is found in the foothills and mountains on each side of the Santa Clara Valley, California. The burrow is comparatively long, simple, with a simple trap door.

Family AVICULARIIDÆ (A-vic-u-la-ri'i-dæ)

The bird-spiders resemble the preceding and the following subfamily in having the postabdomen situated immediately above the spinnerets, in that the furrow of the chelicerae is well marked and in that the median furrow of the thorax is transverse or elliptic; but they differ from both of these subfamilies in having the tarsi of the legs furnished with a bundle of terminal tenent hairs and in having the third claw wanting or obsolete.

These are running tarantulas; a fact that might be inferred from the armature of their tarsi; which are furnished with scopulæ and terminal tenent hairs.

To this subfamily belong the larger tarantulas, those that are most feared on account of their size. In South America there is a species having a body two inches in length and whose legs

Superfamily Avicularioidea

expand more than seven inches. The habits of this spider were described by Madam Merian two hundred years ago. She stated that it destroyed small birds; this suggested the name of the typical genus, *Avicularia*, and also the popular name Bird-spiders. (See p. 185, 186.)

The Aviculariinæ rarely dig true burrows; but make use of natural cavities in the ground or in trunks of trees. Correlated with this fact is the absence of a rake of the cheliceræ. Simon states that they line the cavities in which they live with a close, but light and transparent web, which is always less extended than that of the Diplurinæ, and is without a tube-shaped retreat. They are nocturnal, watching in the evening for their prey, which consists almost always of large beetles. Simon also states that their eggs are numerous, and are enveloped in a cocoon of white tissue, which certain species carry in the cheliceræ till the young emerge.

Our known species represent four genera, which can be separated by the following table.

TABLE OF GENERA OF THE AVICULARIINÆ

A. Tibiæ and metatarsi of the third and fourth pairs of legs with very few spines.
  B. Anterior eyes subequal, in a very strongly procurved line; anterior tibia of male furnished with one spur at the apex. P. 244.   AVICULARIA
  BB. Anterior eyes unequal, median eyes nearly twice as large as the lateral; anterior eyes in a slightly procurved or nearly straight line; anterior tibia of male with two spurs at the apex. P. 244.   TAPINAUCHENIUS
AA. Tibiæ and metatarsi of the third and fourth pairs of legs with many spines.
  B. Metatarsus of the forelegs thickly scopulate to the base and usually without basal spines. P. 241.   EURYPELMA
  BB. Metatarsus of the forelegs with scopula not reaching the base; with basal spines. P. 240.   RHECHOSTICA

Genus RHECHOSTICA (Rhe-chos'ti-ca)

The scopulæ of the anterior metatarsi do not reach the base of the segment, covering only about two thirds of the segment;

the base of the anterior metatarsi is armed with spines; the apex of the posterior metatarsi is minutely scopulate; the posterior median eyes are broadly removed from the anterior median and are much smaller.

Only one species has been described.

Rhechostica texense (R. tex-en'se).— The male is black and measures one and one half inches in length; the four anterior eyes are nearly equal in size; the posterior median eyes are evidently smaller than the anterior median, and are barely separated from the posterior lateral eyes.

The female has not been described.

Nothing is known of the habits of the species; it was described from specimens collected in Texas.

Genus EURYPELMA (Eu-ryp'el-ma)

The tibiæ and metatarsi of the third and fourth pairs of legs are armed with many spines. The metatarsi of the fore legs are thickly scopulate to the base and are usually without spines.

To this group belong the largest spiders that occur within the limits of our fauna. They are the tarantulas which, in the South and the West, most often attract attention on account of their size, and which are greatly feared on account of the supposed deadly nature of their bite. One of them is represented natural size by Fig. 224.

Fig. 224. EURYPELMA HENTZII

They are nocturnal, hiding during the day in the cracks of trees, under logs, stones, stumps, or other rubbish, and coming forth in the evening and lying in wait for their prey.

Several species have been described from the United States; but in most cases the descriptions have been based on an examination of but one sex. In fact, notwithstanding the large size of these spiders and their abundance in the regions where

Superfamily Avicularioidea

they occur, it is not possible in the present state of our knowledge to accurately define the species.

The following table, which is a modification of one published by Simon, will aid in the separation of our better known species:

- A. With a large velvety brown spot on the otherwise light-coloured abdomen. *E. steindachneri*
- AA. Abdomen not light with a brown spot.
    - B. Anterior metatarsus shorter than the tibia. All tarsi long, not much shorter than the anterior metatarsus, the posterior only one third shorter. Adults small, the male measuring only four fifths inch in length; body entirely black. *E. marxi*
    - BB. Anterior metatarsus as long as or longer than the tibia. Anterior tarsus about one third shorter than the metatarsus, posterior not more than half as long. Adults large.
        - C. Anterior median eyes hardly more remote from each other than from the lateral eyes and the space between the lateral eyes not much less than the diameter of an eye. Posterior median eyes round. *E. helluo*
        - CC. Anterior median eyes evidently more remote from each other than from the lateral eyes and the space between the lateral eyes barely half the diameter of an eye. Posterior median eyes elongate and straight, not oblique.
            - D. Hairs silky olivaceous. Legs long. Anterior metatarsus curved and evidently longer than the tibia. Posterior metatarsus more than twice as long as the tarsus. *E. rusticum*
            - DD. Hairs ferruginous. Legs shorter. Anterior metatarsus about as long as the tibia. Posterior metatarsus about twice as long as the tarsus.
                - E. Posterior median eyes evidently smaller than the lateral eyes; both median and lateral elongate. Tibia of the pedipalp usually armed on the inside with five or six reddish spines. *E. californicum*
                - EE. Posterior median eyes elongate, lateral subrotund and not larger than the posterior median. Tibia

of the pedipalp armed on the inside with a subbasal spine, two submedian spines, and an apical spine. *E. hentzi*

Eurypelma steindachneri (E. stein-dach'ner-i).— This is the most easily recognized of all of our tarantulas, being characterized by a large velvety brown spot on the otherwise light-coloured abdomen. It is a large species, the adult measuring from one and one half to two inches in length. There are specimens of both sexes in the collection of Cornell University, and in both the brown spot is present. It is evident, therefore, that what was described by Ausserer as the female of this species belongs to some other species, probably to *E. californicum*. The species is widely distributed in the Southwest. Figure 225 is of a young individual which I collected in Texas.

Eurypelma marxi (E. mar'xi).— This species can be distinguished from our other species of *Eurypelma* by its comparatively small size; the adult measuring only about four fifths inch in length, and the body is entirely black. The tibia of the pedipalp is armed on the inside with only two submedian spines. This species is found in the Southwest.

Fig. 225.
EURYPELMA STEINDACHNERI

Eurypelma helluo (E. hel'lu-o).— This is our largest species, the male measuring two inches in length. The distinctive characteristics are given in the table above. It occurs in the Southwest.

Eurypelma rusticum (E. rus'ti-cum).— See table above for distinctive characteristics. This species inhabits the Southwest.

Eurypelma californicum (E. cal-i-for'ni-cum).— This is the most common of the large tarantulas found in California, Arizona, and Texas. See table above for distinctive characteristics.

Eurypelma hentzi (E. hent'zi).— This is one of the most common tarantulas of the Southern and Southwestern States. It can be distinguished from the preceding species by the characteristics given in the table above.

Superfamily Avicularioidea

Genus AVICULARIA (A-vic-u-la'ri-a)

The group of eyes is barely one third wider than long; the anterior eyes are quite small, nearly equal, and in a strongly procurved line. The posterior sigilla are small and marginal. The anterior tibia of the male is furnished with one spur at the apex.

Avicularia californica (A. cal-i-for'ni-ca).— The cephalothorax is a little more than one inch in length and two thirds inch in width. The cephalothorax is densely clothed with yellowish gray hair, rather paler on the sides and brighter in the middle; on the anterior margin there is a fringe overhanging the cheliceræ of long, pale gray hairs with tawny tips. The abdomen is clothed above with fine dark brown hair, and many long, recurved bristles of a tawny colour with pale tips; the lower side of the abdomen is clothed with black hair, and with long tawny bristles. The endites and the furrow of the cheliceræ bear long tawny bristles.

This is the only species of Avicularia found in our fauna. It was described by Mr. Nathan Banks from several specimens collected in the vicinity of San Diego, Cal.

Genus TAPINAUCHENIUS (Ta-pin-au-chen'i-us)

The group of eyes is at least twice as wide as long. The anterior eyes are in a nearly straight line; and the anterior median eyes are larger than the anterior lateral. The posterior sigilla are quite large, ovate, and submarginal. The anterior tibia of the male is furnished with two spurs at the apex.

Two species have been found in the Southwest.

Tapinauchenius cærulescens (T. cæ-ru-les'cens).— This species was described by Simon from an immature female which measured four fifths inch in length. The cephalothorax is densely clothed with very long, silky, dark blue hairs; and the abdomen, which was badly rubbed in the specimen described, is silky blue pubescent above with a few long, yellow hairs, and is black below.

The specimen was collected at Fort Sill, Indian Territory.

Tapinauchenius texensis (T. tex-en'sis).— This species, like the preceding, was described from an immature female. The specimen measured one inch in length. The cephalothorax is densely clothed with yellow hairs; and the abdomen, badly rubbed, with close, long dark hairs, and marked with an obscure longitudinal, median stripe and a transverse zone. The abdomen is black below.

The specimen was from Eagle Pass, Tex.

Family DIPLURIDÆ (Di-plur'i-dæ)

The Funnel-web Tarantulas

The funnel-web tarantulas differ from the bird-spiders in having the tarsi furnished with a third claw instead of a bunch of terminal tenent hairs (claw tufts); and from the trap-door spiders in the absence of the rake of the cheliceræ. The genus *Brachythele* apparently forms an exception to this statement for a poorly developed rake is present in both sexes. The postabdomen is situated immediately above the spinnerets in most members of the family but in the Hexurinæ it is a considerable distance above the spinnerets. The spinnerets are typically very long, particularly the apical joint which may have numerous false articulations, and they are set wide apart.

The habits of our species have not been investigated fully. Certain exotic forms spin large webs of close, cloth-like tissue, terminating in a funnel, like the webs of the grass spiders. The presence of the third claw shows that they live upon webs or sheets of silk; and the absence or weak development of the rake of the cheliceræ indicate that their burrowing activities are limited.

Seven species, representing four genera and at least three subfamilies, have been found in the United States. They can be separated by the following table:

A. With six spinnerets; palpus of male with a conductor of the embolus. (Hexurinæ.) P. 247. HEXURA
AA. With four spinnerets; palpus of male lacking a conductor.
  B. Anterior tarsi and metatarsi scopulate; space between the fore spinnerets twice the width of the spinnerets. (Diplurinæ.) P. 246. BRACHYTHELE
  BB. Anterior tarsi not scopulate; fore spinnerets separated by at least four times width of spinnerets. (Macrothelinæ.)
    C. Tibia of second leg of male incrassated beneath and armed with two robust spines; mature spiders one half inch long; tibia of first leg set with a series of ventral spines. P. 246. EVAGRUS

CC. Tibia of second leg of male normal; first tibia with a strong ventral spine which extends beyond the tip; small species, one sixth inch long. P. 246.
MICROHEXURA

Genus BRACHYTHELE (Brach-y-the'le)

The anterior eyes are of equal size or the anterior median eyes are a little smaller than the lateral. The metatarsi of the first and second pairs of legs are scopulate all the way to the base. Other characteristics are given in the table above.

Brachythele longitarsis (B. lon-gi-tar'sis).—This species is widely distributed in the far west, being found from Idaho to Lower California. The adult female measures from one to nearly two inches in length; the male is somewhat smaller. A closely related species (*B. theveneti*) also occurs in California.

This species digs a deep burrow, and lines only the upper fourth or less with silk.

Genus EVAGRUS (E-va'grus)

The anterior lateral eyes are much larger than the anterior median; and the posterior median eyes are distinctly larger than the anterior median eyes. The tibia of the second pair of legs in the male is furnished beneath with a large apophysis armed with spines. The following is our best known species.

Evagrus comstocki (E. com-stock'i).—The female measures about two thirds inch in length, and is nearly uniform pale brown in color. The male is somewhat smaller. The species is common in southern Texas. Other species occur in the southwestern part of our country.

Genus MICROHEXURA (Mi-cro-hex-u'ra)

This genus resembles the preceding one but is distinct in the secondary sexual characters present in the legs of the male. The first tibia bears a strong ventral spur which extends beyond the tip. The second leg is not modified.

The only known species of this genus, *Microhexura montivagus*, which is probably the smallest member of the Dipluridæ, being only one sixth inch long, was described from North Carolina.

Superfamily Avicularioidea

Genus HEXURA (Hex-u′ra)

The median furrow of the cephalothorax is longitudinal but broad and short. There are three pairs of spinnerets. The group of eyes is twice as wide as long; the anterior median eyes are situated in the middle of the group and are very small compared with the other eyes.

Hexura picea (H. pic′e-a).—The body measures a little more than one fourth inch in length. The chelicerae are long and project forward in a prominent manner; the distal portion is armed with spines in the male. The male palpus has a conductor of the embolus. This species occurs in the northwestern part of our country. A second species has been described from California.

Family ATYPIDÆ (A-typ′i-dæ)

The Purse-web Spiders

The tarantulas constituting the family of Atypidæ are distinguished from the other members of the superfamily by the form of the coxa of the pedipalps, which bears a large conical lobe, the endite (Fig. 226), and by the more complicated form of the

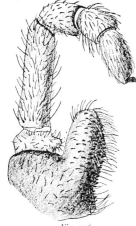

Fig. 226.
PEDIPALP OF ATYPUS

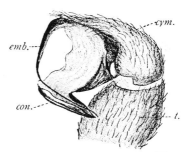

Fig. 227. TARSUS OF MALE OF ATYPUS BICOLOR

palpus of the male, which approaches more nearly the form seen in many true spiders. The bulb of the palpus is of the intermediate type, the apical division consisting of two distinct parts, an embolus and a conductor of the embolus (Fig. 227).

Superfamily Avicularioidea

The presence of an endite does not, strictly speaking, distinguish the Atypidæ from the other tarantulas, for in some of them the coxa bears a small lobe which is really an endite. It is very small compared with the endite of *Atypus*.

The purse-web spiders agree with the Antrodiætinæ and the Hexurinæ in having the postabdomen situated a considerable distance above the spinnerets; and in that the furrow of the chelicerae is indistinct. They possess three pairs of spinnerets.

This is a small family, including only two genera of which but one occurs in the New World.

Genus ATYPUS (At'y-pus)

The sternum bears four pairs of sigilla, of which the fourth pair is much the largest. The hind spinnerets are three- or four-jointed, much longer than the others. The labium is joined to the sternum without any trace of a suture.

Four species are known from our fauna.

Atypus abboti (A. ab-bot'i).—The adult female measures one half inch or more in length. The cephalothorax is brown with a very narrow black margin; and the eye space is dusky. The legs, pedipalps, and cheliceræ are of the same colour as the cephalothorax; but the claws of the cheliceræ are dark. The abdomen is dark brown, with many paler points and small spots. The male is a handsome spider. The carapace and legs are black or nearly so, sometimes with a greenish tinge. The abdomen is iridescent blue or purple above and has a median black hastate marking at the base. The carapace and the sternum are longer than broad in both sexes, and the posterior pair of spinnerets is four-jointed. In our three other species the hind spinnerets are three-jointed.

This species is not uncommon in Florida and also occurs in parts of Georgia. It is rare farther north.

The name purse-web spider, given to this species by Abbot, who discovered it in Georgia one hundred or more years ago, is a happy one for the web spun by it is a long silken tube, which resembles in a striking way the silken purses of our grandmothers.

The purse-web spider lives in a tunnel in the ground, which in those that I studied was almost invariably made at the base of a tree; but sometimes they make their nests among herbaceous plants. This tunnel is lined by a silken tube, which is extended vertically above the surface of the ground for a considerable distance against the side of the tree (Fig. 228).

In the case of the web of an adult spider the portion above the ground measures nearly or quite one foot in length and from one half to three fourths of an inch in width. The part below the surface of the ground is shorter; in the specimens that I have taken, it measures from four to six inches in length.

The outer surface of the purse-web is always protectively coloured to a considerable extent. The colour varies from a light gray to a very dark brown, almost black, corresponding to the colour of the bark of the tree against which the tube is built. The colour is due to a coat made of minute bits of bark, lichens, and moss, which are evidently collected by the spider from the trunk of the tree and fastened to the surface of the web. In the case of a web built by a spider that I kept in confinement, the web was partly covered with grains of sand taken from the soil in the breeding cage, and there are usually grains of sand on the webs built under natural conditions.

Notwithstanding their colouring, the tubes are easily seen when one has learned their appearance. They look like a dead stick or a section of a climbing vine; but present a very characteristic appearance, owing to their being perfectly straight. This is well shown in Fig. 204 which is from a photograph of of a tree trunk bearing both a purse-web and a vine.

The purse-web is attached to the surface of the tree only by its upper extremity, where it is flattened and somewhat expanded into bands of threads by which it is fastened to the tree. Between this point of attachment and its base, where it is firmly anchored in the ground, it is stretched like a fiddle-string. The result is that the walking of an insect over it causes it to vibrate; and thus the presence of the insect is revealed to the waiting spider, who waits in the tube below the surface of the ground with its head upward ready to rush up the tube.

When the tube is disturbed by the passage of an insect over it, the spider rushes to the point where the insect is, and biting through the web, captures it. The web is then slit and the insect pulled inside the tube. After the spider has sucked its prey dry, the remnant is thrown out, the slit is repaired, and the spider waits for another victim.

As a rule one finds only a single web of an adult spider on a

(*Photographed by H. Harold Hume*)
Fig. 228. PURSE-WEB OF ATYPUS ABBOTII ON DOGWOOD TREE, CORNUS FLORIDA, AT LAKE CITY, FLA.

tree; but occasionally two or three tubes are built side by side on the same tree. The young spiders are more gregarious. I have seen as many as seven small tubes side by side near a large tube. In selecting a place to build, these spiders show no preference for either the shady or sunny side of the tree.

In the case of a French species of *Atypus*, *A. piceus*, that part of the purse-web which is above ground rests horizontally on the surface of the soil (Fig. 218, *a*). This nest forms an interesting intermediate type between a simple silk-lined tunnel with no external prolongation, such as is made by many spiders, and the stretched and vibrating tube of *Atypus abboti*.

Atypus bicolor (A. bi'co-lor).—This is a much larger species than the preceding one, the females averaging an inch or more in length. The females are coloured essentially as in *abboti*. The male is a very striking spider with black carapace, abdomen and palpi, and the legs are carmine-red. The tarsus of the male palpus is illustrated in Fig. 227. The posterior spinnerets are three-jointed.

This handsome species has been taken in the vicinity of Washington, D. C., and in the western part of Florida.

Atypus niger (A. ni'ger).—This species is all or nearly all black. Only the male is well known. The posterior spinnerets are three-jointed as in the preceding species. It was described by Hentz from a solitary individual found on newly turned soil, at Northampton, Mass. It is rare in the northern states but is apparently widely distributed, having been reported from Wisconsin and Illinois and south to Ohio.

Atypus milberti (A. mil-bert'i).—This name was applied to a spider found near Philadelphia by Milbert and sent to Walckenaer. Although no examples have been discovered since the description of the species, it is quite probable that when specimens are discovered, it will prove to be a valid species.

CHAPTER VII:

Superfamily ARGIOPOIDEA (Ar-gi-o-poi′de-a)

THE TRUE SPIDERS

To THIS superfamily, which is coextensive with the *Araneæ veræ* of Simon, belong nearly all the familiar spiders found in the Temperate Zone, all except the tarantulas which occur for the most part in the South and the West. The true spiders are wonderfully diversified in structure. The hundreds of genera and thousands of species found throughout the world give ample evidence of their dominance. Their varied habits and interesting peculiarities have made them favored subjects for study.

In the Argiopoidea the cheliceræ project downward (ventrad) or obliquely downward (Fig. 229), and the claw moves more or less directly toward the middle plane of the body. The pedipalps are furnished with distinct endites (Fig. 230). With the exception of the Hypochilidæ, the true spiders differ from the tarantulas in having no more than one pair of book-lungs.

The Argiopoidea may be distinguished from the Avicularioidea as follows:

  A. Cheliceræ paraxial: i.e., the plane of articulation is vertical (Fig. 211); two pairs of book-lungs always present. See Chapter VI: Superfamily AVICULARIOIDEA, P. 222.
AA. Cheliceræ diaxial: i.e., the plane of articulation is horizontal (Fig. 229); not more than one pair of book-lungs except in the Hypochilidæ.   Superfamily ARGIOPOIDEA

TABLE OF FAMILIES OF THE ARGIOPOIDEA

Key to the Series

  A. With two pairs of book-lungs.   SERIES I
AA. With no more than one pair of book-lungs.
    B. Cheliceræ fused together at the base.   SERIES II

Superfamily Argiopoidea

BB. Cheliceræ not fused together at the base.
  C. Cribellum and calamistrum present.   SERIES III
  CC. Cribellum and calamistrum absent.
    D. With a pair of tracheal spiracles opening just behind the genital groove.   SERIES IV
    DD. With a single tracheal spiracle or none.
      E. With three tarsal claws.
        F. Fourth tarsus with spurious claws or with a ventral line of serrated bristles or both.   SERIES V
        FF. Fourth tarsus never with spurious claws, occasionally with serrated bristles, the latter never arranged in a ventral series.   SERIES VI
      EE. With only two tarsal claws.   SERIES VII

Series I

Two pairs of book-lungs are present. This group comprises a single family. P. 256.   HYPOCHILIDÆ

Series II

The cheliceræ are fused together at the base. A cribellum and calamistrum may or may not be present.

  A. Cribellum and calamistrum present; tracheal spiracles advanced in front of spinnerets. P. 292.   FILISTATIDÆ
  AA. Without a cribellum and calamistrum.
    B. Tarsi normal; usually six eyes; labium longer than broad. P. 313.   SCYTODIDÆ
    BB. Tarsi long, flexible, with false articulations; usually eight eyes; labium broader than long. P. 339.   PHOLCIDÆ

Series III

The cheliceræ are free at the base. A cribellum and calamistrum are present, rarely absent in adult males.

  A. Anal tubercle large, two-jointed, fringed with long hairs. P. 289.   OECOBIIDÆ

Superfamily Argiopoidea

AA. Anal tubercle normal.
    B. With two tarsal claws. P. 302.   Zoropsidæ
    BB. With three tarsal claws.
        C. Eyes diurnal, all dark in color.
            D. Posterior median eyes large, directed forward (Fig. 249). P. 272.   Deinopidæ
            DD. Posterior median eyes of moderate size. P. 261.   Uloboridæ
        CC. Eyes all light in color, or at most only anterior median eyes dark.
            D. Eyes all pearly white. P. 273.   Amaurobiidæ
            DD. Anterior median eyes dark, the other eyes light in color. P. 279.   Dictynidæ

Series IV

A cribellum and calamistrum are never present. A pair of tracheal spiracles open just behind the genital groove.

    A. With only two eyes. P. 303.   Caponiidæ
    AA. With six eyes.
        B. Tarsi with three claws. P. 306.   Segestriidæ
        BB. Tarsi with two claws.
            C. Claws pectinate in a single row; first coxæ elongated. P. 303.   Dysderidæ
            CC. Claws pectinate in a double row; all coxæ subequal in length. P. 309.   Oonopidæ

Series V

A single tracheal spiracle is placed near or advanced in front of spinnerets. The tarsi bear three claws. The fourth tarsus is supplied with spurious claws or with a ventral line of serrated bristles (Fig. 323) or both.

    A. Anterior tibiae and metatarsi armed with a series of very long spines, and a series of shorter spines between each two long ones (Fig. 564). P. 531.   Mimetidæ
    AA. No such armature of spines on first two legs.
        B. Labium not rebordered; fourth tarsi with a ventral row of serrated bristles. P. 345.   Theridiidæ

BB. Labium rebordered.
    C. Labium immovable, not thickened at end; only six eyes. P. 318.   LEPTONETIDÆ
    CC. Labium freely movable; eight eyes, rarely six.
        D. Eyes heterogeneous; cheliceræ with stridulating ridges on lateral surface. P. 382.   LINYPHIIDÆ
        DD. Eyes homogeneous; cheliceræ without stridulating ridges. P. 414.   ARGIOPIDÆ

Series VI

The tarsi bear three claws. The fourth tarsus is never supplied with spurious claws, occasionally with serrated bristles, the latter never arranged in a ventral line.

A. Tracheal spiracle advanced far in front of spinnerets; spinnerets in a transverse row. P. 613.   HAHNIIDÆ
AA. Tracheal spiracle immediately in front of spinnerets.
    B. Chelicera without a lateral condyle.
        C. Colulus absent; posterior spinnerets short, sometimes obsolete. P. 336.   ZODARIIDÆ
        CC. Colulus present; posterior spinnerets very long and attenuate. P. 633.   HERSILIIDÆ
    BB. Chelicera with a distinct lateral condyle.
        C. Trochanters not notched. P. 597.   AGELENIDÆ
        CC. At least posterior trochanters notched.
            D. Only fourth trochanters deeply notched. P. 665.
                                                                 OXYOPIDÆ
            DD. All trochanters deeply notched.
                E. Paired claws with numerous teeth; median claw with two or three. P. 615.   PISAURIDÆ
                EE. Paired claws with few teeth; median claw with one or no teeth. P. 635.   LYCOSIDÆ

Series VII

The cheliceræ are free. The cribellum and calamistrum are lacking. The tarsi bear only two claws.

A. Eyes heterogeneous, only the anterior median dark.

Superfamily Argiopoidea

    B. Eyes in two rows; tarsal claws toothed. P. 321.
                                                                    GNAPHOSIDÆ
    BB. Eyes in three rows; tarsal claws smooth. P. 319.
                                                                    PRODIDOMIDÆ
AA. All eyes homogeneous, dark in colour.
    B. Tarsal claws smooth. P. 338.   HOMALONYCHIDÆ
    BB. Tarsal claws toothed.
        C. Legs laterigrade.
            D. Lower margin of the furrow of the chelicera distinct, armed with teeth.
                E. Six eyes in first row. P. 563.   SELENOPIDÆ
                EE. Four eyes in first row. P. 564. HETEROPODIDÆ
            DD. Lower margin of chelicera indistinct, without teeth. P. 534.   THOMISIDÆ
        CC. Legs normal, the front ones directed forward.
            D. Eyes in two rows. P. 571.   CLUBIONIDÆ
            DD. Eyes in three or four rows.
                E. Anterior median eyes smaller than the lateral eyes. P. 568.   CTENIDÆ
                EE. Anterior median eyes (if in four rows, the front eyes) very much larger than the lateral eyes. P. 669.   ATTIDÆ

Family HYPOCHILIDÆ (Hyp-o-chil'i-dæ)

The Four-lunged True Spiders

In the mountains of eastern Tennessee and of neighbouring states there lives a very remarkable spider, which was first described by Dr. George Marx ('88), and for which the family Hypochilidæ was established (Fig. 231). Two other species belonging to this family have been discovered, one in China, and one in Tasmania. These are undoubtedly True Spiders, that is they belong to the Superfamily Argiopoidea, but they differ from all other members of this superfamily and agree with the Tarantulas, the Avicularioidea, in the possession of two pairs of book-lungs.

The second pair of book-lungs is situated near the middle of the ventral aspect of the abdomen, and their spiracles are connected by a prominent furrow. There are eight eyes, of which the anterior median are dark and the others pearly white;

Superfamily Argiopoidea

the eyes occupy the entire width of the head. The claws of the cheliceræ are nearly vertical. The pedipalps are in front of the labium. It was the position of the labium below the pedipalps that suggested the name *Hypochilus*. The palpus of the male is remarkable in that the bulb is borne

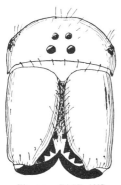

Fig. 229. FACE AND HELICERÆ OF ARANEA

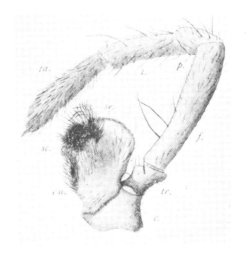

Fig. 230. PEDIPALP OF TRACHELAS
en, endite   *sc*, scopula   *se*, serrula

near the extremity of the tarsus (Fig. 232). It is of the intermediate type, the apical division of the bulb being separated into two parts, the embolus and the conductor of the embolus.

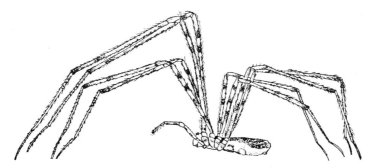

Fig. 231. HYPOCHILUS THORELLII (after Marx)

The embolus is coiled about the conductor. The tarsus bears a branch which supports a prominent bunch of bristles; this may be a rudimentary form of paracymbium.

257

Superfamily Argiopoidea

Genus HYPOCHILUS (Hyp-o-chi'lus)

This genus is distinguished from the genus *Ectatosticta*, to which the two exotic members of the family belong, in having a transverse labium, which is much shorter than long. It contains only a single known species.

Hypochilus thorellii (H. tho-rel'li-i).— This is a spider with an elongate body and very long and slender legs (Fig. 231). It resembles a *Pholcus* in general appearance. The male measures two fifths inch in length; the female, three fifths. The cribellum is semicircular and undivided. The calamistrum is situated near the base of the fourth metatarsus and consists of long slender hairs.

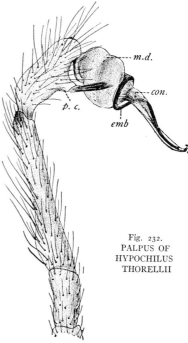

Fig. 232. PALPUS OF HYPOCHILUS THORELLII

It has not been my good fortune to study this spider in the field and very little has been published regarding its habits. But, thanks to the courtesy of correspondents, I am able to give the following account:

Hypochilus constructs its webs on the under surface of projecting cliffs and rocks, and especially in the vicinity of streams. Dr. W. H. Fox has sent me a photograph of a typical locality for this spider (Fig. 233). The picture was taken on Walden Ridge, Tenn., and there were half a dozen or more webs on the under surface of the largest slab.

The web is a meshed one, and is shaped like a lamp shade. Owing to the darkness of the situations in which they are built, it is impracticable to procure photographs of these webs in the field; but Dr. J. Chester Bradley sent me living individuals, from Tallulah Falls, Ga., which I placed in cages in our insectary, and which made webs that I have been able to photograph.

The spider first spins a circular sheet upon the lower surface of the supporting object, which in this case was the roof of the cage. It then spins the foundation of the sides of a downward extending cylinder, which is at first a very open net-work; this is shown in the upper part of Fig. 234. In the completed web the net-work is finely meshed, as is shown in the lower part of Fig. 234. When at rest the spider remains upon the sheet forming the roof of the web, hanging back downward, with its legs extended radiately, the tips of them reaching the edge of the

(Photographed by W. H. Fox)
Fig. 233. TYPICAL LOCALITY FOR HYPOCHILUS

horizontal sheet, so that the claws can touch the descending portion of the web. From this edge the sides of the web extend downward like the sides of a lamp shade.

The framework of the web consists of smooth threads; but a very large part of the meshes is made of a hackled band. The hackled band is of the same type as that of *Amaurobius*, figured later (Fig. 250). The warp consists of four threads. Two of these lie in the central portion of the band; they are straight and parallel. On either side of these two straight threads,

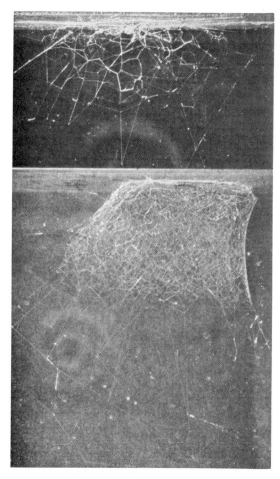

Fig. 234. WEB OF HYPOCHILUS
UPPER FIGURE, BEGINNING OF WEB, NATURAL SIZE;
LOWER FIGURE, COMPLETED WEB

Superfamily Argiopoidea

there is a very much curled thread. And supported by these four threads there is a band of viscid silk, the edges of which are undulating, and at a considerable distance from the curled threads; that is, the woof is much wider than the warp.

Family ULOBORIDÆ (U-lo-bor'i-dæ)

The Uloborids (U-lob'o-rids)

The Uloboridæ includes a small number of very remarkable spiders. They agree with the Argiopidæ and differ from all other spiders in spinning orb-webs. But they differ from the Argiopidæ

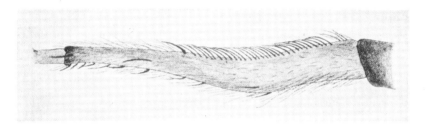

Fig. 235. POSTERIOR METATARSUS OF ULOBORUS

in the possession of a cribellum and a calamistrum; and their webs differ from those of that family in containing a hackled band, instead of the ordinary viscid thread.

Fig. 236. HACKLED BAND OF HYPTIOTES

The eyes are all dark in colour; the lateral eyes on each side are farther apart than are the two pairs of median eyes; and the posterior median eyes are of moderate size. The posterior

Superfamily Argiopoidea

metatarsi are much curved and are armed below with a series of spines (Fig. 235); the calamistrum occupies more than half of the length of the segment.

I have studied the hackled band of our two genera, *Uloborus* and *Hyptiotes*. The structure of it is the same in the two and is quite characteristic (Fig. 236). There is a warp consisting of two straight threads, upon which the woof of viscid silk is borne. The two strands of the warp are strictly parallel, are smooth, and are exceedingly delicate. They do not show well in the figure as they were out of focus when the picture was taken. I found, with the high magnification required, that it was impossible to get both the warp and the woof in focus at once, they being in slightly different planes, the woof resting upon the warp. The woof consists of an exceedingly regular series of overlapping lobes.

It seems probable that the warp is spun from two spigots, one on each of a pair of spinnerets, and that the woof is combed from the cribellum by the calamistrum.

Our two genera can be separated as follows:

TABLE OF GENERA OF THE ULOBORIDÆ

A. Cephalothorax ovate, rounded behind.   ULOBORUS
AA. Cephalothorax wide in the middle, narrowed in front and cut off squarely behind; the sides concave in front, nearly parallel behind.   HYPTIOTES

Genus ULOBORUS (U-lob'o-rus)

The great length and robust form of the front legs of these spiders cause them to present a very characteristic appearance; this is shown in Fig. 237, which represents our most common species. Other species differ in the form of the abdomen, in the arrangement of the eyes, in markings, and in the nature of the clothing of hairs. Figure 238 represents the ventral aspect of the cephalothorax of *Uloborus americanus*.

The spiders of this genus make orb-webs, which resemble in general appearance those of the more common orb-weavers, the Argiopidæ, but differ in that the spiral thread is a hackled band. These webs are almost always horizontal, and usually made in low bushes, or between objects near the ground, or in buildings,

Superfamily Argiopoidea

the spiders preferring the cool and shady places. The webs are never furnished with a retreat, but are often strengthened by a sheeted hub or by a stabilimentum. Certain exotic species are social, a large number of individuals living near together and spinning their orbs on a common foundation.

More than sixty species of Uloborus are known, but only six are found in the United States; most of the species inhabit hot countries; this being true, it seems strange that our northern species should prefer cool situations.

The following are the best-known of our species:

Uloborus americanus (U. a-mer-i-ca′nus).— The female is about one fifth inch in length. It varies greatly in colour and markings. The abdomen is slightly notched in front, is widest and highest at one third of its length from its base; and at this

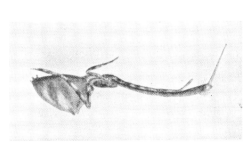

Fig. 237. ULOBORUS AMERICANUS

Fig. 238.
ULOBORUS AMERICANUS
VENTRAL ASPECT

point, it bears a pair of humps. The posterior row of eyes is strongly recurved. The tibiæ of the fore legs bear a bunch of hairs at the tip.

The male is from one tenth to one eighth inch in length. The humps of the abdomen are small or wanting; and the tibiæ of the fore legs lack the brushes of hairs characteristic of the female.

The species has been commonly known as *Uloborus plumipes;* but it was first described under the specific name used here.

This remarkable spider has been found throughout the eastern United States from the Great Lakes to the Gulf of Mexico, and it probably occurs over the entire extent of our country; it is also found in Europe.

Although so widely distributed, it is found, so far as I have

Fig. 239. COMPLETE ORB OF ULOBORUS AMERICANUS ON GOLDEN ROD

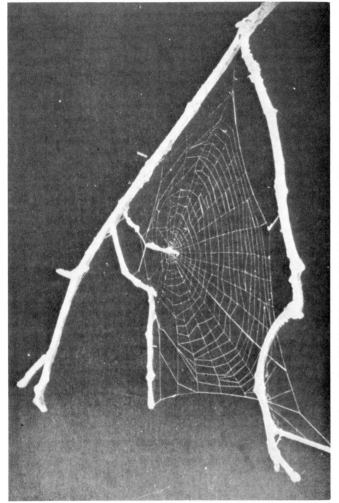

Fig. 240. INCOMPLETE ORB OF ULOBORUS AMERICANUS ON DEAD BRANCH OF HEMLOCK, WITH EGG-SAC SUSPENDED ABOVE IT

Superfamily Argiopoidea

observed, in very limited localities, and only in small numbers. There is a little nook, a few minutes walk from my laboratory, on the south bank of a deep ravine, where but little direct sunlight falls, and where the air is kept moist by a near cataract. Here, upon the low straggling branches of the American yew, in the dense shade of tall hemlock trees, there is established a

Fig. 241. MESHED HUB OF ULOBORUS AMERICANUS

colony of *Uloborus*, which has persisted through several years. By careful search one can find here, at the right season, perhaps a score of webs. But nowhere else in the region about have I been able to find this spider except occasionally an individual; and these have been found in somewhat similar situations.

Although most of the webs that I have studied were made on the yew, it happens that those that I was able to photograph, and consequently those that are used for illustration here, were made on other plants.

The web is usually nearly horizontal. It is sometimes a complete orb (Fig. 239), and sometimes an incomplete one (Fig. 240). The hub is meshed (Fig. 241); there is no clear space, and to the unaided eye, the hackled band, which constitutes the spiral thread, appears like the spiral thread of an *Aranea*. The structure of this hackled band has been discussed on page 263.

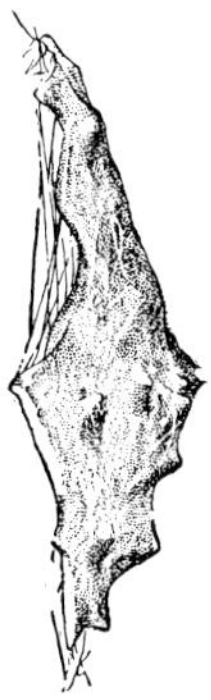

The spiders reach maturity in early summer; their orbs are then from three to five inches in diameter. The spider rests beneath the hub, and resembles a bit of dried leaf. At this time, the egg-sacs are made. They are light brown in colour, are elongate, about one fourth inch in length, and bear several tubercles (Fig. 242). They are suspended in or near the web (Fig. 240). In central New York the eggs hatch in the latter part of July.

On one occasion I had a brood of young of this species in a cage; each one made a stabilimentum across the centre of its web; but the parent spider had none in her web.

Fig. 242. EGG-SAC OF ULOBORUS AMERICANUS

Fig. 243. ULOBORUS GENICULATUS

This species is the *Uloborus plumipes* of Lucas.

Uloborus geniculatus (U. ge-nic-u-la′tus).— The female measures one fourth inch in length; the male one fifth inch. The abdomen bears a single obtuse hump at the end of the first third of its length. The posterior row of eyes is nearly straight. The markings are very conspicuous in living individuals; but the markings of the abdomen are very apt to disappear when the specimen is placed in alcohol. Figure 243 is from a photograph

Superfamily Argiopoidea

of a specimen resting on some threads it had spun in a glass cage; but the threads were not sufficiently well lighted to be photographed. Upon the hump of the abdomen there is a black transverse bar; in front of this bar the dorsum is light gray; back of it, a light brown, with some gray hairs. There are three pairs of light gray spots margined with black; those of the first pair are situated one at each end of the black bar; the third pair is midway between the first pair and the tip of the abdomen.

This is a tropical species, which is found throughout the tropics of both hemispheres; and it extends into the southern part of the United States.

I found many specimens which had built their webs on the ceiling of an old powder magazine at Baton Rouge, La. The room was poorly lighted and was much like a cellar.

I was unable to see the webs distinctly in this dark room at Baton Rouge. But later I found this species common in build-

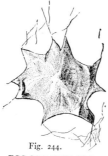

Fig. 244.
EGG-SAC OF ULOBO-
RUS GENICULATUS

Fig. 245. HYPTIOTES CAVATUS, ENLARGED

ings on the Bermuda Islands. In one case a web was in a good light so that I could see it well. There was a sheeted hub upon which the spider was resting; and the hackled band was attached to the radii in the same manner as in the hackled band of *Hyptiotes*; that is, the band did not extend directly across a radius, but followed it for a short distance, as is the case with the spiral thread in the notched zone of the web of some of the Argiopidæ.

The egg-sacs of this species (Fig. 244) are of a pinkish brown colour, and star-like in outline, and measure about one fourth inch in diameter.

Superfamily Argiopoidea

Genus HYPTIOTES (Hyp-ti'o-tes)

The following is the best known of the two species that have been described from the United States.

The Triangle Spider, *Hyptiotes cavatus* (H. ca-va'tus).—This inconspicuous spider will ordinarily be recognized by the form of its web; the spider itself being so well-protected by its form and colouring as to escape observation. It usually rests close to a dead branch and resembles a dried bud of this branch.

The adult female is one sixth inch in length. The outline of the body is shown in Fig. 245. On the back of the abdomen there are four pairs of slight elevations, on which are a few stiff hairs. The male is one twelfth inch in length; the abdomen is of a more slender form and the humps are not so prominent.

The habits of this spider have been made well-known through the writings of Dr. B. G. Wilder, who first described them more than thirty-five years ago in the *Popular Science Monthly* (April, 1875). Its web is most often found stretched between the twigs of a dead branch of pine or hemlock. At first sight it appears like a fragment of an orb-web (Fig. 246); but a little study will show that it is complete. It consists of four plain lines corresponding to the radiating lines of an orb-web, and supported by these a variable number of threads which appear like sections of the spiral line of an orb-web. From the point where the radiating lines meet a strong line extends to one of the supporting twigs.

Each of the transverse lines supported by the four radii is a hackled band consisting of a warp of two threads and a woof of overlapping lobes of viscid silk like that of *Uloborus* (Fig. 236). Each band is fastened to each radius by being applied to it lengthwise for a short distance; this makes the course of each band a zigzag one. In Fig. 246 one can see that in the spaces between the bands there is in each case a short section of the radius not overlapped by the bands, which consequently appears as a more delicate line; this is best shown on the two intermediate radii.

The number of radii in the web of this spider is always four; but the number of hackled bands varies greatly; it is usually about ten, but often less than that number, and sometimes more than twenty.

The spider rests on the single line, upon which the four radii converge, near the point where the line joins the supporting

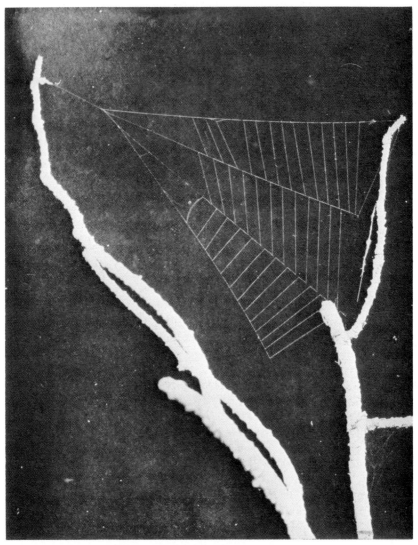

Fig. 246. WEB OF HYPTIOTES CAVATUS

twig. The spider is usually quite close to the twig, so that it appears as a small bud (Fig. 246); but sometimes it rests a small distance from the twig (Fig. 245).

While at rest the spider pulls the web taut, so that there is some loose line between its legs. When an insect becomes entangled in one of the hackled bands, the spider suddenly lets go the loose line so that the whole web springs forward, and the insect is entangled in other bands. The spider then draws the web tight and snaps it again. This may be repeated several times before the spider goes out upon the web after its prey

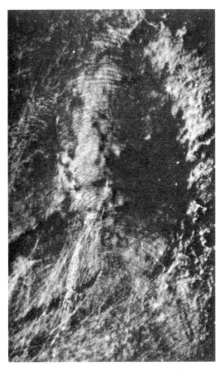

Fig. 247. EGG-SAC OF HYPTIOTES CAVATUS

Fig. 248. PHOTOMICROGRAPH OF COVERING OF EGG-SAC OF HYPTIOTES CAVATUS

The egg-sac of the spider illustrates protective colouring as well as does the spider itself. Although it is a common object in a region where I have collected much, I never saw it until a spider that was in a bottle in my laboratory made one on the cork stopper of the bottle; having seen one specimen I was able to find others in the field. Figure 247 represents the most conspicuous one in our collection, somewhat enlarged. It is flat, oval, and closely

Superfamily Argiopoidea

applied to a twig. The egg-sac proper is about one fourth inch in length; but this is covered by a somewhat larger sheet, which by its gray colour serves to conceal it. This covering layer is well worth study with a microscope. It consists of a sheet of dirty white silk, upon which are transverse parallel lines near together of crochet-work in black silk (Fig. 248).

This species is widely distributed in the eastern United States. A closely allied species, *Hyptiotes gertschi*, has been described from the western part of our country.

Family DEINOPIDÆ (Dei-nop'i-dæ)

The Ogre-faced Spiders

The enormous size of one pair of eyes, the posterior median, gives the face of these spiders a very unusual appearance (Fig. 249). It was probably this characteristic that suggested the name of the typical genus, *Deinopis;* from the Greek *deinos*, terrible, and *opsis*, appearance. And for the same reason I suggest for the family the common name used above. The family is represented in our fauna by a single genus.

Genus DEINOPIS (Dein'o-pis)

The body is slender and the legs very long. The eyes are all small except the posterior median, which are enormously developed and project forward (Fig. 249). The cribellum is transversely elongate, and not divided by a septum. And the calamistrum occupies less than half of the length of the metatarsus. The endites are divergent and curved on the outside. The name of this genus is often written *Dinopis;* but the form adopted here is the original one.

Deinopis spinosus (D. spi-no'sus).— This is a slender spider, measuring two thirds of an inch in length and less than one eighth inch in width. The legs are long, the first two pairs being more than twice as long as the body. The abdomen is greenish yellow, with a lancet-shaped,

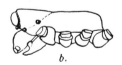

Fig. 249. DEINOPIS *a*, face *b*, lateral view of cephalothorax (after Marx)

brownish folium, and many round black dots. The species is rare; it has been found in Florida and Alabama; and is the only representative of the family reported from the United States. Mr. Banks informs me that he has taken this species at Washington, D. C.

Very little is known regarding the habits of the spiders of this family. They are said to build a horizontal web, which resembles that of the grass spider, *Agelena*, except that it has no tubular retreat; the spider stands in the centre of the web while waiting for its prey. Nothing is known concerning the structure of the hackled band or of the part it plays in the building of the web.

Family AMAUROBIIDÆ (Am-au-ro-bi'i-dæ)

The Amaurobiids (Am-au-ro-bi'ids)

To this family belong a small number of spiders which resemble in a striking way the members of the Agelenidæ. They differ from the agelenids in having a cribellum and a calamistrum. The species are for the most part of average size. They may be distinguished from the Dictynidæ, of which group they were long regarded as constituting only a subfamily, in having all the eyes light in colour and in the structure of the genital organs. The legs are provided with strong spines, whereas in the dictynids only in a few cases are true spines present. The tarsi of the legs bear three claws. The cribellum is divided into two parts. The calamistrum consists either of a single or a double row of curved setæ. Adult males sometimes lack the cribellum and the calamistrum, but the former is usually discernible, although reduced in size, even in this sex.

The amaurobiids construct irregular webs consisting of a framework of plain threads supporting an irregular net-work of the hackled band. Sometimes the supporting threads radiate from the opening of a retreat, with a certain degree of symmetry, giving the web a somewhat regular appearance.

The structure of the hackled band is most easily seen in the webs of *Amaurobius*, on account of its coarser nature here than in the webs of other genera. It is a comparatively easy matter to determine the arrangement of the parts of this band with a microscope; but it is difficult to secure a perfectly satisfactory photomicrograph of it, owing to the impossibility of getting all

Superfamily Argiopoidea

portions of the curled threads in focus at once when greatly magnified.

The warp consists of four threads (Fig. 250). Two of these lie in the central portion of the band; they are straight and parallel. The other threads extend, one along the middle of each lateral half of the band, and are curled. These four threads, consti-

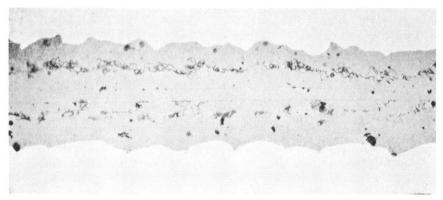

Fig. 250. HACKLED BAND OF AMAUROBIUS

tuting the warp of the band, support a sheet of viscid silk, the woof. The woof has a wavy outline, but does not consist of a regular series of lobes as in the Uloboridæ.

The genera at present known from our fauna can be separated by the following chart. Some of the described species are of uncertain position, particularly those from the western United States.

TABLE OF GENERA OF THE AMAUROBIIDÆ

A. Calamistrum consisting of a single row of curved setæ.
  B. Anterior median eyes much larger than the lateral eyes.
     P. 279.   HESPERAUXIMUS
  BB. Anterior median eyes smaller than the lateral eyes.
    C. Eyes of median quadrangle subequal in size or nearly so. P. 278.   TITANŒCA
    CC. Anterior median eyes very much smaller than the posterior median eyes. P. 278.   CALLIOPLUS
AA. Calamistrum consisting of two parallel rows of curved setæ.
    P. 275.   AMAUROBIUS

Superfamily Argiopoidea

Genus AMAUROBIUS (Am-au-ro'bi-us)

The sternum is not prolonged between the posterior coxæ (Fig. 251); and the cribellum is divided into two parts (Fig. 252). This genus includes the larger representatives of the family, some of them being nearly one half inch in length. Seven species have been described from the United States.

These spiders prefer cool, moist, and poorly lighted situations. They live in cracks in cliffs, in cellar walls, in stumps, in hollow

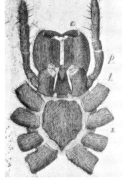

Fig. 251. AMAUROBIUS, VENTRAL ASPECT

Fig. 252. CRIBELLUM OF AMAUROBIUS

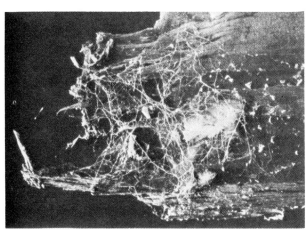

Fig. 253. WEB OF AMAUROBIUS

logs, and under stones. Some of them spin a loose, irregular web, in which there appears to be no definite plan (Fig. 253). Others that live in a retreat in a crack in a cliff or wall spin a sheet of silk about the entrance to the retreat (Fig. 254). Some

275

Superfamily Argiopoidea

of the lines composing this sheet radiate from the central part of it and support a network of hackled bands.

I have observed the egg-sac of *A. bennetti*. This is a loosely woven, flat sac, attached to a stone or other object and covered with an irregular mesh of threads. It is made near the web,

Fig. 254. WEB OF AMAUROBIUS

and the young, after they leave it, are attended by the mother. In this species the young emerge in the latter part of the summer.

The following are our more common species.

Amaurobius bennetti (A. ben-net′ti).— This is the most common species in the North (Fig. 255). The females are two fifths of an inch long; the males, one third of an inch. It is of a brownish black colour marked with yellowish white. On the basal half of the abdomen there are two parallel, longitudinal, light bands; each of these is continued in a zigzag course to the tip of the abdomen, or there may be several light chevrons on the hind part of the abdomen, the light markings varying greatly in size and form in different individuals. These markings are much more distinct after a specimen has been kept in alcohol.

The epigynum of the female consists of three lobes, a small median lobe which is pointed behind, and two larger lateral

Fig. 255. AMAUROBIUS BENNETTI

Fig. 256. EPIGYNUM OF AMAUROBIUS BENNETTI

Fig. 257. TIBIA OF THE PALPUS OF MALE OF AMAUROBIUS BENNETTI

lobes which meet behind the median lobe (Fig. 256). The tibia of the palpus of the male is of the form shown in Fig. 257.

Amaurobius ferox (A. fe'rox).— This is a larger species, adult females often measuring a half inch in length. It is dusky brown in colour. In well-marked individuals, there are, on the base of the abdomen, three yellowish white, longitudinal bands, and on the hind part, four pairs of light spots; but there are great variations in the markings. The most distinctive characteristics are the following.

The lateral lobes of the epigynum do not meet behind the middle lobe (Fig. 258); and the tibia of the palpus of the male is of the form shown in Fig. 259.

This is a domestic spider, being found in cellars, under the

Superfamily Argiopoidea

floors of dwelling houses, in outhouses, and in other buildings. It is also found under logs and stones in the field.

Genus TITANŒCA (Ti-ta-nœ'ca)

The calamistrum is uniseriate, consisting of a single row of curved setæ. The anterior median eyes are somewhat smaller than the anterior lateral and are about the same size as the posterior median eyes. The following is our best known species.

Titanœca americana (T. a-mer-i-ca'na).—The body is one

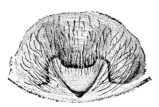

Fig. 258. EPIGYNUM OF AMAUROBIUS FEROX

Fig. 259. TIBIA OF PALPUS OF MALE OF AMAUROBIUS FEROX

fourth inch in length, and of a deep black colour, except the cephalothorax, which is of a dull orange colour. The abdomen is usually without markings, but in some individuals there are a few light gray spots in pairs on the abdomen.

Emerton, who first described this species, states that it lives under stones in the hottest and dryest places.

Genus CALLIOPLUS (Cal-li-op'lus)

The species of this genus resemble *Amaurobius* in coloration and general appearance, but all the known species are very much smaller in size. The calamistrum consists of a single row of curved setæ. The anterior median eyes are very much smaller than the posterior median eyes. Several species have been described of which the following is the best known one.

Callioplus tibialis (C. ti-bi-al'is).—The female is about one third inch long, and the male is somewhat smaller. The cephalothorax is orange, smooth and shining. The abdomen is gray, marked with rows of pale spots. The tibial apophysis of the male consists of a very long curved spur. The species is not uncommon in the northeastern part of the United States.

Superfamily Argiopoidea

Genus HESPERAUXIMUS (Hes-per-aux'i-mus)

This genus is easily distinguished from the preceding ones by the anterior median eyes which exceed the anterior lateral eyes in size. The calamistrum is uniseriate. A single species has been described from California.

Hesperauximus sternitzkii (H. ster-nitz'ki-i).—The carapace is dark reddish brown. The abdomen is dusky brown, with a basal darker maculation and a row of black spots on each side. Females are one half inch long, and the male is of nearly equal size. The sexes are alike in coloration.

Family DICTYNIDÆ (Dic-tyn'i-dæ)

The Dictynids (Dic-tyn'ids)

To this family belong the greater number of our species that are furnished with a cribellum and a calamistrum, the Dictynidæ including several times as many species as all other families of hackled-band weavers taken together.

With the dictynids the median furrow of the cephalothorax is longitudinal; the posterior metatarsi are not armed below with a series of spines; the cheliceræ are robust, and are furnished with a lateral condyle; the anterior median eyes are dark in colour, the others pearly white; and the lateral eyes of each side are contiguous or near together; the tarsi of the legs bear three claws; a cribellum and a calamistrum are present; and the fore and hind spinnerets are of about the same length. Tactile hairs are present on the tarsi and metatarsi of most of the genera. In *Dictyna* they are not present on the tarsi.

These spiders may be distinguished from the amaurobiids by the fact that the anterior median eyes are dark while the others are pearly white. The sternum is prolonged between the posterior coxæ. The cribellum is divided only in rare cases, notably in the subgenus *Ergatis* (of which *Dictynina* and *Mallos* are seemingly synonymous) of the genus *Dictyna*.

The numerous members of this family are for the most part small spiders which live on plants, in leaf mold, or under debris on the ground. Our largest species measures about one fourth inch in length. The dictynids construct irregular webs consisting of a

Superfamily Argiopoidea

framework of plain threads supporting an irregular net-work of the hackled band.

Seven or eight genera have been described from the United States. Five of these are rather well known and may be separated by the following chart. Most of our species belong in *Dictyna*.

TABLE OF GENERA OF THE DICTYNIDÆ

A. Fourth tarsus without row of tactile hairs (trichobothria); clypeus usually much wider than the anterior eyes; head very convex, often subgibbose. P. 281. DICTYNA
AA. Fourth tarsus with a row of long tactile hairs; clypeus not or scarcely wider than anterior eyes; head lower.
  B. With only six eyes. P. 289. SCOTOLATHYS
  BB. With eight eyes.
    C. Anterior median eyes much smaller than the posterior median, equal at most to their radius. P. 280. LATHYS
    CC. Eyes of the median quadrangle subequal in size.
      D. Lower margin of the furrow of chelicera with two teeth. P. 288. ARGENNA
      DD. Lower margin with four teeth. P. 288. TRICHOLATHYS

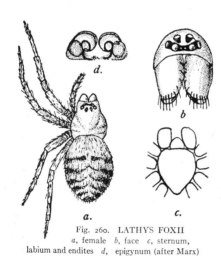

Fig. 260. LATHYS FOXII
a, female   *b*, face   *c*, sternum, labium and endites   *d*, epigynum (after Marx)

Genus LATHYS (La'thys)

This genus is related to *Scotolathys* from which it differs in having the anterior median eyes present, although small. The posterior median eyes are much larger than the anterior median. The head is relatively broad and only moderately elevated. The tarsi of the fourth pair of legs have a single row of trichobothria.

Three or four species of this genus have been described from the United States. The following one is best known.

Superfamily Argiopoidea

Lathys foxii (L. fox'i-i).—The length of the body is less than one twelfth inch. The cephalothorax is yellowish. The abdomen is grayish white with dark gray markings, consisting of about five transverse lines. The accompanying figures (Fig. 260) are by Dr. Marx ('91) who described the species from specimens collected in Tennessee. It occurs generally in the northeastern part of our country. This is the *Prodalia foxii* of Marx.

Genus DICTYNA (Dic-ty'na)

The sternum extends between the hind coxæ; the clypeus is much wider from before backward, than the diameter of the anterior eyes; and the endites are moderately long and convergent (Fig. 261).

The most familiar of the hackled-band weavers belong to this genus, which is a very large one, including nearly one hundred species, of which thirty or more occur in our fauna.

Fig. 261.
STERNUM OF DICTYNA

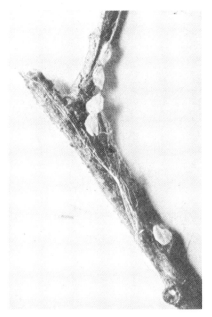

Fig. 262. EGG-SACS OF DICTYNA

These spiders are of small or moderate size; and the different species show marked differences in habits. The species described below as illustrating the genus have been selected with the view of showing the more striking of these differences. One of them usually builds its webs on the walls of buildings, one in the heads of plants, and one on the surface of leaves.

Superfamily Argiopoidea

A single female makes several egg-sacs; these are lenticular in form, snowy white in colour, and are usually made in or near the web. They are placed side by side (Fig. 262) or in an overlapping series. Nineteen species of *Dictyna* are recognized as occurring in our fauna; of these the following are the more common.

Dictyna sublata (D. sub-la′ta).— The adult female is one sixth inch in length. The cephalothorax is reddish brown, clothed with white hairs, which are arranged upon the head in five longitudinal lines. The abdomen is brown above, with a central, longitudinal dark band, with irregular edges, and more or less nearly broken in the middle of its length. The shape of the abdominal marking varies greatly in different individuals, and especially on the hind half of the abdomen. The form of the

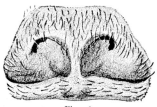

Fig. 263.
EPIGYNUM OF DICTYNA SUBLATA

Fig. 264. TIBIA OF PALPUS OF MALE OF DICTYNA SUBLATA

epigynum is shown in Fig. 263. The tibia of the palpus of the male (Fig. 264) bears a short apophysis at the base; this apophysis is much shorter than the diameter of the tibia and is bifid at the tip.

The sheet-like webs of this species are very common on the sides of buildings. The spider builds a retreat in some small opening or corner, and makes its web about the opening of this retreat. Frequently some of the threads forming the foundation of the web extend in a radiating manner from the small, circular opening of the retreat, giving the web a certain degree of regularity.

The web represented in Fig. 265 was made on the outside of a window sash. The retreat was in the angle of the sash near the centre of the web but it does not show in the picture. And the plain threads forming the foundation of the web are also invisible; but the zigzag courses of the hackled band resting on these threads are well-shown. This is especially the case in the lower left corner of the web.

Fig. 265. WEB OF DICTYNA SUBLATA

Superfamily Argiopoidea

I have found similar webs very common on the rough bark of the trunks of elm trees on the Cornell Campus. In these cases, the retreat of the spider was beneath a scale of the bark.

There is in our collection a web, which I believe to be of this species, and which was made on a dead branch on a larch tree

Fig. 266. WEB OF DICTYNA SUBLATA

(Fig. 266). In this web the characteristic form of the entrance to the retreat is well-shown.

The adult spiders can be found in their webs in early summer. It is evident that they reach maturity in the autumn or early spring. The egg-sacs are made early in June; and then follows an interval in which no new webs are made. In August the young spiders can be found in their webs.

Dictyna volucripes (D. vo-lu'cri-pes).— The body is about one sixth inch in length. This species resembles the preceding in colour and markings; but can be distinguished by the form of the epigynum (Fig. 267) and by the form of the apophysis on the base of the tibia of the palpus of the male (Fig. 268), which is very much longer than in *D. sublata*.

The web is usually made near the top of some herbaceous plant. In the centre of it is a silken retreat, spun in an angle at the base of a branch or of a leaf stalk. The two examples figured (Figs. 269 and 270) illustrate the most common type. In Fig. 269 the spider can be seen resting on the retreat. A very common feature in these webs is a ladder-like structure formed by stretching a hackled band back and forth between two supporting threads. This is shown in the edges of the upper part of Fig. 269. Old webs are often refurbished by the addition of new ladders.

The adults pair in mid-summer; the male and female live together in their nest till after the female has made her egg-sacs; these are made in the centre of the web, a single spider making several of them. The female continues to use the web after making her egg-sacs; and in the autumn the old female and the young brood live together in the old web.

Dictyna foliacea (D. fo-li-a'ce-a).— The length of the body is about one eighth inch. The cephalothorax is light brown, darker on the sides and light on the head. The abdomen is yellow in the middle and brown, sometimes red, at the sides. The outline of the yellow portion differs greatly in different individuals. The form of the epigynum is shown in Fig. 271. In the male the cheliceræ are very long, almost as long as the cephalothorax; and the palpi are long and large. The tibia of the male palpus is twice as long as wide, with a short, two-spined apophysis near its base.

This species is commonly known as *D. volupis;* but it is evidently the *Theridion foliacea* of Hentz.

The web is made in the hollow of a leaf, the edges of which have been slightly rolled (Fig. 272). No part of the web is formed for a retreat, the spider resting on the surface of the web. The specimen figured here contains four egg-sacs, which are placed in two piles of two each; it was collected in July.

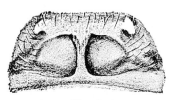

Fig. 267.
EPIGYNUM OF DICTYNA
VOLUCRIPES

Fig. 268. TIBIA OF
PALPUS OF MALE OF
DICTYNA VOLUCRIPES

Fig. 269. WEB OF DICTYNA VOLUCRIPES

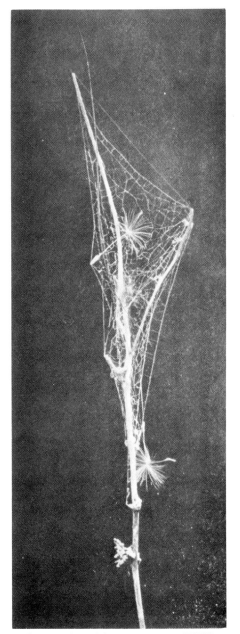

Fig. 270. WEB OF DICTYNA VOLUCRIPES

Fig. 271. EPIGYNUM OF DICTYNA FOLIACEA

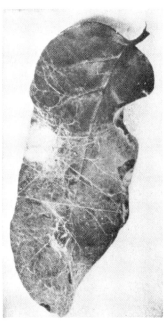

Fig. 272. WEB OF DICTYNA FOLIACEA

Superfamily Argiopoidea

Genus TRICHOLATHYS (Tri-cho-la′thys)

The cephalothorax is longer than broad and moderately high, the head being relatively wide in front. The width of the clypeus is scarcely as great as the diameter of an anterior lateral eye. The anterior median eyes are slightly smaller than the anterior lateral and the posterior lateral eyes. The median ocular quadrangle is wider than long and slightly wider behind than in front. The cheliceræ are stout, the lower margin set with four short teeth. The moderately stout legs are provided with scattered spines, particularly the last two pairs. The oval abdomen is of moderate height. The cribellum is undivided. The posterior tarsi bear a row of trichobothria.

Only two species of this genus have been described, both from the western part of our country. The following species is representative of the genus. It occurs in Utah.

Tricholathys spiralis (T. spir-al′is).—The carapace is light yellowish brown, the margins being somewhat darker. The dusky abdomen is provided above with lighter markings. The palpus of the male has the caudal process of the conductor very long and slender, the distal portion forming a coil.

Genus ARGENNA (Ar-gen′na)

This genus resembles the preceding closely in general appearance and structure. It differs chiefly in the armature of the lower margin of the furrow of the chelicera where only two subequal teeth are present. The clypeus is low and the head is only moderately elevated. Both eye rows are nearly straight, the anterior median eyes being subequal in size to the posterior median eyes. The legs are rather stout and are rarely provided with spines. When weak spines are present, they are confined to the last two pairs of legs. A row of trichobothria is present on the fourth tarsus.

Several species occur in our fauna. The following one, from Utah, will serve as an example.

Argenna aktia (A. ak′ti-a).—This species measures one sixth inch in length. The carapace is brown. The legs are yellowish brown, with dusky shadings, but without distinct annulae. The abdomen is dusky above and marked with faint spots and a median basal streak.

Superfamily Argiopoidea

Genus SCOTOLATHYS (Sco-to-la′thys)

The six-eyed condition of these spiders distinguishes them from the other members of the family found in our fauna. The fourth tarsus is provided with a dorsal row of trichobothria. The legs lack true spines. Three or four species have been described from the eastern United States. The following is an example.

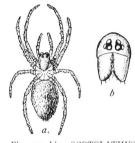

Fig. 272 bis. SCOTOLATHYS PALLIDUS *a*, female enlarged *b*, face (after Marx)

Scotolathys pallidus (S. pal′li-dus).— The body is about one sixteenth of an inch in length. The cephalothorax is pale yellow, the abdomen and the legs yellowish white. The eyes are on black patches (Fig. 272 *bis*). The species is found under stones, in leaf mold, and under debris.

Family ŒCOBIIDÆ (Œ-co-bi′i-dæ)

The Œcobiids (Œ-co-bi′ids)

The family Œcobiidæ is represented by two or possibly three genera. The best known is *Œcobius* which includes a small number of tropical and subtropical species.

In this family the cephalo-

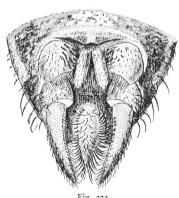

Fig. 273.
END OF ABDOMEN OF ŒCOBIUS

Fig. 274. CALAMISTRUM OF ŒCOBIUS

thorax is broader than long, with the group of eyes nearly in the centre. The eyes are unequal in size and dissimilar in form. The anterior median and the posterior lateral eyes are dark in colour; the others, pearly white. The posterior median eyes are elongate and usually angular; but the arrangement of the eyes varies in

Superfamily Argiopoidea

different species. The cheliceræ are small, without a lateral condyle, but with a comparatively long claw. The endites are strongly inclined and nearly contiguous at the apex. The cribellum is narrow, transverse, and divided into two parts (Fig. 273). The hind spinnerets are two-jointed; the second segment is long and furnished with a comb of long spinning tubes on the inner side (Fig. 273); the spinning tubes are easily broken off and are consequently frequently wanting. The postabdomen is prominent; the terminal segment is furnished with a fringe of long, curved hairs (Fig. 273). The calamistrum (Fig. 274) consists of a single series of long, slender, curved hairs; parallel with the calamistrum, in our species at least, there are two or more rows of similar hairs, which are arranged in a less regular manner, and which perhaps form a part of the calamistrum.

I have not yet succeeded in obtaining a satisfactory photograph of the hackled band.

Two genera have been found within the borders of the United States. They may be distinguished as follows:

- A. Median ocular quadrangle as long as, or longer than, broad. P. 290.   Œcobius
- AA. Median ocular quadrangle very much wider than long. P. 291.   Platœcobius

Genus ŒCOBIUS (Œ-co′bi-us)

This is the best known genus of the family. The cephalothorax is convex and moderately elevated. The legs are proportionately longer than in *Platœcobius*, the metatarsus and tarsus together of each of the legs far exceeding the carapace in length. Three species are known from our fauna and may be identified as follows:

- A. Carapace entirely dusky or black. Occurs in Arizona, California, and Mexico.   *Œcobius isolatus*
- AA. Carapace light with dark markings.
  - B. Posterior row of eyes moderately procurved (part of the medians advanced in front of a line along the caudal margins of the lateral eyes). Common in the southern United States.   *Œcobius parietalis*

Superfamily Argiopoidea

BB. Posterior row of eyes strongly procurved (no part of the medians advanced in front of a line along the caudal margins of the lateral eyes). Known only from Texas.
Œcobius texanus

The following well known species is discussed in more detail. The habits of the other ones are presumably similar.

Œcobius parietalis (Œ. pa-ri-e-ta'lis).—This is a small spider, measuring less than one eighth inch in length, which is found in crevices on the sides of buildings and of walls and within buildings. In fact the spiders of this genus are essentially domestic spiders. This fact suggested the generic name, which is from the Greek *oikobios*, living at home, domestic.

The cephalothorax and legs are a very pale yellowish or greenish white marked with dark bands and spots (Fig. 275); sometimes many of these spots are indistinct or wanting. The abdomen is light brown, marked with dark spots and many smaller white ones.

The webs of this spider are made over cracks in the sides of buildings and in angles; or are stretched over some slightly projecting object, as the head of a nail. The principal part of the web is a sheet of very fine silk, which is usually less than one inch in diameter. This sheet has a warp of direct lines, which support a filmy woof, and is often more or less star-like in outline. Beneath this sheet, there is either a tube within which the spider waits, or a second sheet, somewhat smaller than the outer one, upon which the spider rests. When the spider is disturbed, it runs with exceeding rapidity.

This species is the *Thalamia parietalis* of Hentz.

Fig. 275. ŒCOBIUS PARIETALIS

Genus PLATŒCOBIUS (Plat-œ-co'bi-us)

The only known species placed in this genus may be distinguished from species of *Œcobius* by its flatter carapace and stouter appendages. The eyes of the median quadrangle form a figure which is very much broader than long. The metatarsus and tarsus of each of the legs together are about the same length as the carapace.

Platæcobius floridanus (P. flor-i-da′nus).—A few examples of this species have been taken in Florida. The male has never been described. The females measure about one twelfth inch in length. The cephalothorax is pale grayish, black on its edges. The abdomen is grayish, blackish on the sides, and is not marked with silvery spots.

Family FILISTATIDÆ (Fil-is-tat′i-dæ)

The Filistatids (Fil-is-tat′ids)

In this family the cephalothorax is oval, longer than broad. The eyes are massed in a small group, which is hardly wider than long; the anterior median eyes are dark in colour and round, the others are pearly white and oval or angular (Fig. 276). The cheliceræ are small and lack a lateral condyle; they are chelate, the short claw being apposed by a prolongation of the basal segment (Fig. 277). The palpus of the male is the most simple found among spiders.

The filistatids comprise a small group of spiders which are tropical or subtropical in distribution. The familiar genus *Filistata* includes the larger members of the family and one of them, *Filistata hibernalis*, is a common house spider in the southern states. Two other genera, the representatives of which are much smaller, have been recorded from the United States. Our genera may be distinguished as follows:

A. Legs armed with numerous robust spines, those beneath the tibiæ and metatarsi paired. P. 294.   FILISTATA
AA. Legs almost completely devoid of spines, occasionally with a single ventral or lateral spine on the tibiæ.
  B. Carapace scarcely longer than broad; clypeus not fully as high as the length of the eye group; legs annulate in black. P. 301.   FILISTATINELLA
  BB. Carapace much longer than broad; clypeus higher than the length of the group of eyes; legs not distinctly annulate in black. P. 301.   FILISTATOIDES

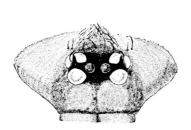
Fig. 276. EYES OF FILISTATA

Fig. 277. CHELICERA OF FILISTATA

Fig. 278. CALAMISTRUM OF FILISTATA

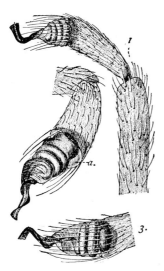
Fig. 279. PALPUS OF MALE OF FILISTATA

Fig. 280. FILISTATA HIBERNALIS

Superfamily Argiopoidea

Genus FILISTATA (Fil-is-ta'ta)

The more important characteristics of this genus are given in the above description of the family, to which may be added the following. The calamistrum is near the base of the fourth metatarsus and is very short (Fig. 278). In the adult male the calamistrum is wanting. The palpus of the male is comparatively

Fig. 281. THE HOME OF FILISTATA

simple in structure (Fig. 279); it is described in detail on page 108. The following is our best known species.

Filistata hibernalis (F. hi-ber-na'lis).—The larger individuals of this species (Fig. 280) measure from one half to five eighths of an inch in length. The legs are long, especially the first pair,

Superfamily Argiopoidea

which are about twice as long as the body. The colour of the body is usually a dark brownish black without markings. But I collected many specimens under stones at Austin, Tex., that appeared velvety black in some lights, in other lights they bore a lead-coloured tinge.

These are sedentary spiders which live under stones, in crevices about buildings, and in other similar situations. The

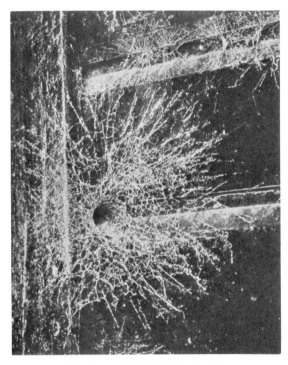

Fig. 282. A DETAIL FROM Fig. 281

spiders themselves are rarely seen, except by the collector, but their webs are often very conspicuous, especially in the extreme south. These webs are frequently built upon the sides of buildings and are more or less circular in outline, surrounding the opening of the retreat of the spider. Figure 281 shows the characteristic form and location of these webs; it represents the side of a neglected building at Miami, Fla. In Fig. 282 some of the same webs are shown less reduced. The webs of this spider

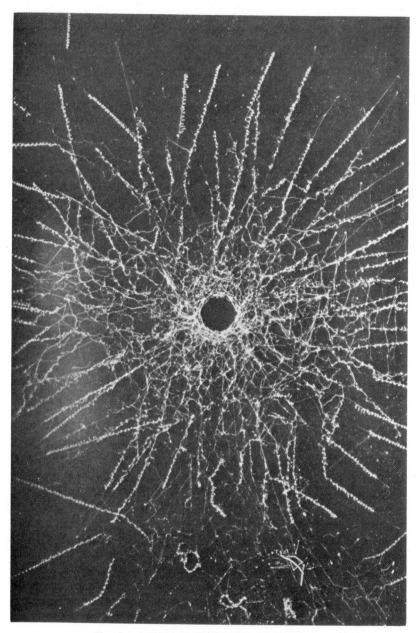

Fig. 283. A MADE-TO-ORDER WEB OF FILISTATA

Superfamily Argiopoidea

are rendered very conspicuous by accumulations of dust, which is caught and held by the hackled band of which the web is largely composed.

Under natural conditions the web is so quickly injured by insects and obscured by dust that its plan of structure is not easily seen. But this is well-shown in some made-to-order webs that were built by spiders in confinement in my laboratory. Figure 283 is from a photograph of one of these.

A shallow box was made, the cavity of which served as a retreat for the spider. In the centre of one face of the box a hole was made for the egress of the spider; and this face of the box was painted black so that if a web were built upon it it would make a good background for a photograph of the web. The box containing an active spider was then placed under a glass bell-jar. Several cages of this kind were prepared, and in every case where a spider that had been uninjured, by its trip from the South, was caged a web was built on the face of the cage around the opening in it, although, as a rule, several days elapsed before the web was begun.

Figure 284 represents the beginning of one of these webs. Additions were made to the web from time to time, but always in the night, the spider never leaving its retreat during the day-time except to capture an insect that had been caught in its web. Flies placed in the bell-jar were caught by the spider as soon as they touched the web, the spider rushing out from its retreat with great rapidity, and immediately carrying its prey into the interior of the box. Figure 283, which is from a photograph taken seventeen days later, represents this web in its most perfect condition; later the regularity of the radiating lines was destroyed by the insects that were given the spider for food; the web presenting an appearance like that shown in Fig. 282, which is the usual appearance of the webs of this species built under natural conditions.

In building this web an irregular net-work of lines was made about the opening of the retreat; this net-work forms the central portion of the web, and is of such structure that any disturbance of the web sets it in vibration. It consists chiefly of plain threads which are fastened to the supporting surface by a few attachment disks but are not closely applied to this surface, so that they can be readily set in motion. The net-work is continued as a tube

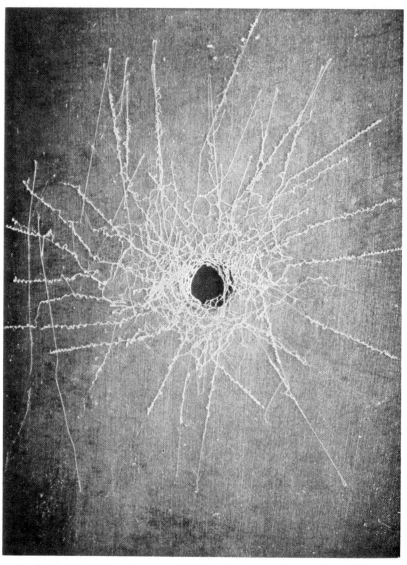

Fig. 284. THE BEGINNING OF A MADE-TO-ORDER WEB OF FILISTATA

Superfamily Argiopoidea

which extends through the hole in the board upon which it is built; but is considerably smaller than this hole, so that it can easily vibrate in it (Fig. 284). On the inner surface of the board the net-work is continued as a sheet resembling somewhat that on the outer surface. It is probably upon this sheet that the spider waits for its prey. Here it is concealed from sight but is in a position to detect any vibration of the web.

But the most characteristic feature of the web is a series of radiating lines, which consist of a doubled plain thread supporting a looped hackled band. In making these lines the spider spins a thread of plain silk, which consists of several parallel strands,

Fig. 285. HACKLED BAND OF FILISTATA

from near the centre of the web to a distant point, where it is fastened by an attachment disk; the spider then returns to the starting point spinning as it goes another similar thread closely parallel to the first. Upon these two threads, which serve as a foundation, are fastened afterward loops of a hackled band. This doubled supporting line and the loops of the hackled band can be seen with the unaided eye, and are shown in Fig. 283. A small section of one of the radiating lines is shown greatly enlarged in Fig. 285. This picture is from a photomicrograph and is not as perfect as could be desired; for with the high magnification necessary to see the details it was impossible to get all parts of a loop in focus at once; hence each loop appears blurred in a part of its course.

Four kinds of silk enter into the formation of this remarkable structure. *First,* — the doubled supporting line; this appears

Superfamily Argiopoidea

as a single thread in Fig. 285; but its double nature is shown in the part not covered with the hackled band near the outer end of each line (Fig. 284); each of the two parts of this double thread consists of several parallel strands; this can be seen by slight magnification of that part where it is flattened near the attachment disk. *Second,* — the primary looped threads; there are two of these, and they form the axis of the hackled band; they are extremely elastic. *Third,* — the secondary looped threads; there is one of these supported by each of the two primary looped threads; each of the secondary looped threads forms a very regular series of loops, each of which is fastened by one end to the primary looped thread; this secondary thread is not looped around the primary thread as it appears to be, but is merely fastened to one side of it by viscid silk. *Fourth,* — the viscid silk; this is an amorphous sheet, which fills the spaces between the loops of the secondary looped thread; it is largely liquid, but when it is highly magnified irregular threads can be seen in it.

It is easy to infer the function of these four kinds of silk: the supporting line not only supports the parts fitted for entangling the prey but communicates to the centre of the web, where the spider is lying in wait, any disturbance of the web; the primary looped threads also have two functions, they support the secondary looped threads and by their elasticity allow an entangled insect to become involved in other threads; I have seen these threads stretch to fifty times their first length; the secondary looped threads support the viscid silk; and the viscid silk clings to anything that touches it.

An interesting problem to be solved is the determination of the source of each of the four kinds of silk in the hackled band and its support, and the method of spinning the hackled band. The doubling of the supporting thread shows that it is spun separately. In the upper part of Fig. 284 are shown several places where the two parts of the supporting thread are widely separated; in these cases the spiders did not return to the starting point after making the attachment disk. It seems probable that when the hackled band is applied to this supporting thread the primary loops are made by a movement of the spinnerets, and that at the same time the secondary loops are formed by the calamistrum. During this operation silk is issuing from four spigots; from two of them comes the elastic silk that forms the

Superfamily Argiopoidea

primary loops, and from two others the silk that is combed by the calamistrum into the secondary loops; in this way a double band is formed, and to each half is applied a sheet of viscid silk. On the inner surface of each fore spinneret there is a series of flattened hairs (Fig. 286) which appear as if they were for the purpose of conducting the silk from the cribellum to the hackled band, the series of each side leading to one of the parts of the double band.

Genus FILISTATOIDES

The appendages are almost completely devoid of spines. The sternum is longer than broad. The clypeus is higher than the length of the eye group.

Fig. 286. SPINNERET OF FILISTATA

Filistatoides insignis (F. in-sig'nis).—This is a very small species, adult females measuring about one sixth inch in length. The cephalothorax is pale yellowish and is marked with a median longitudinal dark stripe. The legs are concolorous with the cephalothorax and are not ringed with black. The abdomen is gray or white, with a median stripe of dark spots and a line of smaller spots on each side.

This Mexican species is not uncommon in southern Texas.

Genus FILISTATINELLA (Fil-is-ta-ti-nel'la)

The appendages are only sparsely provided with true spines, and there are no paired spines beneath the first and second tibiæ and metatarsi. The cephalothorax is slightly longer than broad. The clypeus is narrower than the height of the group of eyes. The anterior median eyes are slightly smaller than the posterior median eyes. The sternum is only slightly longer than broad.

Filistatinella crassipalpus (F. cras-si-pal'pus).—Mature examples of this species are even smaller than the previous one, measuring only one twelfth inch in total length. The cephalothorax and abdomen are brown. The legs are marked with conspicuous black rings. The tibia of the male palpus is incrassated.

This species occurs in Texas and Arizona.

Superfamily Argiopoidea

Family ZOROPSIDÆ (Zo-rop'si-dæ)
The Zoropsids (Zo-rop'sids)

The zoropsids are eight-eyed, cribellate spiders which resemble superficially some of the amaurobiids, but their tarsi bear only two claws. These spiders also resemble in general appearance some of the gnaphosids and clubionids, for example *Syrisca* and *Syspira*, spiders which are not provided with a cribellum and a calamistrum. The tarsi and metatarsi are densely scopulate beneath, particularly the anterior pairs, and the tarsi have terminal claw tufts. The legs are set with strong spines, those beneath the tibiæ and metatarsi paired. The eyes are placed in two lightly procurved rows and differ little in size, the anterior median eyes usually being slightly smaller. The posterior spinnerets are somewhat longer than the anterior pair, two-jointed, the distal segment short and conical. The cribellum is divided.

Three or four species of *Zorocrates*, the only genus of these spiders known from the New World, have been found in the southwestern United States. Other species are known from Lower California and Mexico. Very little is known of the habits of this group of spiders. They are said to live under stones and spin webs which resemble those made by *Amaurobius*.

Genus ZOROCRATES (Zo-ro-cra'tes)

The spiders of this genus are of average to large size. The cephalothorax is longer than broad, convex, and the median groove is longitudinal. The first row of eyes is lightly procurved, the eyes equidistantly spaced, and the median eyes slightly smaller than the lateral eyes. The second row is somewhat wider than the first and the eyes are subequal in size, the median pair somewhat nearer to each other. The lower margin of the furrow of the chelicera is armed with three or four teeth. The legs have four or five pairs of spines beneath the tibiæ.

Species of this genus are known from Texas, New Mexico, and Arizona. The following one will serve as an example.

Zorocrates æmulus (Z. æm'u-lus).—The females are about one half inch long, the males somewhat smaller. The cephalothorax is dusky yellow and the appendages are similar in colour, without dark spots or rings. The abdomen is gray with black markings above. The chelicerae are armed with three stout teeth on the lower margin. The first tibia has four pairs of ventral spines.

This species is not uncommon in the lower Rio Grande River Valley of southern Texas.

Family DYSDERIDÆ (Dys-der'i-dæ)

The Dysderids (Dys-der'ids)

The dysderids are six-eyed spiders which have four conspicuous spiracles near the base of the abdomen, a pair of lung-slits and a pair of tracheal spiracles (Fig. 130). They can be distinguished from the members of the following families, which also have only six eyes and possess four spiracles, by the form of the coxæ of the first two pairs of legs, which are long and cylindrical (Fig. 130).

This family is represented in our fauna by a single species of the genus *Dysdera*, *D. crocata*, a spider which is widely distributed in Europe and Asia as well as in the New World.

Genus DYSDERA (Dys-de'ra)

Our only species of this genus is readily distinguished from the genera of various related families by the fact that the lateral extensions of the sternum completely surround the coxa of each leg. The third claw is lacking on the tarsi and claw tufts are present. The bulb of the palpus of the male (Fig. 99) is of an unusual form. It is described in an earlier chapter (p. 109).

Dysdera crocata (D. cro-ca'ta).—This is an orange-brown species with a pale abdomen. The body is one half inch in length, and the more striking features are represented in Fig. 130. This spider lives under stones and in similar situations. It is widely distributed in the United States but is apparently local. Simon states that the species of *Dysdera* enclose themselves in an oval, depressed sac of close, firm tissue, and that the female deposits her eggs there without enveloping them in an egg sac.

Family CAPONIIDÆ (Ca-po-ni'i-dæ)

The Caponiids (Ca-po-ni'ids)

The members of this family resemble the dysderids and related families in having four spiracles near the base of the abdomen.

Superfamily Argiopoidea

However, both pairs of openings lead into tracheal tubes, no booklungs being present. In this respect they resemble the rare, exotic spiders of the families Telemidæ and Symphytognathidæ. A distinguishing peculiarity of most of the genera of the Caponiidæ is the presence of only two eyes, a condition found in no other members of the order. In two genera from Africa the full complement of eight eyes is present, and in one of the American genera, *Nopsides*, there are four eyes. The labium is joined to the oval sternum, the juncture marked by a shallow transverse depression. The endites are not much longer than the labium and converge around it, the tips nearly touching.

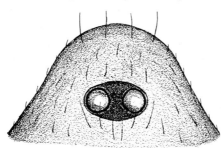

Fig. 287. EYES OF ORTHONOPS GERTSCHI

The legs bear no true spines. The tarsi are provided with two or three claws of which the paired claws are pectinate. The tarsi are provided ordinarily with one or several false articulations. In our genera the anterior metatarsi bear a translucent keel along their ventral line and have a translucent apophysis at the base of the tarsi.

The abdomen is oval and clothed with a light covering of black hairs. Six spinnerets are present. The anterior and middle pairs are placed close together in a transverse row. The palpus of the male (Fig. 288) is similar to that found in the Segestriidæ.

These spiders are found under stones in the arid and semi-arid parts of our southwest. Two genera occur in our fauna.

A. Cephalothorax ovate, only three fourths as broad as long; sternum distinctly longer than broad. P. 304.
ORTHONOPS

AA. Cephalothorax subround; sternum about as broad as long. P. 305.
TARSONOPS

Genus ORTHONOPS (Or'tho-nops)

The ovate cephalothorax is considerably narrowed in front, moderately convex, and is not marked by a median groove. The sternum is three fourths as broad as long and separates the fourth coxæ by their width. The two eyes are set back from the clypeal margin by two full diameters or more. All the coxæ are robust. The legs are of moderate length and are devoid of true spines. The median claw is present on all the tarsi.

Orthonops gertschi (O. ger'tschi).—Females are one half inch in length but the males are smaller. The carapace is bright orange and provided with inconspicuous black hairs.

The eyes are placed on a black spot. The legs are yellow to light orange, unspined, but clothed with black hairs. The abdomen is gray to pale yellow. The bulb of the male palpus is prolonged into a heavy, curved spine. This species was described from Utah.

Genus TARSONOPS
(Tar'so-nops)

The cephalothorax is much broader, nearly as broad as long, narrowed in front, and more depressed than in *Ortho-*

Fig. 288. PALPUS OF ORTHONOPS GERTSCHI

nops. The eyes are set back from the clypeal margin by a little more than their diameter. The sternum is subround, only slightly longer than broad. The tarsus and metatarsus are provided with two or three distinct false sutures.

Tarsonops systematicus (T. sys-te-mat'i-cus).—Females average about one third of an inch in length. The males are somewhat smaller. The cephalothorax and legs are pale yellow and are clothed with black hairs. The abdomen is gray. The palpus of the male is similar to that of *Orthonops gertschi*, but the embolus is a short, stout spur.

This species has been recorded from southern Texas. It was described from Lower California from which locality several other closely related species are known.

Superfamily Argiopoidea

Family SEGESTRIIDÆ (Se-ges-tri'i-dæ)

The Segestriids (Se-ges-tri'ids)

The segestriids were formerly placed in the same family with *Dysdera* and related genera of six-eyed spiders. Four conspicuous spiracles are present near the base of the abdomen, the posterior pair the openings to tracheal tubes. Six eyes are present. They are in three groups of two each, placed in two rows, and are well separated, the median eyes and the posterior lateral eyes forming a straight or recurved row. The coxæ are all robust. The sternum is longer than broad and lacks the lateral extensions characteristic of *Dysdera*. The first three pairs of legs are directed forward. The legs are spinose, the first two pairs armed with numerous stout ventral spines. The tarsi bear three claws.

Six spinnerets are present. The anterior pair is robust, subcontiguous at their bases, and the colulus is distinct.

Three genera have been found in the United States. They may be separated as follows:

  A. Median eyes and posterior lateral eyes forming a strongly recurved line. P. 306.   Segestria
AA. Median eyes and posterior lateral eyes forming a nearly straight line.
    B. Sides of the head with a row of stridulating ridges. P. 307.   Citharoceps
  BB. Sides of the head smooth. P. 307.   Ariadna

Genus SEGESTRIA (Se-ges'tri-a)

This and the following two genera are closely allied in structure and in habits. They differ in the position of the median eyes as given in the table above. Three or four species are known from the United States, all from the Pacific Coast. The following is our best known species.

Segestria pacifica (S. pa-ci'fi-ca).—This species is quite common along the Pacific Coast. The cephalothorax is brown; the abdomen is nearly white, with reddish brown spots on the middle line, and scattered ones on the sides. The length of the body is about one third inch. The male palpus is prolonged into a long, curved spine.

Genus CITHAROCEPS (Cith'a-ro-ceps)

This genus is closely allied to *Ariadna* in structure and closely similar to it in appearance. The head is provided with a series of conspicuous stridulating ridges. The first femur bears a short cusp which presumably is scraped across the patch of ridges. The following, our only known species, was described from California.

Citharoceps californica (C. cal-i-for'ni-ca).—This spider is one third inch in length. The cephalothorax is chestnut brown, the legs orange brown, and the abdomen is purplish gray above. Only females have been described. They live under the loose bark on Eucalyptus and other trees.

Genus ARIADNA (Ar-i-ad'na)

In both this and the preceding genus the body is more elongate and more nearly cylindrical than in *Dysdera*, the result of their living in slender tubes. Another striking peculiarity is that the third pair of legs, as well as the first and second, is directed forward. In this genus the median eyes are situated between the posterior lateral. The palpus of the male closely resembles that of *Loxosceles* (p. 107). The following is our only species:

Ariadna bicolor (A. bi'col-or).— This spider measures about one third inch in length. The cephalothorax and legs are yellowish brown, and the abdomen purplish brown.

The habits of this species are very remarkable. I collected the specimens that I studied at Agricultural College, Mississippi, in March. They were hibernating, each in a long, slender tube in a crack in a boat house. I brought them alive in vials to Ithaca, where they were left unnoticed till the middle of May. I then placed each in a hole in a block made by nailing together several small blocks. The face of the block was painted black, so as to render any silk that might be spun by the spiders more conspicuous. My efforts were rewarded; for each spider made a nest for me the very first night; nests of such marvellous engineering skill that I have never ceased to wonder at them.

Superfamily Argiopoidea

A long, slender tube is built in the hole that serves as a retreat for the spider. This tube is suspended from a framework of threads, built at the entrance of the retreat, in such a way that any disturbance of the exposed parts of the nest is communicated to the occupant of the tube.

A fine sheet of silk is made surrounding the entrance of the tube like a collar (Fig. 289). The tubular part of the nest does not show in the figure on account of the darkness of the retreat.

The most striking feature of the nest, however, is a series of radiating lines, which begin in the framework supporting the outer end of the tube, and extend out a considerable distance

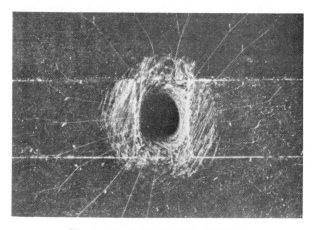

Fig. 289. NEST OF ARIADNA BICOLOR

from the nest. Each of these lines passes over two or more piers, which keep it suspended a short distance from the face of the block. There is a set of piers on the edge of the hole in which the tube is built, these are not well-shown in the figure but are very conspicuous in the specimens; and there is another set a considerable distance beyond the outer edge of the collar of the tube.

The radiating lines being held some distance from the face of the block by the supporting piers, are in position to be disturbed by any passing insect. And this disturbance is communicated to the framework supporting the tube. The spider waits within the tube with six of its eight legs projecting forward ready to make a leap. The touching of one of the trap lines by an insect results like the touching of the spring of a jack-in-the-box. The spider comes forth with amazing swiftness, seizes the unlucky insect, and retreats with it instantly to its lair.

About the first of July one of my specimens removed all of that part of the nest that was on the face of the block, leaving only the tube within the hole. She then laid about fifteen eggs. These were large and were in a spherical mass but not enclosed in an egg-sac.

Family OONOPIDÆ (O-o-nop'i-dæ)

The Oonopids (O-o-nop'ids)

These are six-eyed spiders which resemble the dysderids and segestriids in having four spiracles, a pair of lung-slits and a pair of tracheal spiracles. But in this family the tracheal spiracles are very inconspicuous, difficult to distinguish because of the small size of all the species. The coxæ of all the legs are similar, nearly globose, narrow at their base and at their juncture with the trochanters. The six eyes are usually close together in a compact group, but occasionally they are more widely separated. The position of the lateral eyes and the curvature of the rows are of importance in the determination of the genera. In some members of the family the abdomen is relatively soft, clothed with pale hairs. In other species the dorsum of the abdomen is ornamented with a strongly chitinized, coloured scutum or shield. The venter is often provided with similar hard plates.

Fig. 290. ORCHESTINA SALTITANS

The oonopids are all very small, most of them measuring less than one sixth of an inch. Numerous species occur in the tropics and subtropics. Most of those known from the United States come from the extreme southern parts of the country. Because of their small size and the choice of a habitat in leaf mold and debris, they are collected ordinarily by sifting these materials. Seven genera have been recorded from within our borders. One of them, *Gamasomorpha*, represented by a single species from Florida, *G. floridana*, is not considered in the following key.

Superfamily Argiopoidea

TABLE OF GENERA OF THE OONOPIDÆ

A. Abdomen relatively soft, lacking a more or less extensive hard covering or scutum.
  B. Ocular area subround, the median eyes placed behind the anterior lateral eyes and forming with them a strongly procurved row.
    C. Carapace pale, concolorous with the legs which have weak spines; first femur nearly equal in length to the width of the carapace; abdomen soft, set with long hairs. P. 311. OONOPS
    CC. Carapace brown, the legs much paler, without spines; first femur equal in length to two-thirds the width of the carapace; abdomen lightly chitinized, with short hairs. P. 311. STENOONOPS
  BB. Ocular area transverse, the median eyes widely separating the anterior lateral eyes and with them forming nearly a straight row. P. 310. ORCHESTINA
AA. Abdomen armed above with a hard plate which is usually reddish brown in colour.
  B. Dorsal scutum a small plate which covers the middle portion of the basal half of abdomen. P. 311. ISCHNOTHYREUS
  BB. Dorsal scutum covering the whole abdomen (divided longitudinally in females of *Scaphiella*); ventral scutum well developed.
    C. Anterior lateral eyes contiguous; posterior eyes in a lightly procurved row; dorsal scutum of female divided longitudinally. P. 312. SCAPHIELLA
    CC. Anterior lateral eyes well separated; posterior eyes in a recurved row; dorsal scutum of female entire, not divided longitudinally. P. 312. OPOPAEA

Genus ORCHESTINA (Or-ches-ti'na)

The abdomen is soft, elevated, nearly round in shape. The median eyes widely separate the anterior lateral eyes, forming with them a straight, transverse row. Three species have been described from the United States of which the following one is representative. Two other species occur in Utah.

Orchestina saltitans (O. sal'ti-tans).—This species occurs in the northeastern states and is often found in houses and buildings. It measures less than one twentieth inch in length; the cephalothorax is whitish, with a black marginal line and a black spot around the eyes. The abdomen is purplish above, quite dark near the tip; the venter is pale; and the spinnerets are white. When disturbed the spider leaps backward.

Genus OONOPS (O'o-nops)

The species of this genus are pale yellow spiders with moderately long legs. The legs are often provided with weak spines. The abdomen is soft. The six eyes are placed in a group which is subround, the anterior lateral eyes being at least partially in front of the median eyes. The four posterior eyes form a recurved row.

Several species have been described from Florida and Texas. The following one is representative of the group.

Oonops floridanus (O. flor-i-da'nus).—Both sexes measure about one twelfth inch in length. The cephalothorax is pale yellow, and the appendages are nearly concolorous. The abdomen is white or gray. The whole animal is clothed with fine black hairs. The bulb of the male palpus is incrassated and terminates in a short process.

This species is common in Florida.

Genus STENOONOPS (Sten-o'o-nops)

The spiders of this genus are a little flatter, stouter, and have shorter legs than species of *Oonops*. The eyes are essentially as in that genus, but the posterior row is usually nearly straight. The legs are not armed with spines. The following, our only known species, was described from Florida.

Stenoonops minutus (S. min-ut'us).—The cephalothorax is reddish brown, and the legs are pale yellow in colour. The abdomen is pale yellow, lightly chitinized, but without a dorsal scutum. The posterior row of eyes is gently recurved.

Genus ISCHNOTHYREUS (Isch-no-thyr'e-us)

In this and the following genera the abdomen is provided above with a hard plate or scutum. In this genus the scutum is

Superfamily Argiopoidea

rather small and extends back only half the length of the abdomen. The venter lacks a hard plate. Our single species was found in Florida.

Ischnothyreus barrowsi (I. bar′rows-i).—The female is about one twelfth inch in length. The anterior lateral eyes are contiguous, and the posterior eyes are in a lightly procured row.

Genus OPOPÆA (O-po-pæ′a)

Both the dorsal and ventral surfaces of the abdomen are provided with hard plates. The posterior row of eyes is recurved and the anterior eyes (lateral eyes) are well separated. The dorsal scutum of the female is entire, not divided longitudinally.

Several species are known from the United States of which the following is representative.

Opopœa meditata (O. me-di-ta′ta).—The coloration is bright reddish brown. The abdomen is completely covered above with a convex plate. The plate on the venter is large, covering most of the surface, and the spinnerets are encircled by a narrow, chitinous ring. The posterior eye row is lightly recurved, and the median eyes slightly exceed the lateral eyes in size. The anterior eyes are separated by scarcely their diameter.

This species is found in Texas.

Genus SCAPHIELLA (Sca-phi-el′la)

In the females of this genus the chitinized covering of the abdomen is divided longitudinally above, leaving a narrow soft band which is clothed with hairs. The scutum is continuous on the sides and around the venter. In the males there is a rather narrow dorsal plate the length of the abdomen in addition to the lateral plates which are continuous beneath. The posterior row of eyes is procurved. The following species has been taken in Texas and California.

Scaphiella hespera (S. hes′per-a).—Both sexes measure about one twelfth inch in length. The whole body is bright orange or reddish brown, and is clothed very sparsely with short hairs. The anterior eyes are close together, separated by scarcely their radius. The posterior row of eyes is recurved, the eyes subcontiguous and the median somewhat larger than the lateral eyes. The tarsus

Superfamily Argiopoidea

of the male palpus is greatly incrassated, and the embolus is a thin spine.

These pretty oonopids are relatively uncommon. They occur in detritus and under stones in the arid southwestern states.

Family SCYTODIDÆ (Scy-tod'i-dæ)

The Scytodids (Scy-tod'ids)

The scytodids belong to that portion of the series of six-eyed spiders in which there is only a single tracheal spiracle. They differ from the following family in having the chelicerae soldered together at the base and by the fact that there is not a distinct suture between the labium and the sternum. The eyes are pearly white. This family is often referred to as the Sicariidae.

In two of the genera placed in this family there are eight eyes; but the simplicity of the external reproductive organs which is characteristic of this series, as well as other family characteristics, distinguishes these genera.

Five genera are represented in our fauna; these can be separated by the following table.

TABLE OF GENERA OF THE SCYTODIDÆ

A. With eight eyes.
  B. First femur robust, curved, incrassated, shorter than the carapace, with few or no dorsal spines. P. 313.
                                                         PLECTREURYS
  BB. First femur slender, essentially straight, longer than the carapace, the dorsal spines numerous. P. 314. KIBRAMOA
AA. With six eyes.
  B. Anterior row of eyes in a nearly straight line. P. 314.
                                                         DIGUETIA
  BB. Anterior row of eyes very strongly recurved.
    C. Cephalothorax low or depressed. P. 315. LOXOSCELES
    CC. Cephalothorax high, subglobose. P. 316. SCYTODES

Genus PLECTREURYS (Plec-treu'rys)

These spiders and those of the following genus differ from all scytodids in having eight eyes. These are placed in two, essentially straight rows. The anterior median eyes are smaller and

Superfamily Argiopoidea

somewhat nearer the lateral eyes. All the legs are stout, the first one much more robust and longer than the others. The first femur is thickened and shorter than the carapace. The first tibia of the male has a stout distal spur on the outer side which supports a heavy spine. There are three tarsal claws.

Two species are known from the southwestern United States.

Plectreurys tristis (P. tris'tis).—This species is found in Utah, Arizona, California, and in adjacent Mexico. The cephalothorax and legs are brown or black, while the abdomen is gray, clothed with isolated hairs. The first leg has paired ventral spines and rows of sublateral spines. The tube of the male palpus is a long spine.

Plectreurys castanea (P. cas-ta'ne-a).—This species is found in California. It is paler and somewhat smaller than *tristis*. The first legs are armed with paired ventral spines but lack sublateral spines. The tube of the male palpus is very thick and abruptly angled near the middle.

Genus KIBRAMOA (Ki-bra-mo'a)

The species of this genus differ from *Plectreurys* in the following respects: The legs are proportionately much longer and more slender, the first femur being longer than the carapace, essentially straight, not at all incrassated. The first tibia of the male lacks the conspicuous distal tubercle and stout spine present in *Plectreurys*. Only two species are known, both from the western part of our country. The following will serve as an example of the genus.

Kibramoa suprenans (K. su'pre-nans).—Both sexes are similar in appearance and average about one half inch in length. The cephalothorax and legs are chestnut brown. The abdomen is gray with a greenish cast and marked with a pale median stripe above at the base. The first femur of the female is slightly longer than the carapace. The bulb of the male palpus is thickened, subround, and the embolus is a long, curved spine.

This species was described from California.

Genus DIGUETIA (Dig-u-e'ti-a)

The members of this genus differ from the other six-eyed scytodids in having the anterior eyes in a nearly straight line. In many respects this genus resembles *Plectreurys* more than any of the other scytodids. The palpus of the male is more complicated than usual in the family. A conductor of the embolus is present.

Several species have been found in the southwestern part of our country and others occur in Mexico. They are all very similar in appearance and structure. The best known species is the following one which occurs in the southern Rocky Mountain states and west to California.

Diguetia canities (D. ca-nit'i-es).—The spider is elongate, about three eighths inch long. The cephalothorax is brown, and the abdomen is somewhat paler. Both are thickly covered with white hairs which form distinctive bands. The legs are yellowish and are ringed with brown or black bands. The first legs of the males are often uniform brown or black.

The species was described from specimens taken at San Bernardino, California. Dr. McCook ('89–'93, 11, 135) gives an account of its remarkable cocooning habits. The mother spins a series of flattened disks, which are overlaid one upon another like the tiles upon a roof, and are bound by silken threads somewhat after the fashion of *Metepeira labyrinthea*. This series of cocoons is sometimes three inches or more in length and is covered with leaves from the plant upon which the string is suspended. Along the entire length of one side of the string of cocoons the mother spins a silken tube within which she dwells. The whole is suspended within a maze of threads and is attached above to a strong thread.

Genus LOXOSCELES (Lox-os'ce-les)

These are spiders of medium size, of a yellowish or brownish colour and without conspicuous markings. The cephalothorax is low and depressed (Fig. 292). The anterior row of eyes is strongly recurved (Fig. 291). The palpus of the male is rather simple and is described on page 109. These spiders are found under the bark of trees and under stones on the ground. Simon states that the webs which they spin are quite large and very irregular, resembling those of *Filistata;* and that the threads have the appearance of those spun by spiders having a cribellum and calamistrum. He even suggests that the colulus (Fig. 142, p. 136) may play a role analogous to that of the cribellum. I have observed our species only in winter. At that season I found them common under stones in Texas. The spiders were frequently found in a silken sac, which

Superfamily Argiopoidea

may be a retreat for the winter, like that spun by the jumping spiders. The silk of which these sacs are made appears to be hackled.

Four or five species have been found in the United States. All of these are very similar in appearance and differ chiefly in the eye relations, the comparative leg lengths, and in the palpi and epigyna. The following is a well known species.

Loxosceles unicolor (L. u'ni-col'or).—The sexes are of about equal size and measure one third inch in length. The legs are very long, the fourth one more than five times as long as the carapace in females and more than seven times as long as the carapace in the males. The fourth leg is slightly longer than the first, and the second leg is longer than either of these.

Montgomery ('08) states that a species of this genus makes a large and irregular web beneath logs and stones, usually in drier

Fig. 291.
EYES OF LOXOSCELES

Fig. 292.
PROFILE OF CEPHALO-
THORAX OF LOXOSCELES

Fig. 293. LOXOSCELES RUFESCENS

situations; and he describes the egg-sac, which is discoidal, with a diameter longer than the spider's body. It is sessile, attached to the snare.

Genus SCYTODES (Scy-to'des)

The spiders of this genus are of small or medium size, of a pale yellow or white color, ornamented with black or gray spots. A few species are nearly uniform black in color. The anterior row of eyes is strongly recurved. The cephalothorax is high and subglobose behind and slopes forward (Fig. 294).

These are tropical or subtropical spiders which normally live under stones or rubbish on the ground, where they spin a small irregular web. Several species are common house spiders in the tropics and have extended their ranges northward. A number of species have been observed in the United States. The following are well known.

Scytodes thoracica (S. thorac'i-ca).—This species is not uncommon, even in the North,

Fig. 294. PROFILE OF CEPHALOTHORAX OF SCYTODES

Fig. 295. SCYTODES THORACICA

where it is found in cellars and closets. It is also found in Europe. This spider measures about one fifth to one fourth inch in length. The cephalothorax is light yellow, and the abdomen white, both marked with black spots and bands as illustrated in Fig. 295.

Scytodes intricata (S. in-tri-ca'ta).—This is a larger species, the adult female measuring about one third inch in length. The body is yellow, mottled with brown. The brown markings form a close net-work on the cephalothorax, but on the abdomen the spots are more distinct. On the ventral side of the abdomen there is a prominent V-shaped black marking with the apex behind. In the adult female there is a pair of dark brown chitinous plates which extend back from the epigastric furrow.

This species is common in Texas and also occurs in tropical America. An allied species (*Scytodes longipes*) is a common house spider of the tropics. In the Bermudas where I found it common on the walls of a room, it is known as the dust-spider. It moved slowly, and was easily captured. It has been taken in Florida.

Superfamily Argiopoidea

Several other species are known from the southwestern United States, but they are not well known.

Family LEPTONETIDÆ (Lep-to-net'i-dæ)

The Leptonetids (Lep-to-net'ids)

This family includes small six-eyed spiders, with long legs. Nearly all of the species live in caves or in dark situations. They differ from the scytodids, with which they agree in having six eyes and one tracheal spiracle, in having a distinct suture between the labium and the sternum. The chelicerae are free, and the fourth tarsus is supplied with spurious claws.

Only three genera are known to occur in our fauna. They can be separated as follows:

A. Anterior lateral eye of each side contiguous with an anterior median eye. LEPTONETA
AA. Anterior lateral eyes separate from the anterior median eyes.
 B. Lateral eyes forming two diverging lines. USOFILA
 BB. Lateral eyes not forming diverging lines. OCHYROCERA

Genus LEPTONETA (Lep-to-ne'ta)

This genus is well represented in Europe, but up to the present time only two species have been described from the United States. Fig. 296 represents the eye arrangement of one of them.

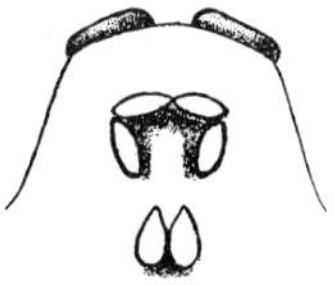

Fig. 296.
EYES OF LEPTONETA
(after Banks)

Leptoneta californica (L. cal-i-for'ni-ca).—Only the female has been described. It measures one tenth inch in length. The cephalothorax and sternum are red brown, both with a black margin. The eyes are on black spots. The abdomen is gray above, marked transversely with blackish behind.

Genus USOFILA (U-sof'i-la)

A single species has been found in this country.

Usofila gracilis (U. grac'i-lis).—This small spider measures but little more than one twenty-fifth of an inch in length. Its form is well shown by the accompanying figure by Dr. Marx ('91), who described the species (Fig. 297). It was found in caves in California.

Genus OCHYROCERA (Och-y-roc'e-ra)

A single species from Olympia, Washington, has been placed in this genus by Mr. Banks, but the validity of this placing has not been verified by recent workers because of the rarity of the species. Figure 298 represents the arrangement of the eyes in an exotic species of this genus.

Ochyrocera pacifica (O. pa-cif'i-ca).—This tiny spider measures about one twentieth of an inch in length. Nothing has been published regarding its habits.

Family PRODIDOMIDÆ (Prod-i-dom'i-dæ)

The Prodidomids (Pro-did'o-mids)

The prodidomids are two-clawed, eight-eyed spiders, with dissimilar eyes in three rows, and very robust cheliceræ, which are furnished with very long and slender claws (Fig. 299). They are very rare spiders. They live under stones and in other dark and dry situations. Our two genera, each represented by a single species, can be separated as follows:

A. Upper spinnerets much stouter than the lower; anterior median eyes larger than the lateral. P. 319. PRODIDOMUS
AA. Upper spinnerets much smaller and more slender than the lower; eyes of first row subequal. P. 321. PERICURUS

Genus PRODIDOMUS (Pro-did'o-mus)

These are small spiders measuring from one twelfth to one sixth inch in length. Our only species occurs in the southern states of Alabama, Louisiana, and Texas.

Prodidomus rufus (P. ru'fus).—The cephalothorax is pale yellow and the abdomen is red or yellow with a pinkish tinge. The anterior median eyes are considerably larger than the lateral eyes as shown in Fig. 299. The interior pair of spinnerets is stout. They are separated by their diameter and contiguous at their bases. This species occurs in cellars and in dark closets in houses in the South. It is also found out of doors on the ground. It makes a tubular web or mesh of threads. For a detailed description of the species see Bryant ('35).

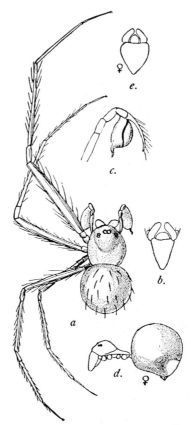

Fig. 297. USOFILA GRACILIS *a*, male, enlarged *b*, sternum, labium, and endites of male *c*, palpus of male *d*, female, enlarged *e*, sternum, labium, and endites of female

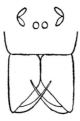

Fig. 298. EYES OF OCHYROCERA (after Cambridge)

Fig. 299. EYES AND CHELICERÆ OF PRODIDOMUS RUFUS (after Banks)

Genus PERICURUS (Per-ic′u-rus)

The eyes of the first row are subequal. The superior spinnerets are less robust than the inferior pair, the latter robust and contiguous.

Pericurus pallida (P. pal′li-da).—The cephalothorax and legs are light yellow. The abdomen is light gray, densely clothed with appressed hair. Immature examples have been taken in Arizona and California. A related species occurs in Lower California.

Family GNAPHOSIDÆ (Gna-phos′i-dæ)

The Gnaphosids (Gna-pho′sids)

The gnaphosids, a family formerly known as the Drassidæ, are those spiders which have eight eyes arranged in two rows and two tarsal claws in which the form of the body and the attitude of the legs are not those characteristic of the crab-spiders, and which differ from the clubionids in having the fore spinnerets widely separated. The tarsi are furnished with bundles of terminal tenent hairs (claw tufts) as in the Clubionidæ.

Most gnaphosids are found under stones or bark or in moss. A few live on the leaves of plants which they roll like the clubionids, but they do not construct a definite tube like that of the clubionids or at the most they spin an irregular retreat at the time of oviposition. Some of the species that live under stones make a silken sac within which they live and in which the egg-sac is made.

Our genera can be separated by the following table which is based almost entirely on one published by Dr. R. V. Chamberlin ('22).

TABLE OF GENERA OF THE GNAPHOSIDÆ

A. Lower margin of the furrow of the chelicera keeled or lobed.
  B. Lower margin of furrow with three contiguous, chitinous lobes. P. 333.   LARONIA
 BB. Lower margin of furrow with a single keel.
  C. Posterior row of eyes much longer than the anterior, strongly recurved, the median eyes usually nearer to each other than the laterals. P. 334.   GNAPHOSA
  CC. Posterior row of eyes but little longer than the anterior, the eyes equidistant or the medians farther from each other. P. 335.   CALLILEPIS

Superfamily Argiopoidea

AA. Lower margin of the furrow of the chelicera unarmed or with one to several ordinary teeth.
  B. Posterior row of eyes very strongly procurved, semicircular or nearly so.
    C. Lower margin of furrow of chelicera unarmed. P. 327. MEGAMYRMECION
    CC. Lower margin of furrow with one tooth. P. 328. SCOPODES
  BB. Posterior row of eyes not thus strongly procurved.
    C. Fourth tibia with two, or less commonly with three or five, median dorsal spines.
      D. Lower margin of furrow of chelicera with a single tooth. P. 325. GEODRASSUS
      DD. Lower margin with two or more teeth.
        E. Eye rows widely separated, the laterals on each side separated by a diameter or nearly so; posterior median eyes much nearer to each other than to the lateral eyes. P. 324. DRASSODES
        EE. Eye rows near together, the laterals on each side separated by no more than their radius; posterior median eyes well separated, only a little farther from the laterals than from each other.
          F. Fourth tarsi flexible, marked with false articulations; tarsal claws very long. P. 327. NEOANAGRAPHIS
          FF. Fourth tarsi and tarsal claws normal.
            G. Eyes of median quadrangle subequal in size. P. 329. SOSTICUS
            GG. Anterior median eyes smaller than the posterior medians. P. 328. RACHODRASSUS
    CC. Fourth tibia with a single dorsal spine or with none.
      D. Lower margin of the furrow of the chelicera with a single tooth or unarmed; bulb of male palpus with no apophyses, at most with one or several small teeth near base of embolus.
        E. Posterior row of eyes more or less recurved. P. 329. POECILOCHROA
        EE. Posterior row of eyes more or less procurved, sometimes straight.
          F. Lower margin of furrow of chelicera unarmed.

Superfamily Argiopoidea

      G. Eyes of first row subequal or the medians larger. P. 333.   LIODRASSUS
      GG. Anterior median eyes smaller than the lateral eyes. P. 328.   NODOCION
   FF. Lower margin of furrow of chelicera with one tooth; eyes of first row subequal.
      G. Posterior median eyes much farther from each other than from the laterals; body with conspicuous longitudinal black and white stripes. P. 333.   CESONIA
      GG. Posterior median eyes not at all or but little farther from each other than from the laterals; body not striped as above.
         H. Posterior median eyes well separated, smaller than or at most as large as the laterals. P. 332.   HERPYLLUS
         HH. Posterior median eyes close together and larger than the lateral eyes. P. 332.   LITOPYLLUS

DD. Lower margin of furrow of chelicera with two or three, or rarely only one, distinct teeth; bulb of male palpus with one or more apophyses.
   E. Upper margin of furrow of chelicera with three teeth, the lower with two, all well developed.
      F. Posterior median eyes large, oblique, typically close together but well removed from the laterals. P. 326.   HAPLODRASSUS
      FF. Posterior median eyes circular, their diameter or more apart. P. 326.   ORODRASSUS
   EE. Upper margin of furrow of chelicera with from four to six teeth, the lower margin with two or three, or rarely one.
      F. Posterior row of eyes straight or but little procurved with the eyes typically nearly equidistant, the median small or at most but little larger than the lateral eyes. P. 330.   ZELOTES
      FF. Posterior row of eyes procurved, the medians close together or contiguous, larger than the laterals, usually much so, oblique. P. 330.   DRASSYLLUS

Superfamily Argiopoidea

Genus DRASSODES (Dras-so'des)

The lower margin of the furrow of the chelicera is armed with two teeth. The eye rows are well separated, the laterals of each side being separated by a distance about equalling or else exceeding their diameter. The posterior median eyes are much nearer to each other than to the subequal lateral eyes.

The species of *Drassodes* are quite uniform in colour, varying from reddish brown to pale yellow, and they are clothed with soft, white, or yellow hairs. Sometimes the abdomen is marked with faint spots or chevrons.

These spiders are found under stones, more rarely under bark and in fissures of rocks.

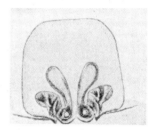

Fig. 300.
DRASSODES NEGLECTUS
FEMALE

Fig. 301. EPIGYNUM OF
DRASSODES NEGLECTUS

The members of this and several related genera were formerly placed in *Drassus*, a name which is now placed as a synonym of *Gnaphosa* and is thus unavailable for this or any other group. Three or four species of *Drassodes* are known from the United States, and others are known from Europe. Because of a close similarity in general appearance and structure, they are difficult to identify. The palpi of the males show little variation, the specific differences being in the finer details of the tibial apophyses.

The following is our best known species.

Drassodes neglectus (D. neg-lec'tus).—The adult female is nearly one half inch in length. It is light gray, with short fine hairs all over its body. The cephalothorax is very narrowly margined with black. The abdomen is either without markings, or with four spots on the muscle-impressions, between which there may be a longitudinal band, and on the hind half there may be faint chevrons (Fig. 300). The most distinctive characteristic is the form of the epigynum (Fig. 301).

The male is smaller and more slender. Figure 302 represents the palpus, as figured by Emerton; the characteristic features of this palpus are the great length of the tibia and the shape of the apophysis at the distal end of the tibia.

This is one of the most common members of the family. It lives under stones and other objects lying on the ground. It makes a large transparent bag of silk in which it lives, and within which the egg-sac is made. The female stays in this bag with the egg-sac until the spiderlings emerge from it. Emerton states that early in the summer a male and female often live together in the nest, even before the female is mature. I found a female in a bag with an egg-sac on July 20th; in this case the young emerged August 21st.

Fig. 302. PALPUS OF DRASSUS NEGLECTUS (after Emerton)

This is the *Drassus neglectus* of Keyserling and Emerton. Several closely related species are known, most of them from the far west.

Genus GEODRASSUS (Ge-o-dras'sus)

This genus is closely allied to *Drassodes*. The lower margin of the furrow of the chelicera is armed with a single small tooth. The eyes of the first row are subequal in size. The palpus of the male is very short, and the tibia completely lacks an apophysis.

Our best known species is the following.

Geodrassus auriculoides (G. au-ri-cu-loi'des).—Females are about one half inch long, and the males are nearly as large. The carapace is yellowish brown. The median eyes are suboval in shape and are set close together, removed from the lateral eyes by nearly twice their diameter. The abdomen is dusky white.

This uncommon species is found in the eastern part of the United States. It was described from Ohio.

Superfamily Argiopoidea

Genus HAPLODRASSUS (Hap-lo-dras'sus)

The fourth tibia lacks spines on the upper surface. Two teeth are present on the lower margin of the furrow of the chelicera, and there are three on the upper margin. The posterior median eyes are large and oblique, set close together, well removed from the lateral eyes. The epigynum of the female is typically large and presents a prominent chitinous ridge on each side (Fig. 303). Several species are known from our fauna. The one described below is a boreal form common in Canada and across the northern United States. It is also found in Europe.

Fig. 303. EPIGYNUM OF HAPLODRASSUS SIGNIFER

Haplodrassus signifer (H. sig'-ni-fer).—The adult female measures from one third to nearly one half inch in length. It is most easily recognized by the form of the epigynum (Fig. 303). This is light colored in the middle with a curved dark ridge on each side. The male is much smaller than the female. The tibia of the male palpus has an apophysis at the distal end that is long and laminate.

Genus ORODRASSUS (O-ro-dras'sus)

This genus resembles *Haplodrassus* closely in structure and general appearance. The posterior median eyes are circular and are separated by at least one full diameter. The epigynum of the female is large and has a conspicuous median septum which is attached to the floor of the atrium.

Several species have been described from the western United States. The following one is common in the Rocky Mountains.

Orodrassus coloradensis (O. col-o-ra-den'sis).—Both male and female are about one half inch in length, the female ordinarily somewhat larger. The cephalothorax and appendages are light to dark brown, and the abdomen is dusky brown. The tibia of the male palpus has an apophysis on the outer side which is short and bicornuate.

This is the *Drassus coloradensis* of Emerton and the *Teminius continentalis* of Keyserling.

Superfamily Argiopoidea

Genus NEOANAGRAPHIS (Ne-o-a-na-graph'is)

The fourth tibia has two dorsal spines on the midline. The lower margin of the furrow of the chelicera is armed with two teeth. All the legs are very long, the last pair five times as long as the carapace. The fourth tarsi are flexible, marked with false articulations in the distal half. The tarsal claws are very long and are armed only at the base with a few teeth. The eyes of the first row are subequal in size.

Neoanagraphis chamberlini (N. cham-ber-lin'i).—This, the only known species of the genus, is found in the White Sands area of New Mexico. The cephalothorax and legs are light yellowish brown, and the abdomen is gray. In general appearance this species resembles *Drassodes neglectus*. The palpus of the male bears a short curved apophysis on the tibia.

Genus MEGAMYRMECION (Meg-a-myr-me'ci-on)

The cephalothorax is ovate, with the front narrow, and with a long median furrow. The anterior eyes are close together, in a strongly procurved line, and the median eyes are larger than the lateral ones. The posterior eyes are in a very strongly procurved, semicircular line (Fig. 304) which is not at all longer than the anterior line. The chelicerae are of moderate size, with the lower margin of the furrow unarmed, and with two minute teeth on the upper margin. The fore spinnerets are very long and are furnished with nine or ten large spinning tubes placed in a semicircle.

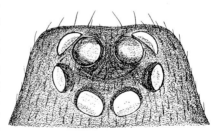

Fig. 304. EYES OF MEGAMYRMECION

This genus is essentially tropical and subtropical in distribution. Two or three species have been taken in the extreme southwestern United States.

Megamyrmecion californicum (M. cal-i-for'ni-cum).—The female measures from one fourth to one half inch in length. It is pale yellow or grayish in colour. The embolus of the male palpus is very long, arising near the base and passing distad along the inner margin of the cymbium. It occurs in California.

Superfamily Argiopoidea

Genus SCOPODES (Scop'o-des)

The single known species placed in this genus resembles *Megamyrmecion* in eye relations and general appearance. The lower margin of the furrow of the chelicera is armed with a single, minute tooth. The bulb of the male palpus has a stout median apophysis.

Scopodes catharius (S. ca-thar'i-us).—The cephalothorax and legs are light yellowish brown. The abdomen is gray, and there is a small yellow scutum at the base. Only the male of this species has been described. It came from Claremont, California.

Genus NODOCION (No-do'ci-on)

The lower margin of the furrow of the chelicera is completely unarmed as in *Liodrassus*. The anterior median eyes are considerably smaller than the lateral eyes. The median ocular quadrangle is narrowed in front, the anterior median eyes smaller than the posterior median. The cheliceræ are armed in front with a thick covering of stiff bristles.

Four or five species are known from the western United States, most of them from California. The following species is an example for the group.

Nodocion barbaranus (N. bar-bar-an'us).—The cephalothorax and legs are dark brown, and the abdomen is dusky gray to black. The posterior eye row is nearly straight, and the eyes are about equal in size and set close together.

This species is rather common along the Pacific Coast.

Genus RACHODRASSUS (Ra-cho-dras'sus)

Two spines are present above the tibia of the fourth leg. The lower margin of the furrow of the chelicera is armed with two teeth. The eyes of the first row are in a nearly straight line, the median eyes smaller than the lateral. The posterior eye row is gently recurved, essentially straight.

Three species have been described. The following is the best known one.

Rachodrassus echinus (R. e-chi'nus).—Females are about one fourth inch in length. The cephalothorax and legs are cloudy yellow. The abdomen is yellow with a dorsal pattern of obscure broken chevron marks. These spiders resemble superficially the clubionids of the genus *Agroeca*.

This species was described from Kentucky.

Genus SOSTICUS (Sos'ti-cus)

Two spines are present on the dorsal surface of the fourth tibia. The lower margin of the furrow of the chelicera is armed ordinarily with two teeth, but three are sometimes present. The posterior row of eyes is in a straight line. The anterior median eyes are only slightly smaller than the anterior lateral eyes. The median ocular quadrangle is somewhat longer than broad, and the front eyes are as large as or larger than the posterior median eyes. Two species have been described.

Sosticus continentalis (S. con-ti-nen-tal'is).—This is a dark spider and resembles the species of *Zelotes* in size and general appearance. The epigynum is elongate, and there is a prominent median lobe projecting caudad into the depression which is free for most of its length.

This spider occurs in the midwestern states.

Genus POECILOCHROA (Poe-ci-loch'ro-a)

The cephalothorax is narrow, oblong, slightly convex, and a little narrowed in front. The anterior eyes are close together, in a straight line, and with the median larger than the lateral eyes. The posterior eyes are small, in a slightly recurved line. The median ocular area is longer than broad, and narrower in front than behind. The cheliceræ are of moderate size, with the margins of the furrow unarmed.

About a dozen species are known from the United States. The following ones are well known.

Poecilochroa montana (P. mon-ta'na).—The female is about one third inch in length. The cephalothorax and legs are dark brown. The abdomen is black with a pair of white spots near the front end and another pair across the middle nearly united. The male measures about one fifth inch in length. The cephalothorax is dark brown, covered with white hairs. The abdomen is black with a narrow white band across the front end and a white band at the caudal end over the spinnerets.

This species occurs across the northern United States.

Superfamily Argiopoidea

Poecilochroa variegata (P. va-ri-e-ga'ta).—This brightly coloured species is easily recognized by the markings of the abdomen. The cephalothorax is bright orange, a little darker toward the eyes. The abdomen is black with three transverse white stripes and a T-shaped white mark between the first and second stripes. The female is one fourth inch in length.

This species is widely distributed in the East.

Genus ZELOTES (Ze-lo'tes)

The cephalothorax is ovate, very much narrowed in front, and furnished with a median furrow. The anterior eyes are near together, in a procurved line, and equal in size or with the median eyes a little smaller. The posterior eyes are in a straight or nearly straight line. The eyes of this row are typically nearly equidistant, the median relatively small or at most not much larger than the lateral eyes. The lower margin of the furrow of the chelicera is usually furnished with two or three teeth.

This is the largest genus of the family Gnaphosidæ, including several hundred described species. About a score have been found in our fauna. All of them are closely allied and agree in general appearance with the species described below.

Zelotes subterraneus (Z. sub-ter-ra'ne-us).—The female is nearly one third inch in length, and the male is nearly equal in size. This is a deep glossy black species without markings. The palpus of the male lacks ventral or lateral apophyses but has two chitinous ridges near the distal end of the bulb.

This, our most common and widely distributed species, is also found in Europe.

Genus DRASSYLLUS (Dras'syl-lus)

The species of this genus are closely allied to *Zelotes*. They differ chiefly in the details of the posterior row of eyes. These eyes are in a procurved row, and the medians are close together or contiguous, larger than the lateral eyes. Some of the more than fifty known species resemble *Zelotes*, but many of them are more brightly coloured. The following are well known.

Drassyllus rufulus (D. ru'fu-lus).—The length of the body varies from one fourth to one third inch in length. The colour is light reddish brown without markings, the abdomen paler than the cephalothorax. The form of the epigynum is shown in Fig. 305. Emerton states that the form of the epigynum varies in shape, that in some individuals the front of it is nearly straight. He has also figured the palpus of the male (Fig. 306).

Mr. Banks, who described this species, states that the egg-sac is attached to the under side of stones; and that it consists of two circular sheets of silk between which are placed the eggs. The outer sheet is often covered with dirt or mud so as to resemble the stones.

Drassyllus frigidus (D. frig'i-dus). —The cephalothorax is brownish yellow with the margin black; abdomen above and below blackish, with the two lines on the venter nearly parallel;

Fig. 305. EPIGYNUM OF DRASSYLLUS FRIGIDUS

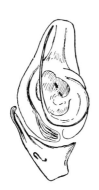

Fig. 306. PALPUS OF MALE OF DRASSYLLUS RUFULUS (after Emerton)

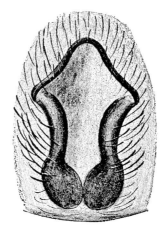

Fig. 307. EPIGYNUM OF DRASSYLLUS FRIGIDUS

the body is covered with black hairs. The form of the epigynum of the female is shown in Fig. 307. The length of the body of both male and female is a little less than one fifth inch.

Superfamily Argiopoidea

Genus HERPYLLUS (Her-pyl'lus)

In this genus, which is closely allied to *Zelotes*, the two rows of eyes are nearly straight and widely separated; the eyes of each row are quite evenly spaced. The anterior middle eyes are larger than any others. The posterior median eyes are a little smaller than the posterior lateral eyes.

Several species have been described from the United States; but only one of them is widely distributed.

Herpyllus vasifer (H. vas'i-fer).— The body is black with white or pinkish markings along the middle of the back (Fig. 308). In alcohol the light stripe on the cephalothorax turns to brown. The abdominal markings consist of a band on the basal two thirds and a spot near the tip. The band is slightly narrowed in the basal portion, and greatly and suddenly narrowed near the hind end. The lower side of the abdomen is dark at the sides and light in the middle; there are two narrow, parallel, faint lines extending from the epigynum almost to the spinnerets. The legs except the femora, are somewhat paler than the body.

Fig. 308. HERPYLLUS VASIFER

The adult female is one third inch in length, the male is much smaller, but is similarly marked.

This species is found under stones and rubbish on the ground, between boards, and in crevices in dark places. It runs with exceeding rapidity. I found its egg-sacs under boards in a barn; they are flat and snowy-white, resembling those of *Gnaphosa gigantea*.

This species is widely distributed throughout the United States; it is the *Herpyllus ecclesiasticus* of Hentz.

Genus LITOPYLLUS (Li-to-pyl'lus)

This genus is related to *Herpyllus* as indicated by the characters in the table. The median eyes of the posterior row are much larger than the lateral, nearer together, and the row is more strongly procurved. The lower margin of the furrow of the chelicera is armed with a single, very small tooth.

Litopyllus rupicolens (L. ru-pi-co'lens).—The female is one third inch in length, and the male is somewhat smaller. The cephalothorax and legs are pale yellowish brown, and the abdomen is nearly uniform gray. The tibial apophysis of the male palpus is distal in position and is very short.

This species occurs in the eastern United States.

Genus LIODRASSUS (Li-o-dras'sus)

This genus is closely related to *Litopyllus* but is easily distinguished by the fact that the lower margin of the furrow of the chelicera is completely devoid of teeth. The known species are pale spiders and lack any strongly contrasting markings. The following species is representative of the group.

Liodrassus utus (L. u'tus).—This is a pale spider similar in size and appearance to *Litopyllus rupicolens*. The eyes of the posterior row are subequidistantly spaced and subequal in size. The tibial apophysis of the male palpus is a long, stout spur. The species was described from Utah.

Genus LARONIA (La-ro'ni-a)

The lower margin of the furrow of the chelicera is furnished with two contiguous lobes, and there is a long keel on the upper margin. The posterior eye row is recurved, the eyes equal in size and nearly equally spaced. The labium is two thirds as long as the endites. The species of this genus resemble those of *Gnaphosa* and *Callilepis* in general appearance.

Laronia bicolor (L. bi'co-lor).—A single specimen of this genus, a male, has been reported from within our borders. This is the *Eilica bicolor* of Banks and was described from Florida. Other species are known from Mexico and tropical America.

Genus CESONIA (Ce-so'ni-a)

The cephalothorax is low, very much narrowed in front, and bears a slender and short median furrow. The two rows of

Superfamily Argiopoidea

eyes are widely separated and nearly straight. The median eyes of each row are farther apart than they are from the lateral eyes. The clypeus is at least twice as wide as the anterior eyes. The fore spinnerets are longer than the hind ones and stout.

The following is the only known species.

Fig. 309.
CESONIA BILINEATA

Cesonia bilineata (C. bi-lin-e-a'ta). — This is a very easily recognized species on account of its markings. The female measures about one fourth inch in length. The body is white with two broad, black, longitudinal stripes extending nearly the whole length of the cephalothorax and abdomen above (Fig. 309), and with a black stripe near each lateral margin of the lower side of the abdomen. The legs are gray with white hairs. The spinnerets are long.

The species is widely distributed throughout the Atlantic region. It makes a snowy-white flat egg-sac, which it loosely fastens to the lower side of a stone. A female with her egg-sac was taken in this situation at Ithaca, N. Y., in August.

Genus GNAPHOSA (Gna-pho'sa)

This and the following genus differ from all other members of the family found in our fauna in having the lower margin of the furrow of the chelicerae armed with a broad keel or lobe (Fig. 310). This genus can be distinguished by the fact that the posterior series of eyes is much wider than the anterior series and is strongly recurved.

Fig. 310. CHELICERÆ OF GNAPHOSA

The posterior lateral eyes are not much larger than the posterior median eyes.

The two following species are our most common representatives.

Gnaphosa gigantea (G. gi-gan'te-a).— This is a robust species

Superfamily Argiopoidea

measuring one half inch in length. It is of a rusty black colour; in alcohol the cephalothorax and legs are dark reddish brown and the abdomen gray. The whole body is covered with fine black hairs. The middle and hind pairs of spinnerets are either greatly reduced in size or are entirely wanting, in which case the spider has only two spinnerets; most of the specimens of this species in our collection are in this condition. The structure of the epigynum of the female is shown in Fig. 311.

This is a widely distributed species in the North; its range extending from the Atlantic to the Pacific. It lives under stones and leaves. Its egg-sac is snowy-white and flat;

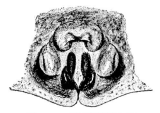

Fig. 311. EPIGYNUM OF GNAPHOSA GIGANTEA

Fig. 312. GNAPHOSA GIGANTEA AND EGG-SAC

its diameter is as great or greater than the length of the spider.

On July 20th I found at Ithaca, N. Y., several females, each with an egg-sac, under stones in a dry pasture. In each case the female was resting on the egg-sac with some of her legs wrapped around it (Fig. 312). From one of these egg-sacs the spiderlings emerged August 21st.

Gnaphosa sericata (G. ser-i-ca'ta).— This is a smaller species, the length of the body being a little less than one fourth inch in both sexes. The cephalothorax and legs are rufous; the abdomen, bluish black. There is a small black ring around each eye.

This is a widely distributed southern species. It is common, in the region where it occurs, on the ground or under stones and leaves. It runs with great rapidity.

Genus CALLILEPIS (Cal-lil'e-pis)

The lower margin of the furrow of the cheliceræ is armed with a broad keel or lobe (Fig. 313) as in *Gnaphosa;* but this genus

Superfamily Argiopoidea

differs from *Gnaphosa* in that the posterior row of eyes is barely wider than the anterior row and is straight or only slightly recurved. The posterior lateral eyes are plainly larger than the posterior median eyes.

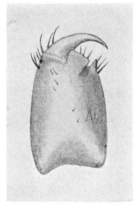

Fig. 313. CHELICERA OF CALLILEPIS IMBECILLA

The spiders of this genus are very active; they live under stones where they spin a slight irregular web. The egg-sac is planoconvex and resembles that of *Zelotes* but is always white.

Our best known species is the following:

Callilepis imbecilla (C. im-be-cil′la).— Although this is a small spider, measuring only about one fourth inch in length, it is striking in appearance owing to its strongly contrasting colours.

The cephalothorax is bright orange-brown and the abdomen blue-black. The armature of the chelicera is shown in Fig. 313.

Family ZODARIIDÆ (Zod-a-ri′i-dæ)

The Zodariids (Zo-da′ri-ids)

The Zodariidæ is barely represented in our fauna, only two rare species having been recorded. The legs are relatively short and stout, and all of them are nearly equal in length. The tarsi of our species bear three claws, but in some of the exotic genera two claws and claw tufts are present. The anterior spinnerets are usually considerably longer than the other ones. In many cases the median and posterior spinnerets are greatly reduced in size, sometimes obsolete or virtually so, when they are represented by inconspicuous tufts of hairs. The chelicerae are smooth or at most armed with a single small tooth on the lower margin. The eight eyes are placed in two or three rows. In our species they are in two nearly straight rows, the medians of each close together. The posterior median eyes are widely separated from the posterior lateral eyes.

These spiders are found under stones on the ground.

Superfamily Argiopoidea

Representatives of two genera have been described from our fauna. They can be separated as follows:
A. With only two well developed spinnerets, the others reduced in size or aborted. P. 338. LUTICA
AA. With six well developed spinnerets. P. 337. STORENA

Genus STORENA (Sto-re′na)

A striking feature of this genus is the great length of the fore spinnerets, and the shortness of the other spinnerets. There are three tarsal claws. This is a widely distributed genus; but only a single species occurs in this country.

Storena americana (S. a-mer-i-ca′na).— Only the female has been described. It measures only one third inch in length. The cephalothorax is reddish; the abdomen is greenish yellow marked

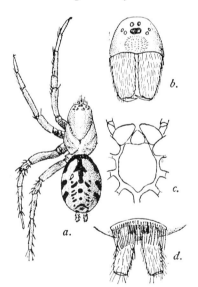

Fig. 314. LUTICA MACULATA *a*, female enlarged *b*, face *c*, sternum, labium, and endites *d*, spinnerets (after Marx)

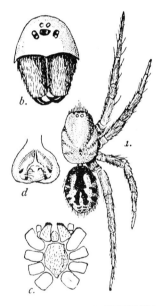

Fig. 315. STORENA AMERICANA *a*, female, enlarged *b*, face *c*, sternum, labium, and endites *d*, epigynum (after Marx)

with black, as shown in Fig. 315, which is a copy of one by Doctor Marx ('91), who described the species. The species was found in Georgia.

Superfamily Argiopoidea

Genus LUTICA (Lu'ti-ca)

The most striking characteristic is the possession of only two spinnerets, the hind pair; the other four are aborted, and only indicated by tufts of hair. Only one species is known.

Lutica maculata (L. mac-u-la'ta).—The species was described from a specimen received from Lake Klamath, Oregon. It can be easily recognized by the accompanying figures (Fig. 314) which were drawn by Marx ('91) who described the species. Mature females are about one half inch long. The male is unknown.

A few examples of this rare spider have been found recently in California.

Family HOMALONYCHIDÆ (Ho-ma-lon-ych'i-dæ)

The Homalonychids (Ho-ma-lon-ych'ids)

The genus *Homalonychus*, formerly regarded as belonging in the Zodariidæ, has been placed recently in a family of its own. These curious spiders have two tarsal claws which are similar and smooth, completely lacking denticles. The tarsi are provided with a brush of terminal tenent hairs (claw tufts) and a pair of spurious claws. The cephalothorax is nearly as broad as long, flat, the pars cephalica not much higher than the pars thoracica. The clypeus is high. The eye arrangement is suggestive of that of some of the ctenids and of *Zora*, a clubionid. The eight eyes are placed in two rows of four each, the last row very strongly recurved, and much wider than the first row. The chelicerae lack teeth on the margins. The sternum is broad, and the posterior coxae are widely separated. The epigynum (Fig. 316) presents a broad median piece which is bordered by lobes or ridges. The tibia of the male palpus has a short, bifid apophysis.

This family, the precise relationships of which are still problematic, is represented by a single known genus.

Genus HOMALONYCHUS (Ho-ma-lon'y-chus)

The generic characters are the same as those given above for the family. The few known species are found in Lower California, western Mexico, and in the southwestern portion of the United States. Specimens are rarely encountered within our borders.

Superfamily Argiopoidea

Homalonychus selenopoides (H. se-len-o-poi'des).— This is a Mexican species which extends into the southwestern part of our country. It was originally described by Doctor Marx ('91),

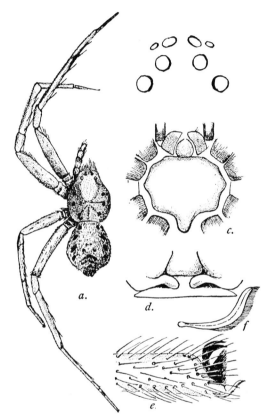

Fig. 316. HOMALONYCHUS SELENOPOIDES *a*, female, enlarged *b*, eyes seen from above *c*, sternum, labium, and endites *d*, epigynum *e*, tarsus *f*, auxiliary claw (after Marx)

and can be easily recognized by his figures of it (Fig. 316). It measures about two thirds of an inch in length.

Family PHOLCIDÆ (Phol'ci-dæ

The Pholcids

The pholcids are spiders with very long legs, which build irregular webs in dark places, in which they hang with the back

Superfamily Argiopoidea

downward. The tarsi of the legs are furnished with three claws. There are usually eight eyes, but in one of our genera there are only six and in another one pair is represented by very minute vestiges. There is a group of three eyes on each side; each group consists of an anterior lateral and two posterior eyes, all of which are pearly white; the anterior median eyes, when present, are isolated, smaller and dark in colour. The claw of the cheliceræ is short and is opposed by a tooth-like projection of the basal segment (Fig. 317). The endites, in all of our genera, are convergent and contiguous at

Fig. 317. CHELICERA OF PHOLCUS PHALANGIODES

Fig. 318. ENDITES OF PHOLCUS

the extremity (Fig. 318). The abdomen varies greatly in form.

Seven genera are represented in our fauna. They can be separated by the following table:

TABLE OF GENERA OF THE PHOLCIDÆ

A. With only six eyes, or with the vestiges of one pair, the anterior median exceedingly minute.
  B. Eyes situated on a very prominent eminence, and not arranged in two widely separated groups. P. 341.
                                                       MODISIMUS
  BB. Eyes not situated on a prominent eminence but arranged in two widely separated groups of three each. P. 341.
                                                       SPERMOPHORA
AA. With eight distinct eyes.
  B. Abdomen elongate. P. 342.   PHOLCUS
  BB. Abdomen globose.
    C. Femur of the first legs not twice the length of the

Superfamily Argiopoidea

cephalothorax, and shorter than the femur of the fourth legs. P. 343.   PHOLCOPHORA
CC. Femur of the first legs twice as long as the cephalothorax.
  D. Posterior row of eyes slightly procurved. P. 344.
                                                PSILOCHORUS
  DD. Posterior row of eyes slightly recurved.
    E. Anterior eyes straight. P. 343.   PHYSOCYCLUS
    EE. Anterior eyes recurved. P. 344.   ARTEMA

Genus MODISIMUS (Mo-dis′i-mus)

The anterior median eyes are reduced to mere vestiges so that there are apparently only six eyes. These, the lateral eyes and the posterior median, are not arranged in two groups as is usually the case in this family. The eyes are situated on a very prominent eminence.

These are tropical or subtropical spiders; only one species has been taken in our fauna.

Modisimus texanus (M. texa′nus).— The cephalothorax is pale yellowish, with a broad median black stripe, tapering a little behind (Fig. 319). Eyes on black spots, but the middle of the eye-tubercle is pale. Abdomen pale with many black and white spots. The epigynum projects forward in a sharp point.

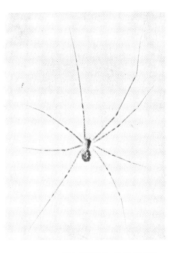

Fig. 319. MODISIMUS TEXANUS

I collected several specimens of this species at Austin, Tex., in March. One of these, a female, was carrying a bundle of eggs. The species proved to be a new one; and was subsequently described, at my request, by Mr. Banks.

Genus SPERMOPHORA (Sper-moph′o-ra)

The number and arrangement of the eyes will serve to distinguish this genus from the other pholcids occurring in our

Superfamily Argiopoidea

fauna. There are only six eyes, the anterior median being absent. The eyes present are not situated on a prominence, and are arranged in two widely separated groups of three each. The abdomen is short and rounded.

Only one species has been described from the United States.

Spermophora meridianalis (S. me-rid-i-a-na'lis).— The body of the adult female measures about one twelfth inch in length; it is white with a pair of pale gray spots on the thorax and two or three pairs on the abdomen. The legs are long and slender.

This is a house-spider living in closets, in dark corners, and under furniture. The female carries her mass of eggs about, clinging to them with her cheliceræ.

Genus PHOLCUS (Phol'cus)

The anterior median eyes are much closer to each other than to the anterior lateral eyes. The median ocular area is trapeziform and not at all or not much wider than long. The abdomen is elongate.

This genus is represented in our fauna by three species; the following one is widely distributed.

Fig. 320. PHOLCUS PHALANGIODES

Pholcus phalangioides (P. pha-lan-gi-oi'des).— This species is easily distinguished from all other pholcids found in the United States by its larger size and the elongated form of the abdomen (Fig. 320). The body is one quarter of an inch long, and the longest legs two inches. The colour of the body is pale brown.

The great length of the legs causes this spider to resemble, somewhat, a harvestman (*Phalangium*), and suggested the specific name. It is a common house-spider, especially in the warmer regions; but it is also found in the North. It prefers cellars and other dark locations.

It spins a very large, loose web. The spider hangs in its web with the abdomen directed upward; and when alarmed shakes its web violently or swings itself around rapidly. The egg-sac is exceedingly thin, in fact it is invisible except on close examination. The eggs are carried by the cheliceræ, as shown in the figure.

Genus PHOLCOPHORA (Phol-coph'o-ra)

All of the eight eyes are present; the anterior row is procurved; the anterior median eyes are but little smaller than the anterior lateral; the posterior row of eyes is moderately recurved.

Only two species have been described.

Pholcophora americana (P. a-mer-i-ca'na).— The length of the body is a little more than one twelfth inch. The cephalothorax is pale yellowish, darker on the head, and with black spots around the eyes. The abdomen is pale beneath and dark gray above. The body is clothed with large stiff bristles, which are most numerous on the abdomen.

This species was discovered in Colorado.

Genus PHYSOCYCLUS (Phy-soc'y-clus)

The eight eyes are all present; the anterior median eyes are several times farther removed from the posterior median eyes than they are from the anterior lateral; the posterior row of eyes is slightly recurved; the femur of the first pair of legs is shorter than that of the fourth.

At least four species have been found in this country, three of them restricted to the southwestern States. The following species is common in the tropics.

Physocyclus globosus (P. glo-bo'sus).—The length of the body of the female is about one sixth inch. The cephalothorax is broader than long; the eyes are borne on a prominent elevation; the abdomen is as high as long. The cephalothorax is yellow with a brownish middle line; the abdomen is yellowish brown with many small black flecks. The epigynum of the female is a chitinous plate which is produced into a single point in front. The cheliceræ of the male are armed in front with a roughened process.

This species is widely distributed in the tropics of the New World. It is a house-spider and is not uncommon in Florida. I found it in a web like that of *Theridion* in corners of outbuildings in Florida. It has not been shown to occur in the southwestern part of our country, where it is replaced by other species.

Genus PSILOCHORUS (Psil-o-cho′rus)

The four anterior eyes are nearly contiguous and are in a procurved line. The posterior row is also moderately procurved. The clypeus is very broad and is inclined forward. The chelicerae of the males are armed in front, usually near the base, with a long, curved tooth or horn, which is characteristic in shape and length for each of the species. The epigynum of the female usually consists of a chitinized ridge or series of ridges which may be ornamented with one or two spurs.

Several species, all of them similar in appearance, have been described from the United States. The following is the only one which occurs in the southeastern part of our country. The others are found in the west.

Psilochorus pullulus (P. pul′lu-lus).—The length of the body is about one tenth inch. The cephalothorax and legs are pale yellow, and the eyes are surrounded with black. There is a forked black marking behind the eyes on the thoracic grooves. The abdomen is pale gray and is often marked with green spots. The epigynum of the female does not bear strong spurs. The tooth on the chelicera of the male is long and curved, set near the base. The abdomen is very greatly arched and projects far behind the spinnerets.

This spider, which is relatively rare, lives in dark situations where it spins a rather small, tangled mass of threads.

Genus ARTEMA (Ar-te′ma)

The carapace is very broad, marked above by a very deep, round excavation at the median furrow. The four anterior eyes are well separated, in a recurved row, the median smaller than the lateral eyes. The posterior row is moderately recurved. The sternum is continued behind between the posterior coxæ, which are separated by no more than half their width, as a bluntly rounded piece. In other pholcids the sternum is broadly truncated behind and the coxæ are more widely separated. The chelicera of the male is armed in front with a large keeled apophysis which is set with short teeth. The abdomen is elevated as in *Physocyclus* and the spinnerets are placed near the base.

A single species is known from the United States.

Artema atlanta (A. at-lan'ta).—This is one of the largest of all the pholcids, the female very often measuring fully one half inch in length. The male is slightly smaller. The legs are very long and are marked with conspicuous, brown annulæ. The carapace is white or pale yellow, with a median dark band in the median groove which continues around the pars cephalica, and a median dusky stripe on the clypeus. The abdomen is gray with an indistinct pattern of large bluish spots.

This large pholcid is widely distributed in tropical America. It has been found within our borders only in southern Arizona, where it spins its webs in caves and ruins.

Family THERIDIIDÆ (Ther-i-di'i-dæ)

The Comb-footed Spiders

Fig. 321. THERIDION TEPIDARIORUM

The most common of all house-spiders, the one that most often spins a tangled maze of threads in the corners of neglected rooms, is *Theridion tepidariorum*, a representative of the typical genus of this family; this spider will serve well, therefore, as an example of the family (Fig. 321).

The comb-footed spiders are, with some exceptions, sedentary spiders which spin webs to catch their prey and in which to place their egg-sacs. Their webs are composed of threads extending

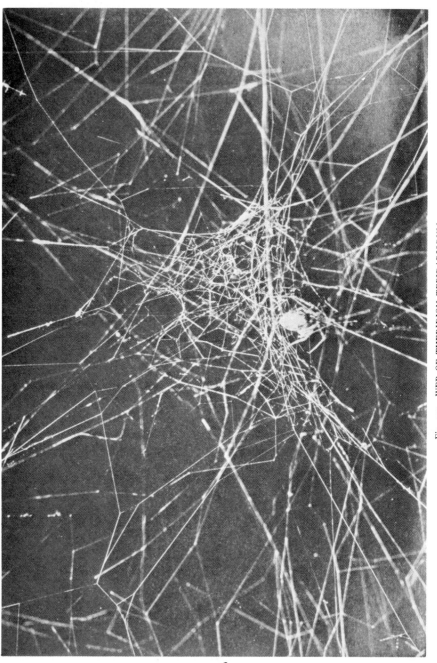

Fig. 322. WEB OF THERIDION TEPIDARIORUM

Superfamily Argiopoidea

in all directions with no apparent regularity (Fig. 322); and the spider hangs in its web with its back downward.

These spiders have eight eyes and three tarsal claws. They are distinguished from other eight-eyed and three-clawed spiders, in fact from all other spiders, by the presence, on the tarsus of the fourth pair of legs, of a distinct comb, consisting of a row of strong, curved, and toothed setæ. Usually this comb is very distinct (Fig. 323); but in some forms, as in *Argyrodes* (Fig. 324) it is considerably reduced.

The comb is used for flinging silk, often in a quite liquid state, over the entangled prey. As the presence of this tarsal comb distinguishes these spiders from all others, I propose the term comb-footed spiders as a popular designation of the family.

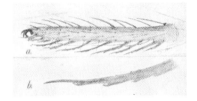

Fig. 323. COMB OF THERIDION TEPIDARIORUM

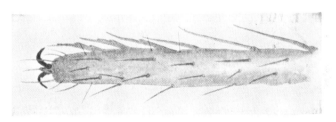

Fig. 324. COMB OF ARGYRODES TRIGONUM

It should be noted that the tarsal comb is quite distinct from the calamistrum of the cribellate spiders, that organ being borne by the metatarsi.

The comb-footed spiders are also distinguished by the fact that the cheliceræ lack a lateral condyle, and that the tarsus of the male palpus is deprived of a paracymbium, which is rarely wanting in the allied families following.

In certain genera the males possess a stridulating organ (Fig. 325). This is composed of a chitinous, curved border,

Superfamily Argiopoidea

armed with teeth, on that part of the abdomen that overhangs the thorax, and of a plate marked with numerous, fine, transverse striæ on each side of that part of the thorax overlapped by the abdomen. The abdomen is freely movable up and down; and by rubbing the teeth on the abdomen across the striæ on the thorax a feeble sound is produced.

While several species live in houses and other protected places, a larger number live on plants in the fields. Most species build irregular webs of the type built by the house-spider mentioned above; but some (*Argyrodes*) live as commensals in the webs of other spiders; and a few (*Euryopis*) do not live in webs and spin very little, but are found under stones, or in moss and leaves, and run with great rapidity.

Fig. 325. STRIDULATING ORGANS OF ASAGENA

One genus of this family, *Latrodectus*, is of especial interest as to it belong those spiders which in all countries where they occur are greatly feared on account of the supposed deadly nature of their bite; two species of this genus occur in the Southern States.

Most of the silk spun by the comb-footed spiders is of the plain type; but occasionally a viscid thread is used. A specimen of *Steatoda borealis* which I had in a cage spun threads upon which the viscid drops could be seen with a hand lens. It has been shown by Apstein ('89) that the Theridiidæ agree with the Argiopidæ in the possession of the aggregate silk glands which are supposed to be the source of the viscid silk.

But the most characteristic silk of the theridiids is that used for enveloping their prey; this is thrown over their victims by the comb on the hind tarsi, and is supposed to be derived from the lobed silk glands, which have been found only in this family. These glands open through a spigot without a tip on the hind spinnerets (Apstein '89).

TABLE OF GENERA OF THE THERIDIIDÆ

A. Lateral eyes of each side widely separate.
  B. Abdomen flattened, and broad behind. P. 357. EPISINUS

  BB. Abdomen globose. P. 372.   Latrodectus
AA. Lateral eyes of each side contiguous or nearly so.
  B. Posterior median eyes fully three times the diameter of one of them apart. P. 355. Spintharus
  BB. Posterior median eyes rarely more than twice the diameter of one of them apart.
    C. Anterior median eyes larger than the posterior median eyes, and much wider apart.
      D. Abdomen pointed behind. P. 357. Euryopis
     DD. Abdomen more globose, broadly rounded behind. P. 371. Dipœna
   CC. Anterior median eyes rarely larger than the posterior median eyes, when they are larger they are not wider apart than the posterior median eyes.
     D. Cephalothorax with a transverse furrow in the middle; abdomen usually prolonged either above or behind the spinnerets.
      E. Abdomen very long and slender, vermiform. P. 351. Rhomphæa
     EE. Abdomen much shorter. P. 353. Argyrodes
    DD. Cephalothorax without transverse furrow; abdomen not greatly prolonged.
      E. Sternum pointed or rounded behind; the hind coxæ either contiguous or more or less widely separated by the end of the sternum.
       F. Anterior median eyes much larger than the anterior lateral eyes. P. 375. Steatoda
      FF. Anterior median eyes not much if any larger than the anterior lateral eyes.
        G. Lateral eyes of each side slightly but distinctly separate.
         H. Clypeus not wider than the area occupied by the eyes. P. 377. Lithyphantes
        HH. Clypeus much wider than the area occupied by the eyes. P. 377. Asagena
       GG. Lateral eyes of each side contiguous.
         H. Lobes of the stridulating organ of the abdomen very long; a small spider measuring less than one tenth inch in length. P. 376. Coleosoma

Superfamily Argiopoidea

- HH. Lobes of the stridulating organ of the abdomen, when present, of moderate length.
  - I. Abdomen dark brown with two white spots across the middle. P. 377.
    ASAGENA
  - II. Abdomen not marked as in *Asagena*.
    - J. Labium long and pointed, more than half as long as the endites. P. 376.
      TEUTANA
    - JJ. Labium transverse, not more than half as long as the endites.
      - K. Endites nearly straight and parallel or slightly convergent.
        - L. Larger species, males measuring at least one sixth inch and females one fourth inch in length.
          - M. Lower margin of the furrow of the cheliceræ conspicuously toothed; abdomen moderately high in front. P. 379.
            ENOPLOGNATHA
          - MM. Lower margin of the furrow of the cheliceræ with a very small tooth or with none; abdomen very high in front. P. 359.
            THERIDION
        - LL. Smaller species.
          - M. Males only; tibia of palpus enormously developed. P. 369.
            THERIDULA
          - MM. Females, and those males in which the tibia of the palpus is not enormously developed. P. 359. THERIDION
      - KK. Endites curved and strongly convergent at the tip.
        - L. Abdomen wider than long with a hump on each side in the middle of its length. P. 369. THERIDULA

  LL. Abdomen without a hump on each side.
    M. First pair of legs longer than the fourth pair; legs usually long. P. 359. THERIDION
    MM. Fourth pair of legs longer than the first pair, all short. P. 380. ROBERTUS
 EE. Sternum broad and truncate behind. All small spiders measuring one twelfth inch or less in length.
  F. Abdomen furnished with several prominent humps. P. 371. ULESANIS
  FF. Abdomen not furnished with humps.
    G. Cephalothorax with numerous, small, crescent-shaped elevations, each at one side of a puncture. P. 374. CRUSTULINA
    GG. Cephalothorax without such elevations.
     H. Cuticle of abdomen soft.
      I. Median ocular area narrowed in front. P. 381. THEONOE
      II. Median ocular area not narrowed in front. P. 381. MYSMENA
     HH. Abdomen hard, with a shield or sigilla.
      I. Clypeus concave; first femur of male with a ventral series of stout spines. P. 382. PAIDISCA
      II. Clypeus convex. P. 382. ANCYLORRHANIS

Several genera of obscure theridiids, which are rare or have been little studied, are not included in the above table or in the following account of the family.

Genus RHOMPHÆA (Rhom'phæ-a)

The lateral eyes of each side are close together; the middle eyes are widely separated, those of each side being near to the lateral eyes, thus forming a group of four eyes on each side. The cephalothorax has a deep transverse furrow near the middle, and the abdomen is elongate.

These spiders are remarkable for the slender form and great

Superfamily Argiopoidea

length of the abdomen, which is extended in a worm-like prolongation far beyond the spinnerets (Fig. 326). The following species is the only representative of the genus known to occur in our territory.

Rhomphæa fictilium (R. fic-til′i-um).—This species is light yellow and silvery white in colour, with three darker bands on the cephalothorax and one on the middle of the abdomen. The legs are very long and slender. The body varies from one fifth to one

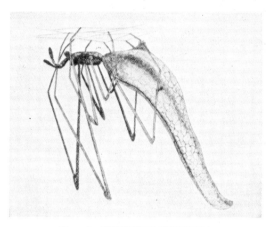

Fig. 326. RHOMPHÆA FICTILIUM

third inch in length; the part behind the spinnerets greatly exceeds the remainder of the body in length. The species is distributed from New England to the Gulf of Mexico, but is rare.

Hentz, who first described the species, states that the spider makes a web like that of *Theridion*, and remains motionless in an inverted position; and that the projection of the abdomen is capable of bending over nearly double. F. O. P. Cambridge writing of the "worm-like appendage" of the abdomen of spiders of this genus states: "This, as I have myself observed in Brazil, is wriggled to and fro, looking like a small caterpillar. But of what service to the spider this accomplishment may be is not easy to guess; for on the one hand it seems likely to attract the attention of grub-eating wasps and ants, though on the other it may attract, within striking distance, gnats and small flies who become curious to ascertain what the wriggling phenomenon may portend."

Superfamily Argiopoidea

Genus ARGYRODES (Ar-gy-ro'des)

In this genus as in the preceding there is a transverse furrow extending across the middle of the cephalothorax. Here too the abdomen is of strange form, but is not worm-like as in *Rhomphæa*. In our more common species the abdomen is greatly elevated above the spinnerets, being triangular when seen from the side, and in several of our species it is furnished with tubercles.

These spiders spin but little silk. They are sometimes found in small webs that are evidently their own; but more often

Fig. 327. ARGYRODES TRIGONUM

they are found in the webs of other and larger spiders, where they nourish themselves with prey which on account of its smallness escapes their host.

This is a large genus, eighty species or more have been described; of these thirteen are found in the United States. The two following are about the most common ones.

Argyrodes trigonum (A. tri-go'num).—This is a small, yellow, triangular species with a high, pointed abdomen. Large females measure an eighth of an inch from the head to the spinnerets and nearly as much from the spinnerets to the tip of the abdomen, which is two-lobed (Fig. 327). In the female, that part of the head bearing the eyes is slightly raised, and the eyes are far removed from the front edge of the clypeus. The colour is light yellow, sometimes with a metallic lustre. On the back of the cephalothorax there are three light brown stripes, and sometimes

Superfamily Argiopoidea

there are dark spots at the sides of the abdomen and over the spinnerets.

The male is smaller than the female, is darker coloured and has a smaller and less angular abdomen. It can be easily recognized by the remarkable form of the head (Fig. 328), which

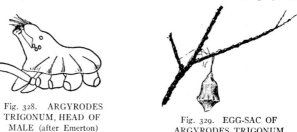

Fig. 328. ARGYRODES TRIGONUM, HEAD OF MALE (after Emerton)

Fig. 329. EGG-SAC OF ARGYRODES TRIGONUM

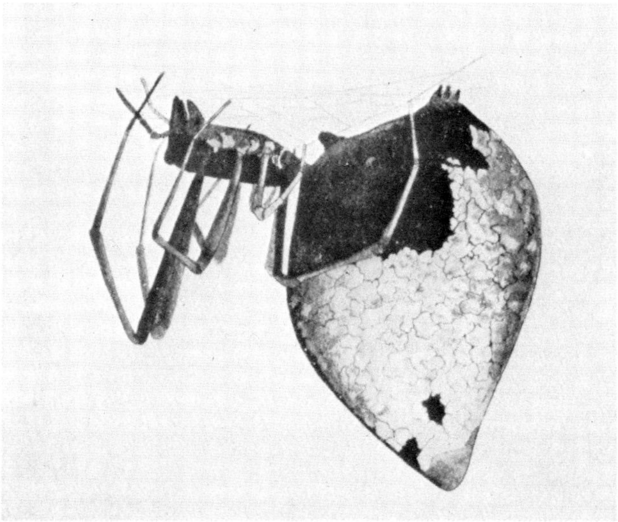

Fig. 330. ARGYRODES NEPHILÆ, FEMALE, SIDE VIEW

bears two projecting horns each tipped by a bunch of toothed hairs; the median eyes are at the base of the hinder horn.

The web is like that made by *Theridion;* it is built among the branches of shrubs or in the outer part of the web of some large spider, with which the *Argyrodes* lives as a commensal, feeding on the smaller insects caught in the web but neglected by its host. The form and colour of the spider is protective, causing it to appear like a little, triangular scale, shed by some bud of a hemlock, and caught in the web.

The egg-sac is a beautiful vase-shaped object; and is suspended by a thread in the web (Fig. 329). When first made it is white, but later it changes to a brown colour. On several occasions we have found two egg-sacs in a single independent web, one of which was brown and the other white; indicating that the spider had lived in this web a considerable time, and had made a second egg-sac. The egg-sacs are made in mid-summer, and the spiderlings emerge in late summer or early autumn.

Argyrodes nephilæ (A. neph'i-læ).— This resident of the South resembles the preceding species in size, and in having a triangular abdomen (Fig. 330); but differs strikingly in appearance, owing to the fact that a large part of the upper portion of the abdomen is silver-white; so that it appears like a drop of quicksilver. The cephalothorax is dark brown or black, and the lower side of the abdomen is black; there is also a black stripe on the middle line of the abdomen above.

Fig. 331. ARGYRODES NEPHILÆ, HEAD OF MALE

In the female the tip of the abdomen is rounded, not split; and the clypeus is nearly vertical.

The head of the male bears two horns, as in the preceding species; but in this species the two pairs of median eyes are borne by the hinder horn. (Fig. 331.)

Like *A. trigonum* this spider sometimes leads an independent existence and sometimes lives as a commensal in the webs of larger spiders, and especially in the webs of large orb-weaving species. The specific name was suggested by the fact that it sometimes lives in the webs of *Nephila*. It is found in the South and is distributed from the Atlantic to the Pacific.

Genus SPINTHARUS (Spin-tha'rus)

This genus can be distinguished from all other comb-footed spiders that occur in our fauna by the arrangement of the eyes. The lateral eyes of each side are contiguous, as is the case with most members of the family, but the posterior median eyes are

Superfamily Argiopoidea

very widely separated, being three or four times the diameter of one of them apart. This is a small genus; only a single representative of it has been found in this country.

Spintharus flavidus (S. flav'i-dus).— This remarkable spider can be easily recognized by the accompanying figures. The female (Fig. 332) measures from one sixth to one fourth inch in length. The cephalothorax and legs are pale yellow; the abdomen reddish brown. On each side of the abdomen there is a white or yellow stripe. The area on the dorsal side of the abdomen, between these stripes, is marked with black, red, and yellow; the distribution of these colours

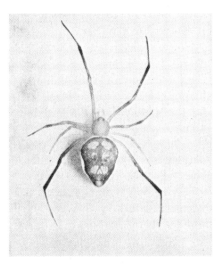

Fig. 332.
SPINTHARUS FLAVIDUS, FEMALE

Fig. 333.
SPINTHARUS FLAVIDUS, MALE

varies greatly in different individuals. This dorsal area is often bordered by a black line or series of dots, inside of which there is a red line, and inside of this two or three pairs of yellow or white spots.

The male measures about one eighth inch in length. It has longer legs and a more slender abdomen than the female. (Fig. 333.) It also varies greatly in markings.

The species is found from New England to Florida. It is common on the lower surface of leaves of bushes. The spider rests near the edge of the leaf as shown in Fig. 334. At first sight it appears to be resting on the leaf; but a careful examination reveals the fact that each foot is supported by a thread. The web, however, is so delicate that it is practically invisible.

Genus EPISINUS (Ep-i-si'nus)

This is one of two genera of the comb-footed spiders occurring in our fauna in which the lateral eyes of each side are widely separated (Fig. 335). This genus is distinguished from the other (*Latrodectus*) in having the abdomen flattened and broad behind. It contains but few species, of which only two have been described from the United States.

Episinus amœnus (E. a-mœ'nus).—The female of this species measures about one sixth inch in length; the male, one tenth inch. It is easily distinguished by the form of the abdomen (Fig. 336), which is narrow and bi-lobed at the base, and gradually widened to near the posterior end; at the widest part of the abdomen there is on each side a tubercle, behind which the abdomen tapers rapidly to a point. This tapering portion is so short that the abdomen has a truncate appearance, which doubtless suggested the specific name.

This species was thought to be the same as the European *Episinus truncatus*, but it is quite distinct. It occurs in Virginia, Tennessee, North Carolina, and presumably in adjacent states. In habits and structure it is closely allied to *Spintharus*.

Genus EURYOPIS (Eu-ry-o'pis)

In this genus the anterior median eyes are much larger and much wider apart than the posterior median eyes (Fig. 337); and the abdomen is pointed behind.

These spiders are of peculiar form, resembling certain crab-spiders more than they do other comb-footed spiders (Fig. 338). They appear to spin but little, for their web is unknown. They are found under stones, and in moss and lichens, running with great rapidity; and they may be shaken from bushes, on which they are probably stalking their prey.

Our species are of small or moderate size; and black with white or silvery markings. They can be distinguished by the

Fig. 334. SPINTHARUS FLAVIDUS, ON LEAF

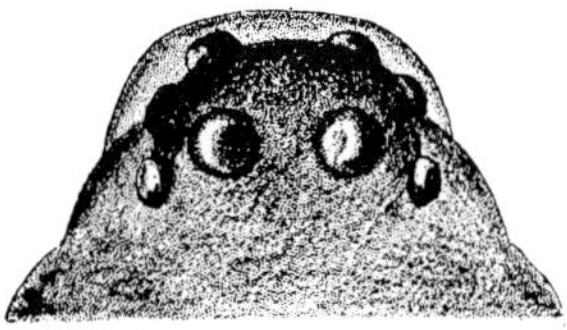

Fig. 337. EYES OF EURYOPIS

Fig. 335. EYES OF EPISINUS

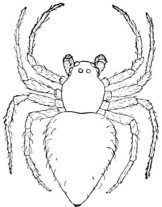

Fig. 338. EURYOPIS FUNEBRIS

Fig. 336.
ABDOMEN OF EPISINUS

Fig. 339. FACE OF THERIDION

markings of the abdomen. The more common ones are the following:

Euryopis funebris (E. fu-ne'bris).— The hind part of the abdomen is bordered with a silvery white stripe. The length of the body of the female is about one eighth inch. This is our most common and most widely distributed species, occurring over the larger part of the United States.

Euryopis scriptipes (E. scrip'ti-pes).— This is a larger species than the preceding, the female being one fifth inch in length and the male one sixth inch. Like that it has a silvery margin on the abdomen; but this completely surrounds an irregular, triangular, dark spot in the middle of the back. The specimens before me are from Colorado, and have been given the above specific name by Mr. Banks.

Euryopis argentea (E. ar-gen'te-a).— There are five or six pairs of white spots along the middle of the abdomen and others at the sides. Only immature individuals have been described. The species is distributed from New England to Florida.

Euryopis quinquemaculata (E. quin-que-mac-u-la'ta).— There are five spots on the abdomen above, a pair near the base, a pair near the middle, and a single one at the tip just above the spinnerets. The length of the body is a little less than one eighth inch. The species has been found at and near Washington, D. C.

Genus THERIDION (The-rid'i-on)

The lateral eyes of each side are contiguous (Fig. 339). The sternum is longer than wide, except in some small aberrant species, and ends in an obtuse, generally narrow point between the hind coxæ, which, as a rule, are but little separated, but there are many exceptions to this. The endites are more than twice as long as the labium, nearly parallel, being but slightly convergent at the tip. The lower margin of the furrow of the cheliceræ is furnished with a very small tooth or with none. Stridulating organs are absent.

This is the largest of all genera of spiders. Simon states that nearly 320 species have been described; and forty are known to occur in North America; this is more than one third of our representatives of the family Theridiidæ. Among our species is our most familiar house-spider, and several that are very com-

Superfamily Argiopoidea

mon in shrubs and trees. Only a few of the more common species can be described here.

The Domestic-spider, *Theridion tepidariorum* (T. tep-i-da-ri-o'rum). — Of all the spiders that inhabit our dwellings this is the most familiar, and consequently best merits the title of *the* domestic-spider. Its tangle of threads can be found in almost any neglected room, throughout the length and breadth of our country; and the species is not limited to our country for it is almost a cosmopolite.

This is an exceedingly variable species in colour and markings. The beginning student of spiders is apt to collect many

Fig. 340. THERIDION TEPIDARIORUM, DORSAL VIEW

Fig. 341. THERIDION TEPIDARIORUM, CAUDAL ASPECT

specimens of it and to think that they represent several species. It is well therefore to become familiar with it in its various guises as soon as practicable.

The female when full-grown may measure more than one fourth inch in length; but many adults are smaller; the male is about one sixth inch in length. The female varies in colour from dirty white with a few dark spots to almost black. Figure 321, p. 330, is a side view of an individual in which the markings are distinct; Fig. 340 is a dorsal view; and Fig. 341 represents the caudal aspect of the same specimen. The most characteristic feature is the presence of several dark chevrons above the tip of the abdomen (Fig. 341). The male differs in being smaller and in having

Superfamily Argiopoidea

the abdomen more slender. In the lighter individuals of both sexes the markings are very indistinct. In cases of doubtful determinations an examination of the external reproductive organs should be made.

The palpus of the male is represented by Fig. 342; the embolus is comparatively short, and is supported by a prominent terminal apophysis; this is roughened on the outside by numerous, crescent-like elevations.

The epigynum of the female (Fig. 343) has a single large oval opening.

The web is an irregular net-work of threads built in a great variety of situations, but usually beneath some object which serves as a protecting roof, as in the upper angles of rooms, in the upper corners of window frames and of doorways. When living in the

Fig. 342. THERIDION TEPIDARIORUM, PALPUS OF MALE

Fig. 343.
EPIGYNUM OF THERIDION TEPIDARIORUM

open air it is most apt to make its webs beneath overhanging cliffs. Frequently in the webs of this spider there is a more densely woven portion forming a sort of a tent beneath which the spider rests (Fig. 322, p. 331). The egg-sacs are brownish and pear-shaped with a dense outer coat. They are suspended in the web, and several of them are made by one spider. Often a brood of spiderlings that has just emerged from its egg-sac can be found.

Theridion fordum (T. for'dum).— This large species, which resembles the domestic-spider in size and habits, is found in the South. It is distributed from Florida to California.

Superfamily Argiopoidea

The abdomen is marked by irregular, narrow, yellow lines (Fig. 344).

The spider makes a large irregular web, like that of the domestic-spider. These webs occur both in houses and on bushes in the field. In the field, there is usually a retreat in the centre of the web, made of one or more dried leaves.

Theridion rupicola (T. ru-pic'o-la).— This species closely resembles the well-marked individuals of *Theridion tepidariorum*, but is much smaller, the females measuring from one tenth to one eighth inch in length and the males about one twelfth inch. It is easily distinguished from *T. tepidariorum* by the fact that the abdomen as seen from above ends in a pointed hump, which is really about midway between the base of the abdomen and the spinnerets (Fig. 345).

It lives under stones and in the midst of rubbish. Emerton states that its web often contains grains of sand which look as if placed there by the spider. I found an individual in August which had a loose web built between two pieces of timber, which were about three inches apart.

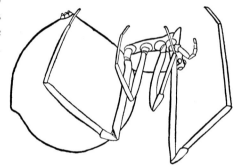

Fig. 344. THERIDION FORDUM   Fig. 345. THERIDION RUPICOLA

In the centre of the web there was a retreat one half inch long made of bits of wood, and in the form of an inverted cup. The spider was in the open mouth of the retreat, and within the retreat there was an egg-sac.

Theridion frondeum (T. fron'de-um).— One of the more common of the species of *Theridion* found in the fields is this little white and black spider, which frequently attracts attention on account of its conspicuous colours (Fig. 346). It is exceedingly variable in its markings, so that a dozen individuals may appear to represent half as many species. The general colour is white

Superfamily Argiopoidea

with black markings, which may be very prominent or almost entirely wanting; every gradation between these two extremes occur (Fig. 347). The cephalothorax is yellowish white with a median longitudinal black line or band; this black mark may be very narrow or may cover the greater part of the cephalothorax; this line is sometimes forked on the head; the cephalothorax may have a narrow marginal line. The abdomen is often marked with three snow-white or yellow longitudinal bands separated by transparent spaces; in the transparent spaces between the median and lateral white bands there may be black spots; sometimes these spaces are entirely black; in another type the centre of the basal part of the abdomen is black.

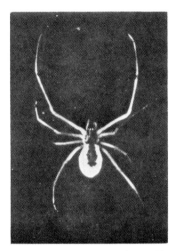

Fig. 346. THERIDION FRONDEUM

The female measures from one eighth to one sixth inch in length; the male, one eighth inch. The front legs are long; those of the female from one third to two fifths inch in length, while those of the smaller male equal those of the female in length.

On each chelicera of the male there is a pointed projection in front near the base. In the female the epigynum has a single opening, which is turned forward. The female is frequently found with her egg-sac in a partly folded leaf on bushes. The egg-sac is snowy white in colour and loose in texture. Sometimes a loose sheet of silk is spun across the space between the two parts of the folded leaf so as to keep the leaf folded and make a retreat for the spider. The spider stays with her egg-sac till the young emerge. We found the egg-sacs in July, and spiders with young during the latter part of August.

Theridion globosum (T. glo-bo'sum).— The male of this species measures only six hundredths of an inch in length, the female eight. The abdomen is very high, suggesting the specific name. The cephalothorax is orange-brown, with the eye-space black. The hind part of the abdomen is white, with a

Superfamily Argiopoidea

large black spot in the middle; the front upper part is yellowish gray.

Theridion unimaculatum (T. u-ni-mac-u-la′tum).— This also is a small species, measuring about one twelfth inch in length. It can usually be recognized by its white abdomen with a black spot in the centre of the back, and a black ring around the base of the spinnerets; sometimes the black spot is wanting, and the ring about the spinnerets is incomplete. The cephalothorax is orange with a black spot around the eyes; this extends back in a point as far as the dorsal groove.

Theridion kentuckyense (T. ken-tuc-ky-en′se).— The cephalothorax is yellow or brownish. The abdomen is white, very thickly

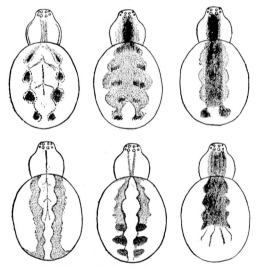

Fig. 347. THERIDION FRONDEUM. VARIATIONS IN MARKINGS

flecked with black. On the middle line of the abdomen above there is a white band, which begins a short distance back from the pedicel; the basal part of this band consists of three quadrangular spots separated by narrow transverse lines, and bounded on each side by a solid black patch; behind these spots the band is indistinctly separated from the flecked sides. Length of body of female one eighth inch, of male, barely cne tenth inch. The species was described from Kentucky; but it occurs as far north as Ithaca, N. Y.

Theridion punctosparsum (T. punc-to-spar'sum). — This species measures one eighth inch in length. The cephalothorax is dark yellow-brown, with a darker stripe in the middle and on each side. The abdomen is dark gray with white spots. The usual stripe is indicated by a large white spot in front and irregular lines of small white spots where the edge of the stripe is in other species. The epigynum has a large oval opening outside near the edge. (Fig. 348).

Emerton states that it is common in the neighbourhood of Salem and Boston, under stones, in stone walls and similar shady places, with a small web. It has also been found south to Florida and west to Colorado.

Theridion zelotypum (T. ze-lot'y-pum).—The female measures one sixth inch in length. The cephalothorax is orange with a distinct dark stripe in the middle, and dark edges. There is a light stripe along the middle of the abdomen above; from each side of this stripe there extend several lateral stripes separating a series of black spots (Fig. 349). The middle of the abdomen is bright red. On the ventral side, there is a black spot in front of the spinnerets, and the epigynum is brown. The epigynum has a single opening outside some distance from the edge.

Fig. 348. EPIGYNUM OF THERIDION PUNCTOSPARSUM (after Emerton)

Emerton, who first described the species from Maine, states that the webs were large, supported between the branches of spruce trees by threads running upward to the branches above and furnishing lodging for numerous specimens of *Argyrodes*. The spider had usually a tent covered with dry spruce leaves, under which it hung with cocoons and young.

Theridion studiosum (T. stu-di-o'sum). — This is an exceedingly common species in the South, and it extends northward to Pennsylvania and New Jersey. The cephalothorax is reddish yellow, with a more or less distinct central band and lateral margins brown. The abdomen is greenish brown or brownish gray on the sides, with a dark median band above, bounded on each side by a white wavy stripe (Fig. 350). On the ventral side there is a dark longitudinal band, which extends the entire length of the abdomen and behind surrounds the spinnerets, which are

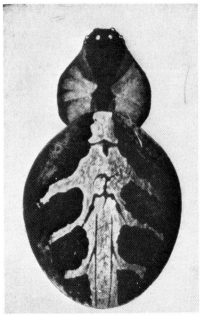

Fig. 349.
THERIDION ZELOTYPUM

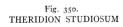

Fig. 350.
THERIDION STUDIOSUM

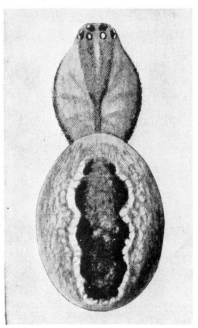

Superfamily Argiopoidea

of the same colour. The male is nearly one eighth inch in length, the female one sixth inch.

This spider is remarkable on account of its social habits, comparatively few social spiders being known. My attention was first attracted to it on the University Campus at Baton Rouge, La., where on various shrubs and trees there were unsightly masses of dead leaves tied together with silk, and extending from them a sheet resembling the sheet made by *Agelena*, but without a funnel. The mass of leaves was found to serve as a retreat; and in each retreat, there were several individuals of this species, evidently the common owners of the retreat and the sheet-web.

A closely allied species, *Theridion eximius* of South America, lives in very large colonies, hundreds or even thousands of individuals uniting to form a common web. Simon states that they sometimes cover an entire coffee tree; and F. O. P. Cambridge says that he has seen their webs spun up to a height of fourteen or fifteen feet amongst the foliage, being at the same time a yard or more across.

This is the *Anelosimus (Adelosimus) studiosum* of Banks's Catalogue.

Theridion differens (T. dif'fe-rens).— We have three common species of *Theridion* that agree in size, the female being about one eighth inch in length, the male a little less, and that resemble each other in having a distinct dorsal band on the abdomen, which is brightly coloured, often red or reddish in the middle and white or yellow at the edges. These species resemble each other so closely that they can be distinguished with certainty only by an examination of the palpus in the case of the males and of the epigynum in the case of the females. The specific differences were first pointed out by Emerton who described the three species.

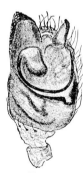

Fig. 351.
THERIDION DIFFERENS, PALPUS OF MALE

In the male of *Theridion differens* the dorsal stripe is obscure and the whole abdomen dark reddish brown. The bulb of the palpus terminates in a projecting point which bears numerous teeth (Fig. 351); the terminal apophysis is of the form shown in the figure; and the embolus is of moderate length. The

Superfamily Argiopoidea

cephalothorax is darker in the middle, but there is no distinct stripe and there are no distinct markings on the lower side of the body.

In the female the dorsal stripe of the abdomen is often very brightly coloured — white at the edges and red in the middle; in the male there is no distinct stripe on the cephalothorax and no distinct spots on the lower side. The openings of the spermathecæ are beneath the plate, so that they are not visible from the outside (Fig. 352).

Emerton states that the web of this species is found on low plants of all kinds, usually two or three feet from the ground. There is sometimes a small tent, often hardly deep enough to

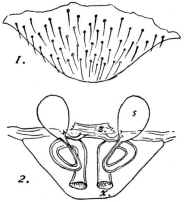

Fig. 352. THERIDION DIFFERENS, EPIGYNUM

Fig. 353. THERIDION MURARIUM, PALPUS OF MALE

cover the spider, from which the web spreads two or three inches according to the shape of the plant. The egg-sac is white, nearly as large as the spider, and is attached in the nest.

Theridion murarium (T. mu-ra′ri-um).— See the introduction to the account of *T. differens* above. In *T. murarium* the cephalothorax is pale with a dark line in the middle and one on each side, the middle line sometimes divided into two near the eyes. The sternum is pale with a black edge and a black stripe in the middle.

The palpal organ of the male is of the form shown in Fig. 353. It differs from that of *T. differens* in lacking the toothed projecting point of the bulb in the form of the terminal apophysis, and in the presence of a sickle-shaped black spine on the inner margin; the embolus is of moderate length.

Superfamily Argiopoidea

The epigynum of the female is represented in Fig. 354; the two openings of the spermathecæ are wide apart, and near the thickened edge of the plate.

Theridion spirale (T. spi-ra'le).— See the introduction to the account of *T. differens* above. The cephalothorax is orange-brown above and below, with an indistinct dark stripe above as

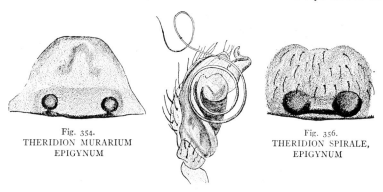

Fig. 354.
THERIDION MURARIUM
EPIGYNUM

Fig. 356.
THERIDION SPIRALE,
EPIGYNUM

Fig. 355.
THERIDION SPIRALE, PALPUS OF MALE

wide as the eyes in front and narrowed behind. The middle stripe of the abdomen is sometimes reddish, but oftener gray, with a dark spot near the front end.

The male palpi are very large, and differ markedly from those of the two preceding species by the great length of the embolus (Fig. 355).

The epigynum of the female is represented by Fig. 356; the two openings of the spermathecæ are about the diameter of one of them apart.

Genus THERIDULA (The-rid'u-la)

The anterior row of eyes is procurved; the abdomen, in the female at least, is wider than long with a hump on each side in the middle of its length (Fig. 357); the tibia of the palpus of the male is enormously developed, concave on the inner side, and overlaps the tarsus half its length; while the bulb itself is very simple (Fig. 358).

These are small spiders, measuring less than one eighth inch in length. They are found on bushes by sweeping, but their

Superfamily Argiopoidea

web has not been described. Simon states that they carry their egg-sacs attached to the spinnerets like *Lycosa*. Two species occur in this country.

Theridula opulenta (T. op-u-len′ta).— The female of this species (Fig. 357) is easily recognized by its portly abdomen, which is high and wider than long; it is yellowish gray, with a

Fig. 357. ABDOMEN OF THERIDULA OPULENTA, FEMALE

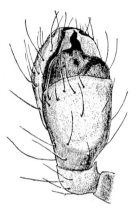

Fig. 358. THERIDULA OPULENTA, PALPUS OF MALE

greenish white spot in the middle and a black spot on a slight elevation on each side; the cephalothorax is yellow with a wide black stripe in the middle. The length of the body is one tenth inch.

In the male the abdomen is not so greatly widened and the markings are indistinct; but the cephalothorax bears a conspicuous, broad, dark, longitudinal band as wide as the eye-space. The length of the body is one twelfth inch.

This is a very widely distributed species. It is the *Theridion sphærula* of Hentz.

Theridula quadripunctata (T. quad-ri-punc-ta′ta). — This species is found in the extreme South. The cephalothorax is marked with a broad band as in the preceding species; the abdomen is black with four oval, white or yellowish spots above. It measures one tenth inch in length. I have seen only the female.

Superfamily Argiopoidea

Genus ULESANIS (U-le-sa'nis)

These are very small spiders in which the dorsal cuticle of the abdomen is firmly chitinized, and raised into a series of humps. The abdomen is higher than long and extends forward over the thorax to the head. Only one species has been found in the United States.

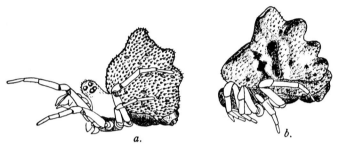

Fig. 359. ULESANIS AMERICANA *a*, male *b*, female

Ulesanis americana (U. a-mer-i-ca'na).— This spider measures only six hundredths of an inch in length and resembles a seed or lump of dirt. It can be recognized by the accompanying copy of a figure by Emerton, who described the species (Fig. 359).

Genus DIPŒNA (Di-pœ'na)

The members of this genus agree with *Euryopis* in having the anterior median eyes larger and much wider apart than the posterior median eyes; but they differ from that genus in having a more globular abdomen, which is broadly rounded behind.

The spiders are found on bushes or on the low branches of trees, especially Conifers, where they spin an irregular web, similar to that of *Theridion;* some species are found in moss; and others under stones.

Four species have been described from the eastern United States. These can be distinguished by the following brief characterizations. Two other species have been described from the Pacific coast.

Dipœna lascivula (D. las-civ'u-la).— Abdomen light yellow. Length of body of female one twelfth inch.

Dipœna buccalis (D. buc-ca'lis).— Abdomen light brown;

opening of the epigynum nearly round; length of body of female nearly one fifth inch.

Dipœna crassiventris (D. cras-si-ven'tris).— Abdomen dark brownish gray, with a light bent line behind; opening of the epigynum heart-shaped. Length of body of female about one seventh inch.

Dipœna nigra (D. ni'gra).— Abdomen black. Length of male six hundredths inch, female one tenth inch.

Genus LATRODECTUS (Lat-ro-dec'tus)

This is one of the two genera of the Theridiidæ occurring in our fauna in which the lateral eyes of each side are widely separated. This genus is distinguished from the other (*Episinus*) in having the abdomen globose. They are comparatively large spiders, the females being the largest of the Theridiidæ.

This genus, as has been well stated by F. P. Cambridge "comprises those very interesting spiders which, under various local names, have been notorious in all ages and in all regions of the world where they occur on account of the reputed deadly nature of their bite." It may be added that this belief has not been shared by many students of spiders. However, sufficient data have been presented by careful investigators to show that this spider is a truly venomous species and one to be avoided. Two well known species occur in our fauna.

The Black Widow, *Latrodectus mactans* (L. mac'tans).— This is a coal-black spider marked with red or yellow or both (Fig. 360). It varies greatly in its markings; the most constant mark is one shaped like an hour-glass on the ventral aspect of the abdomen (Fig. 361). The female, when full grown, is often one half inch in length, with a globose abdomen, marked with one or more red spots over the spinnerets and along the middle of the back; these spots, however, vary greatly in number and size and may be wanting entirely. The male is much smaller than the female, measuring about one fourth inch in length, and is even more conspicuously marked, having in addition to the marks of the female four pairs of stripes along the sides of the abdomen (Fig. 362). It is a curious fact that immature females are often marked like the males.

This species is very common and widely distributed in the

South. It is found under stones and pieces of wood on the ground, about stumps, in holes in the ground, and about outbuildings. It spins an irregular web like that of *Theridion* but

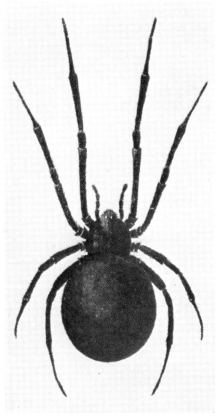

Fig. 360. LATRODECTUS MACTANS, FEMALE

Fig. 361.
LATRODECTUS MACTANS,
VENTER OF FEMALE

Fig. 362.
LATRODECTUS
MACTANS, MALE

of much coarser silk; in fact its web can be recognized, in most cases, at a glance by the coarseness of the thread.

This species is known to occur from Canada to Tierra del Fuego. Although it is essentially a southern species within our country, it is found in almost all the States. There are a few records from Canada, particularly from the western Provinces. In certain areas of the south and the west it is extremely abundant. *Latrodectus mactans* is subject to considerable variation in coloration, and several geographic forms or races have been described.

Typical *mactans* is widely distributed in the region east of the Rocky Mountains. The western form has been dubbed *hesperus*. A third form in which the bright colour markings on the abdomen persist even in adult females is found in Texas and the extreme southwestern States, *L. mactans texanus*. A fourth form, *L. mactans bishopi*, has been described from Florida. In this latter form the legs are bright reddish brown.

The belief in the venomous nature of the bite of this species is very widespread. An intelligent Negro, who saw me collecting the spider in Mississippi, told me that its bite is poisonous. And Dr. C. Hart Merriam in his volume, *The Dawn of the World, Myths and Weird Tales told by the Mewan Indians of California*, ('10) states that the Northern Mewuk say "Po'ko-moo the small black spider with a red spot under his belly is poison. Sometimes he scratches people with his long fingers, and the scratch makes a bad sore." Doctor Merriam adds, "All the tribes know that the spider is poisonous and some of them make use of the poison."

The effects of the bite of this spider on man have been rather carefully studied in recent years. The venom, which is chemically an albumen, is strongly neurotoxic. The symptoms consist ordinarily of severe abdominal pains, a rise in the blood pressure, profuse perspiration, nausea, and a general systemic distress. The severe symptoms usually disappear by the end of twenty-four hours. No accurate estimate of the number of deaths caused by this spider can be made on the basis of the present data. The percentage is certainly very small.

Latrodectus geometricus (L. ge-o-met'ri-cus).—This is a gray species in which the anterior median eyes are distinctly larger than the anterior laterals. It is found in southern Florida.

Genus CRUSTULINA (Crus-tu-li'na)

The base of the abdomen in these spiders is furnished with a horny ring around the insertion of the pedicel. The males have stridulating organs. This genus differs from our other genera of the Theridiidæ that have stridulating organs in having the sternum truncate behind. In our common species the cephalothorax is conspicuously marked with numerous, small, black, crescent-shaped elevations, each at one side of a puncture. Four species have been described from the United States, of which the following is the most common:

Crustulina guttata (C. gut-ta′ta).— There are two varieties of this species differing in the colouring of the abdomen; but these varieties intergrade. In both the cephalothorax is dark brown; in one the abdomen is bright yellow or orange without markings; in the other the abdomen is brown and marked on the dorsal aspect with a median band on the basal part, a curved line around the anterior part, and four spaces about the four muscle impressions of a paler colour; and with several silvery white spots, usually two on each side and one or two in the middle line. The body is about one twelfth inch in length. This spider is common under stones at all seasons.

Genus STEATODA (Ste-a-to′da)

This is one of several genera of comb-footed spiders in which the sternum is pointed behind; it can be distinguished from the others that occur in our fauna by the fact that the anterior median eyes are much larger than the anterior lateral eyes. It is represented in the United States by the following common, and widely distributed species; three others are found in the Far West, and one in New England.

Fig. 363. STEATODA BOREALIS

Steatoda borealis (S. bor-e-a′lis).— This is a dark reddish brown spider measuring when full grown about one fifth inch in length. The cephalothorax is orange-brown, the abdomen, chocolate-brown. Usually there is a light longitudinal band in the middle of the basal half of the abdomen above, which joins a semicircular band on the margin of the front half of the abdomen (Fig. 363). Sometimes these light markings are wanting. The male has stridulating organs.

Superfamily Argiopoidea

This spider makes an irregular, more or less sheet-like web, supported by threads extending in all directions. It is found in buildings or in protected places.

Genus TEUTANA (Teu-ta'na)

The posterior median eyes are fully as large as the anterior median eyes; and the labium is long and pointed, more than half as long as the endites (Fig. 364). The male has well-developed stridulating organs. Two species occur in our fauna. The following is very common.

Teutana triangulosa. (T. tri-an-gu-lo'sa). — The cephalothorax is orange-brown; the legs

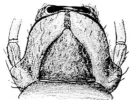

Fig. 364. TEUTANA TRIANGULOSA LABIUM AND ENDITES

Fig. 365. TEUTANA TRIANGULOSA, FEMALE

are light yellow, and very long; the abdomen brown, ornamented with three series of yellow spots, one central and one on each side (Fig. 365). The length of the body of the female is one fifth to one fourth inch; of the male, one eighth inch.

I have found this spider common in the basements of buildings, where usually its web is built in the lower angle of a window in this respect contrasting strongly with the web of *Theridion tepidariorum*, which is usually built in an upper angle. The principal part of the web is an imperfect sheet; a very common feature of it is a number of vertical guy-lines extending down from the sheet. The egg-sacs are made of white silk, loosely woven, without a dense covering. The eggs are plainly visible through the walls of the sac.

Genus COLEOSOMA (Co-le-o-so'ma)

The males of this genus differ from all others of the family in the great development of the bifid plate which forms the ab-

Superfamily Argiopoidea

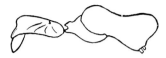

Fig. 366. OUTLINE OF BODY OF COLEOSOMA (after Keyserling)

dominal part of the stridulating organ (Fig. 366). No female has been described.

Coleosoma floridana (C. flor-i-da'na). — This is the only representative of the genus yet found in the United States. It is a small species, the male measuring less than one tenth inch in length. It has been taken in Florida.

Genus LITHYPHANTES (Lith-y-phan'tes)

The slight but distinct separation from each other of the lateral eyes of each side distinguishes this genus from the closely allied genera, except perhaps *Asagena*, in which there is a slight but less distinct separation of these eyes. *Lithyphantes* differs from *Asagena* in that the point of the sternum extends about halfway between the posterior coxæ, and in the narrower clypeus. The males possess stridulating organs. Five species occur in our fauna; they live under stones; the two following are the most common.

Lithyphantes corollatus (L. cor-ol-la'tus).— The cephalothorax is dark brown; in the more common forms the abdomen is yellowish above with four or five, more or less connected, transverse, brown bands (Fig. 367), and dark below with three narrow, yellow lines, which are connected behind; but the species is exceedingly variable. The length of the female varies from one fifth to one fourth inch. This species is very widely distributed in the North.

Lithyphantes fulvus (L. ful'vus).— The cephalothorax and legs are reddish; the abdomen is brownish yellow, with two white spots on each side, a median band or a series of white spots above the spinnerets, and a spot on the ventral aspect of the abdomen of the same colour (Fig. 368). The length of female varies from one fifth to one fourth inch. This is a Southern species which occurs from Florida to Texas and northward.

Genus ASAGENA (As-a-ge'na)

The lateral eyes of each side are slightly separated, but to so slight a degree that they often appear contiguous; hence this

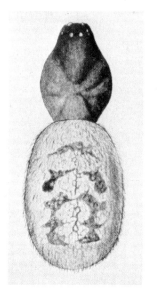
Fig. 367.
LITHYPHANTES COROLLATUS

Fig. 368. LITHYPHANTES FULVUS

Fig. 369. ASAGENA
AMERICANA, MALE

Fig. 370. ENOPLOGNATHA
MARMORATA, MALE

genus is placed in two divisions of the table on page 333. This is a widely distributed genus; but it includes only a few species, of which only one has been found in the United States. The males possess stridulating organs. (Fig. 325, p. 333.)

Asagena americana (A. a-mer-i-ca′na).— The cephalothorax is dark reddish brown; the legs, yellowish brown. In the male the legs bear two rows of teeth under each femur. The abdomen is dark brown with two white spots across the middle (Fig. 369); these white spots are sometimes indistinct. The length of the body is about one sixth inch. This spider is usually found under stones.

Genus ENOPLOGNATHA (En-o-plog′na-tha)

The lateral eyes of each side are contiguous; the lower margin of the furrow of the cheliceræ is toothed, and in the males the cheliceræ are furnished with long teeth. The males possess stridulating organs. This is a very large and widely distributed genus of which several species occur in the United States.

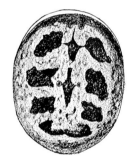

Fig. 371. ENOPLOGNATHA MARMORATA (after Emerton)

Enoplognatha marmorata (E. mar-mo-ra′ta). — The cephalothorax and legs are yellowish brown; the legs are covered with fine hair. The abdomen is whitish, marked with dark spots and lines; there is usually an oblong dark spot covering the greater part of the dorsal aspect, the middle of which is lighter with a central dark stripe (Fig. 370). "In some individuals the dark markings are broken up into four pairs of black spots partly connected with a middle line" (Fig. 371). "It lives under stones and leaves at all seasons and, occasionally on bushes" (Emerton). The length of the body is about one fourth inch. It occurs over a large part of the United States.

Enoplognatha tecta is a closely allied species described from Colorado. The abdomen is brown streaked with black on the sides, with a median dark stripe above, and with two little round white flecks on the side below.

Enoplognatha rugosa (E. ru-go′sa).— This species was recently discovered by Emerton in New Hampshire. It is about

Superfamily Argiopoidea

half as large as *E. marmorata*, measuring a little more than one eighth inch in length. The colour is pale and less yellow than in *E. marmorata;* the abdomen has an indistinct pattern consisting of a broken middle line and, on each side of this, a row of spots.

Genus ROBERTUS (Ro-ber'tus)

The fourth pair of legs are a little longer than the first pair, and all the legs are comparatively short and robust; the chelicerae are robust, being thicker than the femora of the first pair of legs; the lower margin of the furrow of the chelicerae is toothed; and the sternum does not extend between the posterior coxae.

Five species have been found within the limits of the United States; of which the following is the most common.

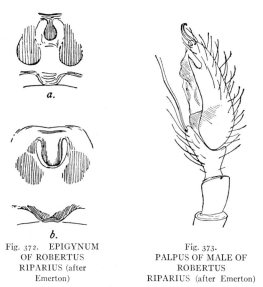

Fig. 372. EPIGYNUM OF ROBERTUS RIPARIUS (after Emerton)

Fig. 373. PALPUS OF MALE OF ROBERTUS RIPARIUS (after Emerton)

Robertus riparius (R. ri-pa'ri-us).—The total length of the body is about one sixth inch. The cephalothorax and abdomen are about equal in length. The cephalothorax is smooth and shining and darkened a little toward the head. The abdomen is gray, generally lighter than the cephalothorax, and covered with dark gray hairs. The guide of the epigynum is pear-shaped in some individuals (Fig. 372, *a*), oblong in others (Fig. 372, *b*).

The cymbium of the palpus of the male is narrow (Fig. 373); near the tip it has a notch on the upper side, and two stiff hairs (Emerton).

This species was first described by Keyserling from Lake Superior; but Emerton states that it is one of the most common spiders under leaves all over New England.

Genus MYSMENA (Mys-me'na)

There are several genera of small theridiids that agree in having the sternum broadly truncated behind, between the widely separated coxæ, as shown in a species of *Histagonia* (Fig. 374). The tarsal claws are completely devoid of denticles. In this and the following genus the abdomen is soft, lacking a hard cover on the dorsum. The median ocular area is as wide as or wider in front than behind and the eyes are subequal. Two or three species are known from our fauna. The following species is not uncommon in the eastern United States.

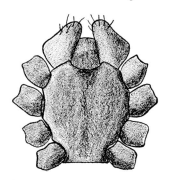

Fig. 374. STERNUM OF HISTAGONIA

Mysmena guttata (M. gut-ta'ta).—These spiders measure about one twentieth of an inch in length. The cephalothorax is yellowish, narrowly margined with black. The legs are pale and are strongly banded with black. The gray abdomen is marked above with four pairs of white or silvery spots.

Genus THEONOE (The-o-no'e)

The abdomen lacks a hard dorsal plate. The anterior eyes are subcontiguous, the median eyes smaller than the lateral. The posterior row is lightly recurved, and the median eyes are nearer the lateral. The median ocular area is wider behind.

Theonoe stridula (T. stri'du-la).—This is a very minute spider, measuring one twenty fifth of an inch in length. The cephalothorax is dusky yellow, and the abdomen is black. This spider was described from Missouri.

Superfamily Argiopoidea

Genus ANCYLORRHANIS (An-cyl-or-rha'nis)

The presence of an abdominal shield and of stridulating organs distinguish these spiders from those of the previous two genera. The clypeus is convex, the eye tubercle evenly rounded, not protruding forward.

Ancylorrhanis hirsutum (A. hir-su'tum).—The abdomen is large and globular. In the male the entire dorsum of the abdomen bears a shield; in the female there is a small plate over the pedicel. The femora of the front legs of the male are armed beneath with a series of stiff bristles.

Fig. 375. EYES OF PHOLCOMMA (After Simon)

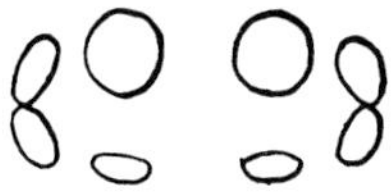

Fig. 376. EYES OF ANCYLORRHANIS

Genus PAIDISCA (Pai-dis'ca)

This genus resembles the preceding one in general appearance. The clypeus is concave, the eyes placed upon a tubercle which, particularly in the males, projects forward.

Paidisca marxi (P. marx'i).—The abdomen of the male is covered by a yellow shield. The abdomen of the female is hard above, dusky white to pale yellow, marked with a dark stripe at the middle and black dashes on each side.

Family LINYPHIIDÆ (Lin-y-phi'i-dæ)

The Sheet-web Weavers

This family includes a large number of common species; but most of them are of such small size and lead such secluded lives that they rarely attract attention. A few of the species, however, are of larger size and build more or less conspicuous webs. The webs made by these are of various forms; but as they usually contain one or more sheets of silk, I propose the term sheet-web weavers as the popular name of the family.

The sheet-web weavers belong to the series of three-clawed, eight-eyed, sedentary spiders. In one genus, *Anthrobia*, which

Superfamily Argiopoidea

inhabits caves, the eyes are wanting. The Linyphiidæ differ from the Theridiidæ, with which they were formerly classed, in lacking the comb on the tarsi of the fourth pair of legs, which is characteristic of that family (Fig. 377); and they differ from

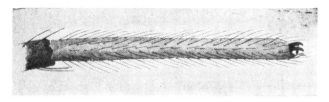

Fig. 377. TARSUS OF LEG OF LINYPHIA

the Argiopidæ, with which they are classed by some later writers, in usually having more or less distinct organs of

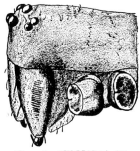

Fig. 378. CHELICERA OF LEPHTHYPHANTES NEBULOSUS, SHOWING THE FILE

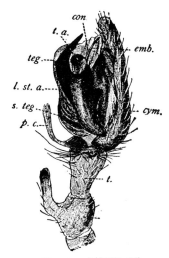

Fig. 379. PALPUS OF LINYPHIA PHRYGIANA
pc, paracymbium

stridulation on the external side of the basal segment of the cheliceræ (Fig. 378), a nd from nearly all argiopids in having dissimilar eyes and in lacking the lateral condyle of the cheliceræ.

The palpi of the males of the Linyphiidæ differ from those of the argiopids in that the paracymbium is closely applied to the bulb, except during the turgesence of the hæmatodocha. The cuticle of the paracymbium is hard and smooth, but its margin is often furnished with hairs or spines, usually in a single series.

Superfamily Argiopoidea

In form the paracymbium is more or less crescent-shaped or horse-shoe-shaped (Fig. 379, *p. c.;* and Fig. 386, *p. c.*).

The most striking difference, however, between these spiders and the Argiopidæ is in web-building habits, the type of web known as the orb-web being characteristic of the Argiopidæ, while the members of the Linyphiidæ, as already stated, build sheet-webs. Several of these are figured in the following pages.

The family Linyphiidæ is represented in North America by two subfamilies, the Erigoninæ and the Linyphiinæ. Each of these contain many genera and species. It is not easy to distinguish these subfamilies; the most available characters for separating them are given below in tabular form.

TABLE OF SUBFAMILIES OF THE LINYPHIIDÆ

Males

A. Tibia of pedipalps furnished at its distal end, either above or on its external angle, with an apophysis; the paracymbium of the tarsus of the pedipalp usually small, P. 384.   ERIGONINÆ

AA. Tibia of pedipalps dilated at the distal end but without a true apophysis; the paracymbium of the tarsus of the pedipalp well-developed and its border often furnished with a series of spines or granulations. P. 388.   LINYPHIINÆ

Females

A. Pedipalps without tarsal claws; epigynum comparatively simple, without a projecting ovipositor. P. 384. ERIGONINÆ

AA. Pedipalps with tarsal claws except in the cave inhabiting genera (*Anthrobia, Phanetta,* and *Troglohyphantes*); epigynum either comparatively simple or with a prominent appendage, the ovipositor. P. 388.   LINYPHIINÆ

Subfamily ERIGONINÆ (E-rig-o-ni′næ)

The Erigonids (E-rig′o-nids)

The erigonids are all very small spiders; for this reason, they are seldom observed except by the more careful collectors.

They are usually found on or near the ground, where some of them build sheet-webs; these are so delicate as to be invisible except when covered with dew.

Many species are found among dead leaves, in grass, and in moss. The best way to collect them is by putting the material in the midst of which they live into a coarse sieve and shaking it over a sheet of cloth or paper.

In the autumn many of the erigonids become aeronautic spiders, migrating by means of silken threads, which buoy them up in the air and transport them long distances.

The males of the Erigoninæ are distinguished from those of the Linyphiinæ by the presence of an apophysis on the tibia of the pedipalps. This apophysis is situated at the distal end of the tibia, either above or on its external angle, and varies greatly in form; it either projects or lies on the base of the tarsus.

In the female, the palpi are without tarsal claws and the epigynum is usually very simple, consisting of a small plate without a projecting ovipositor.

This subfamily is a very large one. Several hundred species and scores of genera are known from the United States. It is unfortunate that up to the present time no adequate key has been published for the separation of the genera. This is true mainly because the characters for differentiating the genera are in the genitalia of the males and are difficult to define and are best expressed by figures. The student is referred to the selected papers by Crosby and Bishop as given in the bibliography for full accounts of many of our genera. The following species are representative.

Ceraticelus fissiceps (Cer-a-tic′e-lus fis′si-ceps).—These small spiders measure only one sixteenth inch in length; their colour is light orange. A shield covers the greater part of the dorsal aspect of the abdomen; this is darker than the remainder of the abdomen. The head is black or dusky about the eyes; and has a transverse furrow back of the anterior median eyes; this furrow is shallow in the female, but very deep in the male (Fig. 380).

Fig. 380. CEPHALOTHORAX OF CERATICELUS FISSICEPS

This is a very common species; it is found on low bushes in the summer, and can be obtained by sifting leaves in the winter.

Ceraticelus lætabilis (C. læ-tab′i-lis).—This resembles the

Superfamily Argiopoidea

preceding species in size, being one sixteenth inch in length. It differs in lacking the transverse furrow on the head, in being of a darker colour, and in the female in usually lacking the shield on the dorsal aspect of the abdomen. The cephalothorax is dark brown, the legs orange. The dorsal aspect of the abdomen of the male is covered with a dark orange-brown shield, in the female this shield is usually reduced to four dots. In both sexes there are shields on the ventral aspect, one in front of the spinnerets and one on each side at the base of the abdomen.

Fig. 381. HEAD OF GNATHONAGRUS UNICORN

Gnathonagrus unicorn (Gna-tho-nag'-rus u-ni-corn').—The cephalothorax and legs are yellowish, the abdomen, olive-gray. The male is easily recognized by the presence of a long horn projecting from the middle of the clypeus (Fig. 381); this horn is clothed with stiff hairs at the tip. Another striking feature is the great length of the apophysis of the tibia of the male palpus, which is sickle-shaped and longer than the body of the segment. The length of the female is one sixteenth inch; of the male, a little less.

Cornicularia directa (Cor-ni-cu-la'ri-a di-rec'ta).—The cephalothorax is brown, the abdomen, olive-gray; the length of the body is one twelfth inch. In the male there is a horn projecting forward from the eye-space and a smaller horn below this (Fig. 382).

Fig. 382. HEAD OF CORNICULARIA DIRECTA

Origanates rostratus (Or-i-ga-na'tes ros-tra'tus).—This is one of a group of species in which the males have cavities in the head opening by holes near the eyes. In the male of this species there is a prominent hump on the head which bears the posterior median eyes; and on each side at the base of this hump and just behind the lateral eyes there is a hole (Fig. 383). The length of the body is one twelfth inch.

Œdothorax montiferus (Œ-do-tho'rax mon-tif'e-rus).—The cephalothorax is dark yellowish brown; the abdomen, dark gray; and the legs, orange-brown. On the cephalothorax of the male

Superfamily Argiopoidea

(Fig. 384) there is a hump half as large as the rest of the cephalothorax; this hump contains large cavities, which open by holes on either side, as large as the eyes. These holes are connected by a deep crease, which extends around the front of the hump. The anterior middle eyes are near together in the middle of the head in front of the hump; the other eyes are in two groups at the

Fig. 383. CEPHALOTHORAX OF
ORIGANATES ROSTRATUS

Fig. 384. CEPHALOTHORAX OF
ŒDOTHORAX MONTIFERUS

extreme corners of the head. In the female the back of the head is considerably elevated, and the posterior middle eyes are farther apart than usual. The length of the body is one twelfth inch.

Eperigone maculata (E. ma-cu-la′ta).—This is probably the most common member of Erigoninæ. It is a small species, measuring one sixteenth inch or a little more in length; and is not striking in appearance, both sexes being of ordinary form. The cephalothorax is yellowish brown; the legs, dull yellow; and the abdomen gray with five or six pairs of obscure yellowish markings. The colour varies, some individuals being almost black and others very pale.

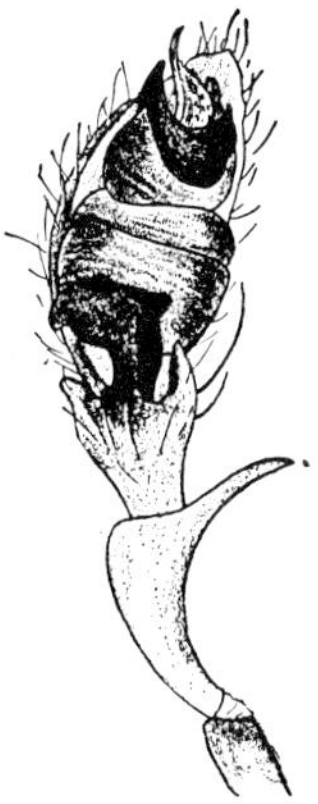

Erigone autumnalis (E-rig′o-ne au-tum-na′lis).— In the palpi of the males of the genus *Erigone* the patella is very long and is armed at

Fig. 385. PALPUS OF ERIGONE AUTUMNALIS

the tip with a large apophysis (Fig. 385); the cheliceræ are armed with teeth on the front side; and there are small teeth on the margin of the cephalothorax. Fourteen species have been catalogued from the United States. *E. autumnalis* is a tiny spider, measuring only one twentieth of an inch in length. It can be

Superfamily Argiopoidea

distinguished from the other species by its light colour and bright yellow head.

Ceratinopsis interpres (Ce-rat-i-nop'sis in-ter'pres).— This is a representative of a group of species which differ from all other Erigoninæ in having longer legs, which are also more delicate toward the extremity; the metatarsi are not at all or barely shorter than the tibiæ; the sternum is broad and ends behind between the broadly separated hind coxæ in a truncated and inflexed point. In *C. interpres* the colour is bright orange with a little black around the eyes and around the spinnerets. The length of the body is about one tenth inch.

Subfamily LINYPHIINÆ (Li-nyph-i-i'næ)

The Linyphiids (Li-nyph-'i-ids)

To this subfamily belong the larger and better known of the species of the sheet-web weavers; some of them are exceedingly common and build elaborate webs, the type of which varies greatly according to the species.

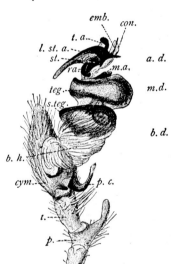

Fig. 386. PALPUS OF LINYPHIA PHRYGIANA

There is no easily seen character by which the linyphiids can be distinguished from the erigonids. In the males of the Linyphiinæ the tibia of the pedipalps lacks a true apophysis at its distal end, although it is sometimes enlarged at the tip or may even bear a tooth; and the paracymbium is more developed than in the preceding subfamily, it is more or less sickle-shaped or horseshoe-shaped. Figure 379 represents the bulb of the male of *Linyphia* with the paracymbium (*p. c.*) slightly removed from it; and Fig. 386 represents an expanded bulb. This is described in detail in an earlier chapter (p. 112).

In the female the pedipalps are furnished with tarsal claws

except in the three cave-inhabiting genera (*Anthrobia, Phanetta,* and *Troglohyphantes*). The epigynum is sometimes comparatively simple; but it is often furnished with prominent appendages, the ovipositor. This ovipositor in the more highly specialized forms (Fig. 387) is remarkable in that it is composed of two projections: one, the *scape* arising in the usual position in front of the

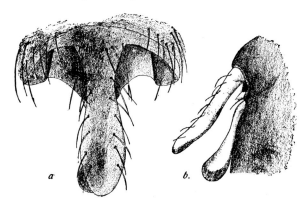

Fig. 387. EPIGYNUM OF BATHYPHANTES NIGRINUS

opening of the oviduct; and the other, the *parmula*, arising behind this opening. The second projection is concealed by the first, except when the organ is viewed in profile. Each of these projections is grooved on the face next its fellow, the two grooves forming a tube.

The following table of genera includes the greater part of our known forms; but there are some of doubtful position that have not been included.

TABLE OF GENERA OF THE LINYPHIINÆ

A. Endites of the pedipalps as wide at the base as they are long. Anterior metatarsi not longer than the tibiæ. Legs unarmed or very sparingly spined; tibiæ always without lateral spines.
  B. Eyes wanting. P. 391.   ANTHROBIA
  BB. Eyes present.
    C. Anterior median eyes very small, nearly obsolete; median quadrangle greatly narrowed in front, less than half as wide there as behind. P. 391. PHANETTA

Superfamily Argiopoidea

  CC. Anterior median eyes often greatly reduced in size, but always well developed; median ocular quadrangle less strongly narrowed in front, more than half as wide there as behind. P. 392.   MICRONETA

AA. Endites longer than their width at the base. Anterior metatarsi not shorter than the tibiæ. Legs armed with spines; the tibiæ almost always furnished with dorsal and lateral spines.

 B. Posterior median eyes much nearer to the posterior lateral eyes than to each other; anterior eyes in a strongly recurved line; median ocular area large. P. 413.
              TAPINOPA

 BB. Posterior eyes nearly equidistant; anterior eyes in a nearly straight line.

  C. Cheliceræ furnished in front with three or four spines. P. 394.   DRAPETISCA

  CC. Cheliceræ unarmed in front.

   D. Posterior eyes near together. Sternum heart-shaped, not at all longer than wide.

    E. Median ocular area large, longer than wide, and barely narrower in front than behind; all eyes subequal. P. 397.   LABULLA

    EE. Median ocular area not longer than wide, and much narrower in front than behind; anterior median eyes smaller than the others.

     F. Metatarsi furnished with one or more spines. P. 393.   LEPHTHYPHANTES

     FF. Metatarsi unarmed. P. 392.   BATHYPHANTES

   DD. Posterior eyes distant.

    E. Eyes greatly reduced in size, evanescent, without pigment, the anterior median often obsolete or nearly so; posterior median eyes much nearer each other than the lateral eyes; small, pale, cave spiders. P. 396.   TROGLOHYPHANTES

    EE. Eyes normal, all well developed and fully pigmented, the anterior median often reduced in size but never obsolete; eyes of the posterior row subequidistantly spaced, the median sometimes larger than the posterior lateral eyes; not cave spiders. P. 398.   LINYPHIA

Superfamily Argiopoidea

Genus ANTHROBIA (An-thro′bi-a)

The eyes are completely wanting. The paired claws are armed with six or more short teeth, and the median claw has a single small denticle near the base. The pedipalps of the female have no tarsal claws. The chelicerae are armed in front with four teeth. The palpus of the male has a long process on the outside of the tibia which ends in a sharp hook.

A single rare species is known.

Anthrobia mammouthia (A. mam-mouth′i-a).— The adult measures six hundredths of an inch in length. It is pale brownish yellow; the abdomen is almost white, with brown hairs; the ends of the pedipalps and the epigynum are reddish brown.

This spider was first described from the Mammoth Cave; but it has since been found in other caves. Its eyeless condition is an excellent illustration of the loss of an organ through disuse. The species is doubtless descended from an eyed form, and has lost its eyes as a result of living in the dark for many generations.

Genus PHANETTA (Pha-net′ta)

This genus like the preceding includes a single cave-inhabiting species; but this is not blind. The eight eyes are present, although the anterior median eyes are greatly reduced in size. The pedipalps of the female have no tarsal claw; in this respect this genus agrees with *Anthrobia* and *Troglohyphantes* and differs from all other members of the Linyphiinæ. It differs from *Troglohyphantes* in having the posterior median eyes about as close to the posterior lateral eyes as to each other, and in that the anterior median eyes are barely the diameter of one of them apart.

Phanetta subterranea (P. sub-ter-ra′ne-a).—This is the only known species of the genus. The length of the body is about one twentieth of an inch. The cephalothorax and legs are pale yellowish brown, in some specimens reddish. The abdomen is nearly white and clothed with pale hairs. The chelicerae are armed in front with six or seven small teeth. The eyes are often evanescent, lacking pigment, and reduced in size. The epigynum is a triangular finger, about as broad as long, which is directed ventrad. The tibia of the male palpus is armed with spurs as in the Erigoninæ, in which subfamily this species may ultimately be placed.

This spider occurs in caves in Kentucky, Tennessee, North Carolina, and Alabama, and probably in caves elsewhere.

Genus MICRONETA (Mi-cron′e-ta)

The members of this genus are small spiders of slender form with long and slender legs. They agree with the preceding genus in having nearly square endites (Fig. 388) and in that the tibiæ are without lateral spines; but differ in that the clypeus is more or less depressed below the eyes, the posterior eyes are close together, the lateral eyes are on slight tubercles, and the legs are longer and more slender.

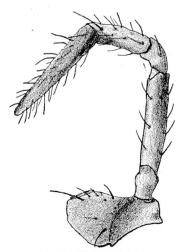

Fig. 388. PEDIPALP OF MICRONETA UNIMACULATA

A score of species are known from the United States; but as most of the species have been collected by sifting dead leaves, comparatively little is known regarding their habits. The species are mostly black or reddish or greenish brown. The following species is well known both in Europe and in the United States.

Microneta viaria (M. vi-a′ri-a).—Both sexes average about one sixth inch in length. The cephalothorax and legs are yellowish brown, and the abdomen is gray or brown.

The common spider is found under leaves in woods.

Genus BATHYPHANTES (Bath-y-phan′tes)

With this genus begins the series of genera of the Linyphiinæ in which the legs are long and are furnished, at least on the tibiæ, with superior and lateral spines; and in which the metatarsi are as long as or longer than the tibiæ.

In *Bathyphantes* the sternum is heart-shaped, the median ocular area is not longer than wide, and the metatarsi are unarmed.

These are small spiders; they are found under stones or leaves, and at the base of plants, where they spin a very delicate sheet.

Nearly thirty species have been found in the United States; of these the following are the more common.

Bathyphantes concolor (B. con'co-lour).— The length of the body is about one twelfth inch in both sexes. The cephalothorax is yellowish brown; the legs, yellow. The abdomen is gray without any markings. The ovipositor (Fig. 389) is long and slender; it reaches nearly or quite to the middle of the abdomen. The ovipositor consists of two pieces; the second piece, the parmula, is much shorter than the first, the scape, and is concealed by it except when viewed in profile. The cymbium of the male palpus is long and tapering.

This is a common species throughout the North. We have taken adults of both sexes in September; they pass the winter under leaves.

Bathyphantes nigrinus (B. ni-gri'nus).— This is a little larger than the preceding species, measuring one tenth inch in length. The cephalothorax and legs are light yellowish brown; and the abdomen is dark gray or black with five or six transverse light bands. The ovipositor is long and slender, and differs from that of the preceding species in that the second piece extends as far back as the first (Fig. 387). The cymbium of the male palpus is short and truncated, and the embolus is twisted in a circle on the end of the bulb.

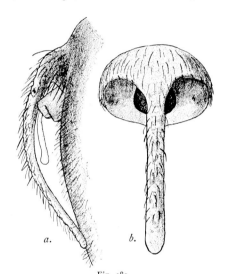

Fig. 389.
EPIGYNUM OF BATHYPHANTES CONCOLOR
a, lateral view   *b*, ventral view

Genus LEPHTHYPHANTES (Leph-thy-phan'tes)

This genus differs from *Bathyphantes* in having the spines of the legs longer and more numerous; the metatarsi, however, usually bear only one spine, the femora of the fore legs one or two, and the femora of the hind legs none. The stridulating striæ on the outer face of the chelicer æ are very conspicuous (Fig. 378).

More than a score of species have been described from our fauna of which the two following are the most common.

Lephthyphantes nebulosus (L. neb-u-lo'sus).—Most of the species of this genus are of small size; but this one is larger, measuring one sixth inch in length. The colour is light brownish yellow with gray or blackish markings. Figure 390 represents the type of the markings of the abdomen; but there is a great variation in the size of the spots. The ovipositor is folded back under the atriolum, so as to be almost concealed except when seen in profile (Fig. 391).

This species lives in damp and shady places and is often found in cellars. It makes a large, flat, sheet-web beneath which the spider hangs. It is a widely distributed species, occurring in Europe and in a large part of this country.

Lephthyphantes leprosus (L. le-pro'sus).—This species is smaller than the preceding, measuring only one eighth inch in length, and differs in being dark gray with light markings (Fig. 392). The epigynum is of the same type, the ovipositor being folded under the atriolum.

This species also lives in damp and shady places and is often found in cellars.

Genus DRAPETISCA (Drap-e-tis'ca)

There is on the front surface of each chelicera three or four spines in an oblique row (Fig. 393); and on the tibia and tarsus of the pedipalps of the female some larger, divergent spines. In the pedipalps of the male the cymbium is furnished with a large, curved apophysis at the base in addition to the paracymbium. The epigynum of the female is furnished with a large ovipositor.

Two species of this genus have been described *D. socialis* of Europe and *D. alteranda* of the United States. Until recently our species has been considered identical with that of Europe and has been described under the name *D. socialis*.

Drapetisca alteranda (D. al-te-ran'da). — The cephalothorax is white margined with black, the eyes are on black spots, there is a black band in the middle of the thorax above, between this

Fig. 390. LEPHTHYPHANTES
NEBULOSUS

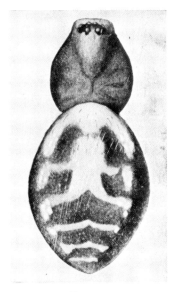

Fig. 302.
LEPHTHYPHANTES LEPROSUS

Fig. 391. EPIGYNUM OF
LEPHTHYPHANTES NEBULOSUS

Fig. 393. CHELICERA
OF DRAPETISCA

Superfamily Argiopoidea

and the black margin there is a series of more or less distinct black spots (Fig. 394, male; Fig. 395, female), the abdomen is white mottled with black or dark gray. The length of the body is one eighth inch.

In its habits this species differs greatly from the more typical sheet-web weavers, for so far as has been observed, it makes no web. It is found under leaves on the ground and on the trunks of trees, where it pursues its prey. Its colour and markings are protective, resembling that of bark and lichens. Our adult specimens were taken in August and September.

Fig. 394. DRAPETISCA ALTERANDA, MALE

Genus TROGLOHYPHANTES
(Trog-lo-hy-phan'tes)

These are cave-inhabiting spiders, which, however, are not blind. Their eyes are greatly reduced in size, especially the anterior median eyes; these in some specimens appear to be wanting. In this genus, as in *Anthrobia* and *Phanetta*, the pedipalps of the female have no tarsal claws. The presence of eyes distinguishes this genus from *Anthrobia*; and it differs from *Phanetta* in the arrangement of the eyes, the posterior median eyes being much closer together than to the posterior lateral eyes, and the

Fig. 395.
DRAPETISCA ALTERANDA, FEMALE

anterior median eyes are much more than the diameter of one of them apart. This is the genus *Willibaldia* of Keyserling.

Two species are known from this country and one from Europe.

Troglohyphantes cavernicolus (T. cav-er-nic′o-lus).— The female is about one tenth inch in length; the male, less than one twelfth. The cephalothorax, legs, and pedipalps are light or brownish yellow; the abdomen, white or light gray. The upper margin of the furrow of the cheliceræ bears two quite large teeth, the lower margin, one small one.

This spider has been found in caves in various parts of the United States. I am indebted to Mr. Cyrus R. Crosby and to Mr. Paul Hayhurst for specimens from a cave at Rocheport, Mo. These were taken on little sheet-webs on the walls and on the roof of a very damp and perfectly dark chamber of the cave. Mr. Crosby collected also the egg-sacs. These are white, and were attached to the lower side of little projections of the wall of the cave.

Troglohyphantes incertus (T. in-cer′tus).— I have not seen this species, but from the published descriptions it appears to resemble the preceding very closely in size, colour, and in habits. The upper margin of the furrow of the cheliceræ bears seven teeth.

Genus LABULLA (La-bul′la)

The anterior eyes are large, equal in size, and in a slightly recurved line; the median ocular area is longer than wide and about as wide in front as behind. The bulb of the pedipalps of the male is large and furnished with a long embolus, rolled in a circle. Three species have been found in our fauna. The following one was described from the State of Washington.

Labulla altioculata (L. al-ti-oc-u-la′ta).— The body measures about one third inch in length. The cephalothorax is light yellow; the abdomen light gray, with a large triangular black spot at the base and several transverse black bands between this and the tip of the abdomen.

Simon states that the spiders of this genus are nocturnal; that they live in woodlands, under large stones, between the roots of stumps, and in similar places. The web is large, built near the ground, and resembles that of *Tegenaria*.

Superfamily Argiopoidea

Genus LINYPHIA (Li-nyph'i-a)

The genus *Linyphia* includes those sheet-web weavers that are most apt to attract attention. The species, although of moderate size, are large compared with the majority of the members of the family, and several of them build more or less conspicuous webs.

The sternum is longer than wide. The endites of the pedipalps are longer than their width at the base; squarely or obliquely truncate, the lateral margin forming a more or less acute angle (Fig. 399). The posterior eyes are widely separated, the median subequal to or larger than the posterior lateral eyes.

Fig. 396. LINYPHIA CLATHRATA

The generic name *Linyphia* is used here in the broad sense of a supergenus to designate a group of species which are quite similar in structure and habits, but which, in more modern classification, have been broken up into a number of separate genera. Our representatives of this group have been studied intensively by Blauvelt ('36). Advanced students are referred to that revision for fuller details regarding the genera and species. The genera are based to a large extent on the structure of the genital organs, which in many cases in this complex group give the only accurate index of the position of the species. Several of our species are Holarctic in distribution, being well known in Europe.

This is a very large genus. It is represented by more than a score of species in the United States, in fact I have nearly twenty species before me as I write; but only a few of the more common ones can be described here.

Linyphia clathrata (L. cla-thra'ta).—The body is about one eighth inch in length. The cephalothorax is yellowish brown, and the eyes are on black spots. The abdomen of the female is pale

Superfamily Argiopoidea

brown, thickly spotted with white and marked with brownish black bars (Fig. 396). The male is darker; in this sex the abdomen is sometimes almost all black with a white spot on each side.

The web is built among grass near the ground; it is a flat sheet, and the spider hangs at one side of it.

Linyphia coccinea (L. coc-cin′e-a).—When alive this spider is crimson or red, with the last three segments of the pedipalps, the area occupied by the eyes, and the tip of the tubercle of the abdomen black. In alcohol, the red colour is lost to a greater or less extent. A very distinctive feature is the presence of a more

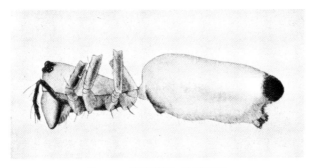

Fig. 397. LINYPHIA COCCINEA

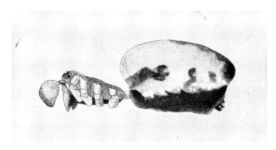

Fig. 398. LINYPHIA VARIABILIS

or less pronounced tubercle at the end of the abdomen above the spinnerets (Fig. 397).

This is a widely distributed species, but it is apparently more common in the South than in the North. I have not seen its web, having taken the spider only by sweeping.

Linyphia variabilis (L. var-i-ab′i-lis).—A striking characteristic of this species is the form of the abdomen, which is compara-

Superfamily Argiopoidea

tively high and ends in a more or less pronounced rounded projection, which is situated a considerable distance above the spinnerets (Fig. 398); there is, however, considerable variation in the degree of prominence of this projection. The cephalothorax is yellowish brown; the venter of the abdomen is dark reddish brown; on each side of the abdomen there is a row of small silvery spots, and above this a row of blackish spots; the dorsal aspect of the abdomen is reddish or yellowish with a few small silvery spots; there is a dark spot on the rounded tip of the abdomen, and in front of this a double row of spots which vary greatly in size. The length of the body is about one eighth inch; the specimens measured were not quite mature.

This species occurs in the northeastern States.

Fig. 399. PEDIPALP OF LINYPHIA PHRYGIANA, FEMALE

The Bowl and Doily Spider, *Linyphia communis* (L. com-mu'nis).—The female of this species measures about one sixth inch in length, usually a little less, but sometimes more. At first sight it resembles the filmy dome spider; but it differs in having the cephalothorax of a uniform, light, brownish yellow, and in that the dorsal band of the abdomen extends its whole length as seen from above; but on the hind end of the abdomen, a short distance before the spinnerets, it is either reduced to a narrow line or is entirely cut in two by the light colour.

The male is smaller than the female, measuring from one tenth to a little more than one eighth of an inch in length. The markings on the abdomen are much less distinct than in the female.

The central feature of the web of this spider (Fig. 400) is a fingerbowl-like cup beneath which is stretched a nearly horizontal sheet. I, therefore, propose as a common name for the species the term the bowl and doily spider.

The webs are built on low bushes, sometimes quite near the ground, in other cases, three or four feet above the ground. In the cavity of the bowl and extending several inches above it is a maze of threads. The maze and the bowl are suspended by strong foundation lines extending to neighbouring twigs. The

Fig. 400. WEB OF LINYPHIA COMMUNIS

Superfamily Argiopoidea

lower sheet or doily is much more nearly horizontal than the bowl, but is somewhat concave. It is not a discarded bowl, but is composed of as fresh and delicate tissue as is the bowl. The spider rests in an inverted position near the centre of the lower surface of the bowl and is protected from attacks from below by the doily. Insects that fly against the thread of which the maze is composed are apt to fall to the bottom of the bowl, in which event they are seized by the spider and pulled through the sheet as is done in the case of the filmy dome spider.

On one occasion I carried a bush bearing the web of this species three or four miles to our insectary. It was a very windy day and the bowl of the web was blown entirely away; but the spider clung to the upper surface of the lower sheet. The insectary was reached at 5 P.M. At 10 P.M. of the same day, I found that the spider had begun to spin a new bowl; and when examined on the following morning, the web was completely restored.

I have found this species in only a few localities in New York but in the South it is one of the most common of the sheet-web weavers.

This is the *Frontinella communis* of some writers.

Linyphia insignis (L. in-sig'nis).— The specific name *insignis* is not a very fortunate one for this species, as the distinguishing marks that probably suggested it are frequently wanting. The cephalothorax and mouth-parts are light orange-yellow. The abdomen varies from dark gray to white and is either without markings or with gray stripes across the back and on the sides (Fig. 401); the stripes on the back are often angular. The epigynum projects in a finger-like process, which reaches the middle of the abdomen and has openings in the end (Fig. 402). The tibia of the palpus of the male has a short pointed process extending directly outward from the side. The length of the body is one eighth inch.

I have taken many mature specimens of both sexes in October by sweeping, but failed to observe the web. Emerton states that the spider lives in flat webs among low plants.

Linyphia lineata (L. lin-e-a'ta).— This species is easily distinguished from our other common species of *Linyphia* by the three rows of black spots on the dorsal aspect of the abdomen (Fig. 403); there are also irregular black spots on the sides of the abdomen. The ground colour of the body is light yellowish gray;

Fig. 401. LINYPHIA INSIGNIS

Fig. 403. LINYPHIA LINEATA

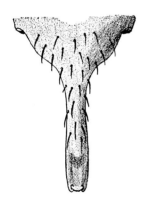

Fig. 402.
EPIGYNUM OF LINYPHIA INSIGNIS

Fig. 404.
LINYPHIA MARGINATA

the cephalothorax is marked with three dark lines, and the legs are prominently ringed with black. The male is nearly one fifth inch in length; the female, about one fourth inch.

This species is found under stones and logs. I have not observed its web; but Menge in his work on Prussian Spiders, for this is a common European species, states that it makes a horizontal net-like web in the grass or heath at the base of trees. And I have taken a female with her egg-sac in thick grass; this was in April.

The male represented by Fig. 403 was taken in August; but was immature. It illustrates well the phenomenon of reproduction of lost parts. This individual had evidently lost, when young, his left hind leg; and a new leg was half grown at the time he was captured.

This is the *Bolyphantes bucculenta* of some authors; but it is the *Aranea lineata* of Linnæus.

The Filmy Dome Spider, *Linyphia marginata* (L. mar-gi-na′ta).— The marvellous delicacy and peculiar form of the web of this spider leads me to suggest the above popular name for it.

The adult spider of either sex measures about one sixth inch in length. The cephalothorax is yellowish brown margined on each side with a light stripe. The abdomen is yellowish white heavily marked with dark bands and stripes (Fig. 404); in the middle of the upper side there is a broad band which consists of three parts united by narrower portions, and at the tip of the abdomen there is a spot which is usually connected with this band by a very narrow line. This dorsal band and apical spot usually include two series of more or less distinct, lighter spots, frequently two pairs of spots in each division of the band. On the sides of the abdomen there are several dark stripes; those on the basal half of the abdomen are longitudinal, while those on the hind half are vertical. These lateral stripes are less distinct in the male.

The web of this spider is usually found on herbs or low bushes in cool moist places, as in the borders of a woodland path, or on shrubs fringing a shady stream. It may be very common and yet not attract attention; for it is so delicate that it is often invisible except when the light falls upon it in the most favourable manner. And even then its marvellous beauty and delicacy can be appreciated only when seen against a dark background. It is

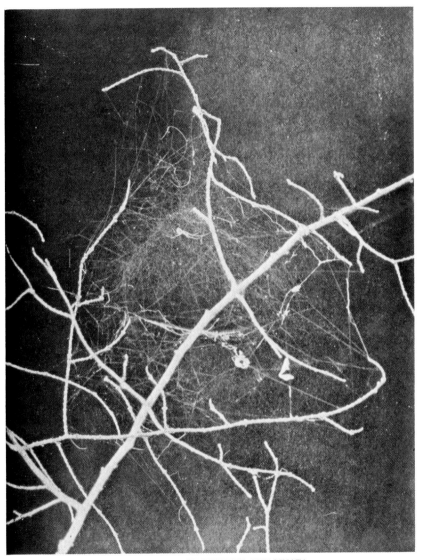

Fig. 405. WEB OF LINYPHIA MARGINATA

Superfamily Argiopoidea

my practice, when showing students this web in the field, to hold a piece of black velvet cloth behind the web. With such a background and with the sunlight falling upon the web, the observer is sure to be filled with enthusiastic admiration.

The characteristic form of the web is well-shown in Fig. 405. There is a maze of threads extending in all directions, and in the centre of this maze a dome-like sheet from three to five inches in diameter. When at rest, the spider hangs beneath the apex of the dome. Here it waits till some insect whose flight is impeded by the maze of threads falls upon the dome; the insect is then seized, pulled through the wall of the dome, and destroyed. When the dome becomes badly injured it is pulled down and a new one built. In the figure there is shown a dense mass of threads extending horizontally a short distance below the dome; this is the remnant of an old dome.

This is a very common and widely distributed species, both in this country and in Europe. The spiders mature in early summer, and the young can be found in tiny webs in August and September.

The Sierra Dome Spider, *Linyphia litigiosa* (L. li-tig-i-o′sa).— In the Sierra Nevada Mountains there is found a species of *Linyphia* that makes a web similar to that of the filmy dome spider of the East except that the dome is much larger and slightly flatter (Fig. 406). Although this spider is widely distributed on the Pacific Coast, and occurs in the Coast Range as well as in the Sierras, I suggest the popular name, the Sierra dome spider, on account of its great abundance in the Sierras. Excepting a species of *Agelena*, this is the most abundant spider found over a very extensive area in these mountains.

The adult spider measures one fourth inch in length. The cephalothorax is yellow, with a narrow median black line, and near each lateral margin there is a wider but less distinct dark band; the sternum is black. The abdomen is silvery white above marked with dark brown or black; there is a median dark band, from which extend a variable number of more or less distinct oblique lines; near the caudal end of the abdomen, this median band is broader and encloses from one to three pairs of white spots; the lateral aspect of the abdomen is dark marked with three or four oblique white lines on the caudal half and a longitudinal white band on the basal half; the venter is dark marked with white dots

(Photographed by G. K. Gilbert)
Fig. 406. WEB OF LINYPHIA LITIGIOSA

Superfamily Argiopoidea

and two more or less distinct white bands. Mature spiders were collected during the first week in August.

The web is built among shrubs, between the branches of small trees, in stumps, and against the sides of logs. The dome is from one foot to two feet in horizontal diameter and from five to eight inches in vertical diameter. It is of the same delicate structure as that of the filmy dome spider, but is more conspicuous on account of its much larger size. I saw thousands of these domes along the trails in the sugar pine belt of Tuolomne County, Cal. It seemed very appropriate that the architects of these large domes should build beneath these magnificent trees. In the higher altitudes i. e., above 7000 feet, I did not find the spider.

Fig. 407. LINYPHIA PHRYGIANA

The species was first described by Keyserling from specimens collected in Washington State; and Mr. R. W. Doane has sent me specimens collected near Stanford University. It is evident, therefore, that the species is widely distributed on the Pacific Coast.

The Hammock Spider, *Linyphia phrygiana* (L. phryg-i-a'na). — This is one of the most common species of *Linyphia* in the eastern half of the United States; and is one that is easily recognized by its characteristic markings. The cephalothorax is light yellow, narrowly margined with black, and with a central dark line on the thorax and two closely parallel lines on the head. The abdomen is yellowish with a dark brown or reddish herringbone stripe in the middle (Fig. 407). The length of the body is one fifth inch.

The web of this spider (Fig. 408) is a netted hammock-like sheet, which is more often quadrangular in outline than otherwise, but the shape depends on the nature of the support. It is

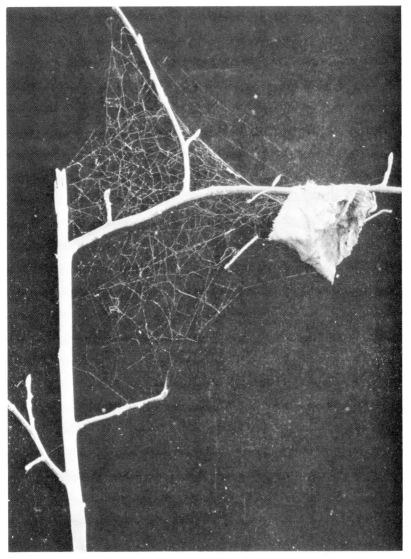

Fig. 408. WEB OF LINYPHIA PHRYGIANA

Fig. 409. WEB OF LINYPHIA PHRYGIANA

Superfamily Argiopoidea

found chiefly on herbaceous plants, although it occurs also on shrubs, the lower branches of trees, and even on fences. I have observed it most often on the borders of woodlands, in gorges, and in valleys near streams. In almost every web that I have examined there was a curled leaf that was used as a retreat by the

Fig. 410. WEB OF LINYPHIA PUSILLA

spider. The relation of this retreat to the web varied greatly, the leaf being either at one end, on one side, or near the middle of the web.

If, however, there is not a curled leaf available for a retreat, the spider will build a tent of silk. This was done in the case of the web represented in Fig. 409. Here the retreat is composed

Superfamily Argiopoidea

entirely of silk, and is furnished with a small opening at one side. Just before taking the photograph for this picture I jarred the web slightly, whereupon the spider rushed forth and assumed the attitude shown in the picture.

This species continues active late in the autumn after most of the orb-weavers have practically disappeared.

The Platform Spider, *Linyphia pusilla* (L. pu-sil'la).— The term platform spider which I apply to this species was suggested by the form of its web (Fig. 410).

Fig. 411. LINYPHIA PUSILLA   Fig. 412. LINYPHIA PUSILLA

The body is about one sixth inch in length. The cephalothorax is dark orange-brown, and the legs a lighter shade of the same colour. The abdomen is dark brown, often almost black, with several white spots, usually two across the front end and several others on the sides (Fig. 411); these sometimes form a complete white margin. In some individuals the upper part of the abdomen is brown with a series of transverse dark bars (Fig. 412); in others it is almost entirely black. Every gradation between these two forms occurs.

In the male both the cephalothorax and the abdomen are long and narrow; and the cheliceræ are more than half as long as the cephalothorax.

Superfamily Argiopoidea

The web (Fig. 410) is a horizontal platform, from three to six inches across, built between blades of grass, and usually from two to six inches from the ground. Above the sheet forming the platform, there is a labyrinth of delicate threads.

The spider usually hangs below the centre of the platform, but is sometimes found in the labyrinth above it. It is exceedingly shy, dropping to the ground on the slightest disturbance.

I have found the adults in April. These may have matured the previous autumn.

Genus TAPINOPA (Tap-i-no'pa)

The posterior median eyes are much nearer to the posterior lateral eyes than to each other; the anterior eyes are in a strongly

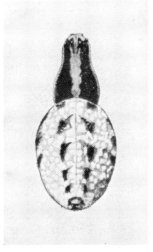

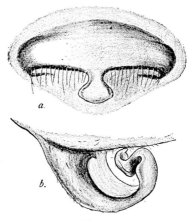

Fig. 413. TAPINOPA BILINEATA

Fig. 414.
EPIGYNUM OF TAPINOPA BILINEATA
a, ventral view *b*, lateral view

recurved line; the median ocular area is large; the anterior median eyes are larger than the posterior median eyes.

This genus is represented in the United States by a single described species; three or four species are known in Europe.

Tapinopa bilineata (T. bi-lin-e-a'ta).— The cephalothorax is pale with a broad black stripe on each side; these stripes do not extend to the lateral margins, but they cover the greater part of the dorsal aspect. The abdomen is pale, grayish brown, blotched with white, and has two rows of four or five spots each above, and a few chevrons at the tip (Fig. 413); the sides are marked with some oblique stripes, and the venter is almost wholly black. The epigynum of the female projects in a prominent manner and has the ovipositor coiled back under the atriolum (Fig. 414). The cymbium of the male palpus is longer than the femur and is furnished on the dorsal side at the base with a long, backward projecting apophysis. The length of the body is one sixth inch.

Mr. Banks, to whom I am indebted for specimens of this species and who first described it, states that it lives among grass or leaves close to the ground. It was first found on Long Island; and it is common in the vicinity of Washington, D. C. It has also been taken at Ithaca, N. Y.

Family ARGIOPIDÆ* (Ar-gi-opi'-dæ)

The Orb-weavers

These spiders are most easily recognized by their web-building habits; all of the species that make webs build what is known as an orb-web, and this type of web is built by no member of any other family except *Uloborus* of the family Uloboridæ; and in this case the nature of the spiral thread is very different from what it is in the Argiopidæ.

The orb-weavers are three-clawed, eight-eyed, sedentary spiders. In nearly all of the genera the eyes are similar; and the lateral condyle of the chelicerae is usually present. The tarsi are more or less clothed with hairs; but they lack the comb characteristic of the Theridiidæ, and the peculiar arrangement of spines that distinguishes the Mimetidæ.

The family Argiopidæ includes seven subfamilies; the following table will aid in the separation of these:

TABLE OF SUBFAMILIES OF THE ARGIOPIDÆ

A. Eyes dissimilar. P. 415.   THERIDIOSOMATINÆ
AA. Eyes similar.

---

*This family has been designated the Epeiridæ by most writers; but as it has been found that the generic name *Epeira* is not tenable, the name of the family has been changed to Argiopidæ by later writers.

B. Epigastric plates not marked by transverse furrows. Lateral condyle of the cheliceræ wanting or rudimentary.
  C. Epigastric furrow between the spiracles procurved. P. 419   TETRAGNATHINÆ
  CC. Epigastric furrow nearly straight. P. 429.
          METINÆ
BB. Epigastric plates marked by transverse furrows. Lateral condyle of the cheliceræ distinct.
  C. Spinnerets not tubulated (See CC below).
    D. Labium longer than broad. P. 440.   NEPHILINÆ
    DD. Labium broader than long.
      E. Posterior row of eyes strongly procurved; legs relatively longer; metatarsi and tarsi together longer than the patellæ and tibiæ. P. 447.
          ARGIOPINÆ
      EE. Posterior row of eyes barely, if at all, procurved; legs relatively shorter; metatarsi and tarsi together rarely longer than the patellæ and tibiæ. P. 457.   ARANEINÆ
  CC. Spinnerets elevated on a very large projection and occupying a circular space limited by a thick flange in the form of a tube or ring. P. 525.
          GASTERACANTHINÆ

Subfamily THERIDIOSOMATINÆ (The-rid-i-o-som-a-ti′næ)

The Ray-spider Family

The best-known representative of this subfamily in our fauna can be most easily recognized by the form of its web, which is described below. These are small spiders in which the eyes are dissimilar in colour; the lateral condyle of the cheliceræ is wanting; the tarsi of the fourth pair of legs are clothed beneath with numerous serrated bristles, but not with a single series of stout serrated spines as in the Theridiidæ. The middle spinnerets are situated between the hind pair, the four forming a straight transverse line; the fore spinnerets are longer than the hind ones; the colulus is distinct.

Only one genus occurs in America north of Mexico.

Superfamily Argiopoidea

Genus THERIDIOSOMA (The-rid-i-o-so'ma)

The abdomen is globular; the posterior median eyes are slightly closer to each other than to the posterior laterals.

The following is our best-known species; a second species, *T. argentatum*, occurs in the South.

The Ray-spider, *Theridiosoma radiosa* (T. ra-di-o'sa).— The length of the body of the female is one tenth inch; of the male, one twelfth inch. The cephalothorax is cordate, truncate at base, the head is much elevated. The abdomen is rounded, oval and highly arched. The external reproductive organs are very prominent; the epigynum of the female being large and vaulted; and the bulb of the palpus of the male being very large. The abdomen varies from straw-yellow to black, but all are marked with many small silvery spots which give the spider a shining appearance.

The web of this spider represents a unique type. As this is the only species of this subfamily whose web has been observed, we are unable to say whether its pecularities are distinctively characteristic of the species or are shared by other members of the subfamily.

The web was first described by Doctor McCook in 1883 and no subsequent account of it has been published. But owing to the common occurrence of the spider at Ithaca, N. Y., I have had abundant opportunities for observing it.

The spider prefers damp situations. It is most often found in the vicinity of streams or in damp forests. It often makes its webs on the face of a cliff over water. Although I have seen large numbers of the web, I have never found one in the field sufficiently well lighted and with a suitable background to enable me to photograph it. The accompanying figures are from photographs of webs made by spiders in confinement.

A remarkable feature of this web is that there is no hub (Fig. 415). The radii converge upon a small number of lines radiating from a point at or near the centre. These lines are termed *rays* by Doctor McCook; and the entire web a *ray-formed web*. There are usually four or five of these main divisions or "rays"; there is a large free zone which occupies about one third of the diameter of the web; there is no notched zone; and, in most of the webs that I have observed, less than a dozen turns of the

Fig. 415. WEB OF THERIDIOSOMA RADIOSA

Superfamily Argiopoidea

viscid spiral. The diameter of the orb is from two and one half to five inches. It is usually vertical or slightly inclined; in a few cases it is horizontal.

The rays converge upon a trapline, which usually extends perpendicular to the plane of the orb; but a convenient point of

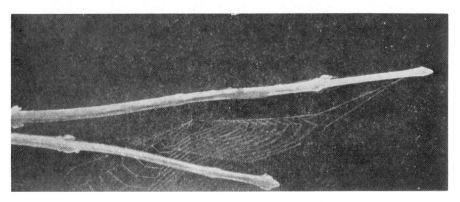

Fig. 416. WEB OF THERIDIOSOMA RADIOSA

attachment may lead the spider to stretch this line in another direction. The spider rests upon the rays at the centre of the web, with its dorsal aspect uppermost and its head away from the orb, and, by pulling on the trapline with its fore feet, pulls the orb into the shape of a cone or funnel (Fig. 416). It is in this position that the spider waits for its prey. When an insect becomes entangled in the web, the spider releases its hold on the trapline allowing the web to spring back. This springing of the web, as in the case of the triangle spider, increases the probability of the insect becoming more firmly ensnared.

I have seen a ray-spider finish its web, the web being partly made when the observations began. There was a spiral guy-line as in the web of orb-weavers; but the spider cut out the whole of the guy-line leaving no hub. In one case the spider fastened three radii together so as to form a "ray," each radius at first extending clear to the centre.

The egg-sac (Fig. 417) is light brown, pear-shaped, and about one eighth inch in its transverse diameter. It is suspended by a thread, which is usually forked. The pointed end of the egg-sac to which this thread is attached is a separable cap, which is partly pushed off when the spiderlings emerge, as shown in the figure.

Superfamily Argiopoidea

Although this spider usually selects a dark situation for its web, the egg-sacs are made in exposed places. Some of them hang from the face of a cliff; but the favourite position is suspended from dead branches of hemlock and other brush. I have also found many attached to living herbs. They are sometimes very abun-

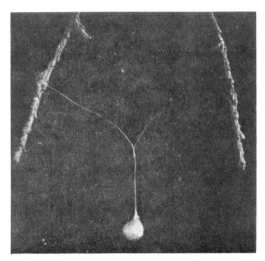

Fig. 417. EGG-SAC OF THERIDIOSOMA RADIOSA

dant; on one occasion I counted twenty-four egg-sacs in one cubic foot of space, among the roots of a tree on the side of a glen. The presence of this remarkable spider in a locality can be most easily determined by the presence of the egg-sacs. The spiderlings emerge from the egg-sac in the latter part of summer.

Subfamily TETRAGNATHINÆ (Te-trag-na-thi'næ)

The Tetragnathids (Te-trag'na-thids)

The most striking characteristic of this subfamily is the large size of the cheliceræ, especially in the males. The endites are also usually quite large. It was probably this fact that suggested the name *Tetragnatha*, four-jawed, for the typical genus. But it would be unwise to speak of these spiders as the four-jawed spiders, as that would suggest that they differed from other spiders in the possession of an extra pair of jaws.

The members of this subfamily are distinguished from other

Superfamily Argiopoidea

argiopids by the simplicity of the structure of their external reproductive organs. The opening of the reproductive organs of the female is furnished with neither an atriolum nor a scape, being merely a transverse slit; and the palpi of the males (Fig. 418) are more simple than in the other divisions of the Argiopidæ; being of the intermediate type. (See page 110 for description.)

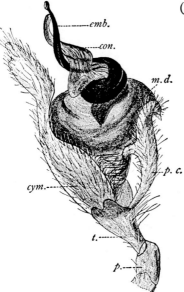

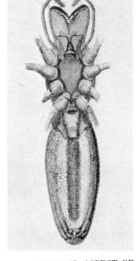

Fig. 418. GENITAL BULB OF PACHYGNATHA, EXTENDED

Fig. 419. VENTRAL ASPECT OF TETRAGNATHA LABORIOSA

The Tetragnathinæ agree with the Metinæ in the absence of transverse furrows on the epigastric plates; but can be distinguished from them by the fact that the epigastric furrow between the spiracles is procurved (Fig. 419), by the absence of a lateral condyle on the chelicerae, and by the fact that the furrow of the chelicerae is armed with numerous teeth.

Four genera of the Tetragnathinæ are represented in our fauna; these can be separated by the following table:

TABLE OF GENERA OF THE TETRAGNATHINÆ

A. Endites more or less convergent and not dilated at the distal end; lateral eyes on each side near together: tarsi without accessory claws.

Superfamily Argiopoidea

B. Posterior spiracle advanced in front of spinnerets.
C. Cheliceræ strongly divergent, both margins set with stout teeth. P. 422. GLENOGNATHA
CC. Cheliceræ subparallel, armed with rather weak teeth. P. 422. MIMOGNATHA
BB. Posterior spiracle situated immediately in front of the spinnerets. P. 421. PACHYGNATHA
AA. Endites parallel, dilated at the distal end; lateral eyes of each side usually distant; tarsi with accessory claws. P. 423. TETRAGNATHA

Genus PACHYGNATHA (Pa-chyg'na-tha)

The species of this genus are of moderate size, and in general appearance resemble certain of the Theridiidæ (*Steatoda*), the abdomen being oval and rounded; but in details of structure they are clearly allied to *Tetragnatha*. The cephalothorax is usually yellow with an obscure border and median band; and the abdomen is ornamented above with a reticulated folium. The endites are convergent and not dilated at the distal end. (Fig. 420).

Fig. 420. MOUTH-PARTS OF PACHYGNATHA TRISTRIATA, MALE

Fig. 421. PACHYGNATHA BREVIS, IMMATURE MALE

These spiders are found on the ground under stones, wood, or leaves, and especially in damp places. I have also swept them from aquatic plants over water in a lagoon. They are not known to spin webs of any kind.

The genus is represented in this country by several widely distributed species, of which the following is the most common.

Superfamily Argiopoidea

Pachygnatha brevis (P. brev'is).—This is our largest species, the body measuring nearly one fourth inch in length, and it is of bright colours. Figure 421 represents an immature male. The net-work of lines on the abdomen is dark brown. In living specimens the folium is brick-red, with central and marginal dark lines; the central line separates two rows of light yellow spots. On each side of the abdomen there is a light yellow stripe bordered with dark.

This species occurs commonly in the eastern portion of the United States.

Genus GLENOGNATHA (Gle-nog'na-tha)

In this and the following genus the posterior spiracle is situated considerably in advance of the spinnerets. In most other respects *Glenognatha* closely resembles the species of *Pachygnatha*. The cheliceræ are strongly divergent and are armed in both sexes with stout teeth.

Glenognatha emertoni (G. em-er-to'ni).—This, our only species, is about one fourth inch in length. The chelicera of the male is armed with two stout teeth on the lower and four smaller teeth on the upper margin. In the female there are four teeth on the lower and three on the upper margin, all of them subequal in size. In addition there is a sharp spur in front above the claw which is lacking in the male. These spiders resemble *Pachygnatha* in colouration and general appearance.

This species is common in Mexico and Central America and occurs more rarely in the southwestern United States.

Genus MIMOGNATHA (Mi-mog'na-tha)

This genus is very closely allied to *Glenognatha* and may not deserve separate generic standing. As in that genus the posterior spiracle is situated a short distance in front of the spinnerets. The cheliceræ are subparallel, relatively weak, and the margins are set with small teeth. In the male there is a short, rounded spur in front above the claw. The male palpus resembles that of Glenognatha, but the elements are proportionately much shorter. The bulb is greatly expanded, suborbicular, and is nearly twice as broad as the length of the tibia and patella together.

Few species of this genus are known.

Mimognatha foxi (M. fox'i).—This, our only species, is a small pink and silver spider sometimes marked with black. Females are one eighth inch in length; the males are about as large. The habits of *foxi* have been investigated by Barrows ('19). He demonstrated that the species belonged in the Argiopidæ and placed it in *Glenognatha*, a genus which it resembles closely in its structure and in habits. A new generic name was later proposed by Mr. Banks for this interesting spider. *Mimognatha foxi* occurs in meadows and waste lands "where it builds its delicate orb web in grass or weeds in rather hot dry situations. Usually the web is placed horizontally about two inches above the ground." "The spiders, both males and females, remain on the under side of the web at its center unless disturbed when they drop to the ground and run rapidly away."

This spider is widely distributed in the United States, particularly in the southern portion, and it is also found in Mexico and Central America.

Genus TETRAGNATHA (Te-trag'na-tha)

The abdomen is long and slender and bears the spinnerets at or near its end. The endites are parallel and dilated at the distal end (Fig. 419); the lateral eyes on each side are usually distant. The chelicerae are without a lateral condyle.

These are orb-weaving spiders; they are common on plants and other objects in the vicinity of water and some of them occur on grass in drier places. They are striking in appearance on account of their slender form, the great length of their legs, and the large size of their chelicerae (Fig. 419). The chelicerae are sometimes of enormous size, especially in males.

The webs (Fig. 422) are either inclined or horizontal. They are of moderate or large size. I have seen horizontal ones, built over a race of running water, that were two feet in diameter. These had an open hub, a notched zone of four or five turns, a broad free space, and from thirty to forty viscid spirals.

In all webs that I have seen made by spiders of this genus the hub was open; but in some there was no free space.

When at rest on a branch the spiders assume a very characteristic attitude; the body is closely applied to the branch, the

Superfamily Argiopoidea

first and second pairs of legs are stretched directly forward, the fourth pair, backward, and the shorter third legs embrace the branch.

When resting on its web the spider stands over the centre with its legs in a somewhat similar position.

The egg-sacs are attached to various objects and present a very characteristic appearance due to their bearing projecting

Fig. 422. WEB OF TETRAGNATHA

tufts of silk, which, in some cases at least, contrast strongly in colour with the body of the sac (Fig. 203, p. 212).

The following descriptions are of adult individuals; in immature ones the teeth on the upper margin of the furrow of the cheliceræ are similar in shape and are regularly spaced, and the claw of the cheliceræ is very short and stout. In the adults the teeth, especially those near the apex of the cheliceræ vary in form,

Superfamily Argiopoidea

size, and spacing in the various species and are fairly constant, slight variations from the normal formulæ sometimes occur.

More than a dozen species have been described from the United States. Most of them were studied by Seeley ('28) in whose paper descriptions and keys to the species may be found. The following species are widely distributed. *Tetragnatha extensa* is also found in Europe.

THE MORE COMMON SPECIES OF TETRAGNATHA

Tetragnatha elongata (T. el-on-ga′ta).— The lateral eyes of each side are not as far apart as the anterior median and posterior median eyes. The chelicerae of the male are longer than the cephalothorax; those of the female are about one tenth their length shorter than the cephalothorax.

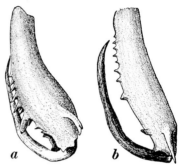

Fig. 423. CHELICERÆ OF TETRAGNATHA
a, *T. straminea*   b, *T. pallidula*

This is the largest of our common species, the body of the full-grown female being often one half inch in length. In the female (Fig. 424) the basal third of the abdomen is usually swollen; the male (Fig. 425) is more slender. It prefers damp situations; the web is often found over running water, and frequently it is perfectly horizontal. This is the *Tetragnatha grallator* of Hentz.

Wishing to obtain a photograph of a web of *Tetragnatha*, and not finding one in the field with a suitable background, I took some spiders of this species to our insectary, and set them free on the edge of a large tank through which water was flowing. Here they remained unconfined, making no effort to wander to the dryer parts of the building. I placed a section of a balustrade upon the top of this tank, and the spiders stretched their webs between its pillars. Figure 422 represents one of these. Before taking the photograph the balustrade was taken from its place over the tank and set in a vertical position before a suitable background.

Tetragnatha extensa (T. ex-ten′sa).— The lateral eyes of each

Superfamily Argiopoidea

side are not so widely separated as are the anterior median and posterior median eyes. The cheliceræ in both sexes are shorter than the cephalothorax, those of the female are only a little more than half as long as the cephalothorax.

This is a smaller species than *T. elongata;* the female measures from one fourth to three eighths inch in length. The abdomen

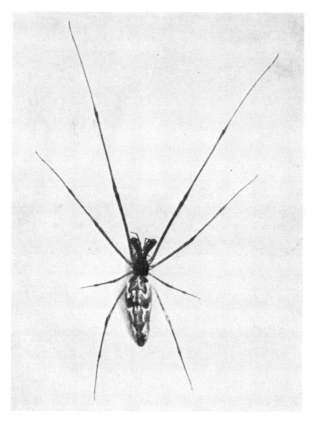

Fig. 424. TETRAGNATHA ELONGATA, FEMALE

is only a little more than twice as long as the cephalothorax; it is proportionately shorter than that of either the preceding or the following species.

Tetragnatha laboriosa (T. la-bo-ri-o'sa).— The lateral eyes of each side are widely separated, being about the same distance apart as are the anterior median and the posterior median eyes;

the second row of eyes is somewhat recurved, but not so markedly so as in the following species. In the male the patella and tibia of the pedipalp are nearly equal in length. In the female the abdomen (Fig. 426) is less than three times as long as the cephal-

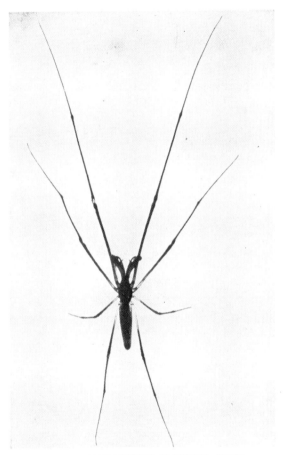

Fig. 425. TETRAGNATHA ELONGATA, MALE

othorax. The full-grown female is about one third inch in length; the male is a little smaller.

This species is common in meadows, where it makes its web between stems of grass. It shows no preference for moist situations, in fact, I have found it more common in dry fields than in the vicinity of water.

Superfamily Argiopoidea

Tetragnatha straminea (T. stra-min′e-a).— This is the most common species in the Northern States of the group of species in which the lateral eyes of each side are more widely separated

[Fig. 426.
TETRAGNATHA LABORIOSA, LATERAL ASPECT OF ABDOMEN

than are the anterior lateral and posterior lateral eyes, the genus *Eugnatha* of some writers. The length of the body is from one fourth to three eighths of an inch. The cephalothorax is light

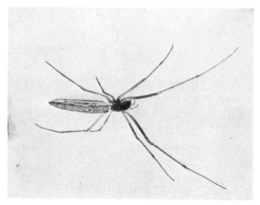

Fig. 427. TETRAGNATHA STRAMINEA

yellow with two parallel gray stripes. The abdomen is usually three times as long as the cephalothorax (Fig. 427); it is silvery white above and dark beneath with two parallel gray stripes.

The male can be easily distinguished from our other two species of this group by the character given in the table above. In the case of the female the distinction between this species and *T. vermiformis* is not so well-marked; but I am unable to point out a more easily recognized means of separating the two.

Tetragnatha vermiformis (T. ver-mi-for′mis).— This species is slightly larger than the preceding, the female measuring about one half inch in length. The male is distinguished by the fact that the tibia of the palpus is not longer than the patella. The female agrees with that of *T. straminea* in that the endite of the pedipalp reaches the tip of the claw of the cheliceræ; but differs in having the outer side of the chelicera not so obviously concave. This is a widely distributed but uncommon species.

Tetragnatha pallescens (T. pal-les'cens).—This is a handsome spider which is widely distributed in the eastern part of the United States. It is relatively rare in the north. It can be easily recognized by the characters given in the table above.

Tetragnatha lacerta (T. la-cer'ta).—This species, appropriately named *caudata* by Hentz, was formerly placed in a separate genus, *Eucta*, because of the fact that the abdomen projects a considerable distance beyond the spinnerets (Fig. 428). The abdomen is very long and slender.

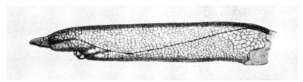

Fig. 428. TETRAGNATHA LACERTA

I have found by observing living individuals that the tail of these spiders can be extended and contracted a considerable distance, so that the length of the abdomen may vary greatly from moment to moment.

I have not seen the web of *lacerta* and can find no description of it. It probably resembles that of typical species of the genus. The species is common on aquatic plants projecting from the water of marshes and in the grassy area bordering ponds and lakes. This spider resembles very closely *Tetragnatha straminea* except in the shape of the abdomen, which is longer and has a tail about one fourth of the length of the abdomen extending backward beyond the spinnerets (Fig. 428).

This curious tetragnathid is widely distributed in the eastern part of the United States, and is particularly abundant in Georgia and Florida.

Subfamily METINÆ (Me-ti'næ)

The Metids (Me'tids)

The metids are closely allied to the tetragnathids and by some writers are classed in the same subfamily. These subfamilies agree

Superfamily Argiopoidea

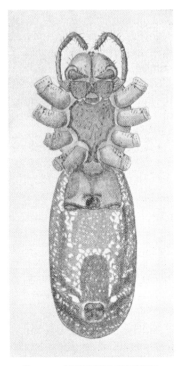

Fig. 429. LEUCAUGE VENUSTA, VENTRAL ASPECT

in having the epigastric plates not marked by transverse furrows; but in the Metinæ the epigastric furrow is nearly straight, (Fig. 429), and not strongly procurved as in *Tetragnatha*. In the Metinæ there is a rudimentary external condyle on the cheliceræ, and the epigynum of the female is often more or less developed.

Some of the metids live in caves and other dark places; others build their webs on bushes and trees, and among these are some of the most brilliantly coloured of our spiders.

The webs of the Metids, except *Hentzia*, resemble those of *Tetragnatha* in being more or less nearly horizontal, and in having an open hub. In some the orb is accompanied by a barrier web; *Hentzia* builds a peculiar web.

Seven genera of the Metinæ are represented in our fauna; these can be separated by the following table.

TABLE OF GENERA OF THE METINÆ

A. Lateral eyes of each side near together.
  B. Eyes nearly equal in size; clypeus narrow.
    C. Posterior femora with a single or double fringe of hairs on the external face of the basal half.
      D. Tibiæ and metatarsi of the first and second legs studded with many triangular thorn-like points. P. 438.   PLESIOMETA
    DD. Tibiæ and metatarsi of the first and second legs not studded with points. P. 435.   LEUCAUGE
  CC. Posterior femora not fringed.
    D. Abdomen with a hump on each side near the base. P. 431.   HENTZIA

DD. Abdomen without basal humps. P. 433.　　META
BB. Anterior median eyes much smaller than the others; clypeus wide. P. 438.　　NESTICUS
AA. Lateral eyes of each side distant.
B. Posterior median eyes small and close together. P. 439.
　　DOLICHOGNATHA
BB. Posterior median eyes equal to the posterior lateral in size and widely separate. P. 439.　　AZILIA

Genus HENTZIA (Hentz'i-a)

The cephalothorax is oval, with the median furrow in the form of a circular pit. The abdomen is cylindrical, much longer than wide, and is furnished with a hump on each side near the base; this characteristic distinguishes the members of this genus from *Meta* which they closely resemble in the characteristics presented by the eyes and mouth-parts.

Only a single species has been found in our fauna.

The Basilica Spider, *Hentzia basilica*, (H. ba-sil'i-ca).— The adult female measures from one fourth to nearly one third inch in length. The cephalothorax is yellow or olive, with a blackish median stripe, and with the margin dark. The cylindrical abdomen projects forward over the thorax and backward beyond the spinnerets. It is yellow striped with blackish brown; the folium extends the entire length of it. On the basal half of the length of the abdomen, the folium is wide, and consists of a median dark line, and on each side two dark lines extending back from the hump; alternating with these dark lines are yellow or yellowish bands. On the caudal half of the abdomen, the folium is narrower and more nearly solid in colour. Figure 430 represents a side view of this spider with the legs removed.

This spider was first discovered by Doctor McCook and its web described by him. He has given an extended account of it in the first volume of his "American Spiders." As I have never seen the species in the field, I make a condensed statement of this account. The web was a composite one, consisting of an irregular net and an open silken dome, suspended within it. Beneath the dome, and from two to three inches distant, there was a light sheet of cobweb, which may have been the collapsed remnant of

Superfamily Argiopoidea

an old dome. The dome consisted of a large number of radii, which were about one sixteenth inch apart at the bottom of the dome, and a spiral line, which was attached to the radii in the same way that the spiral line of the notched zone of an ordinary orb is attached. Doctor McCook does not state whether this line was dry or viscid. His account was based on observations made on a single web, which he found near Austin, Tex. It was hung about two feet from the ground upon a bush, which stood in the midst of a grove of young live oaks.

Later, Dr. Geo. Marx observed several specimens of this species in the shrubbery of some parks in Washington, D. C. He watched the building of the dome. This was, when first observed,

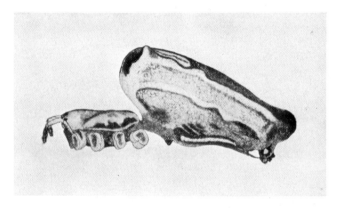

Fig. 430. HENTZIA BASILICA

a horizontal orb, with more than fifty radii, supporting the spiral line. After the completion of the orb its margin was pulled down by means of lines extending from it and its centre was elevated until the dome-shaped form was attained.

It is very desirable that further observations should be made upon this remarkable orb. Is the spiral line dry and inelastic, like the thread of the notched zone of other orbs, or is it viscid and elastic? In the former case the web could be regarded as a specialization of the Linyphia type, as illustrated by *Linyphia marginata;* in the latter, as a modified web of *Aranea*. The former would be a connecting link between the webs of the Linyphiidæ and those of the Argiopidæ; the latter, a highly specialized argiopid web.

Superfamily Argiopoidea

Genus META (Me′ta)

The lateral eyes of each side are near together; the eight eyes are nearly equal in size; the clypeus is but little if at all wider than the diameter of an anterior median eye; the cheliceræ are long, stout at the base, and strongly arched forward near the base,

Fig. 431. META MENARDII, FEMALE

the endites are longer than wide, narrowed at the base and blunt at the apex.

Meta is represented in this country by a single well-known species.

Meta menardii (M. me-nar′di-i).—This is a dark brown spider with translucent and yellowish markings. The female (Fig. 431), when full-grown measures one half inch or more in length; the abdomen is longer than wide, and high in front. The cephalothorax has median and lateral stripes; the abdomen has an

Superfamily Argiopoidea

irregular net-work of light markings as shown in the figure. The adult male (Fig. 432), is nearly three eighths inch in length. The cephalothorax is large and broad; the abdomen, of moderate size. The light markings of the abdomen are more marked near the

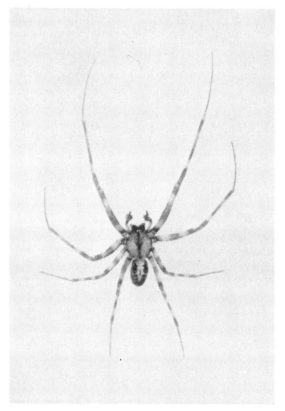

Fig. 432. META MENARDII, MALE

middle line than in the female. The tarsus of the pedipalp bears, on the upper side near its base a very prominent spur, which is hooked at the tip.

This cave-inhabiting species I find common in deep recesses in the cliffs on the shores of Cayuga Lake, and especially in those that are so situated that the sun does not shine into them, and even in such of these as are darkened by a dense growth of shrubbery at the mouth. I have also found them in a culvert under a

Superfamily Argiopo

highway and in a dark tunnel through which water was flow
It is a widely distributed species; Simon states that it is foun
Madagascar as well as in America.

The webs are usually inclined; but they vary from vertical to horizontal. The spider hangs from the hub of the web but when disturbed, retreats quickly by means of a trapline, to the rocks supporting the web. The orbs are of moderate size, from six to ten inches in diameter. The hub is open; the notched zone is narrow, usually consisting of from three to five turns; the clear space is wide; and the viscid spirals are from fifteen to thirty in number.

Meta must be a very patient hunter. I rarely observe insects in its web, which are almost invariably in perfect condition at all times of the day. On the other hand this spider suffers little from competition in its dark and damp retreats. Occasionally a *Theridion tepidariorum* builds a web in the same cave.

Fig. 433. EGG-SAC OF META MENARDII

The egg-sacs are large (Fig. 433), snow-white, and so translucent that the ball of eggs can be seen within. They are suspended from the roof of the cave, near the web, by a short thread.

Genus LEUCAUGE (Leu-cau'ge)

The members of this genus can be separated from *Meta*, to which they are closely allied, by the fringe of hairs on the posterior femora (Fig. 434); and by their much more brilliant colouring, being green and silver-white, with bronze and sometimes coppered markings. These spiders are even more closely allied to the following genus, from which they can be distinguished by the absence of the armature of the first two pairs of legs characteristic

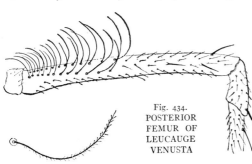

Fig. 434. POSTERIOR FEMUR OF LEUCAUGE VENUSTA

Superfamily Argiopoidea

of that genus. The abdomen is about twice as long as wide, blunt at both ends, and projects far over the cephalothorax. The legs are long and slender, especially the first two pairs.

The genus contains more than one hundred species, but only one has been found in our fauna.

The beautiful Leucauge, *Leucauge venusta* (L. ve-nus′ta).— This is a common and widely distributed species. It is a bright green and silver-white spider, tinged with golden, and sometimes with orange-yellow or copper-red spots. In the *female* the cephalothorax is yellowish with a dark green stripe in the middle and one on each side. The abdomen is egg-shaped twice as long as wide, with the front end rounded and projecting above the cephalothorax. It is silver-white above, with a dark line in the middle, from which extend four pairs of more or less distinct bars, as shown in Fig. 435. On the sides of the abdomen there are two yellowish stripes. Sometimes there are two bright orange-yellow or copper-red spots above near the hind end of the abdomen, and a large spot of the same colour on the ventral side near the middle. None of the specimens that I have taken in New York have the red spots but in all that I have taken in the South they are present. The length of body is one fifth to one fourth inch.

Fig. 435. LEUCAUGE VENUSTA

The *males* are half as large as the females; with longer legs, and similar colours.

The specific name *venusta* or beautiful, applied to this spider is well deserved, for it is one of the most beautiful of all our spiders. It is a very widely distributed species, extending beyond the limits of the United States both north and south.

Unlike *Meta*, this spider lives in open, well-lighted situations. It builds its web on shrubs and trees. The webs are horizontal

Superfamily Argiopo

or inclined, and are sometimes more than one foot across, ʻ very many turns of the spiral thread, presenting a beauuᵤᵤ appearance. One web that I examined carefully had a notched zone of five turns, a wide clear space, and a viscid zone of sixty-two turns. Figure 436 represents the central part of a web of this species.

The web is usually nearly horizontal with a barrier web below;

Fig. 436. CENTRE OF WEB OF LEUCAUGE VENUSTA

sometimes the barrier web is wanting. When present it consists of a few lines extending in all directions like a loose web of *Theridion*.

Sometimes the spider rests in the barrier web or between it and the orb, with a trap line leading to the edge of the hub. But the more usual position is on the orb with the tip of the abdomen at the centre of the open hub, and with the first and second pairs of legs on radii of the clear space and the third and fourth legs on the notched zone.

This species is the *Epeira hortorum* of Hentz.

Genus PLESIOMETA (Ple-si-o-me'ta)

This genus has been recently separated from *Leucauge*, which it resembles in the possession of the characteristic fringe of hairs on the external face of the basal half of the posterior femora (Fig. 434). It differs from *Leucauge* in having the tibiæ and metatarsi of the first two pairs of legs studded with many triangular, thorn-like points. In the single described species, the central part of the epigynum bears a prominent tubercle which projects at right angles to the length of the abdomen.

Plesiometa argyra (P. ar'gy-ra).— This is a Central American species which extends its range into the southern portions of the United States. It bears a striking resemblance to the beautiful *Leucauge*, but is somewhat larger and can be readily distinguished by the characters given in the above generic description.

Genus NESTICUS (Nes'ti-cus)

The lateral eyes on each side are near together. The median eyes of the first row are much smaller than the others, in some cases rudimentary or even absent. In some species all the eyes have been lost. The tarsi are armed with a line of serrated bristles as in the theridiids. The labium is distinctly rebordered as in the linyphiids and argiopids.

These spiders resemble *Theridion* in appearance and are placed in the Theridiidæ by some authors. Others place them in the Linyphiidæ, the Argiopidæ, and even in a separate family, the Nesticidæ. In view of this disagreement, in this book they are left in the place assigned to them by Simon.

The nesticids, while similar in appearance and habits, live in a variety of places, such as caves, mines, tunnels, almost always in dark situations. A few of them live in the corners of buildings, in which case they have the eyes fully pigmented, normal in appearance. In other species from caves, especially from deep inside, they have begun to lose their eyes, the first evidences being the loss of the black pigment and the reduction in size of all of them.

The following is our best known species.

Nesticus pallidus (N. pal'li-dus).—The female measures one seventh inch in length. The cephalothorax and legs are pale orange-brown, and the abdomen is yellowish white with brown hairs. The eyes are all well developed, pigmented, the anterior median eyes smaller than the others.

This species was described by Emerton from specimens taken in a cave in Virginia, among stalactites where there was no daylight. However, it is very often found outside of caves. It is widely distributed in the United States and also occurs in Mexico and the West Indies.

Genus AZILIA (A-zil'i-a)

The eight eyes are all large; the posterior median eyes are a little larger than the posterior lateral; and the four posterior eyes are nearly equally distant from each other. The abdomen is rounded in front, narrowed and sloping behind.

The following is our only known species.

Azilia vagepicta (A. vag-e-pic'ta).— The female measures nearly one half inch in length. It was described from Georgia. No observations have been published regarding its habits.

Simon states that some species of *Azilia*, which he observed in Venezuela, live in the darkest parts of forests, under damp rocks, and spin an orb-web, which has a large free zone between the hub and the viscid spiral.

Genus DOLICHOGNATHA (Dol-i-chog'na-tha)

The lateral eyes of each side are separated by fully the diameter of one of them; the anterior median eyes are larger than the anterior lateral; the posterior median eyes are small, close together and widely separated from the posterior lateral eyes. The anterior metatarsi are armed with a series of setæ below.

The following is our only described species.

Dolichognatha tuberculata (D. tu-ber-cu-la'ta).— The female measures one eighth inch in length. The cephalothorax is nearly

Superfamily Argiopoidea

one half as long as the entire body. The abdomen bears, on the dorsal surface, two pairs of tubercles.

This species was described from Florida. Nothing is known regarding its habits. Simon states that the web of *Dolichognatha* is not an orb-web, but a delicate sheet under which the spider hangs like a *Linyphia*. It is very desirable that the habits of our species be observed; for this is certainly a remarkable variation from the usual habits of the family.

Subfamily NEPHILINÆ (Neph-i-li′næ)

The Silk Spiders

The members of this subfamily are remarkable for the large quantity and the great strength of their silk, which is being used, to a limited extent, for the production of fabrics. For this reason they are here designated as the silk spiders.

The Nephilinæ are distinguished from the preceding subfamilies of the Argiopidæ by the presence of transverse furrows on the epigastric plates, and from the following subfamilies of this family by the greater length of the labium, which is longer than broad. The legs are relatively long; the metatarsi and tarsi together are longer than the patellæ and tibiæ together. The posterior row of eyes are straight.

The subfamily includes only a single genus, *Nephila*.

Genus NEPHILA (Neph′i-la)

The cephalothorax is longer than wide, the eyes are quite small and are nearly equal in size. In the female the eyes are separated into three groups, the lateral eyes being far removed from the median eyes; the lateral eyes of each side are situated on a tubercle. The legs are long. The adult females are of large size. The males are very much smaller than the females; and in this sex the eyes are closer together than in the female.

About sixty species of this genus are known. They occur in the tropics and in the warmer portions of the subtropical regions; only one or perhaps two of them extend into our fauna. The following is well known.

Nephila clavipes (N. clav′i-pes).— The adult female (Fig. 437) measures from seven eighths to one and one tenth inches in length. The abdomen is long with nearly parallel sides. The legs are long; and excepting the third pair are clothed with tufts of hair. The cephalothorax is black above, but covered, except in spots, with silver-coloured hairs. The abdomen is olive-brown variously marked with yellow and white spots and stripes. The

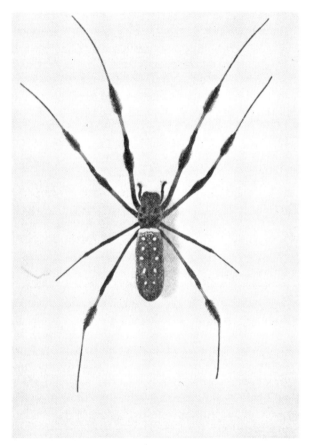

Fig. 437. NEPHILA CLAVIPES, ADULT FEMALE

markings of the young differ somewhat from those of the adult (Fig. 438).

The male is small, the female weighing more than one hundred times as much as the male. The length of the body is one fourth inch, and the legs spread less than one inch in a longitudinal and

Superfamily Argiopoidea

three fourths of an inch in a lateral direction. The general colour of both body and legs is dark brown. The legs lack the tufts of hairs characteristic of the female.

The male, in the adult state at least, spins no web, but lives in the web of the female. Comparatively few specimens of this sex are found in collections; but Mr. Schwartz states that it is just as common as the female, but that it is easily overlooked from its small size and the fact that it always occurs opposite the large body of the female on the other side of the web.

Fig. 438. NEPHILA CLAVIPES, YOUNG FEMALE

This remarkable spider attracted attention very early. It was described by Linnæus in his *Systema Naturæ* (1767). It has been commonly known by American writers under the name *Nephila plumipes*, which, however, according to Simon, is correctly applied to a species from the Islands of the South Sea.

Nephila clavipes is widely distributed through the southern states. It builds large webs frequently two or three feet in diameter in shady forests. The supporting lines of these webs are frequently exceedingly strong and are apt to attract the attention of people who run into them in going through such forests. The webs of this spider differ in several very striking features from those of other orb-weavers. They are slightly inclined. The most striking feature at first sight is the looped nature of the viscid lines (Fig. 439). In the webs of old spiders the loops occupy but little more than one half of a circle. The webs of younger spiders are much more nearly complete orbs (Fig. 440).

Unlike other orb-weavers which rebuild their webs at frequent intervals, this spider makes use of the same web for a long period replacing only the viscid lines. Correlated with this fact are several features which contribute to the permanency of the web. The radii are branched so that the interval between two adjacent radii in the outer portions of the web is not greater than that between two near the centre (Fig. 439). The guy-line, corre-

Fig. 439. NEPHILA CLAVIPES, WEB OF AN OLD SPIDER

Fig. 440. NEPHILA CLAVIPES, WEB OF A YOUNG SPIDER

Superfamily Argiopoidea

sponding to the spiral guy-line of other orb-weavers like the viscid line, is looped back and forth and remains a permanent part of the web. Its attachment to the web is of a much more firm nature than is that of the spiral guy-line of other orb-weavers where its use is limited to the short period of construction of the web. As it crosses each radius it is united with it for a short distance in a

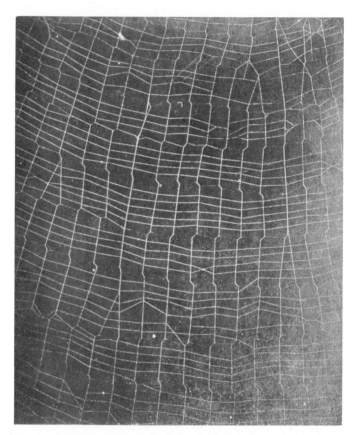

Fig. 441. NEPHILA CLAVIPES, SECTION OF WEB SLIGHTLY ENLARGED

way similar to that in which the spiral guy-line of other orb-weavers is attached to the radii in the notched zone. This is well-shown in Fig. 441. As this guy-line is pulled taut it draws the radius out of its direct course. By reference to Fig. 441 it will be seen that the direction of these notches alternate in the successive turns of the spiral showing that the spider passed back and forth, in making it.

This notched nature of the spiral is doubtless correlated with the fact that the spiral is to remain a permanent part of the web, and is made of dry and inelastic silk as is the case with the few turns of the spiral guy-line constituting the notched zone of an ordinary orb-web.

As the turns of the viscid lines are spun in the spaces between the turns of the guy-line it results that the viscid threads are in groups as shown in Fig. 441.

The viscid silk and the dry silk differ in colour, the former being yellow, the latter white. The viscid drops are frequently so large that they can be easily seen with the unaided eye. According to Doctor Wilder ('66) the yellow silk is spun from the fore spinnerets.

In the case of old spiders where the abdomen is filled with eggs and movement is consequently more laborious, the spider repairs only one half of the web each day. Figure 442 represents a web, one half of which was repaired the night before the picture was taken. The spider that made this web was under observation for a considerable time and it was found that each night she rebuilt one half of the web. During the period that this spider was under observation there was a very severe storm, five inches of rain falling in the course of a few hours. When the web was visited on the following morning it was found that it had been repaired throughout. (Fig. 439.) In repairing the web this large spider walked sideways with her head directed upward and her legs extending over two or three turns of the spiral guy-line. As stated above, the web is slightly inclined and the spider hangs throughout the day from the lower side of the hub.

The silk of the spiders of the genus *Nephila* surpasses in strength and in beauty that of the silkworm; and it is being utilized to some extent. The more important of the investigations which demonstrated the practicability of using this silk were the following: those made in this country by Prof. Burt G. Wilder, with *Nephila clavipes;* those by Pere Camboni, a French Roman Catholic missionary in Madagascar, with *Nephila madagascariensis;* and those by some Chinese at Yun-Nan, with *Nephila clavata.*

Professor Wilder published an account of his experiments

Superfamily Argiopoidea

in the Proceedings of the Boston Society of Natural History, Oct. 1865, and in the *Atlantic Monthly* for August, 1866; but no practical application has been made of them in this country. In Madagascar, however, the French have founded schools for the instruction of the natives in the methods of rearing the spiders, and

Fig. 442. NEPHILA CLAVIPES, WEB OF AN OLD SPIDER; ONE HALF OF THIS WEB HAS BEEN RECENTLY REPAIRED

in winding, spinning, and weaving the silk. I have not at hand information as to what is being done by the Chinese.

The method of obtaining the silk from these spiders is very different from that in which the silk of the silkworm is procured, which is by unwinding the cocoons. It is also different from that used in the earlier attempts to utilize the silk of spiders, which was by carding the silk of the egg-sacs. The silk of *Nephila* is obtained by pulling it directly from the body of the living spider.

The full-grown spider is fastened in a tiny stanchion which fits over the body between the cephalothorax and the abdomen, in such a way that the spider is firmly held without injury, and so that the legs are kept away from the spinnerets. By lightly touching the spinnerets a thread can be obtained, and by slowly pulling this thread it will be constantly lengthened by a flow of silk from the spinning tubes. A thread of silk is drawn in this way from each of a considerable number of spiders at the same time; and all are twisted into a single larger thread, by a mechanical twister, from which it passes to a reel.

This process was shown at the Paris Exposition; and a complete set of bed hangings made from the silk of *Nephila* were exhibited there.

Subfamily ARGIOPINÆ (Ar-gi-o-pi'næ)

The Garden Spiders

Although the members of this subfamily are not so striking in appearance as is the species of *Nephila* described in the preceding pages, in the colder parts of our country, where *Nephila* is not found, some of them are the most conspicuous of our orb-weavers. These have been commonly known in this country as the garden spiders; and I have adopted this as the popular name for the subfamily, although in the Old World it has been applied to certain large and conspicuous species of *Aranea*.

The argiopinæ differ from the preceding and from the following subfamilies in having the posterior row of eyes strongly procurved. They differ from *Nephila* in that the labium is broader than long, and from the Araneinæ in the relatively longer legs, and in having the metatarsi and tarsi taken together, longer than the patellæ and tibiæ.

The web of the garden spiders is a typical orb-web; but it is accompanied by a barrier web, which consists of an irregular network of lines stretched behind the orb. Sometimes a barrier web is built on each side of the orb. The barrier web doubtless serves as a protection to the spider. In it, at the mating season the male is often found. But the immature males resemble the females in their web-building habits. The orb is not provided with a trapline. The spider hangs on the hub of the web, throughout the day. When disturbed it either drops to the ground or runs off from the web upon the supporting plants. If all is quiet for a few minutes, the spider returns to its station on the hub. The web is often provided with a stabilimentum. Our

Superfamily Argiopoidea

common species mature late in the summer, and make their egg-sacs at this season.

This subfamily is represented in our fauna by four genera; the females of these can be separated as follows:

- A. Anterior median eyes nearer to each other than to the anterior lateral eyes; adult females large.
  - B. Vulva of female divided by a septum into a pair of equal concavities.
    - C. Abdomen scalloped or lobate on the sides. P. 456. ARGIOPE
    - CC. Abdomen more or less evenly rounded on the sides. P. 452. METARGIOPE
  - BB. Vulva of female not divided by a septum; but the atriolum of the epigynum is extended into a broad, convex process, with a single cavity beneath it. P. 448. MIRANDA
- AA. Eyes of the anterior row almost equidistant; both sexes small. P. 457. GEA

Genus MIRANDA (Mi-ran'da)

The cephalothorax is flat; the head is small; the second row of eyes is so strongly procurved that the posterior lateral eyes are almost as far forward as the anterior median eyes. The vulva of the female is not divided by a septum; but the atriolum of the epigynum is extended into a broad, convex process, with a single cavity beneath it. The two sexes differ greatly in size, the males being small while the females are very large. The following is a very common species.

The Orange Garden Spider, *Miranda aurantia* (M. au-ran'ti-a). — This is a spider that often attracts attention on account of its large size, bright colouring, and the beauty of its web. The adult female frequently measures an inch or more in length, and is marked with spots and bands of bright orange (Fig. 443). The cephalothorax is covered with silvery white hairs. The abdomen is oval, with a pair of humps at the base. The ground colour is black marked with bright yellow or orange spots. On each lateral margin of the abdomen the yellow spots form an almost continuous band. In the black band between these two rows of spots there are from one to three pairs of yellow spots.

Fig. 443.
MIRANDA AURANTIA, ADULT FEMALE

Fig. 444.
MIRANDA AURANTIA, YOUNG MALE

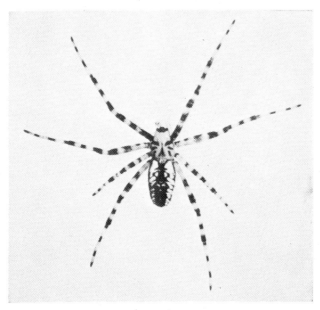
Fig. 445. MIRANDA AURANTIA, YOUNG FEMALE

Superfamily Argiopoidea

The male differs greatly from the female, being only about one fourth as long; the cephalothorax is yellowish brown; the abdomen bears a broad brown band along the middle of the back, and on each side a zigzag band of white; the palpal organ is large.

The young of this species differs much in appearance from the adult, a difference due largely to a banding of the legs; Fig. 444 represents an immature male, and Fig. 445, an immature female.

The genus *Miranda* to which this species belongs was recently separated from *Argiope* by F. O. Pickard-Cambridge in the *Biologia Centrali Americana*. This species has been commonly known under the name *Epeira riparia* given to it by Hentz in 1847; it has also been known under the name *Epeira cophinaria* given to it by Walckenaer in 1837; but it was described and figured by Lucas in 1833, under the specific name used here.

The web of this species is often found upon shrubs; but it is more frequently made upon herbaceous growth in marshy places and upon grass in meadows and pastures. It seems strange that so large a species should choose such feeble supports for its web.

When first made in a protected situation and before it is injured by wind and insects it is a very beautiful structure, resembling the web of the next species shown in Fig. 449, which is of a web made in our insectary. But under natural conditions the symmetry of the web is very soon lost. Figure 446 is from a photograph of one taken in the field.

These webs are large, sometimes two feet in diameter. The hub is sheeted and usually furnished with a stabilimentum; the notched zone is broad, and extends nearly to the viscid spiral; the free zone is therefore limited. The web is inclined and the spider rests upon the hub on the lower side of the web; but it sometimes passes through the narrow free zone to the upper side. The web is usually accompanied by what I have termed a barrier web; this consists of an irregular net-work of lines stretched behind the orb, and probably serves to protect the spider from attacks. Sometimes a barrier web is built upon each side of the orb.

The two sexes have similar web-building habits; but when the males reach maturity they wander in search of the females and are then to be found in the webs of the females, usually upon the barrier web. In the Northern States, this occurs late in August and in September.

During the late summer and in the autumn, grasshoppers

Fig. 440. WEB OF MIRANDA AURANTIA, GREATLY REDUCED. FROM A PHOTOGRAPH TAKEN IN THE FIELD

Superfamily Argiopoidea

form a large part of the food of this and the following species. It is interesting to see how skillfully the spider manages her huge prey. The instant it becomes entangled she rushes to it, and spreading her spinnerets far apart she fastens a swathing band to it; then by a few dexterous kicks she rolls it over two or three times and it is securely swathed in a shroud; a quick bite with her cheliceræ completes the destruction of the victim.

In the autumn, the female makes a pear-shaped egg-sac as large as a hickory nut (Fig. 197, p. 209); this is suspended among the branches of some shrub or in the top of some weed, and is fastened by many ropes of silk so that the storms of winter shall not tear it loose. Within this egg-sac the young spiders pass the winter.

The egg-sacs of this species are frequently infested by Ichneumon parasites and these parasites are preyed upon in turn by secondary parasites. It is easy to rear specimens of both by keeping egg-sacs in a closed bottle.

Genus METARGIOPE (Met-ar-gi′o-pe)

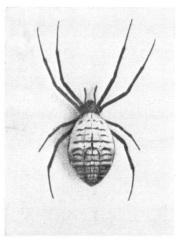

Fig. 447. METARGIOPE TRIFASCIATA, ADULT FEMALE

This genus agrees with *Miranda* in all of the characteristics given for that genus above except that the vulva of the female is divided by a septum into a pair of equal concavities; it differs from the following genus in the form of the abdomen, which is more or less evenly rounded on the sides. Three species have been found in the United States. The following is the only widely distributed one; the other two occur in Arizona, and perhaps elsewhere in the Southwest.

The Banded Garden Spider, *Metargiope trifasciata* (M. tri-fas-ci-a′ta).— Almost as conspicuous as the species just described is a closely allied one, the banded Argiope. This is a slightly smaller species, the adult female measuring from three

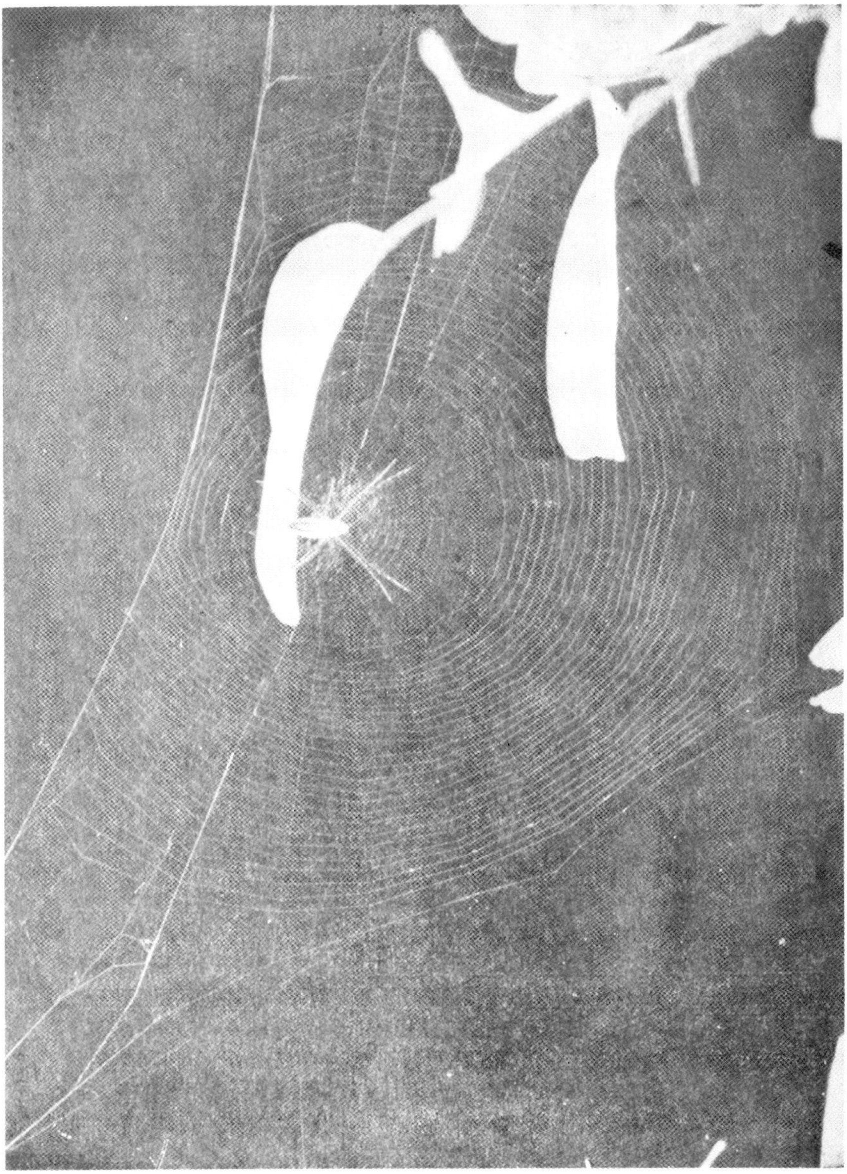

Fig. 448. METARGIOPE TRIFASCIATA. YOUNG FEMALE

Superfamily Argiopoidea

fifths to four fifths inch in length; and it is very differently coloured. The ground colour is white or light yellow; the abdomen is crossed by many, transverse lines (Fig. 447); and there are several longitudinal lines on the caudal half of the abdomen as shown in the figure.

The male is about one fifth inch in length; the legs and cephalothorax are yellowish and the abdomen white.

The young female is silvery white,

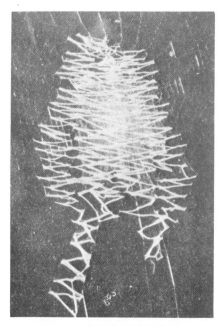

Fig. 450. STABILIMENTUM OF METARGIOPE, NARROW TYPE

Fig. 451. STABILIMENTUM OF METARGIOPE, LACE-LIKE TYPE

due to a covering of white hairs; on the caudal half of the abdomen, the coating of hairs is not continuous but forms three stripes (Fig. 448). The sternum is bright yellow and there are two yellow bands on the lower side of the abdomen.

It is an interesting fact that the hairs forming the silvery coat point toward the head; as the spider hangs head downward this position fits them for shedding rain.

The specific name of the species, *trifasciata*, is appropriate when the immature spider is considered, the three stripes on the caudal half of the abdomen being well-marked; but it is not a fortunate designation, for the bands of the adult are many more than three. The species was named *fasciata* by Hentz and *transversa* by Emerton. Either of these names would be more appropriate, but as *trifasciata* is the older name it must be used.

Fig. 452. EGG-SAC OF METARGIOPE TRIFASCIATA

This is a widely distributed species both in the Old World and in the New; in this hemisphere it extends from the United States to Chili.

The web of the banded Argiope closely resembles that of the preceding species described above. A very perfect specimen that was made in our insectary is represented by the frontispiece. This figure is greatly reduced; the centre of this web is shown natural size at page 198. Sometimes the barrier web of this species is sheet-like, concave, and almost as large as the orb.

There are great variations in the hub of the orb of this

Superfamily Argiopoidea

species. In some it is merely sheeted; in others there is a stabilimentum. The stabilimentum is sometimes a narrow wavy band (Fig. 450) and sometimes a broad sheet of lace (Fig. 451). I have found the more elaborate form only in the webs of immature individuals.

This spider reaches maturity somewhat later in the season than does the orange Argiope, and it is sometimes quite late in the autumn before the egg-sac is made.

The egg-sac is of very characteristic form, not to be mistaken for

Fig. 453. METARGIOPE TRIFASCIATA MAKING HER EGG-SAC

any other in our fauna. It is cup-shaped with a flat top (Fig. 452); and is fastened among the branches of some low shrub or between the leaves of herbs. In building the egg-sac the spider makes the flat side first, and then attaches the mass of eggs to it, and finally covers the mass of eggs with the cup-shaped portion. Figure 453 represents a spider making its egg-sac in one of our breeding cages.

Genus ARGIOPE (Ar-gi'o-pe)

The members of this genus can be distinguished from all other members of this subfamily occurring in our fauna by the form of the abdomen, the margin of which is lobate.

Of the genus *Argiope* as now restricted only a single species occurs within

Fig. 454. ARGIOPE ARGENTATA (after F. O. Pickard-Cambridge,)

the limits of the United States; formerly all of our Argiopinæ except *Gea heptagon* were included in it.

The Silvered Garden Spider, *Argiope argentata* (A. ar-gen-ta′ta).— This is a tropical species whose range extends into the southern part of our territory, being found in the Gulf States and in the Southwest. The adult female measures three fourths inch in length. It can be easily recognized by its striking form (Fig. 454).

Genus GEA (Ge′a)

The cephalothorax is arched; the head is large; the eyes of the anterior row are nearly equidistant. The two sexes are nearly equal in size.

Only one species is found in this country.

Gea heptagon (G. hep′ta-gon).— The adult female measures one fifth inch in length. The general colours of the cephalothorax are dark brown with yellowish and blackish markings.

The dorsal field of the abdomen is brown, relieved with metallic white; there is a black shield-shaped folium in the middle of the apical half of the abdomen. There are three tubercles on each side of the abdomen, one at the base, and two near the middle. It was probably these that suggested the specific name. The male closely resembles the female in size and markings; but differs in that the four anterior tibiæ are furnished, on the internal side, with many, long and strong spines.

The web of our species has not been described. That of an exotic species described by Workman, as quoted by Simon, has no stabilimentum, and the viscid spirals are numerous and close together.

This is a southern species which is found as far north as the District of Columbia.

Subfamily ARANEINÆ (A-ran-e-i′næ)

The Typical Orb-weavers

To this family belong the larger number of our orb-weaving spiders, including as it does more species than the other six sub-families of the Argiopidæ taken together. As the Araneinæ constitute the central and most abundant type of orb-weavers, I give to the subfamily the popular name of the typical orb-weavers.

Superfamily Argiopoidea

In the Araneinæ, as in the two preceding and in the following subfamilies, the epigastric plates are marked by transverse furrows (Fig. 455), and the lateral condyle of the chelicerae is distinct. But these spiders differ from the two preceding subfamilies in having the legs relatively shorter; and in having, except in a few cases, the metatarsi and tarsi together not longer than the patellæ and tibiæ. They differ from the following subfamily in not having the spinnerets elevated on a large tubular projection.

Fig. 455.
EPIGASTRIC PLATE OF ARANEA

The separation of the subfamily Araneinæ into genera has not been made, as yet, in a satisfactory manner. The large number of species involved and the presence of intergrading forms between supposed distinct generic types make the subject an extremely difficult one. Great differences of opinion exist regarding the validity of certain genera that have been proposed. The following classification, therefore, must be regarded as merely provisional.

TABLE OF GENERA OF THE ARANEINÆ

- A. Cephalothorax either as high as long or with horny outgrowths.
    - B. Cephalothorax as high as long. P. 461.   SCOLODERUS
    - BB. Cephalothorax with horny outgrowths. P. 462.
        GLYPTOCRANIUM
- AA. Cephalothorax moderately arched and without horny outgrowths.
    - B. Abdomen with a hump on each side at the base, which bears irregular tubercles. P. 464.   KAIRA
    - BB. Abdomen with or without a hump on each side at the base; but without tubercles on the humps when present.
        - C. Median furrow of the thorax when a narrow longitudinal one slight, not reaching the cervical groove; often a pit with transverse extensions.
            - D. Head and thorax separated by a deep cervical groove in the female at least. P. 464.   CYCLOSA

Superfamily Argiopoidea

DD. Head and thorax not separated by a well-marked complete cervical groove.
  E. Abdomen with a horny shield. P. 469.
    CERCIDIA
  EE. Abdomen without a horny shield.
    F. Abdomen with a median hump or cone at the base, as well as lateral projections. P. 469.
      MARXIA
    FF. Abdomen without median together with lateral projections.
      G. Posterior median eyes nearly or quite as close to the posterior lateral eyes as to each other. P. 471. ZILLA
      GG. Posterior median eyes much closer to each other than to the posterior lateral eyes.
        H. The tibiæ of the first and second pairs of legs without spines above.
          I. Abdomen with spines or humps. P. 474.
            WAGNERIANA
          II. Abdomen without spines or humps. P. 475. METAZYGIA
        HH. The tibiæ of the first and second pairs of legs with some spines above, at least one.
          I. Abdomen as high behind its middle as at its base, and elliptical in outline or broader behind the middle. Small species with short legs. P. 475.
            SINGA
          II. Abdomen highest toward its base, and usually broadest near the base.
            J. Abdomen with cusps or tubercles behind.
              K. Abdomen with several marginal tubercles behind. P. 479. VERRUCOSA
              KK. Abdomen without marginal tubercles behind; but with two round black tubercles on the middle line of the hind part of the abdomen. P. 516.
                ERIOPHORA
            JJ. Abdomen of various forms but without tubercles behind.

Superfamily Argiopoidea

    K. Abdomen greatly elevated in front so that the pedicel is near the middle of its length. P. 480. Wixia
   KK. Abdomen much less elevated in front than in *Wixia*.
     L. Scape of the epigynum greatly elongate extending nearly or quite to the base of the spinnerets. Male with a very long and elbowed embolus. P. 516.
                 Eriophora
     LL. Scape of the epigynum of moderate length. Embolus of male not elbowed as in *Eriophora*.
       M. Metatarsus and tarsus of the first legs together longer than the tibia and patella. Males with no hook on the coxa of the first legs and no groove on the femur of the second legs. P. 476. Metepeira
      MM. Metarsus and tarsus of the first legs not longer than the tibia and patella except in a few species. Males with a hook on the coxa of the first legs and a groove on the femur of the second legs. P. 481. Aranea
 CC. Thorax with a deep median longitudinal furrow, which usually extends forward so as to reach the cervical groove.
   D. Elongate spiders, the abdomen being two or three times as long as wide. P. 520. Larinia
  DD. Abdomen not greatly elongate.
    E. Tibia of the third legs with a cluster of long cilia on the anterior side near the base. Second row of eyes straight or procurved. P. 517. Mangora
   EE. Tibia of the third legs without a cluster of long cilia on the anterior side near the base. Second row of eyes more or less recurved.

Superfamily Argiopoidea

F. Abdomen marked with a broad folium, darker at the edges and bordered by a white line, and enclosing a lanciform stripe bordered in a similar manner. P. 522. ACACESIA
FF. Abdomen not marked as in *Acacesia*.
    G. Females.
        H. Epigynum with the scape directed backward. P. 509. NEOSCONA
        HH. Epigynum with the scape directed forward. P. 523. EUSTALA
    GG. Males.
        H. Patella of the pedipalps with only one apical spine. P. 523. EUSTALA
        HH. Patella of the pedipalp with two apical spines.
            I. Lateral eyes of each side situated on a prominent tubercle.
                J. Abdomen with a bright green patch on the dorsum. P. 516. ERIOPHORA
                JJ. Abdomen not marked as in *Eriophora*. P. 481. ARANEA
            II. Lateral eyes of each side not situated on a prominent tubercle. P. 509. NEOSCONA

Genus SCOLODERUS (Sco-lod'e-rus)

The spiders of this genus are distinguished by a very high and rounded cephalothorax; this region being as high as long (Fig. 456). Both rows of eyes are procurved; and the lateral eyes are placed low down upon the lateral margins of the clypeus.

These are tropical spiders; the range of a single species extends into the southern part of the United States.

Fig. 456. SCOLODERUS TUBERCULIFERUS, SIDE VIEW

Scoloderus tuberculiferus (S. tu-ber-cu-lif'e-rus).— This species is easily recognized by the peculiar form of the cephalothorax and of the abdomen (Fig. 456). The cephalothorax is as high as

Superfamily Argiopoidea

long; the abdomen is cylindrical, with two thick cones at the base whose summits are bifid. Only the female is described; this measures one sixth inch in length. It was found in Florida.

This is the *Carepalxis tuberculifera* of some authors.

Genus GLYPTOCRANIUM (Glyp-to-cra′ni-um)

Our species of this genus can be easily recognized by the form of the cephalothorax, which bears prominent horny outgrowths and many smaller warts. The abdomen is subglobose and is as wide as or wider than long.

Only two species are described from the United States. These were formerly placed in the genus *Ordgarius*.

Glyptocranium cornigerum (G. cor-nig′e-rum).— The female measures one half inch in length. The cephalothorax is red o reddish brown and yellow, with dark brown markings; the abdomen is yellow, with dark or brownish markings upon the basal part. When resting on a leaf it looks exactly like bird-lime; it rests with its legs folded so as to increase this resemblance (Fig. 457), which is chiefly due, however, to the colours of the spider. Figure 458 represents a front view of the cephalothorax. The abdomen bears a very prominent pair of shoulder humps. The male measures only one tenth inch in length; the cephalothorax bears four large tubercles on the hind part of the head; and the abdomen is nearly spherical.

This species is found throughout the southern half of the United States.

Glyptocranium bisaccatum (G. bi-sac-ca′tum).— This species is a little smaller than the preceding, the adult female measuring from one third to four tenths of an inch in length. The cephalothorax is slightly scalloped at the sides. It rises from the eyes backward and has at the highest part behind the middle two large horns. The carapace is covered with conical scattered points. The abdomen is wider in front than long and extends over the thorax as far as the two horns (Fig. 459). The cephalothorax is light brown, darkest in front; the front of the abdomen is light brown with various whitish irregular markings, the back part is yellowish white; the under side of the body is white. The male resembles the female in the general form of the body, but measures only one eighth inch in length.

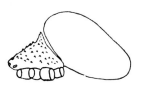

Fig. 459. GLYPTOCRANIUM BISACCATUM (after Emerton)

Fig. 460. EGG-SAC OF G. BISACCATUM (after Emerton)

Fig. 457.
GLYPTOCRANIUM CORNIGERUM ON A LEAF

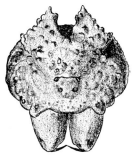

Fig. 458. GLYPTOCRANIUM CORNIGERUM, FRONT VIEW OF THE CEPHALOTHORAX

Fig. 461. KAIRA ALBA (after McCook)

Superfamily Argiopoidea

The egg-sac has been figured by Emerton, who first described the species. It is of the form shown in Fig. 460, of a dark brown colour, and very firm in texture. The egg-sac was found in October in Connecticut and the young emerged in June.

This species occurs in the Northern States.

Genus KAIRA (Kai′ra)

The cephalothorax is longer than broad, moderately arched, and without horny outgrowths. The abdomen has a hump on each side at the base which bears irregular tubercles (Fig. 461).

The following is the only species reported from the United States:

Kaira alba.— The female (Fig. 461) measures about one third inch in length. The cephalothorax is triangular ovate, widest near the base, and yellowish brown in colour. The abdomen is subglobose, with hump on each side at the base; each hump bears numerous conical tubercles. The abdomen is chalky white, mottled with blackish spots, and with an indistinct folium.

This species occurs in the Southern States. It is the *Epeira alba* of Banks's Catalogue.

Genus CYCLOSA (Cyc′lo-sa)

The eyes are subequal; the posterior median are almost in contact; the median ocular area is wider in front than behind; both rows of eyes are recurved; the head and thorax are separated by a deep cervical groove, in the female at least (Fig. 462).

Five species have been described from the United States; these can be separated as follows:

A. Caudal end of abdomen bifurcate. P. 467. *C. bifurca*
AA. Caudal end of abdomen not forked.
  B. Abdomen extremely long. P. 467. *C. caroli*
  BB. Abdomen not extremely long.
    C. Abdomen with five tubercles, a pair near the base, a caudal tubercle, which is divided, and one on each side of this. P. 468. *C. walckenœri*
    CC. Abdomen with less than five tubercles.

Superfamily Argiopoidea

D. Abdomen of female with a pair of dorsal median tubercles and with a slender caudal projection. P. 468. *C. turbinata*

DD. Abdomen without a dorsal tubercle and with the caudal projection stout. P. 465. *C. conica*

Cyclosa conica (C. con'i-ca).— In the Northern States we have only one common species of *Cyclosa*, which is this one. The full-grown female is about one fourth inch in length and is easily recognized by the form of the caudal end of the abdomen which is extended into a prominent hump (Fig. 463); in the male there is only a slight trace of the hump. The cephalothorax is dark

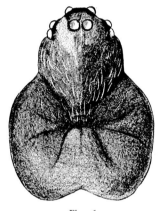

Fig. 462.
CEPHALOTHORAX OF CYCLOSA

Fig. 463.
CYCLOSA CONICA

gray or black; and the abdomen is mottled with gray and white; but there are great variations in colour; some individuals are almost white while others are nearly black.

The web is built upon shrubs, and is common in open woodlands. It is a complete orb and is remarkable for its symmetry and for the fineness of its meshes; there being a large number of radii, and the turns of the viscid spiral being very close together (Fig. 464). The foundation lines are often very few in number; in the web figured here there are only four main ones, which form a quadrangular space in which the orb is stretched, and a short secondary one at each corner of this space. The hub is meshed; there is a distinct notched zone, and a rather wide

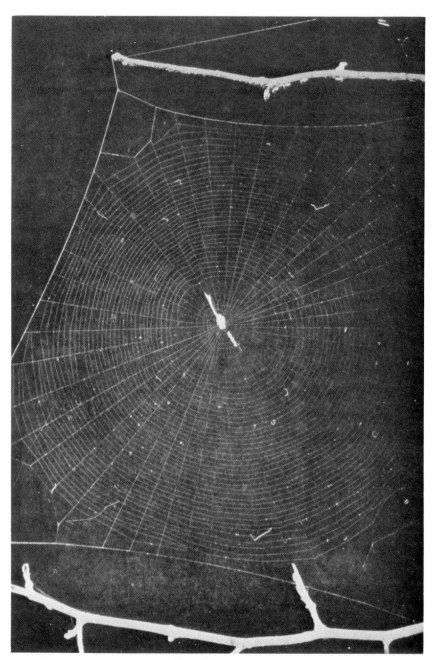

Fig. 464. WEB OF CYCLOSA CONICA

free zone. Across the centre of the web there is stretched a stabilimentum.

The stabilimentum of *Cyclosa* differs greatly from that of a garden spider. It often consists largely of the remains of the insects that the spider has destroyed fastened together and in place with threads of silk. Frequently a cast skin of the spider is woven in with the insect remains; and bits of vegetation, as fallen budscales, are utilized. The stabilimentum in a web that was built near a place where rugs were frequently beaten consisted largely of lint; evidently the spider had removed all of the fibres that had lodged in its web and used them in making its stabilimentum. Sometimes the stabilimentum consists entirely of silk.

There is no retreat, the spider remaining constantly on the web. It rests on the hub in the centre of the stabilimentum, and appears like a part of the rubbish fastened in it.

Emerton states that the egg-sacs are fastened into the stabilimentum in the middle of the summer.

Cyclosa caroli (C. car'o-li).— The characteristic feature of this species is the very long and slender abdomen. The adult female measures from one fourth to one third inch in length.

This spider is remarkable for its self protective habits. These I observed in the jungle near Miami, Fla. The orb of the adult is about six inches in diameter. The female fastens her egg-sacs in a series which extends across the orb from the hub to the upper margin like a stabilimentum, and looks like a dead twig caught in the web. This band of egg-sacs and the spider are of the same gray colour. When disturbed the spider rushes to the band and appears as if it were a part of it. And here it will cling motionless even when the band is removed from the web. A spider which I removed with its egg-sac and placed in a bottle was still in this position on the band ten minutes later.

I also observed smaller individuals shake their webs; these clung to the stabilimentum, projecting the body at right angles to it, and in this position shook the web violently. The evident object was to frighten away the intruder.

The young, when they emerge from the egg-sac, are not tailed.

This is a southern species which has been found as far north as the District of Columbia.

Cyclosa bifurca (C. bi-fur'ca).— The strange form of the abdomen sharply distinguishes this species from the other species of *Cyclosa* in our fauna. The abdomen is long; the basal half bears two pairs of humps; the caudal half is more slender and is bifurcate at the tip (Fig. 465). The body is green mottled with white; there is a light wavy band on the side of the abdomen; and the abdomen is bordered with a black line above, back of the humps. On the ventral side of the abdomen, there is a bright red spot between the epigastric furrow and the spinnerets. The legs are banded with reddish brown.

Fig. 465.
CYCLOSA BIFURCA

I found this to be a common species near Miami, Fla. I first found it in the jungle near the shore of the bay, where it made an orb-web with a string of egg-sacs across it like a stabilimentum. Later I found it to be a pest at the cottage connected with the Sub-tropical Laboratory. There it built its webs on the ceiling of the veranda and there strings of egg-sacs hung from the ceiling by the hundred; and at the lower end of each string of egg-sacs there was a spider. One of these strings is represented by Fig. 466.

Cyclosa turbinata (C. tur-bi-na'ta).— This species of *Cyclosa* is distributed throughout the United States; but I have found it much more common in the South than in the North. The female can be easily distinguished from the other species of *Cyclosa* occurring in our fauna by the shape of the abdomen (Fig. 467). The abdomen bears a pair of dorsal median tubercles and is prolonged into a slender caudal projection. In the male the abdomen is rounded and is marked above at the base with a transverse white band, which is sometimes interrupted in the middle, and a pair of white spots in front of the middle of its length.

Fig. 466. EGG-SACS OF CYCLOSA BIFURCA

This is the *Epeira caudata* of Hentz.

Cyclosa walckenæri (C. walck-e-næ'ri).— This is a tropical species whose range extends into the southern part of the United States. It can be distinguished from our other species by its

Superfamily Argiopoidea

greater number of abdominal tubercles, as indicated in the table above.

Genus CERCIDIA (Cer-cid′i-a)

This genus is distinguished from the closely allied genera by the nature of the cuticula of the dorsum of the abdomen, which is hard and glossy. Two species occur in our fauna.

Cercidia funebris (C. fu′ne-bris).— This is a small spider, the female measuring only one sixth inch in length, which has

Fig. 467. CYCLOSA TURBINATA

been found in Florida. The cephalothorax is reddish brown; the abdomen glossy black, with a median and lateral stripe of chalky white or yellow.

Cercidia prominens (C. prom′i-nens).— The cephalothorax is red; the cheliceræ are red with black spots; the sternum is black; the legs are yellowish with brown rings; the abdomen is brownish above, with a large reddish shield nearly covering the dorsum, there is an indistinct light stripe and behind some transverse black lines. The cephalothorax has above on the median line two prominent spines. The length of the body is one fifth inch.

This species was described from New Hampshire.

Genus MARXIA (Marx′i-a)

The anterior row of eyes is strongly procurved, the median eyes being twice as far from the margin of the clypeus as are the

Superfamily Argiopoidea

lateral eyes. The lateral eyes of each side are placed under a conical tubercle. The sternum is at least one half longer than broad. The abdomen is armed with a median hump or cone at the base as well as with lateral projections. The patella of the pedipalp of the male bears a single spine at the apex. Our species are placed in the genus *Plectana* by some writers.

Four species occur in our fauna; one of these occurs in Florida and one in

Fig. 468. MARXIA STELLATA, IMMATURE MALE

Fig. 469. EGG-SAC OF MARXIA STELLATA

Arizona. The other two are the following:

Marxia stellata (M. stella′ta). — This star-shaped spider is from a quarter to a third of an inch long and nearly as broad. The abdomen is encircled by a series of tubercles, eleven, twelve, or thirteen in number (Fig. 468). The abdomen is grayish brown marked with a light and dark pattern, which varies in different individuals; the thorax is black on the sides and brown above; the sternum is brown bordered by black; and the legs are ringed with dark brown or black.

This spider lives in low bushes and among weeds and grass.

It makes a complete orb from six to ten inches in diameter. The hub of the orb is nearly open, the central space being crossed by comparatively few lines; there is a distinct notched zone, and a free zone, and, usually, from twenty to thirty-five viscid spirals.

The spider is sometimes found on its orb in midday; but usually it rests in a retreat made in the dead head of a plant forming one of the supports of the web. In this case, the brown colour of the body and the spines on the abdomen serve to render the spider inconspicuous.

The mass of eggs is attached to a leaf and enclosed in a mass of loose brown silk (Fig. 469).

This is a common species both in the North and in the South.

Marxia mœsta (M. mœs'ta). — This species is found in the Southern States. It differs from the preceding in its smaller number of abdominal tubercles. "Two shoulder tubercles mark the base well back of the anterior middle point, leaving thus the fore part of the abdomen as a wide subtriangular space sloping toward the front, while the remainder of the dorsum slopes somewhat, though but little, toward the rear. The apex is marked by a prominent rounded tubercle, resembling those upon the shoulders, but smaller; on either side of this is a similar smaller tubercle, and beneath it on the apical wall of the abdomen are two others in a row, of similar character, but somewhat flattened. The colour is yellow, much broken by irregular and lateral black lines upon the sides." (McCook.)

Genus ZILLA (Zil'la)

The abdomen is short oval, rather depressed, and without humps at the base. The posterior median eyes are scarcely more than one and one half times the diameter of one of them apart, and as close to the posterior lateral eyes as to each other; the lateral eyes are slightly separate; and all of the eyes are subequal in size. The epigynum is without a scape in our species.

Four species of *Zilla* are found in the United States, at least three of which are believed to have been introduced from Europe. They are of moderate size, the largest being about three eighths of an inch in length. In their general appearance they resemble spiders of the genus *Steatoda* of the family Theridiidæ. Of the three species found in the East, Emerton states as follows:

Fig. 470. WEB OF ZILLA

Superfamily Argiopoidea

"The colour of all of the species is gray, with sometimes a little yellow or pink in the lighter parts. The cephalothorax has usually, but not always, a dark border at the sides and a middle dark line that widens and becomes lighter toward the eyes. The abdomen has a wide middle stripe like *Aranea*, scalloped at the sides and crossed at the hinder end by two or three pairs of transverse spots. In front it is almost white or tinted with pink or yellow, and narrows almost to a point, with a much darker spot on each side, The sides of the abdomen are marked with oblique

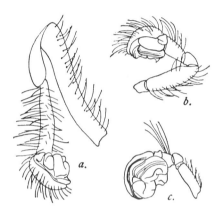

Fig. 471. PALPI OF MALES OF ZILLA
a, *Zilla atrica* b, *Zilla x-notata*
c, *Zilla montana* (after Emerton)

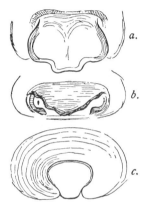

Fig. 472. EPIGYNA OF ZILLA
a, *Zilla atrica* b, *Zilla x-notata*
c, *Zilla montana* (after Emerton)

dark marks that extend underneath. The sternum has a light middle stripe. Under the abdomen is a dark middle stripe, with light each side of it."

The three species resemble each other so closely that it is almost impossible to distinguish them except by an examination of the palpi of the males and the epigyna of the females.

The spiders of this genus usually make an incomplete orb of the form which I have designated as the zilla type of orb web (Fig. 470); but sometimes they make a complete orb.

Zilla atrica (Z. at'ri-ca).— See above for the more general characteristics. The palpi of the males are as long as the whole body, with the femur and tibia both slightly curved and the tarsus and bulb small (Fig. 471, *a*). The epigynum of the female is of the form shown in Fig. 472, *a*.

Superfamily Argiopoidea

Zilla x-notata (Z. x-no-ta′ta).— See above for the more general characteristics. The palpi of the males are as long as the cephalothorax, and the tarsus and bulb small and round (Fig. 471, *b*). The epigynum of the female is of the form shown in Fig. 472, *b*.

Zilla montana (Z. mon-ta′na).— See above for the more general characteristics. The palpi of the males are shorter than in the preceding species, the tibia thicker and the tarsus and bulb larger (Fig. 471, *c*). The epigynum of the female is twice as large as that of either of the preceding species and of the form shown in Fig. 472, *c*.

Zilla californica (Z. cal-i-for′ni-ca).— This species is found in California and in Washington. Only the female has been described. The cephalothorax is whitish, with a black marginal seam, and a large triangular black spot over the head part. The abdomen is grayish, with a broad folium, rather silvery near the middle, black on the edge and with a silvery margin. The sides are finely striped with black. The epigynum shows a dark transverse area, three times as wide as long and with a small projection behind from the middle.

Genus WAGNERIANA (Wag-ner-i-a′na)

The abdomen is armed with prominent humps both in front and behind, but lacks the median hump at the base characteristic of *Marxia*. The posterior median eyes are much closer to each other than to the posterior lateral eyes. The tibiæ of the first and second pairs of legs are without spines above. Two species are found in Florida; the following is the more common one:

Fig. 473.
WAGNERIANA TAURICORNIS

Wagneriana tauricornis (tau-ri-cor′nis).— The female measures about one third inch in length; the male, about one fifth. The species is easily recognized by the prominent humps on the abdomen; these are more prominent in the female (Fig. 473) than in the male. In the male the shoulder humps are single instead of bifid, and there is a single lateral one instead of three as in the female. There are four apical humps in the male and five in the female.

This species is found in the southern portions of the United States, and its range extends through Central America.

I studied this spider at Miami, Fla., and noted that it held its legs closely folded when hanging in its web; and that when in this position, it appeared like a bit of dirt.

Genus METAZYGIA (Met-a-zyg′i-a)

This genus is represented in the United States by a single species, which occurs in Florida and in Mexico, and which I have not seen. It differs from *Wagneriana* in the absence of spines or humps on the abdomen.

Metazygia wittfeldæ (M. witt-fel′dæ).— The female measures a little more than one third inch in length; the male, a little less. The head is dark brown, the thorax yellow, and the abdomen yellow or yellowish green, with a large central scalloped folium, outlined in black.

Genus SINGA (Sin′ga)

The posterior median eyes are closer to each other than to the posterior lateral eyes. The tibiæ of the first and second pairs of legs are armed with some spines above, at least with one. The abdomen is as high behind its middle as at its base, and elliptical in outline or broader behind the middle.

This genus includes small species with short legs; they live among herbage in low open places. The following are our more common species. Seven others are known to occur in the United States.

Singa truncata (S. trun-ca′ta).— This species measures about one eighth inch in length. The head is as high as wide. The cephalothorax is orange-brown except about the eyes where it is black. The abdomen is orange with indistinct blackish markings across the hinder part.

Singa variabilis (S. va-ri-ab′i-lis).— The female measures one sixth inch in length; the male, one eighth. The cephalothorax is light orange except about the eyes where it is black. "The abdomen is usually entirely black, but occasionally has bright yellow markings (Fig. 474). Sometimes there is a wide middle stripe, with narrower ones at the sides and two underneath.

Superfamily Argiopoidea

Sometimes there are only the two lateral stripes, and there are all variations between these markings." (Emerton.)

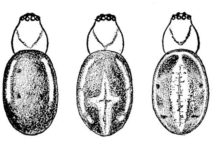

Fig. 474. SINGA VARIABILIS (after Emerton)

Fig. 475. SINGA PRATENSIS, FEMALE (after Emerton)

Singa pratensis (S. pra-ten'sis).— "When full grown the females are a fifth of an inch long, with the abdomen oval and marked with a double white stripe in the middle and a single one on each side (Fig. 475). The cephalothorax is yellow, with a little black between the middle eyes not extending to the lateral pairs. The males are marked in the same way." (Emerton.)

Genus METEPEIRA (Met-e-pei'ra)

The metatarsus and tarsus of the first legs together are longer than the tibia and patella; the lateral condyle of the

Fig. 476. PALPUS OF MALE OF METEPEIRA LABYRINTHEA

cheliceræ is comparatively small. In both of these respects this genus resembles the Metinæ more than the Argiopinæ. In the males the bulb of the palpus bears a median apophysis resembling a double-bladed reaping-hook (Fig. 476). There is no hook on the coxa of the first legs, and no groove on the femur of the second; the patella of the pedipalp bears two spines at its apex. In the female the tip of the scape of the epigynum is recurved.

The following is a common and widely distributed species.

The Labyrinth Spider, *Metepeira labyrinthea*, (M. lab-y-rin'the-a).— This exceedingly common species is most easily recognized by the form of its web, which is described below. The adult female varies from one fourth to two fifths inch in length. The carapace is dark brown with the ocular area yellow; the sternum is brown with a central yellow band; the legs are dull yellow banded with brown; the abdomen bears a well-marked folium, in which there is a central band and two pairs of spots extending laterally from this band (Fig. 477); the sides of the abdomen are mottled with rich reddish brown.

The male measures one sixth inch in length; it resembles the female, but has longer legs.

The web of this species is a composite one, consisting of an incomplete orb and an irregular net (Fig. 478). The irregular net part of the web resembles the webs of some theridiid spiders and like them is a more or less permanent structure. The orb is built in front of and usually slightly below the irregular net. It varies in form from a nearly complete orb in which the viscid line forms many complete spiral turns to an incomplete one in which no turn of the viscid thread passes entirely around the web. On one occasion I saw an orb of this species that was of the zilla type, but after this was destroyed the individual that made it made a complete orb. The hub of the orb is of the meshed type, and from it there extend several traplines to a retreat. This is usually situated in the midst of the irregular net, and ordinarily consists of a bunch of dry leaves enclosing a space lined with silk. I have observed the webs of the young labyrinth spiders in June; They were mostly built among dead branches and the retreat was made entirely of silk; it was usually near a branch, but sometimes it was in the midst of the labyrinth. A little later in the season, early in July. the spiders made their tents near the centre of the labyrinth.

Fig. 477. METEPEIRA LABYRINTHEA, FEMALE

The spiders mature early in the autumn; the males are then to be found in the webs of the females. A little later the females

Fig. 478. WEB AND SERIES OF EGG-SACS OF METEPEIRA LABYRINTHEA

Superfamily Argiopoidea

begin to deposit their eggs. These are enclosed in a string of lenticular egg-sacs (Fig. 479), the formation of which extends over a considerable period of time. The first egg-sac is formed near the entrance of the retreat, and the others are placed successively in line below this one, and are fastened to a strong cord of silk. The cord supporting the egg-sac is stretched between two branches and is made so strong by the addition of many draglines that the egg-sacs are securely held in place through all the storms of winter, long after all other traces of the web are swept away. In walking through woodlands in the spring one often encounters these strong silken cords, each with its string of egg-sacs securely held in place.

This species is found throughout the greater part of the United States.

Genus VERRUCOSA (Ver-ru-co'sa)

In this genus the cervical groove is deeply marked; the head is much elevated above the thorax; the median furrow of the thorax closely resembles that of *Aranea;* the abdomen is flattened above, is subtriangular in outline, and is armed with tubercles behind; the cuticula of the dorsum is somewhat hardened. The scape of the epigynum is elongate.

Fig. 479. EGG-SACS OF METEPEIRA LABYRINTHEA

There is some doubt regarding the distinctness of this genus; several authors class the species placed here in *Aranea (Eperia).* The following is our best-known species.

Verrucosa arenata (V. ar-e-na'ta).— The abdomen is subtriangular in outline and in the female is nearly as wide in front as it is long (Fig. 480); it is narrowed behind but is not pointed. The caudal end bears a number of small tubercles. The most conspicuous feature is a large, triangular, light-coloured spot on the abdomen; this is white, yellow, pink, or greenish, varying in different individuals, and is divided by a mesh of fine, vein-like

Superfamily Argiopoidea

dark lines. The scape of the epigynum is very long reaching halfway to the spinnerets.

In the male the abdomen is less triangular, not exceeding the cephalothorax in width at its base; and the dorsal spot is apt to be broken. The legs are longer and more slender than in the female; the metatarsus of the second pair is strongly curved, and the tibia of these legs bears a long forked spine on the inner face near the tip.

This is a Southern species which occasionally occurs as far north as Philadelphia and Long Island. It spins a large web with a coarse mesh and long guy-lines in forests and woodlands. When alarmed it leaves the web by one of the guy-lines unless immediate danger seems pending when it drops to the ground.

Fig. 480. VERRUCOSA ARENATA, FEMALE

Genus WIXIA (Wix′i-a)

As with the preceding genus, there is some doubt regarding the distinctness of this genus from *Aranea*. But our only representative can be distinguished at a glance by the form of the abdomen, which is greatly elevated in front so that the pedicel is near the middle of its length (Fig. 481). The median furrow is transverse; the posterior median eyes are larger than the anterior median; the median ocular area is wider behind than in front and wider behind than long; the sides of the head are parallel, and the head is not narrowed in front.

Fig. 481. WIXIA ECTYPA

Superfamily Argiopoidea

Wixia ectypa (W. ec′ty-pa).— The female measures a little less than one third inch in length and the male is of same size. The cephalothorax is a long oval; high, truncated, and indented at the base. The abdomen is ovate, greatly elevated in front (Fig. 481) and contracted into a prominence, which is bifid at the top.

This species occurs in the Southern States, and its range extends far south of our border. It is the *Epeira infumata* of Hentz.

Genus ARANEA (A-ran′e-a)

The cephalothorax is moderately arched and is without horny outgrowths. The median furrow of the thorax of the female is transverse, straight or recurved, and prolonged at each end by a stria, which extends backward and outward (Fig. 482); these three lines bound in front a smooth area, which is covered by the overlapping abdomen; this area is sometimes marked by a median longitudinal furrow; in a few cases the median furrow is a circular pit. In the male, of all of the species examined by me, the median furrow of the thorax is a pit with prolongations extending forward, backward, and to each side. The patella of the pedipalp of the male is armed with two spines at its apex. The anterior and posterior median eyes are slightly unequal in size; the median ocular area is not much longer than wide. The lateral eyes of each side are contiguous or nearly so and widely removed from the median eyes. The clypeus is narrower than the median ocular area.

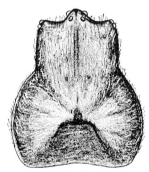

Fig. 482.
CEPHALOTHORAX OF ARANEA

The genus *Aranea* is a very large one, in the United States alone nearly fifty species belonging to it have already been described.*

This genus has been commonly known under the name *Epeira;* this, however, is a much later name than *Aranea*, which was proposed by Linnæus in his *Systema Naturæ* (Ed. X., p. 619). It had

---

*Banks in his Catalogue lists fifty-seven species, under the name *Epeira;* twelve of these species are placed in other genera in this book. See *Kaira, Verrucosa, Wixia, Neoscona, Eriophora,* and *Eustala.*

Superfamily Argiopoidea

previously been named *Araneus* by Clerk; but this form of the name was published one year before the date adopted for the beginning of the binomial nomenclature.

Although a large number of species occur in our fauna, the few described below include those that most commonly attract attention. To facilitate reference to the descriptions of these they are grouped under the following heads:

The Large Angulate Araneas. P. 482.
The Smaller Angulate Araneas. P. 486.
The Large Round-shouldered Araneas. P. 488.
The Three House Araneas. P. 498.
The Smaller Round-shouldered Araneas. P. 504.

THE LARGE ANGULATE ARANEAS

There is a group of species of *Aranea* in which the abdomen bears a pair of prominent humps near its base; this group has been designated *The Angulate Araneas*. The species composing this group can be easily separated into two subgroups, one containing the species of large size; the other, those of moderate or small size.

Of the large angulate Araneas there are four species found in the United States that are sufficiently common to merit mention here. The adult females of these species usually measure one half inch or more in length, and some individuals are nearly one inch in length.

Each of these species exhibits great variations in size, colour, and markings. A careful study of the published descriptions and of a fairly large series of specimens has convinced me that comparatively little use can be made of colours and markings in separating these species; but the females can be easily separated by the form of the epigynum; and it is probable that the males can be separated by the structure of the palpi and the shape and armature of the second legs.

I have not sufficient material to enable me to prepare a table for separating the males of these species; the females can be separated as follows:

A. Scape of the epigynum triangular, as wide at the base as long. P. 486.   *A. gemma*

Superfamily Argiopoidea

AA. Scape of the epigynum longer than its width at the base.
 B. Scape of the epigynum less than twice as long as its width at the base. P. 484. *A. cavatica*
 BB. Scape of the epigynum more than twice as long as its width at the base.
  C. Each lateral half of the atriolum more or less distinctly divided by a furrow; scape long, usually narrow at the base, sometimes slightly widened, but not tapered evenly to the tip. P. 483. *A. angulata*
  CC. Lateral halves of the atriolum not subdivided; scape widest at the base and usually tapered evenly to the tip. P. 484. *A. nordmanni*

Aranea angulata (A. an-gu-la′ta).— This is an exceedingly variable species which is commonly believed to occur both in Europe and in North America. On both sides of the Atlantic several of the varieties of what is regarded as this species have received distinct specific names. It is one of the larger of the four species mentioned here; it lives among trees; and is usually dark coloured like bark. In what may be considered the more typical form (Fig. 483), there is a yellow spot or group of spots on the middle line of the abdomen between or in front of the shoulder humps; and on the hind half of the abdomen there is a distinct folium bordered by a fairly even, undulating dark line. The yellow marks between or in front of the humps vary greatly in shape and may be wanting.

Fig. 483.
ARANEA ANGULATA, FEMALE

In many individuals the dark line bordering the folium is broken into several pairs of very dark, oblique bars; as is the case in the more common form of *A. nordmanni* (Fig. 484).

This species is distinguished by the narrower and longer scape

Superfamily Argiopoidea

of the epigynum and by the fact that each lateral half of the atriolum is divided by a furrow (Fig. 483 *bis*).

The male resembles the female in colour and markings, but is only half as large.

This species often builds its web very high in trees.

In case our form proves to be specifically distinct from that found in Europe, it will be known as *A. silvatica*, the name given to it by Emerton in 1884.

Aranea nordmanni (A. nord-man'ni).— This is a smaller species than *A. angulata;* but the two cannot be separated by size; as some individuals of *A. nordmanni* are as large as the smaller individuals of *A. angulata*. *A. nordmanni* rarely exceeds three fifths inch in length. The same variations in colour and markings are found here as in the preceding species; but in *A. nordmanni* the more common type is that in which the dark line bordering the folium is broken into several pairs of very dark oblique bars (Fig. 484).

In many individuals, the folium, which covers only the hind half of the abdomen, is solid black, and is bordered by a light yellow line (Fig. 485). This variety is represented by both sexes. I have not observed a similar variation in either of the other species. Figure 485 *bis* represents an intermediate variety.

The female of this species can be recognized by the form of the epigynum; the scape is not so long as in *A. angulata;* it is widest at the base and is usually tapered evenly to the tip (Fig. 486); the lateral halves of the atriolum are not divided.

This, like the preceding, is an introduced species; it is widely distributed in the Atlantic region;

Aranea cavatica (A. ca-vat'i-ca).— This spider is dirty white in colour with grayish markings. The abdomen is clothed with numerous whitish or gray hairs, which give it in life a grayish appearance; this is not so marked in alcoholic specimens. The folium is often distinct (Fig. 487); but is usually not so well-marked as in the two preceding species, and is sometimes indistinct. On the ventral side of the abdomen there is a broad black band extending from the epigastric furrow to the spinnerets; the basal half of this band is bordered by two curved yellow lines; and near the middle of its length there is a pair of yellow spots (Fig. 488).

Fig. 483 bis. EPIGYNUM OF ARANEA ANGULATA

Fig. 484. ARANEA NORDMANNI, FEMALE

Fig. 485. ARANEA NORDMANNI, VARIETY

Fig. 485 bis. ARANEA NORDMANNI, VARIETY

Superfamily **Argiopoidea**

Although the scape of the epigynum (Fig. 489) is elongate, it is less than twice as long as its width at the base.

The males before me are remarkable for their size, being about as large as the females.

This species, as its specific name indicates, prefers shady situations. Emerton states that it lives in great numbers about houses and barns in northern New England. I have found it in a tunnel at Ithaca, and on the sides of cliffs in a ravine. Its webs are sometimes very large.

This is the *Epeira cinerea* of Emerton.

Aranea gemma (A. gem'ma).— This is a very large species which is widely distributed in the western half of the United States. The larger specimens that I have examined measure four fifths inch in length and two thirds inch in width. The spider varies greatly in colour; usually the ground colour is yellow with brown or darkish markings. Among some specimens which I collected in California one was almost entirely white, the others were yellowish marked with dark gray. In some the folium is distinct on the hind half of the abdomen; in others it is wanting. The humps on the abdomen are very prominent. On the ventral side of the abdomen, there is a brown band extending from the epigynum to the spinnerets; this band is bordered on each side by a more or less broken yellow stripe. The most distinctive feature is the form of the epigynum, which is small for so large a spider; the scape is short, triangular, as broad at the base as long, and ends in a spoon-shaped tip (Fig. 490).

I have not seen the male. McCook states that it is small compared with the female, measuring only one third inch in length.

THE SMALLER ANGULATE ARANEAS

Among the species of *Aranea* that are characterized by the presence of a pair of humps near the base of the abdomen, there are some that are distinguished from those described above by their smaller size, the adult females usually measuring less than one third inch in length. The more common species of this group are the two following:

Aranea corticaria (A. cor-ti-ca'ri-a).— This is the larger of the two common species of the smaller angulate Araneas, the adult female measuring nearly one third inch in length. The abdomen

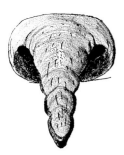

Fig. 486. EPIGYNUM OF
ARANEA NORDMANNI

Fig. 487.
ARANEA CAVATICA, FEMALE

Fig. 488.
ARANEA CAVATICA, VENTER

Fig. 489. EPIGYNUM
OF ARANEA CAVATICA

Fig. 490. EPIGYNUM
OF ARANEA GEMMA

Superfamily Argiopoidea

of the female is about as broad as long; the colour varies from yellowish brown to dark brown, with light markings, on the front part that are often bright red or yellow. The epigynum (Fig. 491) bears a scape of moderate length; and has each lateral half of the atriolum deeply divided by a furrow which nearly cuts it in two.

This is a northern species whose range extends over New England and the Northern Middle States.

Aranea miniata (A. min-i-a′ta).— This species is smaller than the preceding; the adult female measuring less than one fifth inch in length, and the male, about one eighth inch. The abdomen of the female is wider

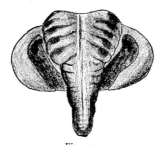

Fig. 491.
EPIGYNUM OF ARANEA CORTICARIA

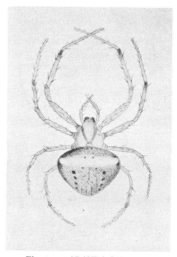

Fig. 492. ARANEA MINIATA

than long (Fig. 492), with distinct shoulder humps; the colour varies from white with reticulated markings to grayish yellow. In some individuals there is an indistinct folium and a light transverse band between the shoulder humps. On the hind half of the abdomen there are often four pairs of brown spots; but these vary greatly in distinctness and may be wanting.

The male resembles the female in colour but the body is more slender.

This is a southern species whose range extends north to New Jersey. It is the *Epeira scutulata* of Hentz.

THE LARGE ROUND-SHOULDERED ARANEAS

There are two common species of *Aranea* that rival in size the large angulate members of this genus described above; these are *A. gigas* and *A. trifolium*. The full-grown females measure from one half to three quarters of an inch in length; but they are easily distinguished from the large angulate species of *Aranea* by the absence of shoulder-humps on the abdomen.

Each of these two species varies greatly in colour and in markings; and as certain varieties of the two species present the same type of markings, it is necessary to make use of structural characteristics in separating them; this can be done as follows:

A. Each lateral half of the epigynum of the female divided by a prominent oblique furrow; ventral side of the abdomen black in the middle with a semicircular yellow spot on each side; the tibia of the second legs of the male thickened and armed with many, short, thick spines. P. 489.
A. gigas

AA. The lateral halves of the epigynum not divided by a furrow; ventral surface of the abdomen usually lacking the semicircular yellow spots; the tibia of the second legs of the male not thickened. P. 493. *A. trifolium*

Aranea gigas (A. gi′gas).— This is an exceedingly variable species in size, colour, and markings. The full-grown female measures from one half to three quarters of an inch in length. The cephalothorax is dull yellow with slightly darker lines in the middle and at the sides. The upper surface of the abdomen varies in colour from light yellow with dark purplish brown markings, through a light grayish brown with markings indistinct, to chocolate brown or dark gray with prominent light markings.

In most cases this species is readily distinguished from the following by the markings of the ventral side of the abdomen, which bears a broad black band in the middle with a semicircular, yellow spot on each side. In this respect this species closely resembles *A. cavatica* (Fig. 488) and many others; but as a rule *A. trifolium* lacks the yellow spots except in the young.

In what must be considered the typical form of this species, being that which was first described, the dark spots are gray or brown with a purplish tint and the light parts are white or yellow; in this form there is comparatively little of the lighter colour, except an irregular spot on each front angle of the abdomen, a broken median line, and a faint outline of the folium (Fig. 493).

Superfamily Argiopoidea

In a very common variety of this species, which has been named the spectacled spider, *Aranea gigas conspicellata*, the abdomen is marked as shown in Fig. 494. There is a brown folium, enclosing a central yellow band or a series of yellow spots; the folium is outlined on each side by a wavy yellow stripe; on each side of the abdomen there are several bright yellow, oblique bars, enclosed more or less completely by darker bands in which there are many small yellow spots.

Fig. 493. ARANEA GIGAS

Fig. 494. ARANEA GIGAS CONSPICELLATA

A third variety is light, grayish brown, lighter on the dorsum, where the muscle impressions are very distinct, and there are darker gray oblique lines on the sides of the abdomen.

The form of the epigynum varies somewhat; but the same variations in form occur in the different varieties of the species described above. The most distinctive features are the division of each lateral half of the atriolum by deep oblique furrows (Fig. 495), and the greater length of the scape than is the case in the following species. In some individuals the scape tapers gradually from base to tip; in others, it is stout throughout the basal two thirds of its length, then suddenly narrowed, and finally slightly enlarged at the tip.

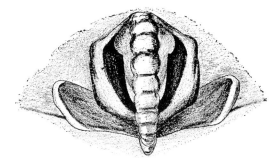

Fig. 495. EPIGYNUM OF ARANEA GIGAS

Fig. 496. TIBIA OF SECOND LEG OF ARANEA GIGAS

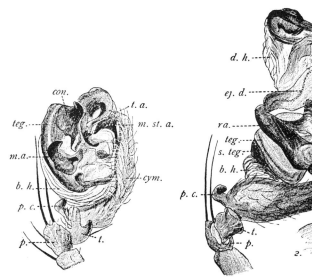

Fig. 497. TARSUS OF MALE OF ARANEA GIGAS

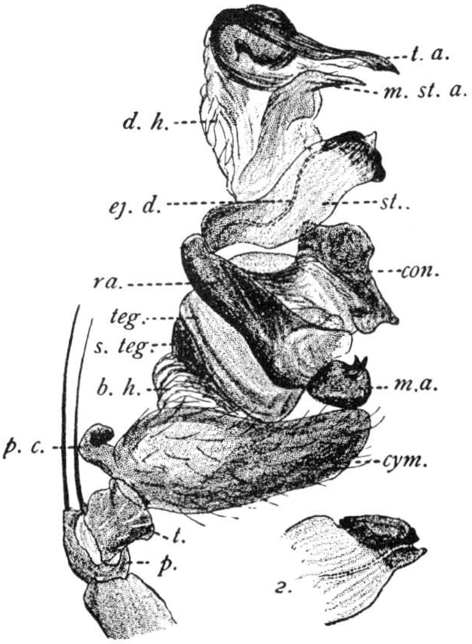

Fig. 498. TARSUS OF MALE OF ARANEA GIGAS WITH THE BULB EXPANDED

Fig. 499. WEB OF ARANEA GIGAS

Superfamily Argiopoidea

The male is much smaller than the female, being about half as long. The tibia of the second legs is thickened and armed on the inner side with many, short, thick spines (Fig. 496). The unexpanded bulb of the palpus is represented by Fig. 497 and the expanded bulb by Fig. 498; these are described in detail in an earlier chapter. (See p. 118.)

The web (Fig. 499) is a complete orb; that of the adult is large, one foot or more in diameter. It is nearly vertical and is usually built in shrubs or among the low branches of trees; but I have seen it on trees high above the ground. The central part of the web of this species is shown in Fig. 185 (p. 191); the hub is meshed, with rather large irregular spaces; the notched zone includes comparatively few turns; and the free zone is narrow or not well-marked; it is probably rarely if ever used by

Fig. 500. EGG-SAC OF ARANEA GIGAS

the spider as a means of passage from one side to the other of the web. The retreat is usually above the orb and at some distance back of it. It is frequently made in a curled leaf or in a bunch of leaves, and is connected with the hub by one or more traplines. Young spiders make their tent entirely of silk.

In the Northern States the spiders reach maturity in August. The egg-sacs are made early in the autumn; they consist of a loose flocculent mass of silk enclosing the ball of eggs (Fig. 500).

This species is the *Epeira insularis* of Hentz.

The Shamrock Spider, *Aranea trifolium* (A. tri-fo'li-um).— The cephalothorax is light with three wide, black stripes. The abdomen varies in colour from almost white without any markings to a gray with an olive tinge or to a dark reddish brown with a purplish tinge. The markings of the abdomen are as variable as is the ground colour. Figure 501 represents a common form.

Superfamily Argiopoidea

In this form there is on the middle line at the base of the abdomen a three-lobed spot resembling, somewhat, a shamrock leaf; in front of each of the four muscle impressions there is a white spot, and there is a third pair of spots farther back and in line with these. On the middle line of the abdomen there may be a row of spots; these are sometimes represented by minute dots, or may be wanting. In addition to these markings there may be a variable number of spots about the margins of the dorsal surface of the

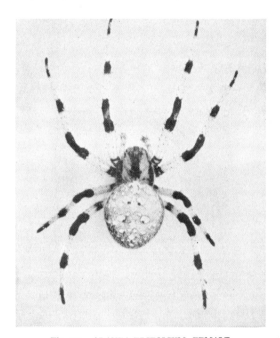

Fig. 501. ARANEA TRIFOLIUM, FEMALE

abdomen. As a rule the folium is not outlined except in young individuals; sometimes, however, it is well-marked (Fig. 502); this form resembles the spectacled variety of the preceding species; but it can be distinguished by the markings of the lower surface of the abdomen, and, more surely, by the form of the epigynum.

The ventral surface of the abdomen is darker than the dorsal and usually lacks the semicircular yellow spots so prominent in *A. gigas* and in certain other species.

The epigynum (Fig. 503) differs markedly in form from that

Superfamily Argiopoidea

of the preceding species; the lateral halves of the atriolum are not divided by furrows, and the scape is comparatively short.

The male (Fig. 504) is much smaller than the female, measuring about one fifth inch in length, and its markings are less distinct. The tibiæ of the second legs are not thickened as in *A. gigas*.

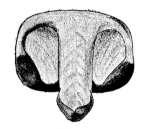

Fig. 503.
EPIGYNUM OF ARANEA TRIFOLIUM

Fig. 502.
ARANEA TRIFOLIUM, FEMALE

Fig. 504.
ARANEA TRIFOLIUM, MALE

Fig. 505.
ARANEA TRIFOLIUM CANDICANS

The three-lobed figure which suggested the common name of this species is not distinctively characteristic; for a similar mark is often borne by individuals of the preceding species.

There is a remarkable female variety of this species, *Aranea*

Fig. 506. WEB OF ARANEA TRIFOLIUM, CENTRAL PART, NATURAL SIZE

Fig. 507. WEB OF ARANEA TRIFOLIUM, REDUCED

Superfamily Argiopoidea

trifolium candicans, which was first described by McCook. In this variety (Fig. 505) the general appearance is strikingly different from that of the typical form, resembling the male in colour and in the shape of the abdomen, but it is much larger than the male. The specimen figured here measures one half inch in length. I collected it near Ithaca, N. Y.

Although *A. trifolium* often builds its web in shrubs, I have found it most abundant on rank herbaceous growth in marshy places. The web is a large, complete orb. The hub of the web is of the meshed type, but it is very open (Fig. 506); there are only a few turns in the notched zone; and the free zone is not very wide or is wanting. A trapline leads from the hub to a retreat, which is usually in a folded leaf or a bunch of leaves above and at one side of the orb (Fig. 507). The spider waits in its retreat for the ensnarement of its prey.

Fig. 508. EGG-SAC OF ARANEA TRIFOLIUM

I have not found the egg-sac of this species in the field; but a confined spider made one which was a very delicate sac attached to a leaf and was so translucent that the contained mass of eggs could be seen through its walls (Fig. 508).

THE THREE HOUSE ARANEAS

Among the round-shouldered Araneas there are three common species that are intermediate in size between the two large species described above and the smaller ones treated later. As these three species are more commonly found about houses, barns, and fences than are other orb-weavers, they were grouped together by Emerton as *The Three House Epeiras;* this is a convenient grouping which I adopt with merely the necessary change in the generic name. It should be remembered, however, that although these are termed house-spiders they frequently build their webs on bushes far from buildings.

The three house Araneas resemble each other in a striking degree in size and in markings. The adult females vary from a little less than one third inch to three fifths inch in length; and the males are somewhat smaller. The colours are various shades of brown; and there is a distinct folium (Fig. 509).

Fig. 509. ARANEA FRONDOSA, FEMALE

Notwithstanding the close resemblance of these three species, they can be easily separated in the adult state by the differences in the form of the epigyna of the females and of the palpi of the males, as indicated in the following table:

A. Females.
   B. Scape of the epigynum finger-like.
      C. Openings of the spermathecæ exposed; the posterior lateral thickenings of the epigynum nearly or quite meeting on the middle line. P. 500. *Aranea sericata*
      CC. With a prominent lobe on each anterior lateral part of the epigynum; the posterior lateral thickenings of the epigynum widely separated; the posterior half of the central portion of the epigynum prominently elevated. P. 501. *Aranea frondosa*

Superfamily Argiopoidea

BB. Scape of the epigynum flat and widened at the tip. P. 503.
Aranea ocellata

AA. Males.
B. The median apophysis of the bulb of the palpus more than twice as long as wide and split less than halfway to its base.
C. The bulb of the palpus with a long blunt terminal apophysis. P. 500. *Aranea sericata*
CC. The terminal apophysis of the bulb of the palpus stouter at the base but ending in a slender spear-like tip. P. 501. *Aranea frondosa*
BB. The median apophysis of the bulb of the palpus nearly as wide as long and split nearly to its base. P. 503.
Aranea ocellata

The Gray Cross Spider, *Aranea sericata* (A. ser-i-ca'ta).—This is the most easily recognized of the three house Araneas. It is darker than the other two species and has lighter abdominal markings (Fig. 510). The most easily recognized distinction is

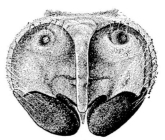

Fig. 511.
EPIGYNUM OF ARANEA SERICATA

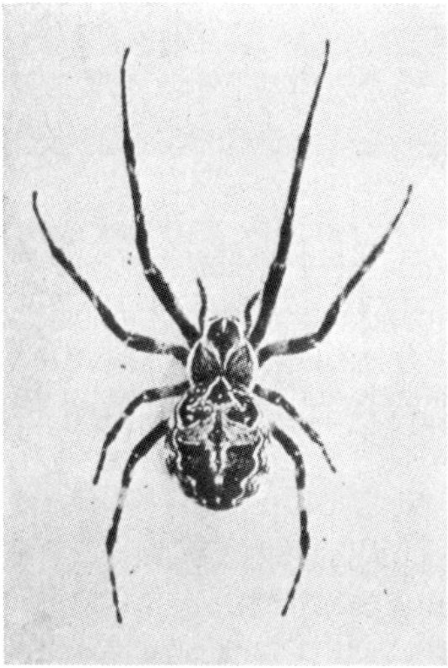

Fig. 510.
ARANEA SERICATA, FEMALE

the fact that the light lines forming the edges of the folium are broken between the first and second abdominal segments; and each part extends inward, more or less distinctly, toward the middle line. At the place where the lines are broken, the folium is crossed by a lighter gray patch, interrupted in the middle. The adult female measures from two fifths to three fifths inch in length; and has a finger-like scape, which is slender and tapers toward the tip (Fig. 511).

The male measures a little less than one third inch in length, and resembles the female in the markings of the abdomen, although the transverse gray patch is usually not as distinct as in the female. The distinctive characteristics of the palpi (Fig. 512) are indicated in the table given above.

This is the most common species of *Aranea* about buildings and other wooden structures; and it is sometimes exceedingly

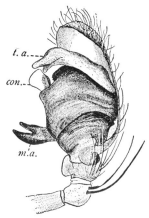

Fig. 512. PALPUS OF MALE OF ARANEA SERICATA
ma, median apophysis *ta*, terminal apophysis *con.* conductor

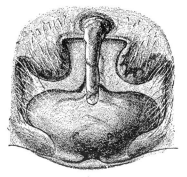

Fig. 513.
EPIGYNUM OF ARANEA CORNUTA

abundant on buildings that are near water. It is rarely found on plants away from houses.

The web is a complete orb of the same type as that of *A. frondosa*. The retreat is usually a dense sheet of silk built across an angle of the supporting structure. The adults are found at all seasons of the year; hence individuals collected at any time will vary greatly in size.

This species has been commonly known under the name *Epeira sclopetaria;* but according to the accepted rules of nomenclature the specific name *sericata* must be used for it. Both names were given by Clerk, but the one adopted here appears first in his book.

The popular name, the gray cross spider, was probably suggested by the transverse gray patch of the abdomen.

The Foliate Spider, *Aranea frondosa* (A. fron-do′sa).— The

Fig. 514.
ARANEA FRONDOSA, MALE

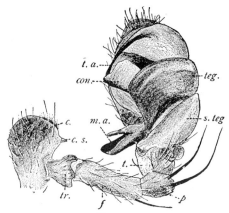

Fig. 515.
PALPUS OF MALE OF ARANEA FRONDOSA
ma, median apophysis *ta*, terminal apophysis

Fig. 516. WEB OF ARANEA FRONDOSA

adult female (Fig. 509) measures about two fifths inch in length. The cephalothorax is reddish brown with a dark stripe on each side and a less distinct one in the middle. The abdomen is light grayish brown, with a darker folium, which includes three or more pairs of indistinctly outlined spots of the lighter shade. This species closely resembles the following one; but can be easily distinguished from it by the form of the scape of the epigynum, which is finger-like (Fig. 513).

The male (Fig. 514) closely resembles the female and is but little smaller. The median apophysis of the bulb of the palpus (Fig. 515) is more than twice as long as wide and is split less than halfway to its base; and the terminal apophysis of the bulb (Fig. 515) ends in a slender spear-like tip. The tibia of the second

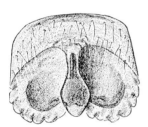

Fig. 517. EPIGYNUM OF ARANEA OCELLATA

Fig. 518. PALPUS OF MALE OF ARANEA OCELLATA

legs is armed with two rows of short stout spines on the inside; these spines differing more markedly from those on the other legs than is the case in either of the other house Araneas.

This is an exceedingly common species, which makes its web about houses, but more frequently on bushes. The spider is usually found in a retreat near the web. Figure 516 is of a web of this species; the spider that made this web had a retreat at the base of the teazle head at the left. Figure 186, p. 194, and Fig. 194, p. 205, represent other webs of this species.

This species is the *Epeira strix* of Hentz and the *Epeira foliata* of Koch.

The popular name doubtless refers to the conspicuous folium.

Aranea ocellata (A. o-cel-la′ta).— This species resembles

Superfamily Argiopoidea

the foliate spider quite closely in appearance; but it can be readily distinguished from it by the form of the epigynum and the form of the bulb of the palpus of the male. The scape of the epigynum is flat and widened at the tip (Fig. 517), and the apophysis of the bulb of the palpus of the male is nearly as wide as long and is split nearly to its base (Fig. 518, *m. a*).

I have found this species much less common than either of the other house Araneas.

This is the *Epeira patigiata* of many authors.

THE SMALLER ROUND-SHOULDERED ARANEAS

Under this head are grouped those species of *Aranea* that are neither angulate nor of large size. This group includes a larger number of species than either of the preceding groups; but we have space to describe only the more common ones.

The Lattice-spider, *Aranea thaddeus* (A. thad'de-us).— The full-grown female spider (Fig. 519) is about one fourth inch long, with a wide round abdomen, which is usually white or light yellow on the upper side, but which varies to the most brilliant purple and pink. There is a dark stripe or a row of dark spots on each side of the abdomen, and a large dark area on the ventral side surrounding a light spot just back of the epigynum. The cephalothorax and legs vary from yellow to orange-yellow and to yellowish brown. The form of the epigynum is shown by Fig. 520. The male is rarely observed; it resembles the female in colour. This species has been found throughout the eastern half of the United States

Like some Oriental ladies this beautiful spider spends the day peering out from behind a lattice, and a wonderfully beautiful lattice it is! It is built on the lower side of a leaf which has been bent and fastened so as to form a tent; sometimes the tent is formed of two or more leaves fastened together. From the apex of this tent a silken tube hangs down. The wall of this tube is not a continuous sheet but is perforated with many openings, which make it appear like lattice-work (Fig. 521). The tube is cylindrical, and in case of the larger specimens one inch or more in length and one half inch in diameter.

Below and slightly in front of this retreat is suspended the orb. This is usually oblong, the vertical diameter being con-

Superfamily Argiopoidea

siderably greater than the horizontal, though sometimes it is nearly circular, as is the one shown in Fig. 522. In the more typical specimens about three fifths of the orb is below the centre of the hub. The hub is meshed; and from the upper edge of it, one or two traplines lead to the lattice-like retreat. There is a more or less distinct free zone. In the webs of full-grown spiders the number of turns of the viscid threads varies from ten to

Fig. 519.
ARANEA THADDEUS

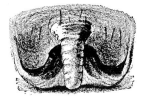

Fig. 520.
EPIGYNUM OF ARANEA THADDEUS

Fig 521. RETREAT OF ARANEA THADDEUS

twenty-five above the hub and from forty to eighty below. Usually the turns are much closer together below the hub than above.

Sometimes this spider builds an orb of the zilla type, the

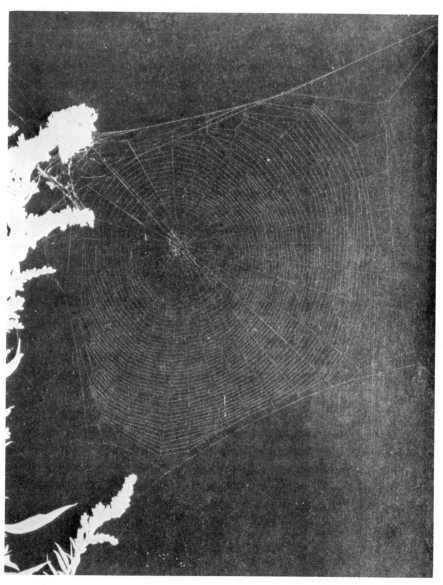

Fig. 522. WEB OF ARANEA THADDEUS, ORDINARY TYPE, REDUCED

Fig. 523. WEB OF ARANEA THADDEUS, ZILLA TYPE

Superfamily Argiopoidea

viscid thread being omitted from one sector of it. Such a web is shown in Fig. 523.

I have found the webs of the lattice-spider most abundant in dense foliage, as in an osage orange hedge, and among the leaves of an ivy on the side of a building. McCook states that this spider may be often found nestled in the angle of a door or window, or other like situations, on the outhouses of farms and rural buildings.

Fig. 524.
ARANEA DISPLICATA, FEMALE

Aranea displicata (A. dis-pli-ca′ta). — This is a small species, the larger individuals being only about one quarter inch in length. The cephalothorax and legs are brownish yellow without markings; the abdomen is oval, and is brightly coloured with light yellow or with crimson; on the hind half there are three small black spots on each side. These black spots make it easy to recognize this spider (Fig. 524). In the male the black spots on the abdomen are larger.

The web is a small one and is usually made among the leaves of shrubs or trees. It is often made in the space enclosed by the bending of a single leaf. The specimen represented in Fig. 525 was made between the lobes of an oak leaf.

Aranea pegnia (A. peg′ni-a). — The adult female measures about one fourth inch in length. The abdomen is globose, being nearly as wide as long. The cephalothorax is yellow, with a very narrow dark line extending from the eyes to the median furrow

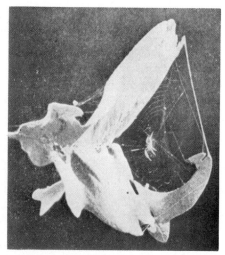

Fig. 525. WEB OF ARANEA DISPLICATA

Superfamily Argiopoidea

and indistinct dark marks on the sides of the head. The general colour of the abdomen is light brownish yellow with a distinct folium. The basal part of the folium is occupied by two pairs of large, white, yellow, or pink spots (Fig. 526) behind these there is a series of pairs of transverse black bars extending nearly to the tip of the abdomen.

The male measures one fifth inch in length. It resembles the female in the markings of the abdomen; but the abdomen is small and more elongate than in the female. The tibia of the second legs is armed with strong spines above.

The web of this species is described by Emerton ('02). It is of a composite type, being an orb-web combined with an irregular net resembling somewhat the web of *Metepeira labyrinthea* (Fig. 478). The orb is incomplete, a segment back of the trapline having few or no turns of the viscid spiral. The retreat to which the trapline extends is a large silken tent; and between the retreat and the orb there is an irregular net.

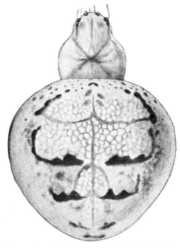

Fig. 526. ARANEA PEGNIA, FEMALE

This is the *Epeira globosa* of Keyserling and the *Epeira triaranea* of McCook.

Genus NEOSCONA (Ne-os-co'na)

This genus includes a group of species which is commonly included in *Aranea*, but which is sharply distinguished from that genus. In *Neoscona* the median furrow of the thorax of the female is longitudinal; and the epigynum is of a very characteristic form, which varies only in details in the different species. There is a circular or elongate atriolum, without depressions, to the hind margin of which is fused a strongly chitinized, more or less elongate, spoon-shaped scape (Fig. 527). The males can be recognized as a rule by the resemblance of the markings of the abdomen to those of the female.

Superfamily Argiopoidea

Although there is a great variation in the colour and markings within the limits of each of the species, there is a characteristic pattern which with modification of details is evident in all of our species (Fig. 528). On the base of the abdomen, in front of the first pair of muscle impressions, there is an irregular, triangular, white patch pointed forward; there is a similar patch in the space between the first and the second pairs of muscle impressions; and behind this is a nar-

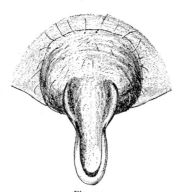

Fig. 527.
EPIGYNUM OF NEOSCONA ARABESCA

Fig. 528.
NEOSCONA ARABESCA

rower band, extending to the tip of the abdomen; sometimes this band is broken into a series of spots; on each side of this band there is a series of four or five conspicuous, black, oblique, oval spots, each of which is usually partly surrounded by a light patch. The widest departure from this type of markings is exhibited by *N. pratensis*, which is figured later.

The following species are the commoner ones which occur within our borders.

Neoscona arabesca (N. ar-a-bes′ca).—This species rarely exceeds one third inch in length. The epigynum is shown in Fig. 527. This spider varies greatly in colour and markings. The most common colour is a mottled brown or brown and red with white or light yellow markings. What may be considered the typical markings is indicated in the generic description above. The basal triangular light patch is often divided into a pair of oblique spots, with more or less distinct dark or red margins, and the narrow band on the hind half of the abdomen is often divided by a central dark line with paired branches, and frequently this band is replaced by a series of more or less distinct spots. The carapace varies in colour from yellow to orange-yellow or to brown; there is a central dark band and on each side a submarginal one; these vary greatly in distinctness.

The male is marked like the female. The tibia of the second legs (Fig. 529) is armed with short stout spines and is usually strongly curved.

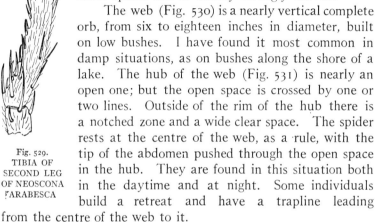

Fig. 529. TIBIA OF SECOND LEG OF NEOSCONA ARABESCA

The web (Fig. 530) is a nearly vertical complete orb, from six to eighteen inches in diameter, built on low bushes. I have found it most common in damp situations, as on bushes along the shore of a lake. The hub of the web (Fig. 531) is nearly an open one; but the open space is crossed by one or two lines. Outside of the rim of the hub there is a notched zone and a wide clear space. The spider rests at the centre of the web, as a rule, with the tip of the abdomen pushed through the open space in the hub. They are found in this situation both in the daytime and at night. Some individuals build a retreat and have a trapline leading from the centre of the web to it.

This species is the *Epeira trivittata* of Keyserling.

Neoscona benjamina (N. ben-ja-mi'na).— This is larger than the preceding species, the adult females measuring from two fifths to three fifths inch in length. The abdomen is ordinarily triangular-oval, being broad at the base (Fig. 532); but sometimes when distended with eggs, it becomes more uniformly oval (Fig. 533). In colour and markings it resembles *N. arabesca* very closely and is almost as variable. It can be most surely distinguished by the form of the epigynum in which the scape is much longer, and the tubercles at the base of the scape more prominent when seen from below (Fig. 534).

I have been unable to distinguish in a satisfactory manner the male of this species from that of *N. arabesca*. According to Emerton the tibia of the second legs of the male of *N. benjamina* (Fig. 535) is not so strongly curved as it is in *N. arabesca;* but I have not found it easy to make use of this distinction.

This is the *Epeira domiciliorum* of Hentz.

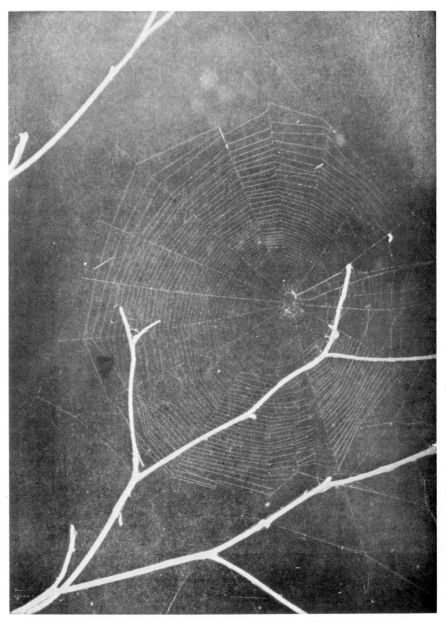

Fig. 530. WEB OF NEOSCONA ARABESCA

Neoscona nautica (N. nau'ti-ca).—This species presents the typical markings of the genus and resembles the preceding species in the triangular-oval form of the abdomen. But it is smaller than *N. benjamina*, the adult females usually measuring

Fig. 531. CENTRE OF WEB OF NEOSCONA ARABESCA

about two fifths inch in length. The most distinctive characteristic is the form of the epigynum (Fig. 536), which is short, not much longer than its width at the base.

This is a widely distributed species; but it is not common in the North.

Fig. 532.
NEOSCONA BENJAMINA

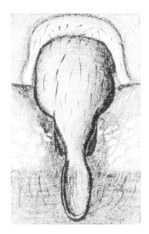

Fig. 534. EPIGYNUM OF
NEOSCONA BENJAMINA

Fig. 533.
NEOSCONA BENJAMINA

Fig. 535.
TIBIA OF
SECOND LEG
OF NEOSCONA
BENJAMINA

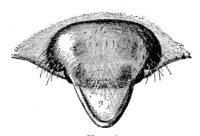

Fig. 536.
EPIGYNUM OF NEOSCONA NAUTICA

Neoscona oaxacensis (N. o-ax-a-cen'sis).— This is a Pacific Coast species whose range extends from California to Panama. The female is one half inch or a little more in length. It resembles the three preceding species in being exceedingly variable in colour and markings, but usually presents the abdominal markings described above as characteristic of the genus. It is easily distinguished from our other species of this genus by the fact that the femora of the first three pairs of legs are armed with a double series of spines beneath.

Fig. 537. NEOSCONA PRATENSIS

The male usually measures a little less than one half inch in length, but varies much in size. The folium resembles quite closely that of the female. It agrees with the following species in having the coxa of the fourth legs armed with a coniform spur beneath.

F. O. P. Cambridge states that the *Epeira vertebrata* of McCook is identical with this species; but Mr. Banks in his catalogue separates the two.

Neoscona pratensis (N. praten'sis). — In this species, which undoubtedly belongs to the genus *Neoscona*, we find the widest departure from the typical markings of the genus (Fig. 537). The adult spiders measure from one third to two fifths inch in length. The colour of the carapace is yellowish brown with darker median and submarginal bands. The colour of the abdomen is yellowish brown, which is lighter on the basal half and darker upon the apical half. There is a broad median brown band; and on each side of this, there is a yellow stripe, which is often more or less broken, especially where it passes over the muscle impressions; and on each side, between the yellow band and the margin of the body, there is a row of six black spots partly or completely surrounded with yellow.

The male is marked like the female and is not much smaller. The tibia of the second legs is curved and armed

Superfamily Argiopoidea

with strong spines. The coxa of the fourth leg has a coniform spur beneath.

This species is distributed over the greater part of the United States. It makes its web among herbage and on shrubs.

Genus ERIOPHORA (Er-i-oph'o-ra)

Several species of spiders that are commonly classed in the genus *Aranea* have been separated from that genus by Pickard-

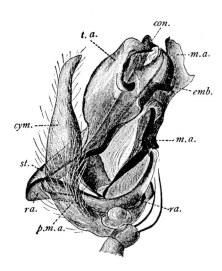
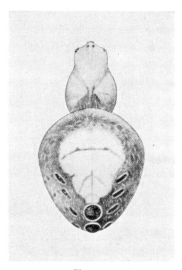

Fig. 538. PALPUS OF MALE OF ERIOPHORA CIRCULATA

Fig. 539. ERIOPHORA CIRCULATA, FEMALE

Cambridge and placed in a new genus, for which he has revived the old name *Eriophora*, first proposed by Simon, but afterward discarded by him.

In the species placed here the scape of the epigynum is greatly elongate, extending nearly or quite to the base of the spinnerets; and the palpus of the male presents an anomalous appearance due to the fact that the genital bulb is twisted so that the embolus appears to arise from its base; the embolus is very long and elbowed (Fig. 538). Other figures of this palpus are given in an earlier chapter, where this palpus is described in detail. (See p. 117.)

Several species of this genus occur in the extreme southern portions of our territory. The following will serve as an example:

Eriophora circulata (E. cir-cu-la′ta).— The adult spiders measure from one third to one half inch in length; the male is nearly or quite as long as the female, but has a smaller abdomen. The cephalothorax is yellow. The abdomen is subtriangular, being almost as wide near the base as long; it is rounded before and obtusely pointed behind (Fig. 539). On the dorsal wall of the abdomen there is large subtriangular patch, which is yellow in alcoholic specimens but which is green in life. At the hind end of this patch there is a round, black tubercle edged with the light colour; and back of this a second similar tubercle. On each side of the hind half of the light patch and of the two tubercles there is a row of four oblique black bars.

This is a subtropical species which is found in the Southern States and on the Pacific Coast as far north as Oregon. It has been mistaken for the *E. bivariolata* of Central America.

Genus MANGORA (Man-go′ra)

The most available character for distinguishing this genus from our allied genera is the fact that the tibia of the third legs bears a cluster of slender hairs on the anterior side near the base (Fig. 540). The cephalothorax is

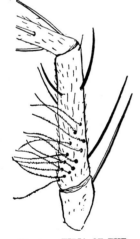

Fig. 540. TIBIA OF THE THIRD LEG OF MANGORA

Fig. 541. MANGORA GIBBEROSA, FEMALE

more or less gibbous or convex posteriorly; the median furrow of the thorax is deep and usually extends forward so as to reach the cervical groove; the median ocular area is narrower in front

Superfamily Argiopoidea

than behind; the posterior median eyes are larger than the lateral eyes; the anterior row of eyes is recurved, the posterior row, straight or procurved.

Three well-known species occur in the United States.

Mangora gibberosa (M. gib-be-ro'sa).— The adult female (Fig. 541) is from one sixth to one quarter of an inch in length, the male is somewhat smaller. The cephalothorax and legs are light greenish yellow; the cephalothorax has a narrow black stripe extending from just behind the eyes back into the median furrow and the legs are armed with prominent black spine-like hairs. The abdomen is white, mottled with yellow and striped and spotted with black. There are three longitudinal stripes on the hind half of the abdomen; the middle one of these extends farther forward than the others; and there are several oblique stripes or spots on each side of the abdomen. The cephalothorax is swollen on each side of the median furrow; this probably suggested the specific name. The abdomen extends forward in a rounded point above the thorax.

This is a very widely distributed species. It is common and makes its web among the stalks of grass and on low bushes. Its web is surprisingly large compared with the size of the spider, the orb varying from six inches to a foot in diameter; but it is so delicate that it rarely attracts attention, even when the spider is very common. The abundance of the spider is frequently made evident first by the appearance of many individuals in the sweeping net. The web (Fig. 542) is either horizontal or inclined and although exceedingly delicate it is very elaborate in structure. The hub is either finely meshed or nearly open, and is surrounded by a notched zone of many turns. Occasionally the hub bears a circular or disk-like sheet and the notched zone consists of many turns of the thread (Fig. 543). The viscid spirals are closely woven and are very numerous; and the number of the radii is also large. The web figured was that of an immature male.

I have most often found the spider resting beneath the hub of its orb; but in some cases I have found it in a curled leaf above the web. If disturbed, when resting on its web, it darts to the ground or runs quickly to a neighbouring leaf. In the North the spiders mature early in August.

Mangora placida (M. pla'ci-da).— The adult is about one sixth inch in length. The cephalothorax is brownish yellow with

Fig. 542. WEB OF MANGORA GIBBEROSA. THE UPPER PART OF THE WEB WAS NOT IN THE SAME PLANE AS THE LOWER, AND HENCE WAS OUT OF FOCUS WHEN THE PHOTOGRAPH WAS TAKEN

Superfamily Argiopoidea

a dark median stripe and dark margins. The abdomen bears a brown stripe which is narrow in front and wide behind (Fig. 544). On each side of this stripe there is a row of black spots; and in the wider portion a pair of white spots. There is a considerable variation in the depth of colour of the brown stripe. The male resembles the female in colour and markings.

This is a very widely distributed species; and, like *M. gibberosa*, it makes a very finely meshed web. The finest meshed web I have ever seen was made by a spider of this species.

Mangora maculata (M. mac-u-la′ta).— The male measures one eighth inch in length; the female, nearly one fifth. This

Fig. 543. HUB OF WEB OF MANGORA GIBBEROSA   Fig. 544. MANGORA PLACIDA

species can be distinguished from *M. gibberosa* by the markings of the abdomen, which consist of several pairs of black spots on the hinder half; sometimes these spots are connected by brownish transverse bands.

Like the two preceding species, this one is widely distributed, but it is much less common.

Genus LARINIA (La-rin′i-a)

These are elongate spiders, the abdomen, in our species, being two or three times as long as wide. The abdomen projects over the cephalothorax in a blunt point; and extends a short

Superfamily Argiopoidea

distance behind the spinnerets. The second row of eyes is nearly straight; the lateral eyes are nearly equal in size and those on each side are close together; the posterior median eyes are near together; the anterior median eyes are widely separated, making the median ocular area twice as wide in front as behind. The patella of the pedipalp of the male is armed with two spines at its apex.

Three species occur in the United States.

Larinia directa (L. di-rec'ta).— This is a common species in the South, where it makes an oblique web in grass or on other herbaceous plants. The adult male measures one fifth inch in length; the adult female, two fifths inch; and in each sex the body is only about one fourth as wide as long. The cephalothorax is yellowish with a narrow, dark, marginal line and also a median, longitudinal line. The abdomen varies greatly in colour and markings (Fig. 545); there is usually a yellowish median stripe above, and often a darker stripe on each side of this. There is also, usually, a series of six pairs of black spots extending the whole length of the abdomen. The spots vary greatly in size in different individuals; and sometimes the spots of the first and third pairs are much larger than the others. The sternum is yellow. On the under side of the abdomen there are two parallel dark stripes, which unite just before the spinnerets. The metatarsus of the first legs is longer than the tibia and twice as long as the width of the body.

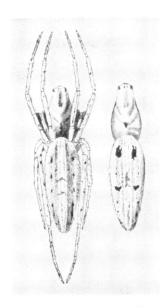

Fig. 545. LARINIA DIRECTA

Larinia borealis (L. bo-re-a'lis).— This species occurs in the North; but is not common. It was described by Banks from specimens taken in the State of Washington, and in New Hampshire. It is not as slender as *L. directa*, the abdomen being only about twice as long as wide. It differs also in that the metatarsus

Superfamily Argiopoidea

of the first legs is not longer than the tibia, and not longer than the width of the body. "The abdomen is gray, with black spots on the sides, above with a pale median stripe, and a row of four black spots on each side near the tip; venter with three narrow black stripes, uniting at base of spinnerets."

Larinia famulatoria (L. fam-u-la-to'ri-a).— Only the female of this species has been described. This was found in Colorado. It measures a little more than one fifth inch in length. The cephalothorax is yellow, with a narrow longitudinal black band extending from the posterior middle eyes to the median furrow; it is one fourth longer than broad, and only one half so broad in front as at the middle. The abdomen is about one third longer than wide; it is bluntly pointed in front and rounded behind.

Genus ACACESIA (Ac-a-ces'i-a)

The median furrow of the thorax is longitudinal and extends forward so as to reach the cervical groove; the abdomen is an

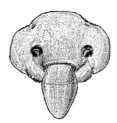

Fig. 546. EPIGYNUM OF ACACESIA FOLIATA

Fig. 547. TIBIA OF SECOND LEG OF MALE OF ACACESIA FOLIATA

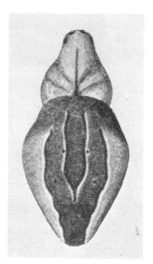

Fig. 548. ACACESIA FOLIATA, FEMALE

elongate rhomboid. The second row of eyes is very strongly recurved; the median ocular area is not much wider in front than behind, is nearly vertical, and is convex and hairy; the posterior median eyes are not close together; the anterior median eyes are farther from the lateral eyes than the lateral eyes are from each other. The legs are armed with

Superfamily Argiopoidea

a very few spines. The scape of the epigynum is short and broad *and is directed backward* (Fig. 546). The epigynum of *Acacesia* closely resembles that of *Neoscona*, but the scape lacks the lateral tubercles, and the openings of the spermathecæ are exposed in the atriolum, whereas in *Neoscona* they are in the dorsal wall at the base of the scape. The tibiæ of the second legs of the male are somewhat thickened and are strongly spined, but the spines are not arranged in a series (Fig. 547); the anterior coxæ of the male are furnished with apical teeth.

The following species, though not common, is distributed over nearly the whole of North America.

Acacesia foliata (A. fol-i-a′ta).— The female is one fourth inch in length; it is easily recognized by the very distinctive markings of the abdomen (Fig. 548); the male is a little smaller but resembles the female in colour and markings.

Genus EUSTALA (Eu′sta-la)

This genus resembles *Acacesia* in the arrangement of the eyes; but differs in that the legs are armed with many spines. The

Fig. 549.
EPIGYNUM OF EUSTALA ANASTERA

Fig. 550.
TIBIA OF SECOND LEG OF MALE OF EUSTALA ANASTERA

Fig. 551.
EUSTALA ANASTERA

scape of the epigynum *is directed forward* (Fig. 549); and in the male the spines on the inner side of the tibia of the second legs

Superfamily Argiopoidea

are arranged in a more or less definite series (Fig. 550). The abdomen is triangular in outline.

The following is our best-known representative of the genus; several other species have been found in our fauna.

Eustala anastera (E. a-nas'te-ra).— This is a common species throughout the United States and southward. The female meas-

Fig. 552. EUSTALA ANASTERA

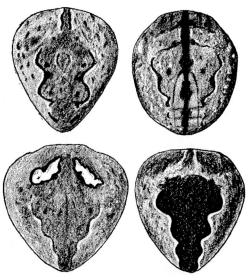

Fig. 553. ABDOMEN OF EUSTALA ANASTERA

ures from one fourth to three eighths of an inch in length. The abdomen is wide in front, bluntly pointed behind (Fig. 551), and about as high behind as in front (Fig. 552); in some individuals the pointed end of the abdomen is turned a little upward, resembling a *Cyclosa*. The markings of the abdomen are exceedingly variable; in Fig. 553 are represented several varieties, which

Superfamily Argiopoidea

are only a small part of those before me. The species is best recognized by the form of the epigynum (Fig. 554).

Fig. 554.
EPIGYNUM OF
EUSTALA
ANASTERA

The forward projecting scape of the epigynum is wedge-shaped, broad at the base and gradually narrowed to a point, which is curved toward the body; it bears many transverse wrinkles; at the base there is a black horseshoe-shaped sclerite, which varies somewhat in form.

The colours of this species are frequently exceedingly protective; the spider closely resembling the bark of the tree or other plant on which it rests; and they act as if conscious of this protection, running only a short distance when disturbed and then crouching down close to the bark. The webs are made in low bushes and are vertical.

This is the *Epeira prompta* of Hentz and the *Epeira parvula* of Keyserling.

Subfamily GASTERACANTHINÆ (Gas-ter-a-can-thi'næ)

The Spiny-bellied Spiders

The striking appearance of most of the members of this subfamily render them easily recognized, the abdomen being armed with prominent spines. Another very striking feature, and one that is distinctly characteristic, is that the spinnerets are elevated on a very large projection and occupy a circular space limited by a thick flange in the form of a tube or ring (Fig. 555). Only two genera are represented in our fauna; these can be separated as follows:

TABLE OF GENERA OF THE GASTERACANTHINÆ

A. Cephalothorax of the female at least as wide as long. P. 525.
GASTERACANTHA

AA. Cephalothorax longer than wide. P. 527.   MICRATHENA

Genus GASTERACANTHA (Gas-ter-a-can'tha)

This genus differs from the closely allied *Micrathena* in the form of the thorax, which is at least as wide as long in the

Superfamily Argiopoidea

female. The very characteristic form of the species make them the most easily recognized of all of our spiders.

Three species are listed as occurring in the United States; but the following is our only common one; the others are found in the extreme South or Southwest.

Gasteracantha cancriformis (G. can-cri-for'mis).— This is a spider whose remarkable shape and conspicuous, strongly contrasting colours are sure to attract attention. The adult female (Fig. 556) measures about one third inch in length and is about as wide as long. The abdomen is leathery and is armed with a fringe of spinose processes. The ground colour is yellow, marked with black spots, which vary in number and shape. This is a southern species, which is

Fig. 555.
MICRATHENA GRACILIS SHOWING THE PROJECTION, ON THE LOWER SIDE OF THE ABDOMEN, THAT BEARS THE SPINNERETS

Fig. 556.
GASTERACANTHA CANCRIFORMIS, FEMALE

found in the Gulf States, where it is common in the more southern portions.

The web is built between the branches of shrubs and trees, and frequently in the tops of tall trees. It is a complete orb, and is either vertical or inclined; the hub is open; the notched zone is narrow; the free zone is wide; and there are many viscid spirals. There is no retreat, the spider resting on the hub with its body over the open space. I have never observed a stabilimentum in a web of this species.

A remarkable feature of the webs of this species is the frequent occurrence in them of series of flocculent tufts of silk attached to either the radii or to some of the foundation lines (Fig. 557). These tufts are composed of a mass of fine threads, like those of which a stabilimentum is made. The only suggestion that I can make as to the use of these tufts is that they may serve as lures for the attraction of midge-eating insects, which in their efforts to capture the supposed midges fly into the orb.

This method of decorating the web has been observed also with certain exotic members of this genus.

Genus MICRATHENA (Mi-cra-the'na)

Our representatives of this genus are of moderate size and are brightly coloured. The cephalothorax is longer than wide; the dorsal aspect of the abdomen is flattened and, with adult females, it bears on its margin several pairs of spines, varying in size and shape. The males resemble the young females in form and colour.

This is the genus *Acrosoma* of most American writers on spiders; but it is believed that *Micrathena* is the older name. The genus is represented in this country by four species.

Micrathena sagittata (M. sag-it-ta'ta).— The arrow-shaped Micrathena is easily recognized by the striking form of the abdomen, which is narrow in front and is terminated behind by two large spreading spines (Fig. 558). The abdomen bears also two other pairs of spines, one near the base, and the other between these and the large caudal spines. The cephalothorax is yellowish brown with white edges. The abdomen is white or bright yellow above spotted with black. The spines are black at the tip and bright red at base. The ventral aspect of the abdomen is darker than the dorsal and is marked with black bands and yellow spots. The adult female measures a little more than one fourth inch in length.

The male (Fig. 559) is about one sixth inch in length. The abdomen is a little widened behind and bears slight humps in the place of the spines of the female.

This species is the *Epeira spinea* of Hentz. It is very common in the South; and Emerton reports it as common in Massachusetts and Connecticut.

The web is made on low bushes and is a very symmetrical, inclined orb, with many radii, and closely placed spirals; that of the adult is about one foot in diameter. The hub is open; there is no free zone, the spiral guy-line being left in the space between the hub and the viscid spiral. Figure 560 represents the central portion of an orb natural size. There is no retreat, the spider

Fig. 557. WEB OF GASTERACANTHA CANCRIFORMIS

Fig. 558.
MICRATHENA SAGITTATA,
FEMALE

Fig. 559.
MICRATHENA SAGITTATA,
MALE

Superfamily Argiopoidea

resting in the open space of the hub in the attitude shown in Fig. 561. A small stabilimentum is usually present above the hub.

Micrathena gracilis (M. grac'i-lis).— The abdomen of this species is armed with five pairs of spines; the first is near the base; the second, nearly midway the length of the abdomen; and the other three pairs at the caudal end. When viewed from above,

Fig. 560. WEB OF MICRATHENA SAGITTATA

the third pair of spines covers more or less completely the fourth and fifth (Fig. 562). The full-grown female measures a little more than one fourth inch in length. The cephalothorax is reddish brown with three dark stripes. The abdomen is spotted with white, yellow and brown; the general colour varies from almost white to nearly black.

529

Superfamily Argiopoidea

I have not seen the male. It is said to have a long narrow abdomen without any humps or spines.

This is a widely distributed species.

Micrathena reduviana (M. re-du-vi-a′na).— This is a smaller species than either of the two preceding, the adult female measur-

Fig. 561. WEB OF MICRATHENA SAGITTATA, SHOWING THE POSITION OF THE SPIDER WHEN AT REST

ing from one sixth to one fifth inch in length; and the spines on the abdomen are much less prominent than on either of them. The cephalothorax is brownish yellow. The abdomen overlaps the hind half of the cephalothorax, and has two pairs of comparatively small spines at the caudal end, the second pair of which are concealed by the first when the spider is viewed from above (Fig. 563). It is white or yellow above; with a dark spot near the base, and another at the hind end.

This species is widely distributed in the Eastern States. It is the *Epeira mitrata* of Hentz.

Micrathena maculata (M. ma-cu-la′ta).— This species, which has been found in Arizona is closely allied to the preceding. The female measures a little more than one fifth inch in length. The cephalothorax is uniform dark brown, about twice as long as broad, broadest in the middle, about as broad in front as

530

Superfamily Argiopoidea

behind, and with a depressed furrow slightly before the middle. The abdomen is about twice as long as broad; the sides are slightly convex, but hardly twice as broad in the middle as at the base; at the basal third above there is a small conical hump or spine

Fig. 562. MICRATHENA GRACILIS, FEMALE

Fig. 563. MICRATHENA REDUVIANA

on each side; and at the apex there are four conical spines. The abdomen is black, marked with yellow spots, a double spot on each side at the base, followed by four spots in a row on each side, and there are other yellow spots.

Family MIMETIDÆ (Mi-met'-i-dæ)

The Mimetids (Mi-met'ids)

The name of the typical genus of this family *Mimetus* is from the Greek *mimetos*, to be imitated or copied, and was sug-

Fig. 564. TARSUS OF FIRST LEG OF MIMETUS INTERFECTOR

gested by a mistaken belief that these spiders build a double web like that of *Theridium* and that of *Aranea* connected.

The members of this family are very easily recognized by

Superfamily Argiopoidea

the armature of the tibiæ and metatarsi of the first two pairs of legs (Fig. 564). These are armed with a series of very long spines regularly spaced, and with a series of much shorter spines between each two long spines; the short spines are curved and the members of each series are successively longer and longer.

These are slow moving spiders; they are found on low plants and bushes, or under rubbish on the ground. They make little use of their thread, as they construct no definite web.

Only about fifty species are known from the entire world. About ten species occur in the United States. They represent two genera which can be separated by the following table:

TABLE OF GENERA OF THE MIMETIDÆ

A. Clypeus much narrower than the ocular area; posterior legs much shorter than the anterior legs. P. 532.
MIMETUS

AA. Clypeus not narrower than the ocular area; posterior legs not much shorter than the anterior legs. P. 533. ERO

Genus MIMETUS (Mi-me′tus)

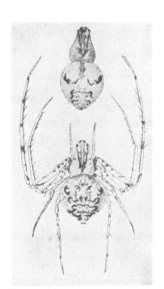

Fig. 565. MIMETUS INTERFECTOR

In this genus the eyes are situated near the front edge of the head, the clypeus being reduced to a very narrow area. These spiders are said to feed exclusively on other spiders. They live in dry and warm places, on bushes and fences and occasionally in houses, where they invade the webs of *Theridion tepidariorum*. According to Hentz the egg-sac is oblong and tapers equally at both ends; they are made in the webs of their victims. Several species have been recognized in our fauna.

Mimetus interfector (M. in-ter-fec′tor).—This is the best known representative of the genus. It is pale yellowish in colour and variable in its markings; two individuals are represented in Fig. 565; there is a double V-shaped band on the cephalothorax; on the dorsal wall of the abdomen there is on each side near the middle of its length a black, more or less projecting point, and extending back from each of these to near the tip of the abdomen, a series of S-shaped spots; fine transverse lines extend between these two series of spots. There may be white spots on the basal part of the abdomen and bright red points scattered over the entire dorsal surface. The legs are conspicuously marked with black points. The length of the body is nearly one fourth of an inch.

Mimetus syllepsicus (M. syl-lep′si-cus).— "Pale green; cephalothorax varied with black; abdomen with a waved line and disk black; feet and palpi very hairy; thighs of first and second pairs of legs with a black ring near the tip." (Hentz.) The species was described from North Carolina. I have not seen it.

Genus ERO (E′ro)

The eyes are much farther from the front edge of the head than in the preceding genus. These spiders are found in damp places under stones and in winter under leaves. About a dozen species are known; only three of them occur in our fauna.

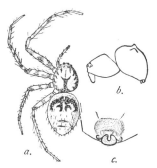

Fig. 566. ERO FURCATA, FEMALE
a, dorsal view *b*, profile
c, epigynum (after Keyserling)

Fig. 567.
EGG-SAC OF ERO FURCATA

Ero furcata (E. fur-ca′ta).— The cephalothorax is light yellow, with a broad dark band on each side, and a narrow median

Superfamily Argiopoidea

line, crossed by a crescent-shaped mark on the highest part. The abdomen is grayish, with brown spots of various shapes (Fig. 566). The abdomen is as high as long, and bears a pair of humps on the highest part.

The egg-sac of this species (Fig. 567) is very characteristic in form. It is nearly spherical and is suspended by a cord of coarse threads, which are continued over the egg-sac as a loose network. I have found it most often attached to cliffs in shady places.

Family THOMISIDÆ (Tho-mis'i-dæ)

The Crab-spiders

The crab-spiders are so called on account of the short and broad form of the body, the crab-like attitude of the legs in most of the species, and the curious fact that these can walk more readily sidewise or backward than forward. In one of our genera (*Tibellus*), however, the body is long and slender.

The first and second pairs of legs in the more typical forms are much stouter and longer than the third and fourth pairs; and in these forms the third pair of legs is directed forward like the first and second pairs (Fig. 568). The tarsi are furnished with two claws. The eyes are small, dark in colour, and arranged in two rows, which are almost always recurved. The lower margin of the furrow of the chelicerae is indistinct and unarmed; the upper margin is either unarmed or furnished with one or two teeth.

Fig. 568. A CRAB-SPIDER

These spiders spin no webs; some species run swiftly and pursue their prey, while others of slower gait depend on their concealing colours and lie in wait for it. They live chiefly on plants and fences, and in the winter hide in cracks and under stones and bark. Most of the species are marked with gray and brown, like the bark upon which they live. Some species conceal themselves in flowers, where they lie in ambush. These are brightly coloured, like the flowers they inhabit, so that insects

Superfamily Argiopoidea

visiting these flowers may alight within reach of a spider before seeing it.

The egg-sac is lenticular in form and is usually formed of two equal valves, united at the border, which presents a little circular fringe. In the subfamily Misumeninæ, the egg-sac is sometimes free and sometimes suspended like a hammock in a retreat formed of rolled or drawn together leaves; in the subfamily Philodrominæ, it is fixed by one of its valves. In most cases after the egg-sac is made, the female quits her wandering habits in order to watch it.

In the palpus of the male, the tibia is armed with apophyses, which vary in form and afford good characters for distinguishing the species. In the male of *Xysticus ferox*, which may be taken as an illustration of the family, the genital bulb, when expanded, is of the form shown in Fig. 569. The subtegulum is well-developed, but does not bear anelli; the tegulum is large; the

Fig. 569.
GENITAL BULB OF XYSTICUS FEROX, EXPANDED

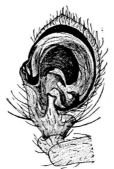
Fig. 570.
GENITAL BULB OF XYSTICUS FEROX, UNEXPANDED

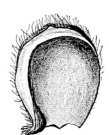
Fig. 571. CYMBIUM OF XYSTICUS GULOSUS, SHOWING THE TUTACULUM

embolus is of the spiral type, and has all of the parts characteristic of that type well-developed; there is a medium apophysis of moderate size; and, beyond the median apophysis, a larger, hooked, terminal apophysis. The development of these apophyses, however, appear to be characteristic of the genus *Xysticus*. I have not observed them in any other members of this family.

The most striking characteristic of the genital bulb is the absence of a conductor of the embolus and, in most of the genera, of a greater or less modification of the edge of the cymbium for

Superfamily Argiopoidea

the protection of the tip of the embolus. To this modified part of the cymbium I have applied the name *tutaculum* (tu-tac'u-lum). The tutaculum is most highly developed in the genus *Xysticus;* here it is formed by an expansion of both the outer face of the cymbium, which part is clothed with hairs, and of an expansion of the lower edge bounding the alveolus, the two constituting a groove in which the tip of the embolus rests in the unexpanded condition of the bulb (Fig. 570). The tip of the edge of the inner part of the tutaculum is often densely chitinized and in these cases may appear in the unexpanded bulb like an apophysis of the bulb. Figure 571 represents the cymbium of *Xysticus gulosus* with the genital bulb removed so as to expose the tutaculum.

The Thomisidæ includes several well-marked subfamilies; but only two of these are represented in our fauna; these can be separated as follows:

TABLE OF SUBFAMILIES OF THE THOMISIDÆ

A. Tarsi of the first and second pairs of legs not scopulate beneath; third and fourth pairs of legs, usually much shorter than the first and second pairs; hairs of the body filiform or rod-shaped and erect; upper margin of the furrow of the chelicerae without teeth. P. 536.

MISUMENINÆ

AA. Tarsi of the first and second pairs of legs scopulate beneath in the females at least; third and fourth pairs of legs as long as or nearly as long as the first and second pairs; hairs of the body pubescent or plumose, and prone, not erect; upper margin of the furrow of the chelicerae with one or two teeth. P. 554. PHILODROMINÆ

Subfamily MISUMENINÆ (Mi-su-me-ni'-næ)

In this subfamily the tarsi of the first and second pairs of legs are not furnished with scopulæ in either sex, though often they are thickly clothed with ordinary hairs; the third and fourth pairs of legs are usually much shorter than the first and second pairs; the hairs of the body are filiform or rod-shaped and erect; and the upper margin of the furrow of the chelicerae is without teeth.

This subfamily includes the majority of our species of crab-

Superfamily Argiopoidea

spiders. The genera occurring in the United States can be separated as follows:

TABLE OF GENERA OF THE MISUMENINÆ

A. Clypeus strongly sloping or horizontal; abdomen high, with a caudal tubercle. P. 537. TMARUS
AA. Clypeus vertical; abdomen broadly rounded behind.
  B. Tubercles of lateral eyes confluent.
    C. Eyes of first row subequal in size; carapace and abdomen almost devoid of spines; first legs with few or no strong dorsal or lateral spines.
      D. Clypeus with a distinct, white, transverse carina; carapace flat. P. 540. MISUMENOIDES
      DD. Clypeus moderately convex, without a carina across the clypeus. P. 538. MISUMENA
    CC. Anterior lateral eyes larger than the median; carapace and legs strongly spinose. P. 542. MISUMENOPS
  BB. Tubercles of lateral eyes broadly separated.
    C. First tibiæ with two pairs of ventral spines. P. 543. OXYPTILA
    CC. First tibiæ with three or more pairs of ventral spines.
      D. Carapace strongly convex; tarsal claws of first legs with six to twelve teeth. P. 553. SYNEMA
      DD. Carapace relatively flat above; tarsal claws of first legs with less than six teeth.
        E. First row of eyes nearly straight; cephalic sutures plainly indicated. P. 544. CORIARACHNE
        EE. First row of eyes moderately recurved; cephalic sutures nearly obsolete. P. 545. XYSTICUS

Genus TMARUS (Tma'rus)

The members of this genus are easily distinguished from other crab-spiders occurring in our fauna by the fact that the abdomen is high and pointed behind (Fig. 572). The lateral eyes of each side are on two distinct tubercles, of which the posterior is the larger.

These spiders live on plants and can run with great rapidity. Their colours are more or less protective. When one of these

Superfamily Argiopoidea

spiders is at rest upon a twig, it clasps it closely with its legs, and the form of the body is such that the spider appears like a bud or a stump of a petiole.

Fig. 572. SIDE VIEW OF THE ABDOMEN OF TMARUS ANGULATUS, FEMALE

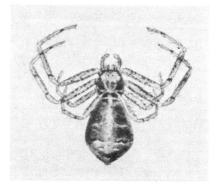

Fig. 573. TMARUS ANGULATUS

Five species have been described from the United States, of which the following is the most common.

Tmarus angulatus (T. an-gu-la′tus).— The colour of the body is dark yellow flecked with five brown spots; the abdomen is crossed by three or four darker bands (Fig. 573). The length of the body is about one fourth inch.

The male differs from the female in having a smaller abdomen. The genital bulb (Fig. 574) lacks apophyses; the embolus is long and curved, and the tip of it is protected by a tutaculum.

This is a widely distributed species; it is the *Thomisus caudatus* of Hentz.

Fig. 574. PALPUS OF MALE OF TMARUS ANGULATUS

Genus MISUMENA (Mi-su′me-na)

The first and second pairs of legs are almost entirely devoid of spines except beneath the tibiæ and the metatarsi. The eyes of the anterior row are equidistant and are in a slightly recurved line. The eyes of the second row are equidistant and in a more or less recurved line. The median ocular area is a little narrower in front than behind; its length and width are equal. The lateral eyes are situated in slightly elevated confluent tubercles.

To this genus belongs a single well known species which is one of the most familiar flower spiders of the United States. It was proved experimentally by Doctor Packard ('05) that white individuals of this species, when placed on goldenrod, changed from white to yellow in the course of ten or eleven days. The species is also very common in Europe.

Misumena vatia (M. va'ti-a).— This is the most commonly observed member of the family, being frequently found in flowers.

Fig. 575. MISUMENA VATIA, FEMALE

The female (Fig. 575) measures, when full-grown, from one third to one half inch in length. It is milk-white or yellow, with, in many cases, a light crimson band on each side of the abdomen, and another in the eye-region. The sides of the thorax are slightly darkened. On the anterior margin of the epigynum (Fig. 576) there is a projecting plate, with a large deep notch behind in the middle, and on each side of this a more or less distinct notch. The openings of the spermathecæ are one on each side behind the point between the central and the lateral notch.

The male is only one eighth or one sixth inch in length. The cephalothorax is darker at the sides than that of the female; the abdomen is marked with two parallel dark marks or lines of spots and has a dark stripe on each side. The embolus, although

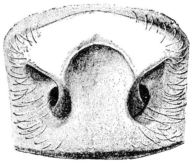

Fig. 576.
EPIGYNUM OF MISUMENA VATIA

coiled, is short. In the unexpanded bulb it arises near the distal end of the alveolus and is coiled backward and outward, so that the tip of it is protected by the margin of the cymbium, which is slightly widened at this point (Fig. 577).

The egg-sac is made upon a leaf and protected by folding a part of the leaf over it and fastening it down with a sheet of silk (Fig. 578). The specimen figured was on a leaf of milkweed

Superfamily Argiopoidea

and was lemon-yellow in colour except the red band on either side of the abdomen. It was taken in August.

This spider is remarkable for the change in colour which takes place in it when it migrates to flowers differing in colour from those previously occupied. In the spring and in the early part of the summer it is most often found in the flowers of Trillium, the white fleabane, and other white flowers. Its ground colour is then white,

Fig. 577.
PALPUS OF MALE OF
MISUMENA VATIA

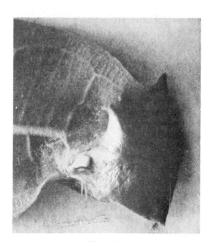

Fig. 578.
MISUMENA VATIA AND EGG-SAC

which protects it from observation by the flower-visiting insects, and enables it more readily to capture them. Later in the season it migrates in large numbers to the flowers of goldenrod, and is then usually yellow.

Genus MISUMENOIDES (Mi-su'me-noi-des)

The genus, represented by the following common species, differs from *Misumena* in having a distinct, white carina on the clypeus and one between the eye rows. The carapace is relatively flatter.

Misumenoides aleatorius (M. a-le-a-to'ri-us).—This is a common white or yellow crab-spider; the female resembles *Misumena vatia*, described above, quite closely; but it (Fig. 580, *a*) does not have the crimson markings at the sides of the abdomen, characteristic of *M. vatia* though it occasionally has dark reddish brown marks in the same places and a double row of dark spots

Superfamily Argiopoidea

in the middle of the back (Fig. 580, b). In this species, there is a pale, transverse band, between the two rows of eyes (Fig. 581),

(Photographed by M. V. Slingerland)
Fig. 579.
SPRAY OF GOLDENROD INHABITED BY MISUMENA VATIA

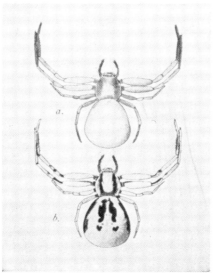

Fig. 580. MISUMENOIDES ALEATORIUS
a, the unspotted form   *b*, the spotted form

which extends from the tubercle bearing the lateral eyes of one side to that of the other side. This band often appears like a ridge; and its presence has led some writers to place this and similarly

Superfamily Argiopoidea

marked species, in the genus *Runcinia*, in which there is a transverse ridge in this position. But the arrangement of the eyes in this species is that characteristic of *Misumena* and not that of *Runcinia*, in which the median eyes of each row are farther from the lateral eyes than from each other; while in *Misumena* the eyes of each row are equidistant.

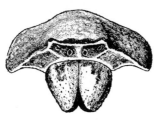

Fig. 581. FACE OF MISUMENOIDES ALEATORIUS

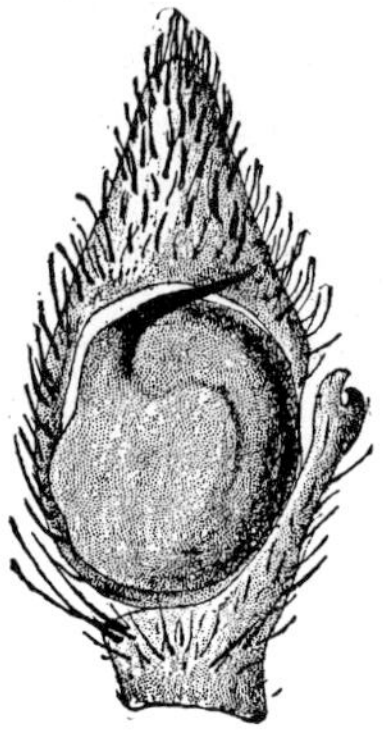

Fig. 582. PALPUS OF MALE OF MISUMENOIDES ALEATORIUS

The epigynum of this species resembles quite closely that of *M. vatia* figured above.

The male measures only one eighth inch in length. The cephalothorax is green with the sides dark brown; the abdomen is bright yellow. The first two pairs of legs are very long and are dark brown in colour. The embolus is short and straight. In the unexpanded bulb, it arises on the distal margin of the bulb in the tip of the alveolus and projects outward and slightly forward (Fig. 582); the tip of it rests in a depression in the face of the cymbium, just outside the margin of the alveolus.

Genus MISUMENOPS (Mi-su'me-nops)

This genus is closely allied to *Misumena* but differs in having large and prominent spines on the femora of the first and second pairs of legs, and on the upper face of the tibiæ of the same legs. The cephalothorax and abdomen are more spiny than in *Misumena*. The tubercles of the lateral eyes are joined by a rounded ridge, and the posterior lateral eyes are not larger than the posterior median eyes.

Eight species of this genus have been found in the United States; they are almost entirely restricted to the South

Superfamily Argiopoidea

and the Far West; but the following species is widely distributed.

Misumenops asperatus (M. as-pe-ra'tus).—The female of this species is one fourth inch in length; it is usually pale yellow in colour with dull red markings (Fig. 583) or the ground colour may be greenish. There is a brownish stripe on each side of the

Fig. 583.
MISUMENOPS ASPERATUS

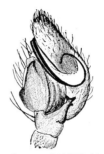

Fig. 584. PALPUS OF MALE OF MISUMENOPS ASPERATUS

thorax; a median light red band on the basal half of the abdomen; two bands or rows of spots on the hind half; and a band on each side.

The male resembles the female in colour and markings but is only about one half as long. The palpus of the male is large; the embolus is long and coiled; in the unexpanded bulb it arises from the distal end of the bulb; it is coiled downward and outward over the lower edge of the cymbium, and the terminal portion of it rests in a depression on the lateral face of the cymbium (Fig. 584).

Genus OXYPTILA (Ox-yp'ti-la)

In this genus the median ocular area is longer than wide. The legs are short and armed with but few spines; the tibiæ of the first and second pairs of legs bear only two pairs of inferior spines (Fig. 585). The cuticle is clothed with setæ of various types, some of them are pointed while others are clavate or spatulate.

A rather small number of species is known from the United States, but the genus is strongly represented in Europe. Several

Superfamily Argiopoidea

of our species are closely related to *Xysticus* of which *Oxyptila* represents a recent offshoot.

The following is the most common of our species:

Oxyptila conspurcata (O. con-spur-ca'ta).— In this species there is only one spine above on the metatarsus of the first legs, and the tibia of these legs is not spotted. The cephalothorax is reddish yellow, paler in the middle, usually with some silvery white lines; the sides of the cephalothorax are either wholly brown or with two brown stripes on each side, the upper one broadest behind and ending in a darker spot. The abdomen is irregularly spotted. The length of the body is one seventh inch.

Fig. 585.
TIBIA OF FIRST LEG OF OXYPTILA

Genus CORIARACHNE (Co-ri-a-rach'ne)

The cephalothorax is strongly depressed and flat. The posterior row of eyes is strongly recurved, with the median eyes smaller than the lateral and farther from the lateral eyes than from each other. The anterior row of eyes is straight or nearly so, with the median eyes much smaller than the lateral; the eyes of this row are usually equidistant. The median ocular area is wider than long. In the more typical species, the lateral eyes on each side are farther apart than are the anterior median and the posterior median eyes; but this is not true of the species described below.

Fig. 586.
CORIARACHNE VERSICOLOR

Four of our species have been placed in this genus. One, *C. brunneipes*, is from the West Coast; another, *C. utahensis*, is common west of the Rocky Mountains; a third, *C. floridana*, is from Florida. The following is widely distributed.

Coriarachne versicolor (C. ver-sic'o-lor).—The female (Fig. 586) is about one fourth inch in length, white or yellowish in colour, and spotted with black and gray. The spots vary greatly in different individuals, and are sometimes so large, especially in males, that the spider is nearly black. This spider closely resembles in appearance a *Xysticus;* but the body is more flattened than in that genus, and the genital bulb of the male is quite different. The female can be most easily recognized by the form of the epigynum (Fig. 587); and the male, by the form of the genital bulb (Fig. 588). The embolus is long and curved. In

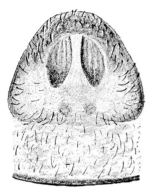

Fig. 587.
EPIGYNUM OF CORIARACHNE VERSICOLOR

Fig. 588.
PALPUS OF MALE OF CORIARACHNE VERSICOLOR

the unexpanded bulb, the terminal part of it is protected by the lower edge of the cymbium, bordering the alveolus, but a well-developed tutaculum, such as is seen in *Xysticus,* is lacking; so also are lacking the prominent apophysis of the bulb characteristic of *Xysticus.*

This is a common species, which is found on fences and under stones; its colours frequently so closely resemble those of the objects upon which it is found that it is seen with difficulty.

Genus XYSTICUS (Xys'ti-cus)

In the genus Xysticus the posterior eyes are nearly equidistant, with the median a little smaller than the lateral; the anterior median eyes are usually a little farther from each other than from the anterior lateral eyes and much smaller than the anterior lateral eyes. The median ocular area is as wide as or wider than long, and is as wide as or a little wider in front than behind. The lateral eyes of each side are situated

Superfamily Argiopoidea

in low subconfluent tubercles. The tibiæ and metatarsi of the anterior legs are usually furnished with more than three pairs of inferior spines. The tarsal claws are furnished with five or six isolated teeth (Fig. 589). The cuticle is clothed with simple, isolated hairs, which are usually pointed, rarely blunt, but never clavate.

These spiders live under stones and leaves or under loose bark; a few live on low plants. They are fawn-coloured or brownish. Usually there is a broad clear band on the cephalothorax enclosing

Fig. 589. TARSUS OF XYSTICUS

in front an obscure triangular spot, and the abdomen is ornamented with a broad, strongly notched band; but there are, however, some unicolorous species.

In the palpus of the males the lower margin of the cymbium bears an appendage for the protection of the tip of the embolus, the tutaculum, and the genital bulb bears a pair of well-developed apophyses.

The genus is a very large one; nearly forty species have been described from the United States, which is about one third of our thomisid fauna. Only a few of our more common and more widely distributed species can be described here; but a large proportion of the species commonly taken are included in this short list.

Xysticus elegans (X. el'e-gans). — The female measures one third inch in length. The ground colour of the cephalothorax is brownish yellow, with a narrow, white, marginal seam; the sides are veined with brownish red; there is a lighter, median, longitudinal band, which is also streaked with red, at least to the beginning of the last third of its length where the red markings end in a blunt point. The abdomen is brown above, somewhat lighter before the middle, and with several narrow, bowed, transverse bands behind; the sides and venter are yellow. The epigynum is represented in Fig. 600.

The female of this species is well illustrated in Fig. 599. It was associated with the male of another species, *X. limbatus*, by Keyserling. The male of that species properly belongs with the female he described as *emertoni*, so the name *limbatus* has been dropped as a synonym of *emertoni*.

The male of *X. elegans* (Fig. 591) measures about one fourth inch in length. It presents a very striking appearance, due to its strongly contrasting colours. The cephalothorax is reddish brown, streaked and flecked with yellow, and with a broad, light, median, longitudinal band, which is bordered on each side with a yellow line; these two lines come together behind in a point; and there is a yellow transverse band between the two rows of eyes. The abdomen is brownish white dotted with brown points, and with four pairs of large brown patches as shown in the figure. The unexpanded palpus is represented in Fig. 592.

This species is widely distributed in the Eastern States.

Xysticus ferox (X. fe'rox).— This species is of medium size, the female (Fig. 593, *a*) measuring about one fourth inch in length, and the male (Fig. 593, *b*) one fifth inch. The cephalothorax is yellowish in the middle and reddish brown on the sides, marked with a net-work of dark lines; at the posterior end there is a small median black spot and a larger one on each side. The abdomen is brownish gray above and smoky white on the sides; there are several small, black spots on the basal part above, and three pairs of transverse black bars bordered in front with white on the hind part. The form of the epigynum is represented in Fig. 594 and that of the male palpus in Figs. 569 and 570 on page 522.

Xysticus formosus (X. for-mo'sus).— This is a beautiful brown and white species, which can be easily recognized by its characteristic markings (Fig. 595). The female measures one fourth inch in length; the male, one fifth. It is found in the Northern States.

Xysticus gulosus (X. gu-lo'sus).— This is a large species, the female (Fig. 596) measuring from one fourth to one third inch in length. It is grayish brown in colour and presents a very distinctively characteristic appearance from the fact that the white ground is largely covered with minute brown specks. The median lighter area of the cephalothorax, and transverse light bands on the hind part of the abdomen are more or less

Fig. 590.
EPIGYNUM OF XYSTICUS EMERTONI

Fig. 591.
XYSTICUS ELEGANS, MALE

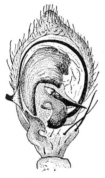

Fig. 592.
PALPUS OF MALE OF XYSTICUS ELEGANS

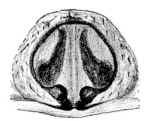

Fig. 594.
EPIGYNUM OF XYSTICUS FEROX

Fig. 593.
XYSTICUS FEROX
a, female   *b*, immature male

distinct, more distinct in immature individuals than in adults. There is a light coloured seam on the lateral margin of the cephalothorax; that part of the thorax that is overlapped by the abdomen is white; and on each side of this white area there is a small white spot on a black patch. On each side of the guide of the epigynum, there is a prominent, dark-coloured, pear-shaped body (Fig. 597).

The male measures one fifth inch in length and is more distinctly marked than the adult female. The two apophyses of the genital bulb are hook-like, with the tips curved toward each other (Fig. 598).

This species is very widely distributed and common in many places; it is usually found under bark or stones.

Xysticus emertoni (X. em-er-to'ni).—The mature female measures about two fifths inch in length. The cephalothorax is reddish brown streaked and marbled with a lighter colour; on the lateral margin there is a white seam; and in the middle a broad longitudinal band, which is bordered with a yellow line and marked with many irregular lines of this colour; the eye space is reddish yellow. The sides of the abdomen are yellowish white, with rows of brown points in oblique wrinkles; the dorsal area is brown marked by transverse lines of a lighter colour. The cavity of the epigynum (Fig. 590) is large and deep. The caudal half of the atrium is filled by an elevated, transverse plate which is emarginated in front. This type of epigynum is quite different from that of most species of the genus.

The male is one third inch in length, and resembles the female in markings; but it is a little darker. The apophysis on the upper and outer face of the tibia of the palpus is very long and slender (Fig. 601).

This species was described from Colorado and Texas; but it is widely distributed in the East.

Xysticus luctans (X. luc'tans).—The full-grown spider measures from one fourth to one third inch in length. The two sexes are quite similar in colour and markings. The cephalothorax is light reddish yellow, with four very distinct, brown, longitudinal stripes, two on each side, one marginal, and one extending back from the posterior lateral eye; there is a small brown spot near the hind end of the cephalothorax, and a pair of similar spots halfway between this and the eyes; the cephalothorax is unusu-

Fig. 595.
XYSTICUS FORMOSUS

Fig. 596.
XYSTICUS GULOSUS

Fig. 597.
EPIGYNUM OF XYSTICUS GULOSUS

Fig. 598.
PALPUS OF MALE OF XYSTICUS GULOSUS

Fig. 599.
XYSTICUS ELEGANS

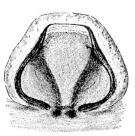

Fig. 600.
EPIGYNUM OF XYSTICUS ELEGANS

ally wide. The abdomen is light gray above, with three indistinct darker longitudinal bands, and three pairs of black spots, one pair near each end and a pair near the middle; besides these there are some smaller marginal spots.

The form of the epigynum is shown in Fig. 602.

This widely distributed species has been commonly known as *Xysticus quadrilineatus*, a name suggested by the four longitudinal bands on the cephalothorax; it is unfortunate that this name must be dropped for an older but less descriptive one.

Xysticus nervosus (X. ner-vo'sus).— The female measures from one fourth to one third inch in length; the male, about one fourth inch. The two sexes are similarly coloured and marked. The cephalothorax is yellow with a narrow black line on the margin; it is veined and marbled with brown on the sides; on the middle of the back there is a broad lighter band (Fig. 603). The abdomen is also light yellowish, and is marbled with white, and spotted with brown; there are three or four pairs of irregular, transverse, brown bars, bordered in front with white, on the hind half of the abdomen, and a pair of small spots near the base.

The central cavity of the epigynum (Fig. 604) is divided by a well-developed guide; and the openings of the spermathecæ are conspicuous and situated some distance behind the central cavity of the epigynum.

In the palpus of the male the terminal apophysis (Fig. 605) is twisted at the end.

This is a very common and widely distributed species.

Xysticus triguttatus (X. tri-gut-ta'tus).— The two sexes of this species differ greatly in appearance. The female (Fig. 606) measures one fifth inch in length; the cephalothorax is brownish yellow, the abdomen, almost white; near the hind end of the carapace there are three black spots, which probably suggested the specific name; from the intermediate of these spots there extends on each side toward the eyes a white band which is wide behind and narrow in front; the eye-space is also white; each posterior median eye is on a black spot; and there are irregular dark markings on each side of the carapace; the abdomen is almost white, with a pair of black spots near the base, and three or four pairs of, more or less broken, transverse, black stripes on the hind half. In the cavity of the epigynum there is a pair of dark coloured plates which probably function as a guide.

Fig. 601. TIBIA OF THE PALPUS OF THE MALE OF XYSTICUS EMERTONI

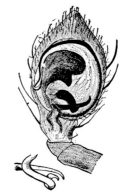

Fig. 605. PALPUS OF XYSTICUS NERVOSUS

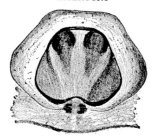

Fig. 602. EPIGYNUM OF XYSTICUS LUCTANS

Fig. 603. XYSTICUS NERVOSUS

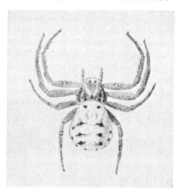

Fig. 606. XYSTICUS TRIGUTTATUS, FEMALE

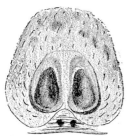

Fig. 604. EPIGYNUM OF XYSTICUS NERVOSUS

Fig. 607. XYSTICUS TRIGUTTATUS, MALE

The male measures a little less than one sixth inch in length. The cephalothorax is dark brown with a lighter median area (Fig. 607); the femora of the first two pairs of legs are dark brown; and the abdomen is white with heavy, brownish black markings, as shown in the figure.

This is a widely distributed species; it lives on grass and on low bushes.

Genus SYNEMA (Sy-ne'ma)

The genus *Synema* is closely allied to *Xysticus* but differs in the following characters: The median ocular area is a little more narrow in front than behind; the anterior eyes are equidistant or even with the median nearer to each other than to the lateral eyes; the tarsal claws of the first two pairs of legs are furnished with more than five or six teeth, and the teeth extend beyond the middle of the claw (Fig. 608); the cuticle is smooth and clothed with scattered long hairs; and there are only three pairs of spines on the lower side of the tibiæ of the first and second pairs of legs.

Fig. 608.
TARSAL CLAWS OF SYNEMA

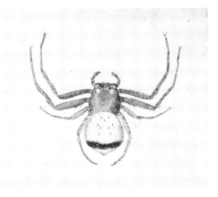

Fig. 609.
SYNEMA PARVULA

The original spelling of the generic name is that adopted here; later Simon changed it to *Synæma*.

This is a large genus. Only four or five species pertaining to it have been found in our fauna.

Synema parvula (S. par'vu-la).—The female measures a little more than one eighth inch in length; the male, a little less. The cephalothorax is brownish yellow, a little darker at the sides and with a marginal, dark brown seam; the lateral ocular tubercles are white, and the median eyes are ringed with white. The abdomen is white or light yellow, with a broad transverse black or brown band near the hind end (Fig. 609). This band usually has a broad deep notch in the middle of the front margin, but it is not always thus notched. There are usually two or three pairs of small brown spots between this band and the base of the abdomen.

This species is widely distributed; it is especially common in the Southern Atlantic States. Hentz, who first described the species, states that it is frequently found on the blossoms of umbelliferous plants.

Synema bicolor (S. bi'co-lor).— The female measures one fifth inch or a little more in length. The cephalothorax is dark brown, almost black, with a lighter line in the middle and a white line on each side near the edge. The abdomen is light gray with indistinct lighter lines at the sides and small light spots in the middle.

This species is distributed from New England to Florida.

Synema obscura (S. ob-scu'ra).— The male measures one seventh inch in length. The cephalothorax is dark red, lighter above, with a yellow seam on the lateral margins. The eye-space is brownish yellow. The abdomen is blackish brown above, with a narrow, white transverse band in front, the hind margin of which is irregularly toothed; the venter is a little lighter brown, and flecked with white, especially in front.

The female has not been described. The male was described from Mount Washington.

Subfamily PHILODROMINÆ (Phil-od-ro-min'æ)

In this subfamily the tarsi of the first and second pairs of legs are furnished with scopulæ beneath, at least in the females (Fig. 610). These scopulæ can be easily distinguished from a

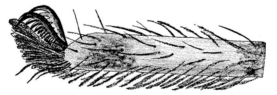

Fig. 610. TARSUS OF PHILODROMUS PERNIX, FEMALE

thick clothing of ordinary hairs, such, for example, as is found in *Xysticus* (Fig. 589, p. 533), by the form of the tenent hairs

(Fig. 611). The third and fourth pairs of legs are as long as or nearly as long as the first and second pairs. The hairs of the body are pubescent or plumose (Fig. 612), and prone, not erect. And the upper margin of the furrow of the chelicerae is armed with one or two teeth.

These spiders are very active and pursue their prey with great rapidity. Most of the species live on plants; when at rest, the body is closely applied to the supporting object, with the legs stretched out separately or in pairs.

Fig. 611.
TENENT HAIR FROM THE SCOPULA OF PHILODROMUS

Fig. 612.
PLUMOSE HAIR OF PHILODROMUS

The following genera include most of the species known from our fauna.

TABLE OF GENERA OF THE PHILODROMINÆ

A. Second pair of legs very much longer than the first pair. P. 560.   EBO
AA. Second pair of legs but little longer than the first pair.
  B. Tibiæ of the first and second pairs of legs armed with five pairs of spines beneath. P. 558.   PHILODROMOIDES
  BB. Tibiæ of the first and second pairs of legs armed with less than five pairs of spines.
    C. Posterior eyes in a slightly recurved line; posterior eyes either equidistant or with the median eyes farther from each other than from the lateral eyes.
      D. Posterior median eyes much farther from each other than from the lateral eyes. Anterior lateral eyes nearer to the anterior median eyes than to the posterior median eyes. P. 556.   PHILODROMUS
      DD. Posterior eyes equidistant. Anterior lateral eyes equidistant from the anterior median and the posterior median, forming with them an equilateral triangle on each side. P. 558.   APOLLOPHANES

Superfamily Argiopoidea

CC. Posterior eyes in a strongly recurved line; posterior median eyes farther from the lateral eyes than from each other.
    D. Cephalothorax not at all or hardly longer than wide. P. 561.   THANATUS
  DD. Cephalothorax much longer than wide. P. 562.
                                                TIBELLUS

Genus PHILODROMUS (Phi-lod′ro-mus)

The second pair of legs are only slightly longer than the first pair; and the third and fourth pairs are but little shorter than the first and second pairs; the body is flat, and the abdomen pointed behind; the posterior median eyes are much farther from each other than from the lateral eyes; and the anterior lateral eyes are nearer to the anterior median eyes than to the posterior median eyes.

Most of the species of this genus live on plants; but a few are found on houses and fences. Some of them are coloured protectively, closely resembling the bark upon which they are found.

The glistening white egg-sacs of some of our species are common and conspicuous objects. They are made upon the branches of the shrubs or trees upon which the species lives; and are often made in the fork of a branch (Fig. 613). In making the cocoon, the spider first spins a disk of silk, the eggs are placed upon this, and are then covered with another disk, and then the whole is covered with the densely woven outer layer, which is stretched very taut. In the case of *Philodromus minutus*, described below, a very different method of caring for the cocoon is practised.

Fig. 613. EGG-SAC OF PHILODROMUS

About thirty species of this genus are known from the United States; among the more common species are the following:

Philodromus pernix (P. per'nix).— This is the most common representative of the genus found on houses and fences; it sometimes occurs also on plants. It measures from one fourth to one third inch in length. It is gray, resembling in colour old unpainted buildings. In the female (Fig. 614), the cephalothorax is darker at the sides; and there is a more or less distinct V-shaped, light band on the hind margin of the head. On the basal part of the abdomen there is a lanceolate median stripe; and on the hind part

Fig. 614. PHILODROMUS PERNIX, FEMALE

a herringbone pattern. The male is coloured like the female; but in all that I have seen the V-shaped band on the head is not so well-marked; the legs are longer, and the abdomen more slender.

This is the *Thomisus vulgaris* of Hentz.

Philodromus minutus (P. mi-nu'tus).— This little spider measures only one eighth inch in length. The cephalothorax is white or yellowish white, the sides reddish brown; the abdomen is dirty white or yellowish; on the basal half of the abdomen there is a median brown stripe, and on the hind half there are two stripes, one on each side.

Although this spider is small, it attracts attention by the curious way in which it cares for its eggs. The egg-sac is made near the tip of a leaf; and then the tip bearing the egg-sac is folded back and fastened down to the body of the leaf by many silken threads. The spider then remains on guard near the folded part of the leaf (Fig. 615). A more common type of nest of this spider is shown in Fig. 616.

This species is found in the Northeastern States.

Philodromus ornatus (P. or-na′tus).— This is a small species, measuring only about one eighth inch in length. It is easily recognized by its colour and markings (Fig. 617). The cephalothorax is yellowish white, with the sides brown. The abdomen is white above and brown or black on the sides; the dark patch of each side is irregular in outline, as shown in the figure. Sometimes there is an indistinct brownish pattern in the middle of the abdomen, and some faint dark chevrons near the hind end.

This species is found in the Eastern States.

Genus PHILODROMOIDES (Phi-lod-ro-moi′des)

The abdomen is twice as long as wide, a little wider behind than at the base, somewhat pointed at the end. The eyes are approximately equal in size; the anterior row much the shorter and slightly recurved; median eyes of this row farther from each other than from the side eyes; posterior row is also recurved; the posterior lateral eyes on larger tubercles than any of the others; posterior median eyes farther from each other than from the posterior lateral eyes. The tibiæ of the first and second pairs of legs are armed beneath with five or six pairs of spines.

A single species has been described from Kansas.

Philodromoides prataria (P. pra-tar′i-æ).— Cephalothorax rusty brown, lighter at the sides and just back of the head; abdomen brown above, mottled and streaked with a lighter shade. The length of the body is one fourth inch, the width one twelfth inch.

Genus APOLLOPHANES (Ap-ol-loph′a-nes)

The cephalothorax is almost circular; the posterior row of eyes is more strongly recurved than in *Philodromus;* the eyes of the posterior row are equidistant. The anterior lateral eyes are equidistant from the anterior median and the posterior median, forming with them an equilateral triangle on each side.

Apollophanes texana (A. tex-a′na).— Cephalothorax yellowish, slightly mottled with brown on sides, and two approximate dark

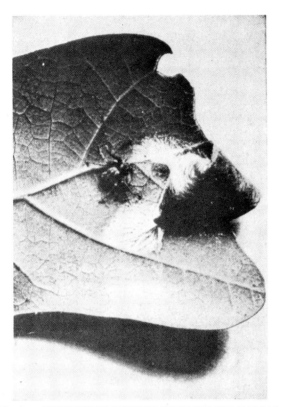

Fig. 615. PHILODROMUS MINUTUS GUARDING HER EGG-SAC

 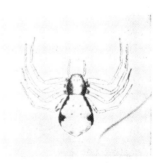

Fig. 616. LEAF BEARING EGG-SAC OF PHILODROMUS MINUTUS

Fig. 617. PHILODROMUS ORNATUS

Superfamily Argiopoidea

marks on base of cephalic part. Abdomen pale, with a basal brown spear-mark, and behind two more or less connected rows of blackish spots. Abdomen about twice as long as broad. The length of the body is about one fourth inch.

This species occurs in the Southwestern States. Two other species have been described from the mountains of Arizona.

Genus EBO (E'bo)

The members of this genus are easily recognized by the great length of the second pair of legs and the almost straight posterior row of eyes. Seven or eight species belong to this genus. The following are examples.

Ebo latithorax (E. lat-i-tho'rax).— This is a small spider measuring only about one eighth inch in length. The cephalothorax is reddish yellow marked with dark brown specks. The abdomen is brownish yellow, flecked with dark on the sides, and quite far up on to the back (Fig. 618), and with several indistinct chevrons on the hind part.

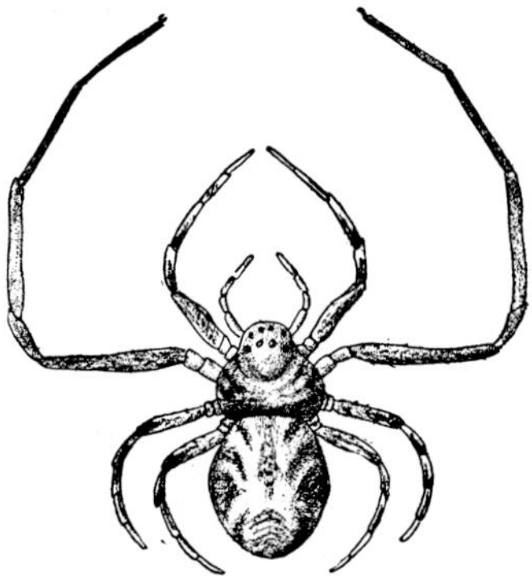

This species occurs in the Southern States.

Ebo oblongus (E. ob-lon'gus). — This is larger than the preceding species, immature specimens measuring nearly one fifth inch in length. It differs also in having the legs feebly spined. It was described from specimens taken in Georgia.

Fig. 618.
EBO LATITHORAX (after Keyserling)

Ebo mexicana (E. mex-i-ca'na).— The body is one sixth inch in length. The cephalothorax is pale, with a large brown spot on each side, not reaching behind to the posterior margin; each spot includes a few white dots. There is a small brown posterior spot, and two elongate brown spots behind the posterior median eyes. The abdomen bears a basal, brown, spear-mark, margined with white.

This species is found in the Southwest.

Superfamily Argiopoidea

Genus THANATUS (Than'a-tus)

In this and the following genus the fourth pair of legs are longer than the first pair; and the posterior row of eyes are more strongly recurved than in the preceding genera. In this genus the cephalothorax is but little if at all longer than wide; and the anterior lateral eyes are closer to the anterior median eyes than to the posterior median eyes.

The species of *Thanatus* resemble each other in general appearance and in colouration; they are all of a yellowish or grayish red, with the abdomen marked with a lance-shaped band. Fifty species are known, several from the United States. The two following are found both in Europe and America.

Thanatus lycosoides (T. ly-co-soi'des).— This species varies considerably in markings. Figure 619 is from a photograph of an unusually well-marked individual. The cephalothorax is reddish yellow with a median, longitudinal, brown band, which is wide in front and tapers to a point behind; the sides of the cephalothorax are streaked with brown and red. The abdomen is light above, with a slender, brown, lance-like spot in front, which reaches beyond the middle, and with an undulating band on each side of the hind part. The anterior median eyes are much smaller than the anterior lateral eyes.

Fig. 619. THANATUS LYCOSOIDES

In many individuals there are no markings on the hind part of the abdomen. In others there are two indistinct longitudinal brown bands in the position occupied by the undulating bands shown in the figure.

The female measures from one fourth to one third inch in length; the male is a little smaller and has longer legs. This is a widely distributed species.

Thanatus coloradensis (T. col-o-ra-den'sis).— This is a Western species which differs from the preceding in that the anterior median eyes are as large as the anterior lateral eyes

Superfamily Argiopoidea

Genus TIBELLUS (Ti-bel'lus)

The members of this genus differ greatly in form from the ordinary type of crab-spiders, the body being long and slender; the cephalothorax is much longer than wide and the abdomen is very long and nearly cylindrical. The posterior row of eyes is strongly recurved, nearly forming a semicircle.

These spiders are found on grass and on bushes; when at rest the legs are stretched out longitudinally, two pairs forward and two pairs backward.

Four species are known from the United States. Two of these also occur in Europe. The following species are representative of this genus.

Tibellus duttoni (T. dut'to-ni). —The colour of the body is light gray or yellow with three longitudinal brown stripes extending the whole length of both the cephalothorax and the abdomen (Fig. 620). There is usually a pair of small black spots on the abdomen near the middle of the second half. Sometimes the lateral bands on the cephalothorax are broken into distinct brown spots; and there may be several dark brown spots along the sides of the abdomen. The male is about one third inch in length; the female, about one half inch. This species is distinguished from the *Tibellus oblongus* by the following characters: The space between the posterior median eyes is considerably less, sometimes only half as great, as the space between one of them and the posterior lateral eye of the same side. In the palpus of the male the short embolus is very much curved. The second legs are more than five times as long as the cephalothorax. The epigyna of the two species are very similar.

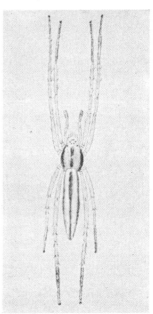

Fig. 620. TIBELLUS DUTTONII

This is a very widely distributed species; and is common on bushes and on grass.

Tibellus oblongus (T. ob-lon'gus).— This species is a little less slender than the preceding and the legs are comparatively shorter, the second legs being not quite five times as long as the cephalothorax. The space between the posterior median eyes is only a little less than that between one of them and the posterior lateral eye of the same side. In the palpus of the male, the short embolus is nearly straight.

This species is widely distributed; and like the preceding is common on bushes and on grass.

Family SELENOPIDÆ (Sel-e-nop'i-dæ)

The Selenopidæ includes a single genus, *Selenops*, of which about a dozen species are known. These are tropical spiders of large or of medium size; and they do not occur naturally within the limits of the United States, except perhaps in the extreme South. A single individual of *Selenops aissus* has been reported from Tortugas Island, Fla.; but this may have been introduced. As specimens of the same species have been taken at Ithaca, N. Y., in a room where bananas were stored, we infer that there is a chance of its being found wherever tropical fruit is taken, and include an account of it here.

Genus SELENOPS (Se-le'nops)

The members of this genus resemble the crab-spiders in the attitude of their legs. They are of large or of medium size and are remarkable for the extreme flatness of the body. They can be easily recognized by the arrangenemt of the eyes, which is unusual in that the posterior median eyes have moved to a position in front of the posterior lateral eyes and in line with the four anterior eyes; the anterior row may be said, therefore, to to consist of six eyes. The following is the only species that has been recorded from the United States.

Selenops aissus (S. a-is'sus).— The adult female measures one half inch in length and nearly one fourth inch in width. The cephalothorax is reddish brown, the abdomen brownish yellow flecked with dark specks; the lateral and the hind margins of the abdomen are dark. When viewed from above only six eyes are visible (Fig. 621); these are the four anterior eyes, which

Superfamily Argiopoidea

are nearly equal in size, the anterior median eyes being slightly smaller than the anterior lateral, and the posterior lateral eyes, which are much larger than the anterior eyes. Also, when viewed from in front only six eyes are visible, the four anterior eyes, and the posterior median eyes, occupying the ends of the

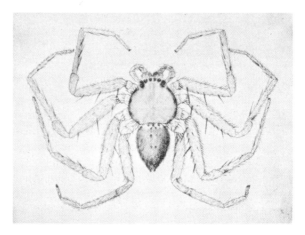

Fig. 621. SELENOPS AISSUS

front row; these eyes are much smaller than the four anterior eyes. The tibiæ of the four legs bear on the lower surface three pairs of spines, and the metatarsi two pairs.

The spiders of this genus are found under the bark of trees, under stones, and in other similar situations; the depressed form of the body enables them to enter thin spaces, or to conceal themselves in slight depressions when lying in wait for their prey; and they run with astonishing rapidity.

Family HETEROPODIDÆ (Het-er-o-pod'i-dæ)

The Giant Crab-spiders

There occur within the limits of our fauna a small number of large spiders, which on account of their size and the fact that they resemble thomisids in the form of the body and the attitude of the legs may be termed giant crab-spiders.

These are really tropical or subtropical spiders whose range extends into the southern part of our territory. One of them, however, is often seen in the North, being brought here with tropical fruits. It is the so-called tarantula found in bunches of bananas, and which periodically gives rise to sensational newspaper stories.

While these spiders resemble the Thomisidæ in the form of the body and the attitude of the legs, they differ from that family and resemble the Clubionidæ in the structure of the cheliceræ, the lower margin of the furrow being distinct and armed. In this respect they resemble the Selenopidæ; but they differ from the members of that family in the arrangement of the eyes. In the Heteropodidæ the eyes are arranged in two rows of four each.

Our three genera may be separated as follows:
A. Spinnerets set upon a distinct, elevated, basal segment. P. 568. TENTABUNDA
AA. Spinnerets normal, not set upon a distinct segment.
B. Anterior median eyes as large or larger than the anterior lateral; clypeus lower than the diameter of an anterior median eye. P. 567. OLIOS
BB. Anterior median eyes smaller than the anterior lateral; clypeus higher than the diameter of an anterior median eye. P. 565. HETEROPODA

Genus HETEROPODA (Het-e-rop'o-da)

This genus includes a large number of tropical species; but the following is the only one whose range extends into our territory.

The Banana Spider, *Heteropoda venatoria* (H. ve-na-to'ri-a). — One often hears of a "tarantula" emerging from a bunch of bananas in a fruit store in the North; but I know of but few instances in which a true tarantula has been found in such a place. The spider that causes consternation among clerks and customers of fruit stores is usually this giant crab-spider; although, more rarely, it may be the *Selenops* described on an earlier page.

The banana spider is a large yellowish spider, with a band of white hairs on the clypeus, and a similar transverse band near the hind margin of the cephalothorax. The male is represented, natural size, by Fig. 622, and the female, also natural size, by Fig. 623.

This species is found in all tropical regions, its range extending clear around the world. It is very abundant in all tropical

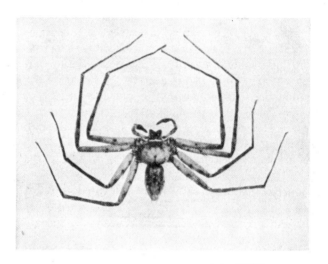

Fig. 622. HETEROPODA VENATORIA, MALE

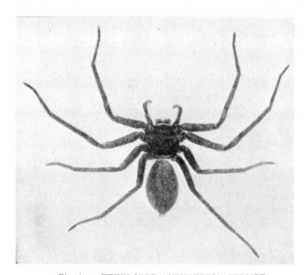

Fig. 623. HETEROPODA VENATORIA, FEMALE

Superfamily Argiopoidea

seaport towns, being transported in trading vessels. Its chief food is cockroaches. Within our territory, it is at home only in the far South; but it is very often found in fruit stores in the North, where it is brought in bunches of bananas.

The female makes a flat, cushion-like egg-sac. The specimens before me are slightly tinged with pink. The egg-sac is three fourths inch wide, but only about one fourth inch thick. It is carried by the female beneath her body (Fig. 624) in a way similar

Fig. 624. HETEROPODA VENATORIA CARRYING HER EGG-SAC

to that of the nursery-web weavers (Pisauridæ). The spiderlings emerge through a slit made in the margin of the egg-sac.

Genus OLIOS (O'li-os)

This is a very large genus, including more than one hundred species, which are distributed in all tropical and subtropical regions. Several species occur in the southwestern portion of our country. The following will serve as an example.

Olios fasciculatus (O. fas-cic-u-la'tus).—This is the largest of our species, often measuring an inch or more in length. The cephalothorax, pedipalps, and legs are reddish brown, but the cephalothorax is somewhat darker than these appendages. The chelicerae are black. The abdomen is yellowish brown, with several brown spots on the middle line, and on the hind part an indistinct lighter chevron.

This species is common in Utah, Arizona, and California.

Genus TENTABUNDA (Ten-ta-bun'da)

The spinnerets are close together in a group and are set upon an elevated, annular segment. The eyes of the first row are subequal in size, and the clypeus is narrow.

Tentabunda cubana (T. cu-ba'na).—This spider is similar in form and general colouration to species of *Olios*. It was described from Cuba and also occurs in Florida.

Family CTENIDÆ (Cten'i-dæ)

The Wandering Spiders

This is a small family of spiders, which is composed of forms closely allied to the Clubionidæ, but which differ, in most cases, from that family in the arrangement of their eyes. These, except in *Titiotus*, are situated in three or four transverse rows (Fig. 625). They also differ from the Clubionidæ in that the truncate end of the endites is entirely clothed with very dense uneven hairs. As a rule they are two-clawed spiders; but in the genus *Cupiennius* a third claw is present.

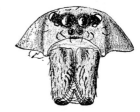

Fig. 625.
HEAD OF CTENUS SHOWING THE ARRANGEMENT OF THE EYES

The members of this family wander about in search of their prey, many of them over the foliage of forests at night. Some of the tropical species are very large. They make no webs for a dwelling, but some species appear to inhabit burrows in the ground.

The genera represented in the United States can be separated as follows:

A. With only two tarsal claws.
  B. First row of eyes only slightly recurved, the eyes subequal in size. P. 570.   TITIOTUS
  BB. First row of eyes very strongly recurved.
    C. Lower margin of the furrow of the chelicera with four or five teeth. P. 569.   CTENUS
  CC. Lower margin with three teeth. P. 570.   ANAHITA
AA. With a third tarsal claw. P. 571.   CUPIENNIUS

Genus CTENUS (Cte'nus)

The eyes are in three rows, the anterior lateral eyes being opposite the posterior median eyes or nearly so (Fig. 625). The anterior lateral eyes are smaller than the anterior median. The tarsi bear only two claws. The lower margin of the chelicera is armed with four, or occasionally five teeth. The labium is longer than broad, rarely slightly broader than long.

Three species occur within our borders.

Ctenus hibernalis (C. hi-ber-na'lis).—This is a tawny species with a median longitudinal yellowish band above extending over both the cephalothorax and the abdomen, and with a darker band on either side. The length of the body of the female is one half inch or more. The epigynum consists of a plate which is broadly rounded on the sides and is truncated behind. The male palpus has a stout tibial apophysis which is directed forward.

This species occurs in Alabama and neighboring states.

Ctenus captiosus (C. cap-ti-o'sus).—This species is similar in size and colour with the preceding. The epigynum of the female is rounded and strongly lobed distally. The tibial apophysis of the male is slender and is directed laterad.

This species was described from Florida.

Ctenus byrrhus (C. byr'rhus).—This species was described by Simon from Mexico as *Leptoctenus byrrhus*, but it belongs in *Ctenus* in the broader usage of that name. These spiders resemble some of the lycosids in general appearance, especially those of the subgenus *Trochosa*, and they are found on the ground in similar situations. Both sexes average about one half inch in length. The general colour is dark reddish brown. The cephalothorax and abdomen are marked above with a median, longitudinal, pale stripe which is flanked by narrow dark stripes. The epigynum of the female consists of a broad, inverted, T-shaped plate, the crosspiece very broad and rounded behind. The tibia of the male palpus has a stout apophysis on the outer side which ends in a slender, curved spine.

This species is quite common in southern Texas.

Genus ANAHITA (A-na-hi'ta)

The eyes are in three rows and agree essentially in arrangement with those of *Ctenus*. The small anterior lateral eyes are placed opposite the large anterior median eyes and form with them a lightly procurved row as seen from in front. The anterior median eyes are much smaller than the posterior median eyes. The labium is broader than long and is not half as long as the subparallel endites. There are three stout teeth on the lower margin of the furrow of the chelicera.

These are ground spiders, found in woods and ravines, that run actively over the leaf mold. A single, common species occurs in the United States.

Anahita punctulata (A. punc-tu-la'ta).—This is the *Ctenus punctulatus* of Hentz. The sexes are similar in appearance and measure about one fourth inch in length. The cephalothorax is pale yellowish brown and is marked with two longitudinal blackish lines and two faint scalloped ones on each side. The legs are pale yellowish brown. The abdomen has "two subobsolete lines of minute white dots becoming more distinct toward the apex, where may be seen a few irregularly placed white dots on the outside of the lines, same colour unspotted beneath."

Genus TITIOTUS (Tit-i-o'tus)

The anterior eyes are in a slightly recurved row, are equal in size and are equidistant. The median ocular area is longer than wide and its sides are subparallel. The clypeus is more than three times as wide as the diameter of an anterior eye.

The following is the only known species.

Titiotus californicus (T. cal-i-for'ni-cus).—The body of the female measures two thirds inch in length. Only the female is known. It was described from California.

Superfamily Argiopoidea

Genus CUPIENNIUS (Cu-pi-en'ni-us)

This genus is distinguished from the other genera of the Ctenidæ by the presence of a third tarsal claw in addition to the tufts of terminal tenent hairs. Figure 626 represents the tip of the tarsus, with one of the two tufts of terminal tenent hairs removed so as to expose the third claw.

Fig. 626. TARSAL CLAWS OF CUPIENNIUS (after Pickard-Cambridge)

Cupiennius sallei (C. sall'e–i).— This is a Central American species which has been found in Florida, and is the only representative of the genus reported from the United States. It is a large spider, measuring from one inch to one and one third inches in length.

Family CLUBIONIDÆ (Clu-bi-on'i-dæ)

The Clubionids (Clu-bi-on'ids)

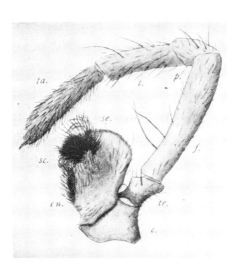

Fig. 627. PEDIPALP OF TRACHELAS

This family includes spiders that live in flat tubular nests on plants, usually in rolled leaves, and others that live on the ground, under stones or rubbish or in moss. As in the Thomisidæ and Heteropodidæ, these spiders have eight eyes arranged in two rows, and have only two tarsal claws. They differ from both of these families in that the form of the body and the attitude of the legs are not those characteristic of crab-spiders; and they differ from the Thomisidæ in that the lower margin of the furrow of the chelicera is distinct and armed with teeth. From the preceding family, the Ctenidæ, they differ in the arrangement of the eyes and in that

Superfamily Argiopoidea

the truncate end of the endite is furnished with a band of hairs (Fig. 627) instead of being entirely clothed with very dense uneven hairs.

The tarsi are usually furnished with bundles of terminal tenent hairs (Figs. 628 and 629). In most of our common light-

Fig. 628. TARSUS OF CLUBIONA

Fig. 629. TIP OF TARSUS OF CLUBIONA

Fig. 630. A TENENT HAIR FROM CLUBIONA

coloured species, these bundles, being dark in colour, form conspicuous cushions beneath the tarsi. These tenent hairs are complicated in structure. Figure 630 represents one of them from the tarsus of *Clubiona*.

There is a striking similarity in appearance between many of the clubionids and certain drassids; but the two families are easily separated in most cases by the fact that in the clubionids the fore spinnerets are contiguous, whereas in the gnaphosids they are widely separated. The few exceptions to this rule need cause little confusion to the general student, for the species concerned are relatively uncommon.

The Clubionidæ includes five subfamilies, all of which are represented in our fauna. The first of these subfamilies, the Anyphæninæ, is sharply distinguished from the others by the position of the furrow of the posterior spiracle. Many workers regard these spiders as belonging to a distinct family. Although there is much in favour of such a disposition, for the purposes of this book the more conservative view is adopted that the anyphænids represent only a subfamily of the Clubionidæ. In the case of the other subfamilies the differences between them are such that a key based upon them must be used with great caution. A preliminary study of the characters indicated for each subfamily will be found helpful in differentiating the groups. The following chart will be found helpful.

TABLE OF SUBFAMILIES OF THE CLUBIONIDÆ

A. Furrow of the posterior spiracle advanced far in front of the spinnerets. P. 574. ANYPHÆNINÆ
AA. Furrow of posterior spiracle near the spinnerets.
  B. Apical segment of the posterior spinnerets conical, often very short but always distinct.
    C. Labium usually much longer than broad, exceeding half the length of the endites; endites narrowed at the middle or at least no broader at the middle than at apex. P. 579. CLUBIONINÆ
    CC. Labium not or scarcely longer than broad, at most half as long as endites; endites broader at middle, narrowed at both ends; first tibiæ often with a series of movable spines. P. 585. LIOCRANINÆ
  BB. Apical segment of the posterior spinnerets always very short, flattened or rounded, usually indistinct.
    C. Endites subquadrate at their apex; sternum not or only slightly margined. P. 592. MICARIINÆ
    CC. Endites rounded at apex; sternum distinctly margined. P. 595. CORINNINÆ

Superfamily Argiopoidea

Subfamily ANYPHÆNINÆ (A-nyph-æ-ni'næ)

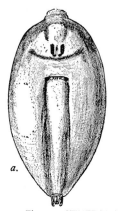 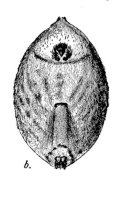

Fig. 631. VENTRAL ASPECT OF ABDOMEN
a, Aysha   b, Anyphæna

This subfamily is sharply distinguished from all other clubionids by the position of the furrow of the posterior spiracle, which is remote from the spinnerets, usually near the middle of the abdomen (Fig. 631). The tufts of terminal tenent hairs of the tarsi consist of a double series of lamelliform hairs; in the other subfamilies the terminal tenent hairs are broom-shaped.

Some writers separate this subfamily from the Clubionidæ, regarding it as a distinct family, the Anyphænidæ.

The anyphænids are spiders of average size which differ little from the species of the Clubioninæ in general appearance, most of them being of pale colouration. They also have similar preferences as to habitat, living for the most part on vegetation. The Anyphæninæ is best represented in terms of genera and species in the Americas.

Four genera are known from our fauna and can be separated by the following table:

TABLE OF GENERA OF THE ANYPHÆNINÆ

- A. Posterior spiracle much nearer the genital groove than the spinnerets (Fig. 631, *a*). P. 575. AYSHA
- AA. Posterior spiracle midway between the genital groove and the spinnerets (Fig. 631, *b*).
  - B. Posterior eye row strongly procurved, the median eyes twice as far from each other as from the lateral eyes; median quadrangle longer than broad. P. 576. GAYENNINA
  - BB. Posterior eye row gently procurved, the eyes nearly equi-

Superfamily Argiopoidea

distantly spaced; median ocular quadrangle about as broad as long.
C. First tibia shorter than carapace. P. 577. ANYPHÆNA
CC. First tibia one and one half times as long as the carapace. P. 576. ANYPHÆNELLA

Genus AYSHA (Ay'sha)

The furrow of the posterior spiracle is situated one third nearer the genital groove than the spinnerets. The anterior median eyes are about equal in size to the lateral eyes or nearly so. These spiders resemble those of *Clubiona* in habits, living in silken tubes in rolled leaves on herbs or shrubs. The following species is the only one found in the north, but also is found in the south where several other species occur.

Aysha gracilis (A. gra'ci-lis).—The two sexes of this common species resemble each other in size and colour. The body is one third inch or a little more in length. The colour is pale yellow; the cephalothorax is darker than the abdomen; in the more distinctly marked individuals there are two longitudinal bands on the cepha-

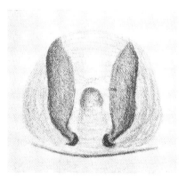

Fig. 632.
EPIGYNUM OF AYSHA GRACILIS

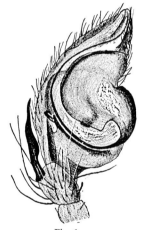

Fig. 633.
PALPUS OF MALE OF AYSHA GRACILIS

lothorax; and on the abdomen two longitudinal series of gray spots irregularly placed, and scattered spots on the sides. The furrow of the posterior spiracle is quite near the epigastric furrow (Fig. 631, *a*).

The epigynum is of the form shown in Fig. 632. The tibia

of the palpus of the male bears a long apophysis at the tip, on the outer lower side, which is more or less notched at the end, and overlaps the tarsus (Fig. 633). The embolus is long, and, in the unexpanded bulb, curves back to the base of the alveolus, and then down and forward, so that the tip rests in the apex of the alveolus, near the tip of the tarsus.

Genus GAYENNINA (Ga-y-en-ni′na)

The spiracle is situated midway between the genital groove and the spinnerets. The first row of eyes is straight, much narrower than the second row which is procurved, the median eyes widely separated. The anterior median eyes are very small. The median ocular area is longer than broad, greatly narrowed in front. The lower margin of the chelicera is armed with two teeth. A single species occurs in the northeastern states.

Gayennina britcheri (G. brit′cher-i).—The carapace is pale yellow, and there are three small submarginal dark spots on each side. The abdomen is mainly gray above but marked with a narrow reddish brown longitudinal stripe.

Genus ANYPHÆNELLA (An-y-phæ-nel′la)

This genus is similar to *Anyphæna* in most respects but may be recognized immediately by the great length of the anterior legs. The tibia of these legs is very much longer than the carapace, and the other joints are correspondingly long. The eyes of the posterior row are gently procurved, subequal in size, and nearly equidistantly spaced. The anterior median eyes are smaller than the lateral eyes.

Several species have been described from the United States of which the following is representative.

Anyphænella saltabunda (A. sal-ta-bun′da).—This species averages about one fifth inch in length. The whole spider is pale yellow, the carapace marked with an indistinct row of spots on each side, the abdomen showing usually a pattern of numerous small black spots. The epigynum (Fig. 638) is wider than long. The male palpus is very characteristic, the tibia (Fig. 639) being three times as long as the patella, and armed with a very long, stout process which is notched at the end.

Superfamily Argiopoidea

Genus ANYPHÆNA (An-y-phæ'na)

The spiracle (Fig. 631, *b*) is situated about midway between the genital furrow and the spinnerets. The eyes of the posterior row are in a lightly procurved line, are essentially equal in size, and subequidistantly spaced. The first legs are of normal length. The anterior median eyes are smaller than the lateral.

The species of this genus resemble each other quite closely in size and markings. The body is from one eighth to one fifth inch in length. The colour is pale yellow or white, with two broken gray stripes on the cephalothorax, and two longitudinal rows of gray spots on the abdomen. The species are best separated by the form of the palpi and epigyna.

This is our largest genus. More than a dozen species are known of which the following are the commonest ones in the northeastern part of our country.

Anyphæna celer (A. ce'ler).—The epigynum (Fig. 634) is longer than wide. The central part of it consists of three sclerites; a transverse one in front; an intermediate one, which is narrower and longer, and tapered toward its hind end; and a smaller one behind. The first two of these sclerites are dark in colour; the third is light. On each side of this series of sclerites, there is a curved one, which is enlarged behind and tapered in front.

The palpus of the male of this species resembles that of *fraterna* in that the tibia is not greatly elongate; but it differs in that the terminal apophyses are comparatively long (Fig. 635).

Anyphæna pectorosa (A. pec-to-ro'sa).—The epigynum (Fig. 636) is wider than long. There is a single, round object behind, on each side of which there is a curved sclerite. The most striking feature of that region is the fact that the posterior edge of the epigastrium is prolonged behind over the epigastric furrow into a plate, which is lobed on each side, and which extends to the lung-slits; the caudal edge of this plate is densely chitinized.

The palpus of the male is similar to that of *Anyphæna fraterna* (Fig. 637), and the coxæ bear apophyses on the coxæ of the third and fourth legs.

Anyphæna fraterna (A. fra-ter'na).—In this species the tibia of the palpus of the male is not greatly elongate. The terminal apophyses of the tibia are comparatively short (Fig. 637). The coxæ of the third and fourth legs bear apophyses beneath.

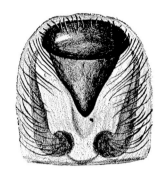

Fig. 634.
EPIGYNUM OF ANYPHÆNA CELER

Fig. 635. PALPUS OF MALE
OF ANYPHÆNA CELER

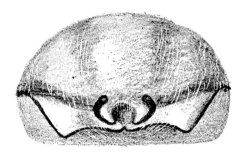

Fig. 636.
EPIGYNUM OF ANYPHÆNA PECTOROSA

Fig. 637. PALPUS OF MALE
OF ANYPHÆNA FRATERNA

Fig. 639.
PALPUS OF MALE
OF ANYPHÆNELLA
SALTABUNDA

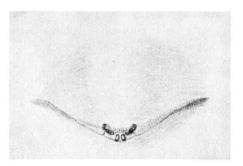

Fig. 638. EPIGYNUM OF A. SALTABUNDA

Subfamily CLUBIONINÆ (Clu-bi-on-i'næ)

In this and the following subfamily the last segment of the hind spinnerets is distinct, conical, often quite long. The labium is usually much longer than wide, and extends beyond the middle of the endites. The endites are narrower at the middle than at the apex and are margined on the inside with incurved hairs. Four genera are represented in our fauna.

A. Posterior row of eyes procurved.
  B. Anterior median eyes much larger than the lateral eyes. P. 579.   MARCELLINA
  BB. Anterior median eyes at most as large as the lateral.
    C. Anterior legs longer than the posterior. P. 583.   CHIRACANTHIUM
    CC. Anterior legs no longer or shorter than the posterior legs. P. 580.   CLUBIONA
AA. Posterior row of eyes recurved. P. 579.   LAURICIUS

Genus MARCELLINA (Mar-cel-li'na)

The posterior eyes are equal in size and equidistantly spaced in a procurved row. The anterior median eyes are very much larger than the anterior lateral and form with the posterior median eyes a quadrangle which is slightly wider in front. The anterior legs are longer than the posterior as in *Chiracanthium*.

Marcellina piscatoria (M. pis-ca-tor'i-a).—This spider resembles our species of *Chiracanthium* in general appearance but is somewhat larger, averaging nearly one half inch in length. The cephalothorax and legs are light yellowish brown, and the abdomen is nearly white. The cymbium of the male palpus is greatly prolonged.

This species is found from New England to Florida, but it is rare throughout its range.

Genus LAURICIUS (Lau-ri'ci-us)

The cephalothorax is relatively flat and broad, and the median groove is distinct. The eyes of the first row are subequal in size. The second row of eyes is distinctly recurved, and the eyes are subequidistantly spaced.

Superfamily Argiopoidea

Lauricius hemiclœinus (L. he-mi-clœ-i'nus).—This species, which measures about one half inch in length, has little resemblance to other members of the subfamily. The carapace and legs are dark reddish brown. The abdomen is dusky yellow, with an irregular black margin and many small dark spots above.

This Mexican species occurs in Arizona and New Mexico.

Genus CLUBIONA (Clu-bi-o'na)

The median furrow of the thorax is present. The posterior eyes are equidistant or the median eyes are farther from each other than from the lateral eyes. The lower margin of the furrow of the cheliceræ is usually armed with two teeth, but sometimes with three, four or five. The posterior legs are longer than the anterior. The tibiæ and metatarsi of the first two pairs of legs are armed beneath with paired spines; there are one or two pairs, usually two, beneath the tibia, and a single pair beneath the metatarsus of each of these legs.

The spiders of this genus are of medium or of small size; they are pale or tawny; the abdomen is clothed with white or pale yellow pubescence, with a silky reflection; most of them are without markings, but some have the abdomen ornamented either with a median line or with a series of chevrons.

These spiders live in silken tubes which they spin under bark or stones or in rolled leaves; these tubes have two openings, and adhere to the object to which they are attached.

The six species described below are our more common ones; they are widely distributed.

Clubiona abbotii (C. ab-bot'i-i).— This is the smallest of our common species, measuring only from one sixth to one fifth inch in length. The cephalothorax is light yellowish brown. The ground colour of the abdomen is lighter than the cephalothorax; but there is a reddish brown lanceolate median stripe on the basal part; and the hind part is mottled with reddish brown. The space between the posterior median eyes is nearly twice as great as that between one of them and the posterior lateral eye of the same side. The epigynum (Fig. 640) is pointed behind, with a notch in the middle; and at some distance in front of it two dark objects, probably the glands of the spermathecæ, show through the skin. The tibia of the palpus of the male bears, on the outside, a very large apophysis, which varies in size and length in different individuals (Fig. 641).

Clubiona riparia (C. ri-pa'ri-a).— This species is easily recognized by the markings of the abdomen (Fig. 642). There is a dark, median, longitudinal stripe, which is narrowed behind and is often broken into a series of spots on the hind half of the abdomen. On each side of this median stripe there is a white or yellow band, with irregular edges. The sides of the abdomen are brown, and are crossed by oblique alternating, light and dark bands of this colour. On the venter, a light median band extends from the epigastric furrow to the spinnerets.

This species makes an interesting provision for the protection of its egg-sac. A leaf of a broad-leaved grass, which grows in marshy places, is folded in the manner shown in Fig. 643. The long three-sided chamber thus formed is lined with silk and contains both the egg-sac and the spider that made it, thus serving as a nursery for the spiderlings and a coffin for the parent. Similar nests are made by folding the leaves of the cat-tail flag.

Clubiona pallens (C. pal'lens).— This is a well-marked species, which can be distinguished from our other common species by the markings of the abdomen (Fig. 644). On the basal half of the abdomen there is a median band, which is distinct in some individuals, and faintly indicated in others. Behind this band, and extending to the tip of the abdomen, there is a series of transverse, more or less arched, dark bars, some of which are frequently broken, especially toward the tip of the abdomen, so as to form a pair of spots. On each side of the abdomen there is a series of oblique dark bands. The ground colour of the abdomen is pale.

Clubiona canadensis (C. can-a-den'sis).— This species measures about one fourth inch in length. The abdomen is yellowish brown, with many light dots, thickly and evenly distributed over the entire surface. In some individuals there is a series of chevrons on the hind half of the abdomen composed of a series of these dots (Fig. 645). A darker lanceolate band on the basal half of the abdomen is usually faintly indicated.

Clubiona obesa (C. o-be'sa).— The length of the body is about one fourth inch. The entire body is pale and almost without markings. On the basal half of the abdomen there

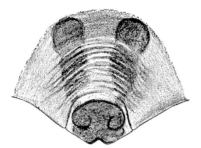

Fig. 640.
EPIGYNUM OF CLUBIONA ABBOTII

Fig. 641.
OUTER FACE
OF PALPUS
OF MALE
CLUBIONA
ABBOTII

Fig. 643.
NEST ENCLOSING
EGG-SAC
OF CLUBIONA
RIPARIA

Fig. 642. CLUBIONA RIPARIA

Fig. 644. CLUBIONA PALLENS

Fig. 645. CLUBIONA CANADENSIS

is a median, longitudinal stripe, which sometimes is only faintly indicated. The females of this and the following species resemble each other closely. The males can be distinguished by the form of the apophysis borne by the tibia of the palpus. In this species this apophysis is of the form shown in Fig. 646.

Fig. 646.
OUTER FACE OF PALPUS
OF CLUBIONA OBESA

Fig. 647.
OUTER FACE OF PALPUS
OF CLUBIONA TIBIALIS

Clubiona tibialis (C. tib-i-a′lis).— This species is of the same size and colour as the preceding. In the case of the females, the differences between the two species have not been determined. The male of this species can be distinguished by the form of the apophysis of the tibia of the palpus (Fig. 647).

Genus CHIRACANTHIUM (Chir-a-can′thi-um)

The cephalothorax is somewhat convex and lacks the median furrow of the thorax. The posterior median eyes are farther from the lateral eyes than from each other. The lower margin of the furrow of the chelicerae is armed with two or three teeth, which are situated some distance from the base of the claw. The anterior legs are longer than the posterior; and the first two pairs of legs are armed with but few and not paired spines.

In colouration these spiders resemble *Clubiona*, and have similar habits, living in silken tubes. Although this is a large genus, only one species has been found in the United States.

Chiracanthium inclusum (C. in-clu′sum).—This species measures about one third inch in length. The colour in life is greenish white, with the chelicerae brown. The cephalothorax

Superfamily Argiopoidea

is a little darker than the abdomen. There is on the abdomen a faintly indicated, median, longitudinal band.

The guide of the epigynum is a large nearly circular plate which nearly fills the cavity of the epigynum (Fig. 648). On the outer face of the palpus of the male there is a slender, curved apophysis extending from the tip of the tibia toward the cymbium, and a slightly larger pointed apophysis extending backward from the cymbium (Fig. 649).

This is a common and very widely distributed species, occurring both in the

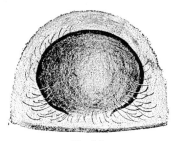

Fig. 648.
EPIGYNUM OF CHIRACANTHIUM INCLUSUM

Fig. 649. OUTER FACE OF TIBIA OF CHIRACANTHIUM INCLUSUM

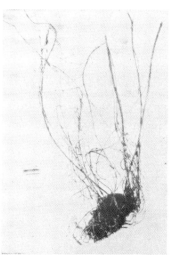

Fig. 650. NEST OF CHIRACANTHIUM INCLUSUM

North and in the South. Hentz, who first described it, states that "This spider was always found in tubes of white silk, the female watching her cocoon, which is covered with a very thin coat of silk; the eggs are loose and not glued together. It probably moves out only at night, as its pale colour indicates. The young are deeper in colour even than the mother."

The nest of this spider is shown in Fig. 650.

Superfamily Argiopoidea

Subfamily LIOCRANINÆ (Li-o-cra-ni'næ)

The tibiæ and metatarsi of the first two pairs of legs are armed below with a definite double series of long spines. The last segment of the hind spinnerets is distinct and conical. In this series the labium is not at all or barely longer than wide, and does not extend beyond the middle of the endites. The endites are not narrower in the middle than at the end.

The following key to our genera does not include *Scotinella*, based on a small spider from North Carolina.

TABLE OF GENERA OF THE LIOCRANINÆ

A. First tibiæ with at most three pairs of ventral spines.
  B. Claws of all legs accompanied by large bundles of tenent hairs (claw tufts).
    C. Posterior row of eyes straight. P. 586. SYRISCA
    CC. Posterior row of eyes recurved. P. 586. SYSPIRA
  BB. Posterior claws without claw tufts; posterior row of eyes procurved. P. 588. AGRŒCA
AA. First tibiæ with four or more pairs of ventral spines.
  B. Sternum prolonged behind between the posterior coxæ.
    C. Posterior row of eyes recurved.
      D. Lower margin of chelicera with one tooth; tarsi with claw tufts. P. 587. GOSIPHRURUS
      DD. Lower margin of chelicera with two teeth; tarsi with two long spatulate hairs. P. 588. APOSTENUS
    CC. Posterior row of eyes procurved, rarely straight.
      D. Anterior median eyes considerably larger than the lateral. P. 590. PIABUNA
      DD. Anterior median eyes usually smaller, but often as large as the lateral.
        E. Carapace pale yellow to brown, with a black marginal seam and dark median stripes or spots; eye rows subequal in width; tibia of palpus with a single spur. P. 589. PHRUROTIMPUS
        EE. Carapace uniformly coloured, without conspicuous contrasting markings; first eye row distinctly narrower than second; tibia of palpus with a double apophysis. P. 590. PHRUROLITHUS

Superfamily Argiopoidea

BB. Sternum not prolonged behind between the posterior coxæ which are contiguous or nearly so.
    C. Anterior median eyes very much larger than the lateral eyes. P. 587.   ZORA
    CC. Anterior median eyes at most as large as the lateral.
        D. Lower margin of the furrow of the chelicera with three teeth. P. 591.   LIOCRANOIDES
    DD. Lower margin with two teeth. P. 591.
                                    DRASSYLOCHEMMIS

Genus SYRISCA (Sy-ris'ca)

In the males the tibiæ of the first legs are armed below with two or three pairs of spines. In the females the tibiæ are usually unspined beneath. The posterior row of eyes is essentially straight and the eyes are well separated. The terminal joint of the posterior spinnerets is as long or even longer than the basal. The front tarsi and metatarsi are heavily scopulate.

Two or three species occur in the United States.

Syrisca affinis (S. af-fin'is).—This is the *Teminius affinis* of Banks. Females measure one half inch in length. The whole animal is dark brown, the abdomen with a median longitudinal pale stripe which is outlined in black and bordered by pale spots.

This species is common in Texas.

Genus SYSPIRA (Sy-spi'ra)

The tibiæ and metatarsi of the front legs are armed below with a few weak spines. The apical segment of the hind spinnerets is moderately long, subacuminate. The posterior eye row is recurved, and the lateral eyes of each side are close together. The anterior median eyes are variable in size, and may be smaller than, equal to, or exceed the anterior lateral eyes in size.

Several species occur in the southwestern states.

Syspira eclectica (S. ec-lec'ti-ca).—The carapace is dusky yellow, with a clearer marginal border on each side and a faintly defined median pale stripe. The legs are yellow with brown rings or broken rings. The abdomen is dusky gray, paler beneath. The anterior median eyes are much larger than the anterior lateral eyes. Females average one half inch in length.

This species is found in Arizona and California.

Superfamily Argiopoidea

Genus ZORA (Zo'ra)

The tibiæ and metatarsi of the anterior legs are armed below with many very long spines and a few elevated ones; the apical segment of the hind spinnerets is short and indistinct; the posterior row of eyes is strongly recurved.

Zora includes small spiders with a yellow or whitish integument ornamented with longitudinal bands or with small spots usually in series; their legs are thick, the two first pairs are often in part black or brown. They live in moss and detritus; they only rarely climb on plants, and they spin neither a tube nor a web. (Simon.)

Only one species is reported from this country.

Zora pumilus (Z. pu'mi-lus).— I have not seen this species and can only copy the description and figure given by Hentz, who first described it; no later description of it has been published. In that part of the figure (Fig. 651) showing the arrangement of the eyes, the posterior row is the upper one:

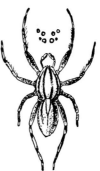

Fig. 651.
ZORA PUMILUS
(after Hentz)

"Livid, testaceous; cephalothorax with two longitudinal bands near the middle, and two curved fillets near the edge, fuscous; abdomen with a line bifurcated anteriorly on the middle, and two lines of minute dots on the sides, fuscous; same colour underneath, with minute fuscous dots."

This is the *Katadysas pumilus* of Hentz. It was found by him under stones in North Alabama.

Genus GOSIPHRURUS (Go-si-phru'rus)

The general structure is similar to that of *Apostenus*. The posterior row of eyes is distinctly recurved and the eyes are subequal. The anterior median eyes are only slightly smaller than the lateral eyes. The lower margin of the chelicera is armed with two small teeth. The sternum broadly separates the posterior coxæ. The front tibiæ are armed with five or six pairs of long, movable spines. There is a thin, chitinous plate on the abdomen of the males at the base.

Superfamily Argiopoidea

The following species is representative of the genus.

Gosiphrurus scleratus (G. scle-ra'tus).—The cephalothorax and legs are reddish brown, and the abdomen is dusky gray. The male measures one sixth inch in length. The palpus is relatively simple and lacks a median apophysis. The tibia is without a conspicuous lateral spur. The species occurs in California.

Genus APOSTENUS (A-pos'te-nus)

In this and in the three following genera the claws of the posterior tarsi are not covered with bundles of spatulate or truncate hairs; but in this genus there are two long, spatulate hairs beneath the claws. The posterior row of eyes is slightly recurved; and the anterior median eyes are smaller than the anterior lateral eyes.

Apostenus cinctipes (A. cinc'ti-pes).—This species, described from a single female taken at Olympia, Washington, has been placed, with considerable doubt, in the present genus. Several species of *Apostenus* are known from Europe. They resemble our species of *Agrœca* but have the posterior eye row recurved and the front tibiæ set with more numerous spines.

Genus AGRŒCA (A-grœ'ca)

This genus is closely allied to *Apostenus* but differs in lacking the two spatulate hairs beneath the claws of the hind tarsi and in having both rows of eyes procurved. The eyes of the anterior row are close together; those of the second row are more widely spaced.

Six species have been described from our fauna. The following is widely distributed across the northern part of our country and occurs far south in our mountain ranges.

Agrœca pratensis (A. pra-ten'sis).—This spider is about one fourth inch in length. The cephalothorax is widest and highest behind the middle, the head a little more than half as wide as the thorax; the abdomen is widest across the hinder third and not much pointed behind. The cephalothorax is light brownish yellow; it has a narrow dark edge on each side and a row of radiating dark lines each side forming two broken dark longitudinal bands. The abdomen has two rows of gray oblique markings on a light ground.

Superfamily Argiopoidea

Genus PHRUROTIMPUS (Phru-ro-tim'pus)

The sternum is large and prolonged behind between the posterior coxæ; the anterior row of eyes is procurved; all of the tarsi lack scopulæ, but are furnished under the claws with very small bundles of from six to ten spatulate hairs; the tibia and metatarsus of the first two pairs of legs are armed, on the underside, with a double row of strong spines.

These are small spiders, varying from one twelfth to one fifth inch in length, which live under stones and among moss or short grass, and are very active. They are sometimes ornamented with bright markings and iridescent scales.

A dozen species are known from the United States. The apophysis of the male palpus is single. The

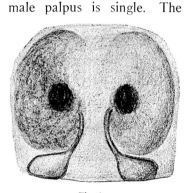
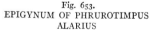

Fig. 652. PHRUROTIMPUS ALARIUS

Fig. 653. EPIGYNUM OF PHRUROTIMPUS ALARIUS

Fig. 654. PALPUS OF PHRUROTIMPUS ALARIUS

males of the different species can be separated by the variations in form of the apophyses of the tibia of the palpus; the females, by the form of the epigynum. The following is our most common species.

Phrurotimpus alarius (P. a-la'ri-us).—The male measures one eighth inch in length; the female, nearly one sixth. The cephalothorax is light yellowish, with a narrow black marginal line, and two light gray stripes. The abdomen is gray, with several light transverse bands or chevrons, which vary in size and shape; it is clothed with iridescent scales. The first and second pairs of legs are stouter than the others. The tibiæ of the first pair are very conspicuous, being black except the tip, which is of a strongly contrasting white (Fig. 652). The form of the epigynum is shown by Fig. 653; and that of the tibial apophysis of the male palpus by Fig. 654.

The egg-sac of this species is often found attached to the lower surface of a stone or of a piece of wood lying on the ground. It is closely applied to the supporting object, flat, circular in outline, and bright red in colour.

Genus PHRUROLITHUS (Phru-rol'i-thus)

This genus is closely related to *Phrurotimpus*. The carapace is typically slightly higher and more convex, and the eyes of the first row are clearly narrower than the second row. The carapace is uniformly coloured, sometimes yellowish brown but ordinarily black, and lacks the pattern of contrasting stripes or spots which characterize *Phrurotimpus*. The tibia of the male palpus is armed with two distinct apophyses, one subventral in position and usually shorter, and a lateral spur or process which is often long and curved. The femur of the palpus has a distinct spur beneath which is set with stiff hairs.

The spiders of this genus are small and live under stones, under forest debris, or in moss. Some of them are found in ants' nests. More than a score of species occur in our fauna. The following one is widely distributed and typical for the group.

Phrurolithus formica (P. for'mi-ca).—Both sexes measure about one sixth inch in length. The carapace is reddish brown, almost unicolorous, but sometimes with indistinct darker markings. The legs are brown with indistinct darker markings. The abdomen is black and the whole dorsum is covered with a shining, chitinous plate.

This species occurs in the eastern United States.

Genus PIABUNA (Pi-a-bu'na)

The few known species of this genus are closely allied to *Phrurolithus*. They are very small spiders, pale yellow or white in colour. The anterior median eyes are considerably larger than the anterior lateral eyes. The front legs have a series of long,

Superfamily Argiopoidea

movable spines beneath the tibiæ and metatarsi. The tibia of the male palpus is armed with two apophyses as in *Phrurolithus*, but the spur on the ventral side of the femur is lacking.

Several species of this genus are known from the southwestern portion of the United States and adjacent Mexico. The following species is representative of the group.

Piabuna longispina (P. lon-gi-spi'na).—This small species, which measures about one twelfth of an inch in length, is almost entirely pale yellow in colour. The male has a thin, chitinous shield on the abdomen. These spiders are found under stones in dry situations. The present species occurs in New Mexico.

Genus LIOCRANOIDES (Li-o-cra-noi'des)

Both rows of eyes are recurved, and all the eyes are nearly equal in size, those of the posterior row nearly equidistant from each other. The tarsi are all lightly scopulate beneath, and all the claws are accompanied by bundles of tenent hairs (claw tufts).

Three species are known from the United States, one from the east and two other from California.

Liocranoides unicolor (L. u-ni-co'lor).—This species was described on the basis of a young example from the Mammoth Cave of Kentucky. Mature females measure nearly one half inch in length. The cephalothorax and legs are yellowish or rusty brown in colour, and the abdomen is gray. These spiders live under stones or debris in dark situations in the forests or mountains. They also occur in caves.

This species occurs in North Carolina, Tennessee, Kentucky, and Alabama.

Genus DRASSYLLOCHEMMIS (Dras-syl-lo-chem'mis)

A single species of this genus, *Drassyllochemmis captiosus*, has been described recently. It is of somewhat uncertain position because of the widely separated anterior spinnerets which ally it with the Gnaphosidæ. The anterior tibiæ are armed below with six pairs of spines. The eyes of the first row are in a straight line, and the anterior median are considerably smaller than the lateral. The lower margin of the chelicera bears two teeth.

This small, pale spider is found in Texas.

Superfamily Argiopoidea

Subfamily MICARIINÆ (Mi-ca-ri-i'næ)

In this and the following subfamily the last segment of the hind spinnerets is very short, frequently indistinct, and subspherical. In this subfamily the apex of the endites is subquadrate, forming an angle on the outside, and the sternum is not at all or barely margined. The front tarsi are long.

Four genera are represented in our fauna.

TABLE OF GENERA OF THE MICARIINÆ

A. Abdomen with a chitinous pedicel armed behind with two stout, erect spines. P. 592. MAZAX
AA. Abdomen usually not pedunculate, if so, without two stout spines near the base.
  B. Posterior row of eyes straight or lightly procurved. P. 592. MYRMECOTYPUS
  BB. Posterior row of eyes at least lightly procurved.
    C. Lower margin of the furrow of the chelicera with two teeth. P. 592. CASTIANEIRA
    CC. Lower margin with one tooth. P. 595. MICARIA

A single representative of *Mazax, M. spinosa,* is not uncommon in southern Texas. The species is also widely distributed in Mexico and Central America. *Myrmecotypus cubanus,* our only species of the genus, has been taken on one or two occasions in southern Florida and in Texas. It was described from Cuba.

Our better known genera are more fully discussed below.

Genus CASTIANEIRA (Cas-ti-a-nei'ra)

In this genus the cephalothorax is ovate, quite convex, and provided with a well-marked median furrow. The lower margin of the furrow of the chelicerae is armed with two small teeth. The body is clothed with plumose hairs, more or less scattered, and often brightly coloured. The tibia of the anterior legs is armed beneath with two or three pairs of spines.

These spiders are of medium size, brown or black in colour, with the abdomen ringed or otherwise marked with white or some other bright colour. As these markings are often due to the colour

Superfamily Argiopoidea

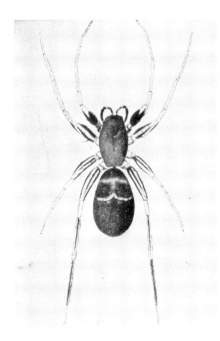

Fig. 655. CASTIANEIRA CINGULATA

Fig. 656. CASTIANEIRA DESCRIPTA

of hairs, which are easily rubbed off, it is frequently difficult to determine the species by their markings. About a score of species occur in our fauna, of which the following are the more common:

Castianeira cingulata (C. cin-gu-la′ta). — This species is easily distinguished by the fact that the femora of all of the legs are striped with black. The male is about one fourth inch in length; the female, one third. The body is dark brown in colour, with two white transverse bands on the abdomen; one of these is near the middle; and the other is between this and the base of the abdomen (Fig. 655).

This very active spider lives under stones and leaves on the ground.

Castianeira descripta (C. des-crip′ta). — The male measures nearly one third inch in length; the female, from one third to two fifths inch. The body is black, with or without red markings on the abdomen. In well-marked individuals there is a basal spot on the abdomen, one near the middle, and a

Superfamily Argiopoidea

series of spots or a band near the tip; but frequently some or all of these spots are wanting (Fig. 656).

This spider is common under stones in pastures. Its egg-sac (Fig. 657) is often seen attached to stones in pastures.

Fig. 657. EGG-SACS OF CASTIANEIRA DESCRIPTA

It is parchment-like and has a metallic lustre. A spider which I had in confinement made its egg-sac August 5th. And egg-sacs which I opened September 19th were filled with spiderlings.

Castianeira longipalpus (C. lon-gi-pal'pus).— The body of the male is one fourth inch or more in length; that of the female about one third inch. The cephalothorax is light brown; the abdomen is dark with several transverse white stripes (Fig. 658). The number of these stripes varies in different individuals. There

Fig. 658.
CASTIANEIRA LONGIPALPULS, MALE

are usually some near the tip of the abdomen, not shown in the figure.

In the banding of the abdomen this species bears more or less resemblance to *Castianeira cingulata;* but the two species are easily separated by the fact that in this species the legs are banded with black, while in *C. cingulata* they are striped.

In habits this species resembles the two described above. Figure 658 is from a photograph of an individual that had lost when young two of its legs, the left front and the right hind one, and these legs were being reproduced.

Genus MICARIA (Mi-ca'ri-a)

The members of this genus are small slender spiders, in which the median furrow of the thorax is wanting or faintly indicated, and in which the lower margin of the furrow of the cheliceræ is armed with only one small tooth. The endites are slightly depressed in the middle. The abdomen, and usually the cephalothorax also, is covered with flattened scales, which are sometimes brightly coloured and iridescent. The anterior legs are usually not armed with spines.

These are exceedingly active spiders, which are found in dry and sandy places, where they hunt their prey, even in the hottest part of the day. Thirteen species have been described from our fauna, of which the following is the most common.

Micaria aurata (M. au-ra'ta).—This species measures from one fifth to one fourth inch in length. It is light yellow-brown in colour, with gray hairs and scales, which have, on the abdomen, green and red metallic reflections. The abdomen is slightly constricted in the middle of its length, and opposite this constriction there is on each side a white bar; these give the spider a more or less ant-like appearance. At the front end of the abdomen there is another pair of less distinct white bars.

Subfamily CORINNINÆ (Cor-in-ni'næ)

In this and in the preceding subfamily the last segment of the hind spinnerets is very short, frequently indistinct, and subspherical. In this subfamily the median furrow of the thorax is distinct; the apex of the endites is rounded, not at all

Superfamily Argiopoidea

angulate on the outside (Fig. 659), and not depressed in the middle; the tarsi of the anterior legs are not unusually long; and the sternum is distinctly margined.

Only two genera are represented in our fauna.

TABLE OF GENERA OF THE CORINNINÆ

A. Posterior row of eyes strongly recurved. P. 596. TRACHELAS
AA. Posterior row of eyes straight. P. 597. MERIOLA

Genus TRACHELAS (Tra-che′las)

In this genus the legs are not armed with spines or with only a few under the tibia of the first pair, and the posterior row of eyes is strongly recurved. Six species have been found

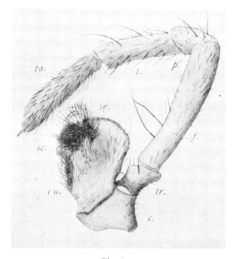

Fig. 659.
PEDIPALP OF TRACHELAS

Fig. 660.
TRACHELAS TRANQUILLA, FEMALE

in our fauna. They are all very closely related in their structural characters and are difficult to distinguish. The following one is common in the eastern United States.

Trachelas tranquilla (T. tran-quil′la).—The male measures one fifth inch or more in length; the female about two fifths inch. The cephalothorax and chelicerae are dark brown in colour; the abdomen is light yellow without markings, except that the four muscle impressions are brown, and there is a darkish median stripe on the basal half (Fig. 660). The first and second pairs of legs are stouter, and usually darker than the other two pairs.

This common species is found in dry and warm places, at the base of plants or under stones or leaves, sometimes on fences.

Genus MERIOLA (Me-ri'o-la)

In this genus as in *Trachelas* the legs are not armed with spines or bear a few under the tibia of the first pair; but this genus is distinguished from *Trachelas* by having the posterior row of eyes straight.

Only two species have been described; one of them, *M. californica*, is from California; the following one is widespread.

Meriola decepta (M. de-cep'ta).— This species appears like a miniature of *Trachelas tranquilla*, resembling that species in form, colour, and the proportions of its legs; the adult female, however, is only one sixth inch in length; the male has not been described.

Mr. Banks, who described the species, found it on the ground in an old meadow; the specimens in my collection were obtained by sifting.

Family AGELENIDÆ (Ag-e-len'i-dæ)

The Funnel-web Spiders

These are three-clawed, almost always eight-eyed, sedentary spiders. They differ from the Clubionidæ in the number of tarsal claws and in lacking scopulæ on the tarsi; and from the two following families in not having the trochanters notched. The eyes may be either silvery white or dark or both types may be present; usually the anterior median eyes are silvery white, the others dark; the eyes are in two rows. The hind spinnerets are very long (Fig. 661).

The members of this family spin sheet-like webs, which are usually furnished with a tubular retreat; this suggests the common name funnel-web spiders for the family. The web of the grass-spider, an exceedingly common species, is a good illustration (Fig. 662); this is described in detail on a later page.

To this family belongs the remarkable aquatic spider, *Argyroneta aquatica*, of Europe, the habits of which have been described

Superfamily Argiopoidea

by many writers. This spider lives among plants at the bottom of clear and quiet ponds; but it breathes air which it brings down from above the surface of the water adhering to its body.

Fig. 661. LATERAL ASPECT OF AGELENA

Fig. 662. WEB OF AGELENA NÆVIA

It constructs a dome of silk among the plants at the bottom of the pond or in a crevice of some kind or in an empty shell and fills it with air; this air-filled dome serves as a home for the spider, a nidus for its egg-sac, and a retreat for passing the winter.

This family is represented in our fauna by nearly a score of genera and more than one hundred species. Some of the genera, such as *Coras* and *Wadotes*, are typically eastern in distribution; others (for example, *Cybæus*, *Cybæota*, and *Chorizomma*) are for the most part restricted to the western part of our country. Because many of our species are still of doubtful generic position, in this book the older, conservative generic arrangement is followed.

The groups and genera may be separated by reference to the table which follows:

TABLE OF GROUPS AND GENERA OF THE AGELENIDÆ

- A. Fore spinnerets contiguous, preceded by a colulus; hind spinnerets one-jointed. P. 599.   Subfamily CYBÆINÆ
- AA. Fore spinnerets separated, the colulus lacking; hind spinnerets two-jointed. P. 600.   Subfamily AGELENINÆ
    - B. Labium longer than wide; posterior coxæ contiguous; hind spinnerets usually with apical segment at least as long as the basal segment. P. 601. Group AGELENEÆ
        - C. Posterior eye row very strongly procurved. P. 601.
              AGELENA
        - CC. Posterior eyes at most slightly procurved, often straight or lightly recurved.
            - D. Anterior median eyes much larger than the anterior lateral eyes. P. 608.   CORAS
            - DD. Anterior median eyes usually smaller, at most as large as the anterior lateral eyes.
                - E. Lower margin of the furrow of the chelicera with two teeth. P. 611.   WADOTES
                - EE. Lower margin with from four to ten teeth.
                    - F. Posterior eye row very lightly procurved. P. 607.   TEGENARIA
                    - FF. Posterior eye row lightly recurved. P. 607.
                          CALYMMARIA
    - BB. Labium as wide as or wider than long; posterior coxæ usually well separated; hind spinnerets with the apical segment much shorter than the basal segment. P. 611.
          Group CRYPHŒCEÆ

Subfamily CYBÆINÆ (Cy-bæ-i'næ)

The spiders of this series differ from other agelenids in having the fore spinnerets contiguous and preceded by a more or less evident colulus. The hind spinnerets consist each of a single segment, and are not longer than the fore spinnerets. The posterior eyes are essentially straight, at most lightly curved. The cheliceræ are very robust and, in the majority of the species at

Superfamily Argiopoidea

least, strongly convex at the base.

At least four genera of this subfamily are known from the United States. The species are restricted for the most part to the western states, where they occur most abundantly in mountainous and wooded areas. They spin irregular webs of the same type as those of *Amaurobius*. The spiders themselves are relatively inactive, semi-sedentary creatures and live under rocks, leaves, and debris on the ground.

In two of the genera, *Cybæina* and *Cybæota*, the front tibiæ are armed with five pairs of long spines, and the cheliceræ are not strongly convex in front. In *Cybæus* and *Cybæozyga* the spines are less numerous. Students are referred to three papers by Chamberlin and Ivie ('32, '33, and '37) for descriptions and figures of the genera and the numerous species.

The following species of *Cybæus* is the more familiar of the two known from the eastern United States.

Cybæus giganteus (C. gi-gan'te-us).—The length of the body is nearly one half inch. The cephalothorax is shining, dark red-brown. The abdomen is covered with short black hairs; the dorsum and sides are dark grayish black, with a short, median, basal, light stripe, not reaching to the middle of the dorsum, and on each side of this two oblique light spots.

This species has been found in New York and in North Carolina and Tennessee.

Subfamily AGELENINÆ (Ag-e-le-ni'næ)

To this subfamily belong all our agelenids except *Cybæus* and related genera. They are characterized by having the fore spinnerets separated and not preceded by a colulus. The hind spinnerets have an apical segment which is always distinct, often very long, and commonly exceeds the basal joint in length. Some of our most familiar spiders belong to this series, the species of which are of average or large size. Their webs are familiar objects in our woods and meadows, and some spin them in our houses.

Fig. 663. EYES OF AGELENA NÆVIA

This subfamily is divided into two groups as indicated in the table above.

Superfamily Argiopoidea

Group AGELENEÆ (Ag-e-le'ne-æ)

The labium is longer than broad. The posterior coxæ are contiguous or nearly so. The apical segment of the hind spinnerets is long, in most cases at least as long as the basal segment.

Genus AGELENA (Ag-e-le'na)

Both rows of eyes are strongly procurved, so much so that the posterior lateral and the anterior median form a nearly straight line (Fig. 663). The cephalothorax is narrowed in front. The apical segment of the hind spinnerets is very long in most species; in some from the west it is shorter than the basal segment. About thirty species have been described from our fauna.

The species of this genus are called grass-spiders. The following common species will serve as an example.

Agelena nævia (A. næ'vi-a).—This is the largest species of the genus, a full grown female often measuring an inch in length. Fig. 664 represents a well marked individual.

Fig. 664. AGELENA NÆVIA, FEMALE

The embolus of the male palpus is greatly elongated, and its form is used to separate the various species. Two of them are illustrated in Figs. 665 and 666.

These spiders are called grass-spiders because their webs are the most common ones found on grass. The several species are abundant throughout our country, and so closely allied that they were thought to represent only a single species. Few people realize what an immense number of webs these spiders spin upon the grass in the fields. But occasionally they are made visible in the early morning by the dew which has condensed upon them. At such

Superfamily Argiopoidea

times we may see the grass covered by an almost continuous carpet of silk (Fig. 667).

The grass-spider lives only one year and passes the winter in the egg state, the old spiders dying soon after oviposition in the autumn. It is not till late in May in the North that the small webs of the young spiders first become observable; and it is much later in the season before they reach their full size. If a spider is not disturbed, it occupies the same web throughout the summer, extending it from time to time until it becomes one foot or more across.

The webs of this spider vary greatly in form and in position; but the typical form is a nearly horizontal, slightly concave sheet, built near the surface of the ground in a grassy place

Fig. 665.
PALPUS OF MALE OF
AGELENA POTTERI

Fig. 666.
PALPUS OF MALE OF
AGELENA PENNSYLVANICA

(Fig. 662); the web is firmly attached to the grass, and there is an irregular open net-work of threads above the sheet supported by stalks of grass that extend above it. The object of this net-work is probably to impede the flight of insects, causing them to fall upon the sheet, where they can be seized by the spider. One side of the sheet is continued into a tubular retreat, which extends downward a greater or less distance, but it is open below so that the spider can escape, by a back door as it were, in an emergency.

(*Photographed by P. B. Mann*)
Fig. 667. A LAWN COVERED WITH WEBS OF THE GRASS SPIDER

Fig. 668. AN OLD WEB OF AGELENA NÆVIA

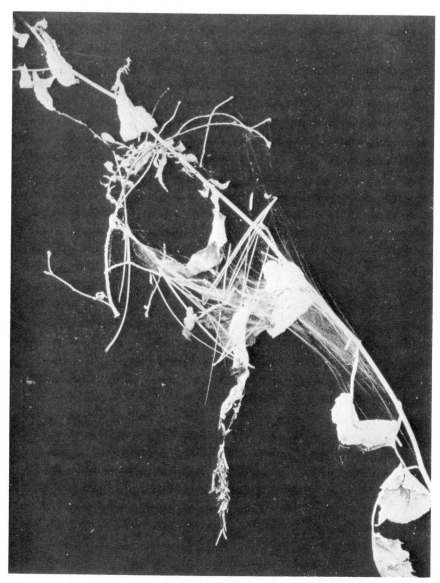

Fig. 669. WEB OF AGELENA NÆVIA ON A SINGLE PLANT

Superfamily Argiopoidea

The spider waits for its prey in its retreat; as soon as an insect alights or falls upon the sheet, the spider rushes forth to seize it.

(Photographed by Glenn W. Herrick)
Fig. 670. FUNNEL-LIKE WEB OF AGELENA NÆVIA

The continued use of a web for several months, the spider spinning a drag-line in each of its trips over it, results in a constant addition to the thickness of the sheet; but this addition is not

Superfamily Argiopoidea

made evenly, but in such a way as to produce a mesh-like structure (Fig. 668).

The agelenids differ from most web-building spiders in that they do not hang from their webs, but run upon them with the dorsal aspect of the body uppermost.

The grass-spider often makes its web in the angles of buildings and frequently in elevated structures. It is also often found upon shrubs or high herbs at a considerable distance from the ground; in such cases the typical form may be greatly modified

Fig. 671. EGG-SACS OF AGELENA NÆVIA

(Fig. 669). Another striking modification of the typical form is shown in Fig. 670; here the retreat was in a stump and the entire sheet was funnel-like.

In the autumn the males wander in search of their mates; and the females leave their webs to seek a suitable place for building their egg-sacs.

The egg-sacs are placed in secluded situations; they are often found beneath loose bark on trees and stumps. They are disk-like, closely applied to the supporting object, and are partly covered with bits of rotten wood or other debris (Fig. 671). It is evident that the female remains near the egg-sac after it is made; for one often finds dead individuals under bark with the egg-sacs.

Genus CALYMMARIA (Cal-ym-ma'ri-a)

The species of this genus were formerly included in *Tegenaria*. The anterior row of eyes is slightly procurved, the medians smaller than the lateral eyes, usually equal to about half their diameter. The posterior row of eyes is very lightly recurved and the eyes are subequal in size. The legs are proportionately much longer than in *Tegenaria*, the femora of all of them far exceeding the carapace in length. The legs are always marked with broad, dark rings. The palpi of the males are characteristic for the genus. One is illustrated in Fig. 672. The cymbium is abruptly narrowed near the distal end of the bulb and prolonged into an appendage as long as the bulb. The tibia is supplied with spurs which differ among the various species.

Fig. 672.
CORAS MEDICINALIS, FEMALE

Several species of this genus are known from the United States, most of them from the Pacific Coast. Only one, *Calymmaria cavicola*, occurs in the east. It was described from a cave in Indiana but also occurs in the mountains of Tennessee and Alabama in wooded areas.

These spiders spin their funnel webs in dark places in wild situations, and are rarely found associated with man as are many of the Tegenarias. They are not uncommon in caves, but the same species occur in comparable places in the open.

Genus TEGENARIA (Teg-e-na'ri-a)

The anterior row of eyes is slightly procurved or rarely straight; the eyes of this row are either equal in size or the median eyes are smaller than the lateral eyes. The clypeus is at least twice as wide as the anterior eyes. The cheliceræ are usually but slightly convex; and the lower margin of the furrow of the cheliceræ is usually armed with from four to six teeth, rarely with only three.

Five species have been described from the United States; three of these occur in the Far West; one, in caves in Indiana; and the following is widely distributed.

Tegenaria derhami (T. der-ha'mi).— This is the best-known representative of the genus. It is a domestic species which inhabits the dwellings of man in all regions of the world from the Arctic Zone to the Tropics. The female (Fig. 673) is two fifths inch in length. The cephalothorax is light yellowish brown with two gray longitudinal stripes. The abdomen is pale with many irregular gray spots, of which there is a series, more or less connected, forming a median band, and many along each side. The markings of the male are similar to those of the female, but the abdomen is smaller.

This species lives almost exclusively in cellars and neglected buildings. Its web (Fig. 674) resembles in its more general features that of the grass-spider; but as with that species varies in form depending on its situation; sometimes instead of its being a flat sheet, as shown in the figure, it is a deep, pocket-like sac.

Genus CORAS (Co'ras)

The rows of eyes are not strongly procurved; and the posterior median eyes are much larger than the anterior lateral. The cheliceræ are robust and strongly convex at the base. The lower margin of the furrow is armed with three teeth. The clypeus is scarcely wider than the anterior eyes. These spiders are closely related to *Tegenaria*, but their form is more thick-set and their legs are shorter. The cymbium of the male palpus does not bear accessory spurs. The female epigynum (Fig. 676) is a convex plate, usually depressed in the middle, which has a spur or tooth on each side. In *Wadotes* the epigynum bears a single elongated appendage at the middle.

These spiders live in hollow trees, in crevices among rocks, and in the angles of buildings. Several species are known from the eastern United States. The following are widely distributed.

Fig. 673. TEGENARIA DERHAMI, FEMALE

Fig 674. WEB OF TEGENARIA DERHAMI

Superfamily Argiopoidea

Coras medicinalis (C. me-dic-i-na′lis).—This is a gray spider measuring about one half inch in length. It is stout and comparatively shortlegged. The carapace is yellowish brown, darkest in front, marked with radiating gray lines which form two longitudinal dark bands on the thoracic part; on the head part there are two shorter bands which meet at the median furrow.

Fig. 675.
HEAD AND CHELICERÆ OF CORAS

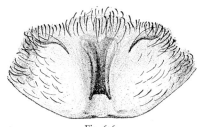

Fig. 676.
EPIGYNUM OF CORAS FIDELIS

Fig. 677.
CORAS FIDELIS, MALE

The web of this spider was formerly believed to be narcotic. Hentz, who first described the species and proposed the name *medicinalis* for it, states that "for some time the use of its web as a narcotic in cases of fever was recommended by many physicians in this country; but now it is probably seldom used."

This species is common in the eastern United States.

Coras fidelis (C. fi-de′lis).—The length of the body is a little less than one half inch. The cephalothorax is yellowish brown, with radiating dark bands. The abdomen is nearly white, marked with many dark spots and lines. The form of the epigynum of the female is shown by Fig. 676, and Fig. 677 represents the male.

Superfamily Argiopoidea

Genus WADOTES (Wa'do-tes)

The species of this genus may be distinguished readily from *Coras*, with which they agree closely in appearance and habits, in having the lower margin of the chelicera armed with only two teeth. The anterior median eyes are much smaller than the lateral eyes. The cymbium of the male palpus is ornamented with one or two spurs, and the epigynum of the female is always characterized by the presence of a median scape which is free and projects caudally over the plate.

Several species are known from the eastern part of our country. The following one is widely distributed.

Wadotes calcaratus (W. cal-ca-ra'tus).—The cephalothorax and legs are yellowish brown. The abdomen is dark gray with a median lighter stripe in front and a double row of pale markings behind. The cymbium of the male palpus has a single long process projecting behind.

Group CRYPHŒCEÆ (Cry-phœ'ce-æ)

In this series of species the labium is clearly broader than long or in some instances at most as long as broad. The posterior coxæ are ordinarily well separated. The terminal joint of the posterior spinnerets is conical, very much shorter than the basal segment. The eye rows are essentially straight, the eyes of the posterior row subequal in size and also subequal to the anterior lateral eyes. The anterior median eyes are subject to considerable variation in size. In *Cicurina* they are in most cases not decidedly smaller, rarely even larger, than the anterior lateral eyes. In the other genera they are often greatly reduced in size and may be completely obsolete. When they are wanting or rudimentary, the lateral eyes are correspondingly closer together, sometimes separated by little more than the radius. Some of the genera include species in which six and eight eyes are present. Most of the species are small or of average size.

Six or seven genera have been found in the United States, most of them from the far west. Several of the groups have been studied by Chamberlin and Ivie ('37). In *Chorizomma, Chorizommoides*, and some species of *Blabomma*, the anterior median eyes are completely obliterated. The genera are hard to define because of the

Superfamily Argiopoidea

complex intergradation and great variability of the characters ordinarily used in the family. The genitalia would seem to be the surest index of the position of each species.

The two genera found in the eastern part of our country are considered below.

Genus CICURINA (Cic-u-ri′na)

The characters are, in general, those of the group. The anterior median eyes are usually well developed and separate the lateral eyes by a distance exceeding the diameter of these eyes. The clypeus is as high or higher than the diameter of an anterior lateral eye. The male palpi of all the known species are very similar to that of *Cicurina robusta* (Fig. 679).

Fig. 678.
EPIGYNUM OF CICURINA ROBUSTA

Fig. 679.
PALPUS OF MALE OF CICURINA ROBUSTA

The species are small, our common ones measuring from one fifth to one fourth inch in length. They are most often found under dead leaves in woods. Simon states that they spin delicate horizontal webs under stones or in the midst of moss; and that the eggs are enclosed in a little white sac, which is covered with bits of earth. About twenty species have been described from our fauna. The following species is widespread in the United States, and is particularly abundant in the north.

Cicurina robusta (C. ro-bus′ta).—This common species measures from one fifth to one quarter of an inch in length. The colour is pale yellowish brown, lighter on the abdomen, which bears

Superfamily Argiopoidea

faint gray markings. The epigynum of the female is illustrated in Fig. 678, and the curious palpus of the male, with the long fine embolus, is shown in Fig. 679.

Genus CRYPHŒCA (Cry-phœ'ca)

The spiders of this genus agree with *Cicurina* in the form of the hind spinnerets, of the labium, and in the number of the eyes; but differ in that the clypeus is not wider than the anterior lateral eyes, and in having the anterior median eyes much smaller than the anterior lateral eyes (Fig. 680).

Only two species have been found in our fauna, one in New England, and one in the State of Washington

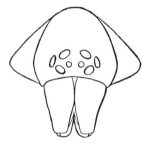

Fig. 680. FACE OF CRYPHŒCA MONTANA, FEMALE
(after Emerton)

Fig. 681. CRYPHŒCA MONTANA, FEMALE
(after Emerton)

Cryphœca montana (C. mon-ta'na).— This species has been described recently by Emerton ('09) from specimens taken on Mount Washington and elsewhere in New England. The male is one sixth inch in length, the female, one eighth. The colours are translucent white and gray; the cephalothorax has a narrow black edge and broken, radiating, dark marks; the abdomen is marked with a series of oblique light spots in pairs (Fig. 681).

Family HAHNIIDÆ (Hah-ni'i-dæ)

The Hahniids (Hah-ni'ids)

The members of this small but interesting family were formerly regarded as constituting only a subfamily of the Agelenidæ. They are sharply distinguished from the agelenids and other spiders by the arrangement of the spinnerets, which are in a single transverse row (Fig. 682). The longer hind spinnerets occupy the ends of

Superfamily Argiopoidea

the row and the fore spinnerets are between them and the middle spinnerets. The fore and hind spinnerets are two-jointed, the middle ones one-jointed. The posterior spiracle is large and is advanced far in front of the spinnerets.

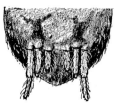

Fig. 682. SPINNERETS OF HAHNIA

These are small spiders which spin delicate sheet-webs near the surface of the ground; the webs are so delicate that they attract attention only when covered with dew; they are not furnished with a retreat. I have often found the webs stretched over slight depressions in the bare ground of country roads and of cultivated fields. In the early morning while the webs are still covered with dew the spiders are to be found under the webs. I infer that they move to the upper surface when the web is dry. The webs are also made among short and thin grass and moss. And the spiders are frequently found under stones and leaves.

Our three genera may be distinguished as follows:

A. Anterior median eyes smaller than the anterior lateral eyes. P. 614. HAHNIA
AA. Anterior median eyes as large or larger than the lateral.
  B. Spiracle midway between the genital furrow and the base of the spinnerets. P. 615. ANTISTEA
  BB. Spiracle twice as far from the spinnerets as from the genital furrow. P. 615. NEOANTISTEA

Genus HAHNIA (Hah'ni-a)

These are small spiders which have the anterior median eyes much smaller than the anterior lateral. The posterior spiracle is situated twice as far from the genital furrow as from the base of the median spinnerets. The labium is broader than long.

Only three species have so far been recorded from the United States. The following one is widespread and common.

Hahnia cinerea (H. ci-ne're-a).—This is a rather small species, measuring only about one twelfth inch in length. The cephalothorax is brownish with dark radiating markings. The abdomen is dark gray with scattered small white spots and a double row of light spots in the middle followed by chevrons.

Genus NEOANTISTEA (Ne-o-an-tis′te-a)

These spiders are probably larger than any other hahniids, some of the species being fully one third inch long. They are stout spiders, and the carapace is nearly as broad as long. The anterior median eyes are larger than the anterior lateral eyes. The posterior spiracle is situated twice as far from the median spinnerets as from the genital groove. The apical joint of the posterior spinnerets is as long as or longer than the basal.

Several species have been described of which the following one is representative.

Neoantistea agilis (N. ag′i-lis).—This species measures about one eighth inch in length. The cephalothorax is bright orange-brown, and the abdomen is light gray with many irregular pale spots. In the middle of the front half of the abdomen there are two orange-coloured spots and behind these there are several pairs of pale oblique spots.

This species is very common in the eastern United States.

Genus ANTISTEA (An-tis′te-a)

This genus resembles *Neoantistea* but the posterior spiracle is situated about midway between the spinnerets and the genital groove and the apical segment of the posterior spinnerets is much shorter than the basal segment. The carapace is nearly as broad as long, and the anterior median eyes are somewhat larger than the anterior lateral eyes.

A single species is known from the northeastern part of the United States and from Canada.

Antistea brunnea (A. brun′ne-a).—The female is one sixth of an inch long. The carapace is light brown, and the abdomen is gray with a dorsal pattern of dusky chevrons. The tibia of the male palpus is armed with a stout, curved spur.

Family PISAURIDÆ (Pi-sau′ri-dæ)

The Nursery-web Weavers

No more striking instance of maternal devotion is to be found among spiders than that exhibited by the nursery-web

Superfamily Argiopoidea

weavers. From the time the egg-sac is made until the spiderlings are ready to emerge, the mother carries about with her, wherever she goes, this great silken ball with its load of eggs or of young. The difficulty of doing this can be seen by a glance at the figure of *Pisaurina mira* and her egg-sac (Fig. 683). The egg-sac is held under the body; and is so large that the mother is forced to run on the tips of her tarsi in order to hold the load clear of

Fig. 683. PISAURINA MIRA AND HER EGG-SAC

obstructions. The specimen figured was resting on my table at the time the picture was taken; but the egg-sac is held free from the ground when the spider runs.

Just before the young are ready to emerge from the egg-sac, or just after they begin to do so, the mother fastens it among leaves at the top of some herbaceous plant or at the end of a branch of a shrub, and builds a nursery about it by fastening the leaves together with a net-work of threads (Fig. 684). She then remains on the outside of the nursery guarding the young.

Sometimes, as is frequently the case with *Dolomedes fontanus*, the nursery is made in an angle between stones and consists only of threads. While this habit of building a nursery is not universal in the family, it is sufficiently common to warrant the use of the term nursery-web weavers as a popular name for the family; they are also termed the pisaurids.

The members of the Pisauridæ are characterized by the presence of a semicircular and bordered notch in the apical margin of the lower side of the trochanters of the legs (Fig. 685). In this respect they differ from the preceding family, the Agelenidæ, but resemble the following family, the Lycosidæ, in which the trochanters are notched in a similar manner.

The pisaurids differ from the lycosids in that the tibia of the pedipalp of the male is furnished with an external apophysis

Fig. 684. PISAURINA MIRA ON HER NURSERY; THE EGG-SAC IS COVERED BY THE FOLDED LEAVES

Superfamily Argiopoidea

(Fig. 686, *ex. a*); the cuticle is almost always furnished with appressed plumose hairs; and in all forms in our fauna the two pieces of the lorum of the pedicle are either united by a transverse suture (Fig. 687), or the anterior piece is furnished with a notch behind into which a projection of the posterior piece fits (Fig. 688).

Fig. 685. TROCHANTER OF A HIND LEG OF DOLOMEDES

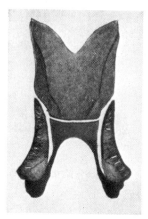

Fig. 687.
LORUM OF PISAURINA

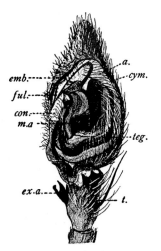

Fig. 686. PALPUS OF THE MALE OF DOLOMEDES FONTANUS

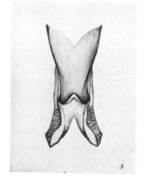

Fig. 688.
LORUM OF DOLOMEDES

The palpus of the male of *Dolomedes fontanus* (Fig. 686) is described in detail on page 118.

If only the species of our fauna be considered, a striking difference in habits between the Lycosidæ and the Pisauridæ is found. The females of the Pisauridæ make an egg-sac composed of a single piece, and carry it under the body, holding it by the cheliceræ (Fig. 683); those of the Lycosidæ make one composed of two valves, and drag it after them, attached to the spinnerets.

Our pisaurids are all wandering spiders which stalk their prey, and make webs only for nurseries for their young. Certain exotic species, however, spin permanent webs, like those of *Tegenaria;* hence this family holds an intermediate position in habits as well as in structure between the Agelenidæ and the Lycosidæ.

I have made careful observations on the method of carrying the egg-sac. A specimen of *Dolomedes urinator* had an egg-sac beneath its body which was so large that the abdomen projected but a short distance beyond it. The tips of the claws of the chelicerae were inserted in the egg-sac; the palpi extended over it in front, the tips of the palpi being closely applied to the lower surface; and the abdomen was closely applied to it behind. The sac was of a dirty brown colour, and was attached by a dragline of clear white silk to the spinnerets. There was an attachment disk on the egg-sac a short distance from the end of the spinnerets and from this disk eight threads extended to the spinnerets, two to each of the fore and middle spinnerets. Similar observations were made on a specimen of *Dolomedes sexpunctatus*. It is obvious that the attachment of the egg-sac to the spinnerets aids the spider in holding it off the ground when she runs.

As to the number of eggs laid by these spiders, one of my students, Miss E. L. Whittaker, reports that during the summer of 1903 she counted the contents of 22 egg-sacs of *Dolomedes*. The average sac was of the size of a small red cherry. The number of eggs in each varied from 236 to 412; but in 18 of the 22 sacs the range of variation was from 275 to 312.

The egg-sac of *Dolomedes fontanus*, and perhaps also those of other species, become mottled in colour a short time before the spiderlings emerge. This appears to be due to a stretching and cracking of the outer layer. Shortly before the opening of her egg-sac a female was observed pulling at the outer casing with her chelicerae, but was not seen to open the sac.

Several females of this species, each with an egg-sac, were kept in confinement. None of these made a nursery until after

Superfamily Argiopoidea

the spiderlings began to emerge; but did so immediately afterward.

It is evident that the female protects her young. When a pencil was inserted into a cage some distance from a female on her nursery she ran toward it and clutched it fiercely, repeating the performance whenever the pencil was put near her.

The adults were not observed to feed the young; but young ones were frequently seen to feed upon other members of the brood.

After the spiderlings had moulted, they left the nursery and migrated. The period that they remained in the nursery varied from three to six days.

Deserted nurseries of *Dolomedes* are common in late summer. Figure 689 represents one of these in which the empty egg-sac can be seen.

Most of our common species of this family belong to the genus *Dolomedes*. Several of these are of large size and often attract attention. Some of them may be found in the dark in elevated dry places; in most cases, however, their favourite haunts are near water, and especially in marshy places. They not only frequent the banks, where they run over the plants growing there, but they also freely run over the surface of the water in pursuit of their prey, and, when frightened, they dive beneath the surface, and hide under floating leaves or other objects. An English species, *Dolomedes fimbriatus*, is said to construct a raft by lashing together floating leaves. "This raft is utilized as a point of departure for raids upon water insects, and as a 'lunch room' in which captured prey are fed upon. It floats upon the fens of England, apparently at the sport of the wind" (McCook).

The usual food of *Dolomedes* is insects, but there is one well-authenticated instance where a spider of this genus measuring three fourths of an inch in length and weighing fourteen grains captured a fish three and one fourth inches long and weighing sixty-six grains. The spider fastened upon the fish with a deadly grip just on the forward side of the dorsal fin, and clung to it till the fish was exhausted and then dragged it out of the water. The observation was made by Mr. Edward A. Spring of Eagleswood, N. J., and is described at length with a figure in Doctor McCook's "American Spiders" (Vol. I. pp. 235–237).

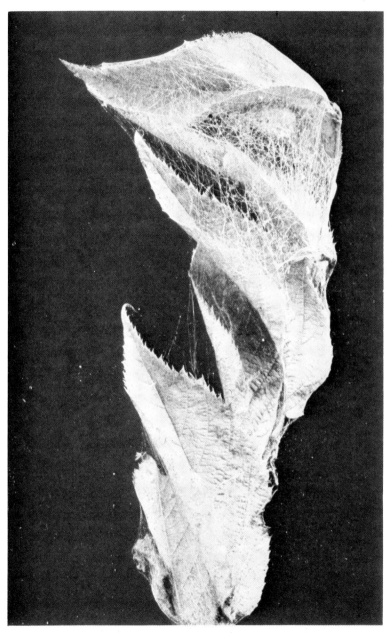

Fig. 689. DESERTED NURSERY OF DOLOMEDES

Superfamily Argiopoidea

This family has been the subject of a splendid paper by Bishop ('24) in which all of our known species have been fully described and figured. Advanced students are referred to that publication for fuller details regarding the genera and species. The genus *Melocosa,* based on a species described as *Lycosa fumosa* by Emerton, has been referred to the Pisauridæ by Gertsch ('37) but it is not discussed here. It may be noted at this time that a few representatives of the genus *Trechalea,* a genus of large pisaurids in which the tarsi are elongated and flexible, have been found recently in the mountains of southern Arizona.

TABLE OF GENERA OF THE PISAURIDÆ

A. With only two eyes in the anterior row.
   B. Median ocular quadrangle (second and third rows) much wider than long. P. 622. THANATIDIUS
  BB. Median ocular quadrangle (second and third rows) at most as broad as long. P. 624. PELOPATIS
AA. With four eyes in the anterior row.
   B. Lower margin of the furrow of the chelicera with four teeth. P. 625. DOLOMEDES
  BB. Lower margin with only three teeth.
    C. Median ocular quadrangle much wider behind than in front. P. 631. TINUS
   CC. Median ocular quadrangle as wide behind as in front or nearly so. P. 625. PISAURINA

Genus THANATIDIUS (Than-a-tid'i-us)

These are elongate spiders which have a considerable resemblance to the crab-spiders of the genus *Tibellus,* in fact Hentz placed them doubtfully in the genus *Thomisus.* The cephalothorax is much broader than long, considerably narrowed in front. The ocular area is correspondingly elongated, about as long as broad, the curvature of the two rows being so great that the eyes are actually in four rows of two eyes each. In this genus the median ocular quadrangle (second and third rows) is broader than long, and as wide behind as the first row (anterior lateral eyes). The lower margin of the chelicera is armed with three teeth. The abdomen is about three times as long as broad.

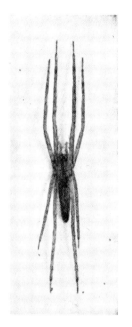

Fig. 690
PELOPATIS UNDULATA

Fig. 692.
PISAURINA MIRA

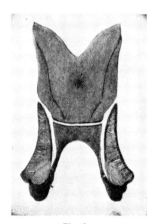

Fig. 691.
LORUM OF PISAURINA

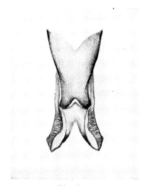

Fig. 693.
LORUM OF DOLOMEDES

Superfamily Argiopoidea

Two species of this genus occur in the southern States.

Thanatidius dubius (T. du'bi-us).—The cephalothorax is pale yellow with a broad median light brown band throughout its length which encloses the eyes in front. The abdomen is pale yellow and is marked by a broad median dark brown band the full length. The legs are pale in colour and very long, in the female more than twice as long as the body. The fully grown female is half an inch long and the male is nearly as large. The tibia of the male palpus bears a short process near the middle of the joint above which there is a shallow excavation. The bulb is relatively simple, and the embolus is evenly curved.

These spiders climb actively over shrubs and low plants near water. They are often taken by sweeping. The species is widely distributed but not abundant.

Thanatidius tenuis (T. ten'u-is).—This species is much rarer than the preceding one. It differs from *dubius* chiefly in the form of the epigynum. It is the *Maypacius floridanus* of Simon, but was previously described as *Thomisus tenuis* by Hentz.

Genus PELOPATIS (Pe-lo'pa-tis)

The species of this genus are similar in general appearance to *Thanatidius* and are easily confused with them. The carapace is much longer than broad, and the eyes are placed in four rows. The median ocular quadrangle (second and third rows) is about as long as broad, only moderately narrowed in front, the posterior median eyes (third row) very much narrower than the first row (anterior lateral eyes). The chelicera is armed with three teeth on the lower margin of the furrow. The abdomen and the legs are very long.

A single species is known from the United States. It occurs commonly in our southern states from Florida to Texas and is much more abundant than the species of *Thanatidius*.

Pelopatis undulata (P. un-du-la'ta).—The cephalothorax is dull yellow and is marked for its full length by a broad dusky band which is divided by a darker line. The legs are light yellowish brown. The abdomen is yellowish above with a dusky longitudinal, continuous with the one on the carapace, which narrows caudally, the edges gently scalloped behind. The shape of the body is well shown in Fig. 690.

Superfamily Argiopoidea

Genus PISAURINA (Pis-au-ri'na)

In this and in the following genus there are four eyes in the anterior row; this genus differs from the following in that the median ocular area is a little longer than wide, the lower margin of the furrow of the cheliceræ is armed with only three teeth, and the two pieces of the lorum of the pedicle are united by a transverse suture (Fig. 691). Only two or three species have been found in the United States; the following is the one best known.

Pisaurina mira (P. mi'ra).— This is an extremely variable species in colour and in size. Full-grown specimens measure about one half inch in length. In the more common type the body is light brownish yellow, with a wide darker and browner band on the middle of both cephalothorax and abdomen; on the cephalothorax the edges of the band are nearly straight, but on the abdomen they are undulating. The band is bordered on each side by a white line (Fig. 692).

This is a common species throughout the eastern part of the United States, and one that frequently attracts attention on account of its beauty. A figure of a female with her egg-sac is given on page 599; and a figure of one resting on her nursery on page 600.

This is the *Micrommata undata* of Hentz.

Genus DOLOMEDES (Dol-o-me'des)

As in *Pisaurina*, in the spiders of this genus there are four eyes in the anterior row; but here the median ocular area is as wide as or wider than long, the lower margin of the furrow of the cheliceræ is armed with four teeth, and the anterior piece of the lorum of the pedicle has a notch behind into which a projection of the posterior piece fits (Fig. 693).

The spiders of this genus frequently attract attention on account of their large size. They are most often observed near water, or in marshy places, but sometimes they are found in cellars or other dark and dry situations.

Notwithstanding their large size and frequent occurrence our species have not been well differentiated by writers on this group, and there is consequently much confusion in the nomen-

Superfamily Argiopoidea

clature of them. Eleven species are recognized by Bishop in his revision of the family, but several are known only from the female sex. Several common species are described below.

The Dark Dolomedes, *Dolomedes tenebrosus* (D. ten-e-bro'sus). — This common species is one of the largest members of the family; the specimen figured here (Fig. 694) is a female and measures seven eighths inch in length. The following description was made from a fresh specimen before it was placed in alcohol.

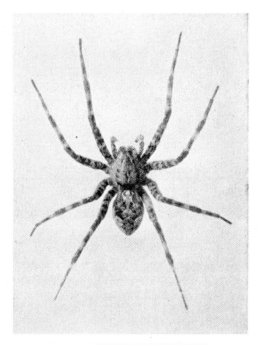

Fig. 694. DOLOMEDES TENEBROSUS

The colours are light brown, gray, and black. The cephalothorax is narrowly margined with black; within this narrow black margin there is a broad band in which the brown hairs predominate; this band extends on each side to the lateral angle of the clypeus, but is interrupted by a black stripe below the eyes. The lower part of this stripe is as wide as the clypeus, the upper part as wide as the first row of eyes is long. The head, back of the eyes is reddish. The central part of the thorax is dark. There is a light spot in the outer end of each radial furrow. There are three well-marked, black, chevrons on the posterior half of the abdomen, with a light brown spot at each end of each chevron. On the basal half of the abdomen there are three pairs of black points.

The epigynum of this species differs markedly from those of the other common species. The guide is very broad in front and does not extend to the hind part of the organ (Fig. 695). It bears a tongue-like projection behind, which in some specimens is strictly continuous with the wide part, as shown in the figure, but in other specimens, which are perhaps shrunken, it is more or less withdrawn into the body and therefore inconspicuous.

This is the *Dolomedes idoneus* of Montgomery.

The Northern Dolomedes, *Dolomedes scriptus* (D. scrip′tus). — This (Fig. 696) is one of the larger and more common species of *Dolomedes;* but it is a little smaller than *D. tenebrosus*. It differs from our other common species in having the sternum blackish with a yellow median band. The cephalothorax is brown with a narrow median light line extending its whole length; from the median side of each posterior lateral eye arises a yellow line which passes backward and is so curved that with its fellow of the opposite side it forms nearly a circle. On the posterior half of the abdomen there are four transverse W-shaped yellow bands.

The epigynum of the female is almost as long as wide; the guide is finger-like, and extends to the hind margin of the organ (Fig. 697). It resembles that of *D. urinator* in the apparent division of the guide.

The apophysis of the tibia of the male palpus terminates in two large teeth and a small one (Fig. 698).

The Whitish Dolomedes, *Dolomedes albineus* (D. al-bin′e-us). — This is a large species closely allied to *D. tenebrosus*. The female is easily recognized by a yellowish longitudinal band edged with black on the ventral aspect of the abdomen; the clypeus bears a white band with black below; and the head is much elevated. I have not seen the male.

This is a Southern species. Hentz states that it does not dwell habitually in caves and cellars, but is usually found on the trunks of trees, yet in dark, shady places.

The Diving Dolomedes, *Dolomedes urinator* (D. u-ri-na′tor). — The female of this species can be distinguished from our

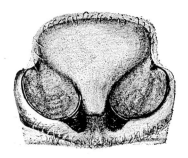

Fig. 695.
EPIGYNUM OF DOLO-
MEDES TENEBROSUS

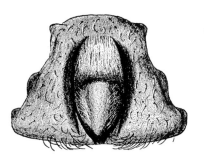

Fig. 697.
EPIGYNUM OF DOLO-
MEDES SCRIPTUS

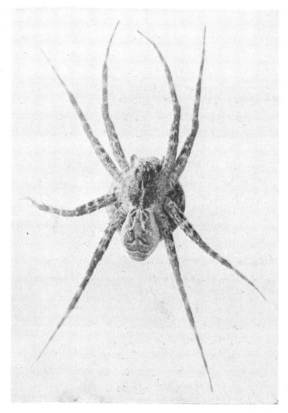

Fig. 696. DOLOMEDES SCRIPTUS, FEMALE WITH HER EGG-SAC

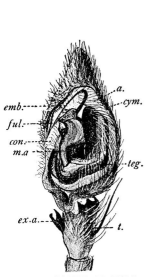

Fig. 698. PALPUS OF MALE OF DOLOMEDES SCRIPTUS

Fig. 699. DOLOMEDES URINATOR, FEMALE

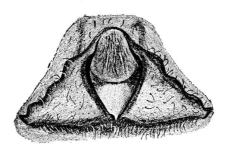

Fig. 700.
EPIGYNUM OF DOLOMEDES URINATOR

Superfamily Argiopoidea

other large common Dolomedes by the markings of the abdomen and the form of the epigynum.

There is a median yellow band on the basal half of the abdomen; and on each side of this at the base of the abdomen, a narrow yellow line each forming an incomplete ring; between these and the tip of the abdomen there are from three to six

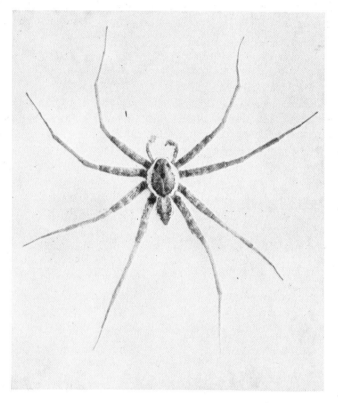

Fig. 701. DOLOMEDES VITTATUS, MALE

pairs of small, white or yellow spots; each pair of spots is connected by a slightly recurved, not W-shaped, black line (Fig. 699). The epigynum (Fig. 700) is distinctly broader than long; the guide extends the whole length of the organ, and has a transverse groove near the middle.

The male of this large species has never been described. It is probably similar to *Dolomedes scriptus*.

The Six-dotted Dolomedes, *Dolomedes triton* (D. tri'ton).—
This is a very beautiful and easily recognized species. It is dark greenish gray in colour, with a white band on each side extending the whole length of the body, two rows of white spots on the surface of the abdomen, and six dark dots on the sternum, three on each side near the coxæ. Two distinct subspecies, differentiated chiefly by the colour pattern, have been recognized. In typical *triton*, which is essentially southern in distribution, the sides of the carapace are marked by a broad, marginal band which is white and clothed with white hairs, and which is continuous across the clypeus. In the northern form, which also occurs quite commonly in some of the southern states, the lateral white band is reduced to a narrow line which is not continued across the clypeus. This is the form which was named *sexpunctatus* by Hentz, the specific name occasioned by the presence of six small dots on the sternum. The female is illustrated in Fig. 702. The male, which is considerably smaller than the female and has much longer legs, is shown in Fig. 704. The disparity in the sizes of the sexes is not so great in typical *triton*. The adult female measures from three fifths to four fifths inch in length. The epigynum is illustrated in Fig. 703.

This beautiful species is common in marshy places. It lives on plants over water, and dives freely when frightened, hiding under floating leaves. It is widely distributed, found in the east from New England to Texas, and is also known from across the northern part of the country in some of our western states and in Canada.

Genus TINUS (Ti'nus)

The lower margin of the furrow of the cheliceræ is armed with three teeth as in *Pisaurina*. The first row of eyes (Fig. 707) is lightly procurved, the median eyes larger than the lateral and closer to them. The posterior eye row is recurved, the eyes equal in size, somewhat larger than those of the anterior row. The median ocular quadrangle is broader behind than in front.

This is a genus of tropical and subtropical Mexico and Central America. Our only species was described in *Thaumasia* but it is congeneric with *Tinus*, a genus proposed by F. Cambridge. It is quite possible that *Tinus* may only represent a subgenus of *Thaumasia*.

Fig. 702. DOLOMEDES SEXPUNCTATUS, FEMALE

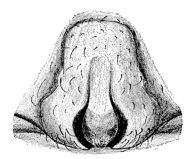

Fig. 703. EPIGYNUM OF DOLOMEDES SEXPUNCTATUS

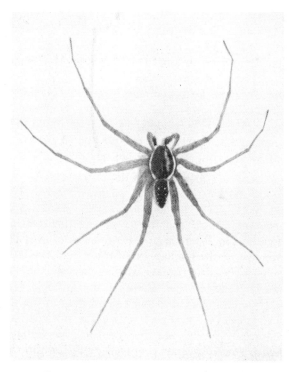

Fig. 704. DOLOMEDES SEXPUNCTATUS, MALE

Superfamily Argiopoidea

Tinus peregrinus (T. per-e-gri′nus).—Large females measure nearly an inch in length. The cephalothorax is light brown above, bordered by a pale marginal band which is continuous across the clypeus. The legs are reddish brown, with indistinct darker bands on the femora. The abdomen is marked with a dull brown band which is margined by narrow light lines. The epigynum of the female is shown in Fig. 708.

This spider is found in the southwestern portion of our country, from Arkansas and Texas to California.

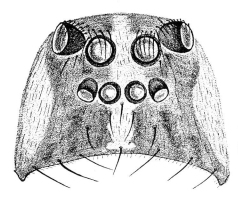
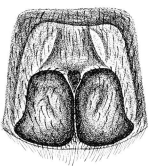

Fig. 707. EYES OF TINUS PEREGRINUS

Fig. 708. EPIGYNUM OF TINUS PEREGRINUS

Family HERSILIIDÆ (Her-si-li′i-dæ)

The Hersiliids (Her-si-li′ids)

The hersiliids are remarkable for the form of their posterior spinnerets which are similar in general appearance to those of some of the diplurids and agelenids. The hind spinnerets are cylindrical, the apical segment greatly elongated and attenuated, equalling the abdomen in length. The inner surfaces of these spinnerets are thickly set with long, fine, setiform spinning tubules (Fig. 709). The front spinnerets are near together between the hind ones, and are separated by a distinct, triangular colulus. The pars cephalica is elevated, and the eyes are in two strongly recurved rows. The abdomen is quite flat, often wider behind than in front. The legs are greatly elongated, especially in the males. In *Tama* the metatarsi are flexible, marked with false articulations.

Superfamily Argiopoidea

In spite of the great length of the spinnerets in this family, the members apparently do not make much use of them for spinning purposes. Some are said to make no webs. The species of *Tama* spin irregular webs similar to those of *Pholcus*.

Genus TAMA (Ta'ma)

This is the only genus known from the Americas. The species are tropical or subtropical in distribution. The metatarsi are

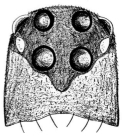

Fig. 710. FACE OF TAMA, SHOWING ARRANGEMENT OF EYES

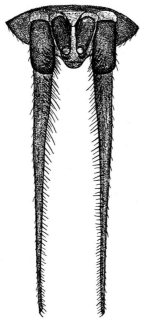

Fig. 709. SPINNERETS OF TAMA

not biarticulate but are weakened just beyond the middle by false articulations. The legs are unequal in length, the third being very much shorter, and the first considerably longer than the others. The apical segment of the hind spinnerets is flagelliform.

Tama mexicana (T. mex-i-ca'na).—The female is about one half inch in length, the male somewhat smaller. The broad cephalothorax is yellowish, narrowly margined with black, and the ocular area is dark. The legs are dusky yellow, with indistinct dark annulæ. The abdomen is white or yellowish above, with a faint longitudinal median dark figure and a black band on each side. The venter and the spinnerets are pale.

This Mexican species has been taken in the lower Rio Grande River Valley of southern Texas. These spiders sit on trees, simulating some of the species of *Philodromus*, and run actively when they are disturbed.

Superfamily Argiopoidea

Family LYCOSIDÆ (Ly-cos'-idæ)

The Wolf-spiders

The Lycosidæ are hunting spiders, which chase their prey. For this reason the typical genus was named *Lycosa*, which is from the Greek word for wolf.

In this family the trochanters of the legs are notched as in the preceding family (Fig. 711); but the lorum of the pedicel of the abdomen is of a different form. In this family the lorum is composed of two pieces of which the posterior one is notched to receive the anterior one (Fig. 712). The cuticle is almost always clothed with simple or squamose hairs; the tibia of the pedipalp of the male is unarmed; and the embolus is short and rarely exerted. The eyes are in three rows, the posterior lateral eyes being situated far behind the posterior median eyes (Fig. 713); the first row consists of four small eyes and the two posterior rows each of two large eyes. The relative proportions are very different from what exists in the jumping spiders, where also the eyes are in three rows.

In the Lycosidæ the palpus of the male is essentially of the same type as that of the Pisauridæ, except that the tibia is not furnished with an external apophysis. Figure 714 *a* represents the expanded bulb of *Lycosa avida*, with the parts lettered as in other figures of palpi; see page 121.

In the unexpanded bulb (Fig. 714 *b*) the embolus is curved so that its tip rests in a furrow in the conductor of the embolus, which is immediately distad of the median apophysis. The part termed the *auricula* by Chamberlin ('08) is evidently the conductor of the embolus.

The wolf-spiders are common; they run through grass or lurk under stones, especially in damp situations. Many species dig tunnels in the earth, and some of these build a turret about the mouth of their tunnel. They all carry their egg-sac attached to the spinnerets by a bundle of threads. After hatching, the young pass to the body of the mother and are carried about by her for a considerable time (Fig. 715). Whether the mother provides nourishment or not to the young during this period has not been definitely determined.

The egg-sac varies in shape, that made by some species is spherical, in other cases it is flattened. It consists of two valves,

Fig. 711.
TROCHANTER OF LEG,
SHOWING NOTCH

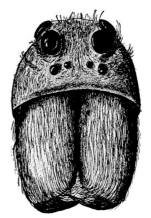

Fig. 713. FACE OF LYCOSA,
SHOWING ARRANGEMENT
OF EYES

Fig. 712. LORUM OF THE
PEDICEL OF LYCOSA

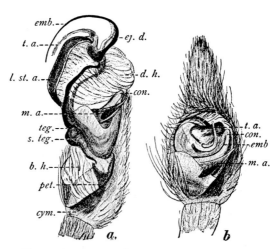

Fig. 714. PALPUS OF MALE OF LYCOSA AVIDA

Superfamily Argiopoidea

one above, the other below; these valves are usually united by a suture of more delicate tissue, which the mother tears in order to let the young escape.

A few members of the family spin webs. In one division of the family, which is represented in this country only by *Sosippus*, the web is said to resemble that of *Agelena*.

Fig. 715. LYCOSA HELLUO, FEMALE WITH YOUNG

It is to this family that the famous tarantula of southern Europe belongs, the bite of which was supposed to cause the dancing madness. This is a large species, resembling in form and colour our *Lycosa carolinensis*.

The Lycosidæ have received the attention of most of the writers on American spiders. The large size of many of the species and the interesting habits of some of them have made them attractive subjects of study. The only paper in which most of our genera and species are fully described is the *Revision of North American Spiders of the Family Lycosidæ* by Dr. Ralph V. Chamberlin ('08). The numerous species that have been described since that revision, most of them from the south and the west, have made it necessary to modify to some extent the nomenclature and the generic limits. Our largest genera are *Lycosa, Arctosa,* and *Pardosa*.

The genera at present known from the United States may be separated as follows:

TABLE OF GENERA OF THE LYCOSIDÆ

A. Posterior spinnerets distinctly longer than the anterior, the apical segment conical, as long or nearly as long as the basal segment; lower margin of the furrow of the chelicera usually with four teeth. P. 639.   SOSIPPUS

AA. Posterior spinnerets at most little longer than the anterior; the apical segment very short, semicircular; lower margin of chelicera armed with no more than three teeth.

Superfamily Argiopoidea

B. Anterior row of eyes strongly procurved, the median eyes much farther from the lateral eyes than from each other. P. 659.   TRABEA
BB. Anterior row of eyes not strongly procurved.
   C. Distal pair of ventral spines of anterior tibiæ never apical in position.
      D. Anterior tibiæ armed below with five or six pairs of long spines. P. 659.   SOSILAUS
      DD. Anterior tibiæ armed below with at most three pairs of spines; cephalothorax with median dark V-shaped marking. P. 653.   PIRATA
   CC. Distal pair of ventral spines of anterior tibiæ apical in position, sometimes small.
      D. Lower margin of chelicera with two teeth.
         E. Cephalothorax with distinct median longitudinal pale band. P. 652.   TARENTULA
         EE. Cephalothorax uniform dusky or black, without a pale band. P. 652.   HESPEROCOSA
      DD. Lower margin of chelicera with three teeth.
         E. At least one, usually two or three spines above on tibiæ of fourth legs.
            F. No true stout spine above at base on fourth tibiæ, replaced by a slender bristle. P. 656.   ARCTOSA
            FF. Fourth tibia with stout spine at base.
               G. Labium wider than long with basal excavations short. P. 660.   PARDOSA
               GG. Labium longer than wide with basal excavations long.
                  H. Cross-piece of epigynum divided at ends; embolus of palpus elbowed; terminal apophysis of bulb usually with an enlarged horn-like process. P. 657.   SCHIZOCOSA
                  HH. Cross-piece of epigynum undivided at ends; embolus of palpus evenly curved, and the terminal apophysis normal. P. 639.   LYCOSA
         EE. No true spines above on tibiæ of the fourth legs. P. 651.   GEOLYCOSA

Superfamily Argiopoidea

Genus SOSIPPUS (So-sip'pus)

The anterior eyes are in a strongly procurved row, the lateral eyes being as large as or larger than the median eyes and set on tubercles. The lower margin of the furrow of the chelicera is armed ordinarily with four teeth, but in some species three teeth seem to be the normal number. The tarsi and metatarsi of the anterior legs are densely scopulate. The last segment of the hind spinnerets is moderately long and conical, usually about as long as the basal segment.

This genus is our only representative of a small group of genera, the Hippasinæ, which, while possessing the structural characteristics of the Lycosidæ, resemble the Agelenidæ in habits. They are sedentary spiders and construct a large permanent funnel-web upon which they run like an *Agelena*.

Three species have been described from our fauna.

Sosippus floridanus (S. flor-i-da'nus).—This is a large spider, some females measuring nearly three fifths inch in length, and the male is only slightly smaller. The cephalothorax is deep reddish brown or reddish black, with a median stripe of yellow hair behind the eye region, and on each side a wider marginal white band. In some examples the pale median band is nearly obsolete. The abdomen is blackish above, with a row of white spots on each side, and narrow transverse white lines on the hind part.

This is a common species in Florida. Mature examples of both sexes may be found during most of the year.

Sosippus mimus (S. mi'mus).—This species resembles *floridanus* closely in appearance but differs in having the lower margin of the furrow of the chelicera armed with only three teeth. It is found from Alabama to Texas.

Sosippus californicus (S. cal-i-for'ni-cus).—This species has the marginal band on the carapace wider than in the above species. It was described from California.

Genus LYCOSA (Ly-co'sa)

The labium is longer than wide and the basal excavations in it are long, usually one third or more of the length of the labium (Fig. 716). The face is much wider below than above and has the sides strongly convex. The anterior tibiæ are armed

Superfamily Argiopoidea

below with three pairs of spines which are but little if any longer than the diameter of the segment.

This genus includes the larger and more familiar members of the family; but some of the species are of moderate size. They are often found running over the ground in damp pastures or lurking under stones or rubbish in fields or at the edges of woods. Some of them live near water upon which they run freely and beneath which they dive when alarmed.

Although some species of *Lycosa* are wandering spiders, resembling the Pardosas in this respect, a large proportion of the species build retreats. The retreat may be merely a shallow excavation under a log or stone, lined with silk, and surrounded by a wall of earth or of sticks and stones. But more often the retreat is a vertical tube in the earth, which in some cases is a foot or more in depth. This tube is often lined with a thin film of silk, especially when made in loose soil; this lining is thicker toward the opening of the tube than in the deeper portion of it.

Some species surround the mouth of their tube with a circular wall of earth and pebbles brought from the burrow or with a turret made of grass and dirt fastened together with silk or of bits of twigs fastened in place in the same way. Figure 717 represents a turret made of grass and earth at the entrance to a burrow of *Lycosa carolinensis*. The specimen figured was taken near Agricultural College, Mississippi. It was one of many found in a bank of earth that had been thrown out of a ditch. These burrows were vertical or nearly so and about six inches in depth; some of them were nearly one inch in diameter. The burrow with the highest turret observed was that of a male; but nests of females were much more abundant.

Fig. 716. LABIUM OF LYCOSA ERRATICA

The material used in the construction of the turret is whatever the spider finds most available in the vicinity of its burrow; consequently the turrets of different individuals of the same species may present a very different appearance. Even the same individuals under changed conditions will vary the nature of its turret. A correspondent sent me an immature male *Lycosa;* it was one of a large number of individuals found near a railroad track, and in each case the spider had built its turret entirely of cinders. The soil in which the burrows were made was quite sandy, and the turrets were in many cases an inch or more in height. When I received the specimen, I put it in a cage made of a glass cylinder placed over a flower pot filled with earth. A hole had been made in the earth near the centre of the cage; into this the spider retreated. A few days later it was observed that the spider had built a low wall of small stones about the entrance of its tube. I then put a quantity of fragments of twigs

Fig. 717. TURRET OF LYCOSA CAROLINENSIS

into the cage, and on the next day was rewarded by seeing that during the night several of the twigs had been placed upon the wall. From time to time, but always in the night, other twigs were placed on the wall, till finally a turret one inch in height had been built (Fig. 718). Each twig was held in place by threads of silk; in fact the turret was lined with a continuous sheet of this substance.

The spider was very shy. I never saw it outside of its turret, although its cage was kept in my office during several months. It was evident that it left its nest only at night, or when

Superfamily Argiopoidea

it was sure that no one was near. But it spent much time during the day perched at the top of the turret, with its head projecting so that it could see the region surrounding its retreat. This, I think, indicates the use of the turret; it is a watch tower from which the spider can see its prey more readily than it could from the surface of the ground. And the fact that the spider used its watch tower during the day indicates that it would leave it to capture prey at this time if it felt it were safe to do so.

Fig. 718. TURRET OF LYCOSA

When stationed at the top of the turret it rested, as it were, on its elbows, with its tarsi inside the turret, so that it could drop into the burrow instantly.

The large Lycosas live two or more years; on the approach of winter they close the entrance to their burrows with débris fastened together with silk.

The egg-sac of *Lycosa* is spherical and usually white; the seam between the two valves is sometimes very conspicuous (Fig. 719), but often it is not well marked. When the spiderlings are ready to emerge the female rips open the sac with her cheliceræ. This I have observed with a large *Lycosa* that I had in a breeding cage. The female rolled the sac and pulled out threads along the equator. I observed the female doing this one day and on the following day the spiderlings were out. The young climb onto the back of the mother and are carried about by her for some time. A female of *Lycosa helluo* with her brood is represented on page 620; the entire body except the head is covered with spiderlings and presents a very unsightly appearance.

Fig. 719. LYCOSA WITH EGG-SAC

Superfamily Argiopoidea

The genus *Lycosa* is a very large one. Even as restricted here, by the removal of the genera *Pirata, Arctosa, Tarentula, Geolycosa,* and *Schizocosa*, it is represented by about thirty known species in the United States and Canada alone. Some of them are among our largest and most interesting spiders. The burrowing ability of many of them is well known, but it is not generally known that certain Lycosas have gone even farther than the camouflage of a turret or canopy for their burrows. *Lycosa lenta* is known to cover the opening with a hinged door which is similar to the wafer type door of some of the trap-door spiders (Ctenizidae).

Many of our lycosids are very widely distributed. The commoner species found in the eastern part of our country may be separated by reference to the following table.

TABLE OF THE COMMON SPECIES OF LYCOSA

A. Cephalothorax with a light median longitudinal stripe which is very narrow or line-like anteriorly, and which extends forward to or between eyes of second row.
  B. Legs yellow or light brown, not annulate, at most with a few dark markings on femora. *L. helluo*
  BB. Legs strongly banded with black, or entirely black.
    C. Anterior row of eyes as wide as or a little wider than the second. *L. aspersa*
    CC. Anterior row of eyes shorter than the second row.
L. riparia
AA. Cephalothorax either without a median band or with a band which is as wide or nearly as wide as third eye-row.
  B. Dorsum of abdomen marked along its entire length by a distinct median dark band.
    C. Sternum pale; dorsal band of abdomen with margins behind dentate, often enclosing pale spots; venter without black spots. *L. rabida*
    CC. Sternum black; dorsal band of abdomen with margins nearly straight; venter with several small black spots.
L. punctulata
  BB. Abdomen not so marked.
    C. Legs pale brown, entirely without darker markings; venter behind epigastric furrow black. *L. lenta*

Superfamily Argiopoidea

CC. Legs with patellæ and often distal end of tibiæ black beneath; anterior femora above and behind with fine longitudinal dark lines; posterior femora with faint dark spots; venter behind epigastric furrow black.
L. baltimoriana

CCC. Not as for *lenta* or *baltimoriana*.

D. All tibiæ black at both ends beneath and the femora black beneath at the distal end; large spiders, the cephalothorax two fifths inch or more in length.
L. carolinensis

DD. Not as above; cephalothorax less than two fifths inch in length, usually much less.

E. Dorsum of abdomen with median light band extending to spinnerets behind where it ends in a point, this band enclosing at base a dark lanceolate marking. *L. avida*

EE. Dorsum of abdomen without such a distinctly limited light band.

F. Females.

G. Lateral depressed areas of epigynum subcircular (Fig. 720 a). *L. avara*

GG. Not so.

H. Guide of epigynum enlarged at the posterior end (Fig. 720 g). *L. gulosa*

HH. Guide of epigynum inversely T-shaped, the arms slender.

I. Epigynum as in Fig. 720 b. *L. pratensis*

II. Epigynum as in Fig. 720 d. *L. frondicola*

FF. Males.

G. Anterior row of eyes as long as or longer than the second row.

H. Tip of embolus not curled. *L. frondicola*

HH. Tip of embolus forming a small loop.
L. pratensis

GG. Anterior row of eyes shorter than the second row.

H. Embolus with several free distal apophyses as shown in Fig. 720 h. *L. gulosa*

HH. Embolus without free apophyses; median apophysis large, curved ventrad. *L. avara*

THE MORE COMMON SPECIES OF LYCOSA

The following notes are merely supplementary to the data given in the table above and are not intended to be sufficient for the determination of species, which can be best done by the use of the table.

Lycosa helluo (L. hel'lu-o).— The colour of the body above is dull yellow or greenish brown. The cephalothorax is marked with a narrow yellow stripe in the middle; this stripe is quite narrow between the eyes and somewhat wider on the thorax; there is also a light stripe along each lateral margin of the thorax. On the basal half of the abdomen there is a longitudinal lanceolate stripe similar to that found in many species, which is wider in the middle and pointed at each end; this stripe is darker at the edges and is bordered by lighter bands. On the hinder half of the abdomen there is a series of indistinct chevrons. The form of the epigynum is shown in Fig. 720, *c*. The female is about three fourths inch in length; the male, one half inch.

This is one of our most common and most widely distributed species; it has been found throughout the East and as far west as Texas and Utah. The female builds a shallow nest, lined with silk, and often surrounded with a low wall of earth or of sticks and leaves, under a stone or other object lying on the ground. They are found in these nests with their egg-sacs early in the summer. This spider is often found in the woods.

This species is known under several different names; the one which has been most used is *Lycosa nidicola* proposed by Emerton in 1885; but the species was described under the name *helluo* by Walckenaer in 1837.

Lycosa riparia (L. ri-pa'ri-a).— The cephalothorax is brown, with a narrow grayish yellow median band, which is widest at the median furrow and reduced to a line between the eyes. The abdomen is grayish brown above, with scattered minute spots of black pubescence, and indistinct outline of a lanceolate stripe on the basal part, and several black chevrons behind. The female is three fourths inch in length.

This is a southern species, its range extending from the District of Columbia southward. It has been found also westward to Kansas and Texas. Hentz, who first described the species states that it "is aquatic in its habits, always found near or on

Superfamily Argiopoidea

water, and diving with ease under the surface when threatened or pursued."

Lycosa aspersa (L. as-per′sa).— The cephalothorax is dark reddish brown, blackish about the eyes; there is a lighter uneven-

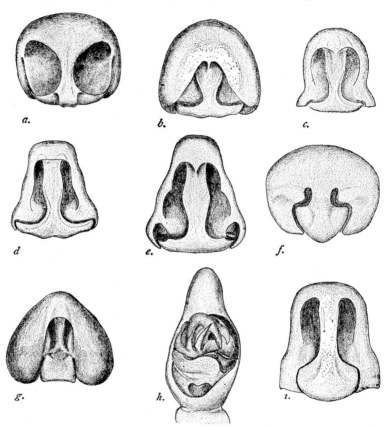

Fig. 720. EXTERNAL REPRODUCTIVE ORGANS OF LYCOSA
a, epigynum of *L. avara* *b*, epigynum of *L. pratensis* *c*, epigynum of *L. helluo* *d*, epigynum of *L. frondicala* *e*, epigynum of *L. erratica* *f*, epigynum of *L. pictilis* *g*, epigynum of *L. gulosa* *h*, palpus of male of *L. gulosa* *i*, epigynum of *L. pikei* (after Chamberlin)

edged marginal band on each side, and a more or less distinct narrow median stripe. The abdomen is very dark above; there is a basal black mark, which is forked behind and followed by a series of chevron marks; these are indistinct in old individuals.

This is a common burrowing species; its known range extends

from New England to the Gulf of Mexico, and west to Kansas. It is easily confused with *Lycosa riparia*, especially in the males, but typical specimens are very much darker. The epigynum is similar to that of *helluo* but is much shorter, the side arms usually being about the same length as the median septum. It has been described several times under different specific names. The synonyms are *Lycosa tigrina*, *Lycosa vulpina*, and *Lycosa inhonesta*. The first description, under the name *Lycosa aspersa*, was by Hentz in 1844.

Lycosa carolinensis (L. car-o-li-nen′sis).—The cephalothorax is clothed with brown and gray hairs and is usually without distinct markings; but in some individuals there is a gray supramarginal band on each side and a similar median one along the

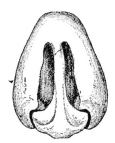

Fig. 721.
EPIGYNUM OF G. FATIFERA

Fig. 722.
TURRET OF G. FATIFERA

dorsum, widening from behind anteriorly. The tegument of the abdomen is light brown covered above with long brown to grayish brown pubescence, except over the dark marks which are clothed and made more distinct by black hair; these marks are as follows: on the basal part of the dorsum a median stripe, which is forked at the hinder end and sends out from its sides several pairs of pointed lines directed backward and toward the sides; behind the basal line there are several chevrons, and often a series of light dots along each side. The female attains a length of nearly one and one half inches; the male, four fifths of an inch.

This spider occurs over the greater part of the United States and is the largest member of the Lycosidæ found in our territory. The female digs a burrow six or eight inches deep and sometimes

Superfamily Argiopoidea

builds a turret around its entrance; but it is also often found running about on the ground or lurking under stones.

Lycosa avida (L. a'vi-da).—The colour of the body varies from light gray to almost black. The cephalothorax is marked with a reddish yellow or reddish brown median band, which is narrowest on the thorax and wider on the head, becoming as wide as the third row of eyes; between the eyes it is reduced to a narrow stripe. There is also a light band on each lateral margin of the cephalothorax; this band is more or less split by a dark line. The abdomen is marked with a longitudinal pale band, which tapers to a point at the spinnerets and encloses a distinct lanceolate stripe in the basal half or two thirds of its length; behind the lanceolate mark there may be several dark transverse lines. The venter is whitish or light yellow and usually marked in specimens from the eastern United States with two black stripes which meet at the spinnerets making a U-shaped figure; in specimens from the Pacific Coast this U-shaped figure is usually wanting in adults. The form of the epigynum is shown in Fig. 720, *e*. The length of the body varies from two fifths to one half inch.

This is a common and very widely distributed spider. It is a vagabond species, which is found running in grass or hiding under stones; so far as is known it makes no burrow. It was described by Hentz under the name adopted here; it was later described by Emerton under the name of *Lycosa communis;* and it has been described under several other names by other authors.

Lycosa rabida (L. ra'bi-da).—The cephalothorax is dark gray with three light longitudinal stripes extending its whole length, and a narrow light line on each lateral margin of the thorax (Fig. 723). The abdomen is marked with a broad dark median band, which is notched on each side in front of the middle of the abdomen, and which encloses several pairs of light spots on the hinder part of the abdomen. At the sides of the median band are narrow light bands, and beyond these fine light and dark oblique lines. The male measures about one half inch in length; the female, four fifths inch or more.

This large species is common and widely distributed; so far as is recorded it has been observed only as a wandering spider.

Lycosa punctulata (L. punc-tu-la'ta).— The cephalothorax is light brown, with each side of the middle a blackish brown stripe, which runs forward over the eyes of the corresponding side; also on each side a very narrow marginal and a wider submarginal blackish line; the median light band is narrower than the dark bands enclosing it; between the eyes of the third and second rows it is narrowed to a line but widens again above the eyes of the first row. The abdomen is marked with a longitudinal black median band, the edges of which are not notched and which does not include light spots as in the preceding species; on each side of the median band is a grayish brown stripe; exterior to the gray stripes the sides are covered with brown and grayish brown intermingled in spots and streaks. The body measures three fifths inch or more in length.

This is a widely distributed species in the eastern half of the United States.

Lycosa frondicola (L. fron-dic′o-la).— The cephalothorax is dark brown above, with a median light brown band, which is widest just behind the eye-space, where it is wider than the third row of eyes. The abdomen is grayish brown above; with the lanceolate stripe faintly indicated, and with indistinct chevrons on the hinder part. The form of the epigynum is shown in Fig. 720 d. The male measures two fifths inch in length; the female, about one half inch.

Fig. 723. LYCOSA RABIDA

The range of this species includes the greater part of the United States. It is common especially in the mountainous portions of the country. It is found most frequently in and at the edges of woods, among fallen leaves and sticks.

Lycosa pratensis (L. pra-ten′sis).— This is a yellowish brown species with indistinct light and dark markings. The cephalo-

Superfamily Argiopoidea

thorax is marked with a median light band, which is widest between the third row of eyes and the median furrow; in this wider portion there are two dark longitudinal lines. On the basal half of the abdomen there is a lanceolate stripe and behind this four or five chevrons. The form of the epigynum is shown in Fig. 720 *b*. The length of the body is about one half inch.

This is a very common species in the northeastern part of the United States and in Canada; it is found under stones.

Lycosa gulosa (L. gu-lo'sa).— The cephalothorax is dark brown with a light gray stripe in the middle; this stripe is strongly constricted in front of the median furrow and less markedly so behind this furrow; it is widest in front of the first constriction, and extends forward to the second row of eyes. The abdomen is grayish brown above, with a darker lanceolate stripe on the basal half; this stripe, however, is sometimes indistinct or wanting. There is a black spot on each side near the base of the abdomen; and usually a row of dark marks extend back on each side from the basal spot to the hind end of the abdomen. The form of the epigynum is shown in Fig. 720 *g*; and the palpus of the male in Fig. 720 *h*. The length of the body is about one half inch.

This is a widely distributed species, its range covering the greater part of the United States. It is common in forests, where its brown and gray colours like those of dead leaves are protective.

This species has been commonly known under the name of *Lycosa kochii*; but this name, according to Chamberlin, should be applied to another species, which is known only from the Far West.

Lycosa avara (L. a-va'ra).—The cephalothorax is marked with a light median band, which begins at the second row of eyes and extends to the hind end of the carapace; it is widest between the third rows of eyes and the median furrow where it is as wide as the third row of eyes; it is somewhat abruptly contracted at the anterior end of the median furrow, and then gradually narrows to the posterior end of the carapace. The abdomen is clothed with gray and brown hairs, and bears an indistinct lanceolate stripe. The form of the epigynum is shown in Fig. 720 *a*. The male measures one third inch in length; the female, one half inch. The known range of the species extends from Massachusetts to Texas. Several closely related species are known from the South and the West.

Superfamily Argiopoidea

Genus GEOLYCOSA (Ge-o-ly-co'sa)

This genus is closely allied to *Lycosa*, in fact it is not recognized as valid by many authors. It is distinguished chiefly by that fact that the posterior tibiæ are completely devoid of dorsal spines in the females. The males have well developed tibial spines and are easily confused with the species of *Lycosa*.

The spiders of this genus burrow into the soil as do many of our other lycosids and often build a turret around the opening. The females rarely leave the burrow, and when outside, remain close to the opening. Several species are known from the United States. The two described below are common in the East.

Geolycosa pikei (G. pi'ke-i).—In the female the cephalothorax is dark reddish brown to blackish, with a median lighter band a little wider than the third eye-row in front, strongly narrowed to the dorsal groove and usually expanding again back of the groove. The abdomen is marked with a dark brown median band, which extends to the spinnerets and which has a broad indentation on each side just in front of the middle and a series of narrow paired indentations behind; the sides of the dorsum are grayish brown. The form of the epigynum is shown in Fig. 720 i. The length of the body is about three fourths inch.

The range of this species includes the Eastern and Middle States and extends to the District of Columbia and to Indiana. It is a burrowing and turret building species; its burrows are often from ten to twelve inches deep and one half inch in width. It was first described by Mr. Scudder in 1877 under the name *Lycosa arenicola;* but this name had already been used for an English species of *Lycosa*.

Geolycosa fatifera (G. fa-ti'fe-ra).—This species, which is the *Lycosa nidifex* of Marx, has been confused with the preceding.

The cephalothorax is dark reddish brown without definite light markings, but the median dorsal portion of the head and the clypeus is paler. The abdomen is light yellowish brown, with a solid black lanceolate mark at the anterior end. The epigynum of the female is of the form shown in Fig. 721; it is relatively longer and narrower anteriorly than that of *L. pikei*, and the furrows are contracted cephalad.

Doctor Marx gave an excellent account of the habits of this species. The turret is shown in Fig. 722.

Superfamily Argiopoidea

Genus TARENTULA (Ta-ren'tu-la)

The lycosids of this genus may be distinguished from allied genera by the armature of the lower margin of the furrow of the chelicera, there being only two teeth present. They differ from the only known species of *Hesperocosa* in having the carapace marked with a pale band. The epigynum of the female is usually as broad or broader than long, the septum often triangular in form.

Tarentula is a boreal genus. All of our species are found in the North and occur within the borders of the United States only on our mountain ranges. The following species is found in Canada and in the northeastern United States.

Tarentula pictilis (T. pic'ti-lis).—The carapace is chocolate-brown, with a median grayish band, which begins at the second row of eyes, extends between the eyes of the third row, and then abruptly bulges on each side, being much wider than the eye-space midway between the eyes and the median furrow, then narrowing to the median furrow, where it is about the same width as the eye-space. The abdomen is dark brown above, with a black spot on each side at the base, which in the female encloses a light spot but not in the male; on each side, between the basal black spot and the middle line of the body, there is a large light spot; on the hinder part of the abdomen, there is a series of chevrons; at the outer end of each of the chevrons is a light spot enclosed by black, and in front of each half of each chevron is a light-coloured spot. The form of the epigynum is shown in Fig. 720, *f*. The male measures nearly two fifths inch in length; the female, about one half inch.

Mr. Emerton, who first described the species, states that it is abundant among the moss and low shrubs on the upper part of Mount Washington, N. H., and the neighbouring mountains.

Genus HESPEROCOSA (Hes-per-o-co'sa)

This genus agrees with *Tarentula* in having the lower margin of the furrow of the chelicera armed with two teeth. It is quite distinct in the genitalia, which are similar to those of the species of *Lycosa*. The median apophysis of the male palpus is sub-lateral in position.

A single species has been described.

Hesperocosa unica (H. u'ni-ca).—This is a small species, the females measuring about one sixth inch in length, the males somewhat smaller. The cephalothorax is dark brown or black, clothed with a longitudinal band of long white hairs, but lacking a pale stripe in the integument. The legs are concolorous with the cephalothorax. The abdomen is dark brown, variegated to some extent with white markings, and clothed with long white hairs. The labium is distinctly broader than long. The epigynum is similar to those of typical species of *Lycosa*.

This little species has been found in the mountains of western Texas, New Mexico, and Arizona.

Genus PIRATA (Pi-ra'ta)

The anterior tibiæ is armed with two or three pairs of spines beneath, but the distal pair is never apical in position, although there may be a subapical pair in the male. The anterior eyes are in a straight or slightly procurved line; they are subequal or the median are a little larger than the lateral. The labium is longer than wide, with the basal excavations short. The epigynum of the female presents no true guide, but usually bears behind two strongly chitnized lobes or tubercles upon which are the openings of the spermathecæ (Fig. 724).

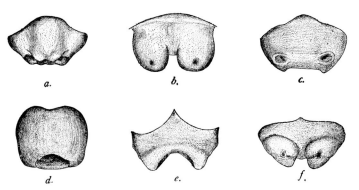

Fig. 724. EPIGYNA OF SPECIES OF PIRATA
*a, P. minutus b, P. insularis c, P. montanus d, P. marxi
e, P. sedentarius f, P. febriculosus* (after Chamberlin)

This genus includes spiders of small or of medium size, which have upon the cephalothorax a pale band which is forked in the

Superfamily Argiopoidea

Fig. 725.
CEPHALOTHORAX OF PIRATA

head-region, where there is also a central pale stripe (Fig. 725). They live in damp fields and in the vicinity of water, upon which they run freely, and beneath which they dive when alarmed. The egg-sacs are white and spherical; the seam between the two valves is less marked than on the egg-sacs of *Pardosa*. The females drag the egg-sacs after them attached to their spinnerets, though when they are at rest they often hold them in their cheliceræ.

THE MORE COMMON SPECIES OF PIRATA

The species of *Pirata* resemble each other to a striking degree in colour and markings. The cephalothorax is brown or blackish, with a pale stripe on each side below the eyes, which extends backward and inward, and merges with its fellow in the region of the median furrow, the two being continued as a single median stripe on the hinder part of the carapace (Fig. 725); in the head-region a pale median stripe divides the darker area enclosed between the two stripes so that this dark area is more or less V-shaped. On each side of the thorax there is usually a submarginal light stripe and frequently a marginal line of white hair. The abdomen may or may not have a basal lanceolate stripe; it usually has a row of white or yellow spots along each side of the dorsum; and there are frequently several chevrons near the caudal end.

Pirata minutus (P. mi-nu'tus).—This is a small species, the male measuring about one eighth inch or less in length; the female, a little more than one eighth inch. The carapace is dark brown, with a forked reddish yellow median band of the usual form and on each side of the thoracic part a marginal light stripe; along this stripe there is in life a marginal line of white hair. The abdomen is black above, with or without a narrow lanceolate mark at base, and clothed, with sparse light brown or grayish hair, with a series of five or six spots of white hair along each side for the entire length and several more or less distinct chevrons on the hinder part. The chelicerae are armed with but two teeth. The form of the epigynum is shown in Fig. 724, *a*.

Pirata montanus (P. mon-ta'nus).—The carapace is deep brown or blackish, with a forked yellow median stripe of the usual form; there are no lighter lateral stripes in the integument, but in life there is on each side a marginal line of white hair. The abdomen is black above, with a reddish brown median lanceolate stripe at base, and on each side of the apex of this stripe a reddish spot; on the hinder part of the abdomen are several chevrons; all of these markings may be indistinct; in life there is on each side a series of about six spots of light yellow hair. The posterior margin of the epigynum is nearly straight. (Fig. 724, *c*.)

This species has been found in the mountains of the Northern States and in Utah.

Pirata marxi (P. marx'i).—The sides of the carapace are brown or grayish black crossed by many radiating lines of black; there is a forked yellow median band of the usual form, and on each side a supramarginal yellow stripe limited below by a black marginal line; there is no marginal stripe of white hair. The abdomen is yellow with markings in black; there is a lanceolate outline at base, and along each side a wavy or zigzag stripe, each united with the lanceolate outline at its base and again at its middle; and on the hinder part of the abdomen there are several short black marks. Figure 724, *d* represents the epigynum. The female measures about three tenths inch in length; the male, about one fourth inch.

This spider occurs in the northeastern United States and south to the District of Columbia.

Pirata insularis (P. in-su-la'ris).—The sides of the carapace are brown crossed by radiating lines of black; there is a median, forked, reddish yellow band of the usual form, and on each side of the thoracic part a yellowish supramarginal stripe limited below by a narrow dark marginal stripe; there is no marginal line of white hairs. The abdomen is blackish above, with a yellow lanceolate median basal stripe; on each side of the apex of this stripe there is a yellow spot; and on the hinder part several yellow chevrons or nearly straight transverse marks; in life there is a row of bunches of white hairs. The form of the epigynum is shown in Fig. 724, *b*. The female measures one fourth inch in length; the male, one sixth.

This species occurs in the Eastern and Middle States, and also has been found in Utah and Colorado.

Superfamily Argiopoidea

Genus ARCTOSA (Arc-to'sa)

This genus is closely allied to *Lycosa* and is still maintained in that genus by some authors. The cephalothorax is glabrous or very nearly so, often smooth and shining, and ordinarily lacks a distinct light-coloured median stripe. In some of our species the cephalothorax is black. The epigynum is simple, presenting no true guide or at most weakly furrowed, and the spermathecæ open posteriorly. The bulb of the male palpus bears an apophysis which is exterior in position. This genus is usually easily separated from *Lycosa* by the fact that the fourth tibiæ do not have a stout spine at the base above. It is replaced by a basally stout, apically slender and pointed bristle.

About twenty species have been described from the United States. The following species are most commonly met with in the east.

Arctosa rubicunda (A. ru-bi-cun'da).—The cephalothorax is dark brown to black above, the clothing of hairs very short which gives the carapace a polished appearance. The legs are dusky, without darker annulæ. The abdomen is gray, light in the middle, and with many closely placed dark dots and dashes on the sides. The first row of eyes is much wider than the second. The tibia and patella of the fourth legs are together shorter than the carapace. The length of the body is about two fifths inch.

This species occurs in the New England States.

Arctosa littoralis (A. lit-to-ral'is).—This is a pale species. The body is clothed with white, gray, and black hairs intermixed in spots and streaks. The abdomen and legs are spotted. The first row of eyes is no wider than the second row. The tibia and patella of the fourth legs are together shorter than the carapace. The length of the body is about three fifths inch.

This common species is found throughout the United States. It lives on beaches and in sandy areas, where its colour and markings possibly afford it some protection. Sometimes it digs a burrow into the sand.

Arctosa funerea (A. fu-ne're-a).—This is a very small species, the female being about one fourth inch in length. The carapace is black, shining, glabrous or nearly so. The femora of the first and second legs are black, the remaining joints of these legs yellow, without annulæ. The abdomen is nearly black and is marked above by a pale longitudinal figure. The first row of eyes is not wider than the second row. The tibia and patella of the fourth legs are together about as long as the carapace.

This species is common in the South and occurs more sparingly north into southern New England.

Genus SCHIZOCOSA (Schiz-o-co'sa)

This genus has been separated from *Lycosa* on account of peculiar features of the external reproductive organs. In the female the lateral extensions of the guide of the epigynum are divided so as to be double (Fig. 726). In the male the embolus is distinctly elbowed and the terminal apophysis of the bulb of

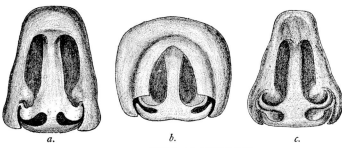

Fig. 726. EPIGYNA OF SCHIZOCOSA
a, *S. ocreata*. *b*, *S. bilineata* *c*, *S. saltatrix* (after Chamberlin)

the palpus is conspicuously elevated and usually more or less produced into a horn-like process extending beyond the front margin of the alveolus.

This genus includes spiders of medium or of small size. In them "the median light band of the cephalothorax widens uniformly from behind forward to the eyes; it is constricted in front of the dorsal groove, but otherwise its margins are nearly straight. The abdomen in all is marked above by a broad light band which is nearly or fully as wide as the dorsum and which extends over its entire length from base to spinnerets; this band encloses at base a lanceolate outline, and behind in some also a series of transverse angular lines of varying degrees of distinctness; sides of the abdomen dark in colour, black at least across anterior lateral angles; the venter is pale. The cocoon is spherical, without any seam at the equator, and is white in colour" (Chamberlin).

Superfamily Argiopoidea

Three species are common in the East. They are separated by Chamberlin as follows:

TABLE OF SPECIES OF SCHIZOCOSA

Females

A. Sternum yellow, with two dark lines or rows of dark spots converging posteriorly. *S. bilineata*

AA. Sternum dark, not marked as in *S. bilineata*.
- B. Septal portion of the guide of the epigynum very broad immediately in front of the transverse arms, narrowing anteriorly where it is not sinuous or bent; the median portion between anterior and posterior divisions of arms very narrow, much narrower than the septum in front of arms (Fig. 726, *a*); sternum usually black except marginally. *S. ocreata*
- BB. Septal portion of guide sinuous or bent near the anterior end; median portion between anterior and posterior divisions of arms wide, wider than septum in front of transverse arms (Fig. 726, *c*); sternum usually reddish brown. *S. saltatrix*

Males

A. First tibiæ clothed with dense hair standing out in brush-like form.
- B. Legs yellow, without dark annuli or markings. *S. bilineata*
- BB. Legs annulate with dark. *S. ocreata*

AA. First tibiæ not so clothed. *S. saltatrix*

Schizocosa bilineata (S. bi-lin-e-a′ta).—This is essentially a Northern species having been found from Connecticut to Kansas, and south to the District of Columbia. The male measures about one fifth inch in length; the female, a little more than one fourth inch.

Schizocosa ocreata (S. oc-re-a′ta).—This species is widely distributed in the Southeastern States; its range extends west to Kansas and north to New York. The length of the body is about one third inch.

Schizocosa saltatrix (S. sal-ta′trix).—This species resembles the preceding in size and in distribution.

Superfamily Argiopoidea

Genus TRABÆA (Tra-bæ′a)

The face is subquadrate, not at all or barely wider below than above; the sides are straight; the anterior row of eyes is very strongly procurved; and the anterior median eyes are much closer to each other than to the anterior lateral eyes.

Only a single species is found in our fauna; and this is one of the smallest of the Lycosidæ.

Trabea aurantiaca (T. au-ran-ti′a-ca).— The female is only one eighth inch or a little more in length; and the male is less than one eighth inch long. The cephalothorax is black or blackish brown, with a yellow median band which begins just back of the second row of eyes and narrows to a point at the median furrow, a yellow spot behind just under the front end of the abdomen, and a narrow yellow stripe on each side. The abdomen is brownish orange with a yellow spot in the middle tapering to a row of smaller spots behind.

This species occurs in New England, New York, and south to the District of Columbia.

Genus SOSILAUS (So-sil′a-us)

This genus differs from all other members of the Lycosidæ found in our fauna in having the anterior tibiæ armed below with five pairs of long slanting spines; and the metatarsi are armed with four pairs of similar spines. The four anterior eyes are subcontiguous and are in a gently recurved row; the anterior median eyes are at least twice as large as the anterior lateral (Fig. 727).

Sosilaus spiniger (S. spin′i-ger). — This is the only known species of the genus and of it only a single specimen is known; this was taken in Louisiana. It is a male and measures a little less than one sixth inch in length. The cephalothorax is "fulvorufous, smooth and subglabrous, a narrow marginal fuscous line and the pars thoracica marked irregularly with short radiating stripes. Abdomen fusco-testaceous, paler in front and below."

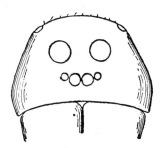

Fig. 727.
FACE OF SOSILAUS SPINIGER
(after Simon)

Superfamily Argiopoidea

Genus PARDOSA (Par-do′sa)

The labium is at least as broad as long and with the basal excavations short, only very rarely more than one fourth of the length of the labium (Fig. 728). The face is subquadrate, straight on each side (Fig. 729). The anterior row of eyes is shorter than the second and is procurved; the eyes of this row are small and subequal or with the median a little larger. The anterior tibiæ are armed with three pairs of spines, of which the basal and median pairs are much longer than the diameter of the segment.

The Pardosas are spiders of small or of medium size and are exceedingly active. They are vagabond spiders, constructing no retreat, and using their silk only in the construction of the egg-sac. They are often found in large numbers in damp fields, a few are semiaquatic like *Pirata*. Almost all of the species live only one year; the young hibernate and become adult in early spring. I have found

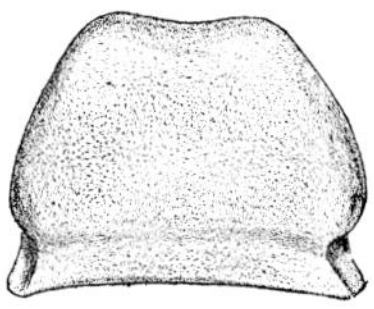

Fig. 728. LABIUM OF PARDOSA ATRA

Fig. 729. FACE OF PARDOSA

Fig. 730. FEMALE WITH EGG-SAC

them with their egg-sacs in May; they were lurking under stones but were very active when disturbed.

The egg-sac is depressed, lenticular in form; it is rarely white, usually yellowish or greenish, sometimes of a deep blue; there is a distinct seam along the edge between the two valves, which is torn open when the spiderlings are ready to emerge. The egg-sac is attached to the spinnerets (Fig. 730) but is held well under the abdomen, more so than is represented in the figure, which is from a photograph of a dead specimen.

The genus *Pardosa* is a large one; nearly fifty species are known from the United States. Those named below are the more common and more widely distributed ones.

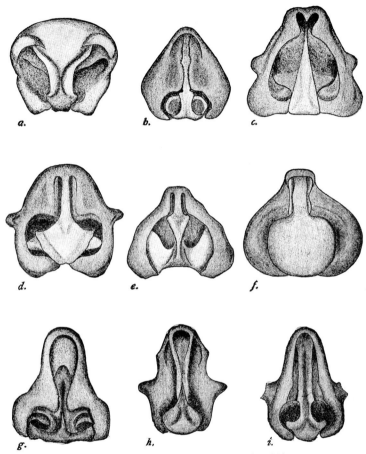

Fig. 731. EPIGYNA OF SPECIES OF PARDOSA
a, *P. xerampelina* b, *P. sternalis* c, *P. grœnlandica* d, *P. glacialis* e, *P. modica*
f, *P. emerton* g, *P. lapidicina* h, *P. saxatilis* i, *P. milvina* (after Chamberlin)

Pardosa saxatilis (P. sax-at′i-lis).—The length of the body is about one fifth inch. The cephalothorax is deep brown or black, with a median reddish yellow band, which is strongly notched on each side midway between the eyes and the median furrow, and narrowed behind the furrow; there is also on each side a yellow stripe near the margin of the carapace; these stripes are

Superfamily Argiopoidea

often obscure in the male. The abdomen is dark gray above, with an indistinct light lanceolate stripe on the basal half, and on each side of the apex of this stripe a light spot with a minute black dot at its centre; on the posterior part of the dorsum there is a series of light cross marks, each formed by the confluence of from two to four spots similar to those at the sides of the apex

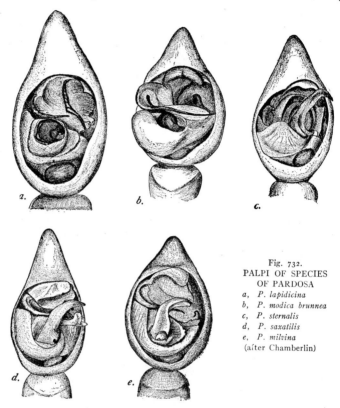

Fig. 732.
PALPI OF SPECIES
OF PARDOSA
a, *P. lapidicina*
b, *P. modica brunnea*
c, *P. sternalis*
d, *P. saxatilis*
e, *P. milvina*
(after Chamberlin)

of the basal stripe. The form of the epigynum is shown in Fig. 731, *b;* that of the palpus of the male in Fig. 732, *d.*

This species is also known under the following names: *Pardosa albopatella, Pardosa minima,* and *Pardosa annulata.* It occurs throughout the eastern half of the United States.

Pardosa milvina (P. mil-vi'na).— This species very closely resembles the preceding in colour and markings but is a little larger, the body being about one fourth inch in length. The two species can be best separated by the differences in the form of the epigynum and the palpus of the male as indicated in the tables. Figure 731, *i* represents the epigynum of this species; and Fig. 732, *e* the palpus of the male.

This species is the *Pardosa nigropalpis* of Emerton. It occurs throughout the eastern half of the United States.

Pardosa sternalis (P. ster-na'lis).— In the female the cephalothorax is black above with a median brown band beginning back of the eyes and narrowed to a line on the posterior declivity of the carapace; there is also a light brown band along each lateral border. The abdomen is blackish above, with a yellow or pale brown median lanceolate stripe at base, a pair of spots near the apex of the lanceolate stripe, and four or five pairs of spots forming chevrons on the hinder part of the dorsum. The form of the epigynum is shown in Fig. 731, *b*.

In the male the cephalothorax is darker, and the median band obscure. The abdomen is entirely black above, without markings. The form of the palpus is shown in Fig. 732, *c*.

The male measures a little less than one fourth inch in length; the female a little more. This species is common throughout most of the United States west of the Mississippi River. In Colorado, Utah, Nevada, and California it is very abundant, being far more common than any other lycosid. It frequents especially open meadows and the grassy borders of streams. It is rare in wooded districts.

Pardosa lapidicina (P. lap-i-di-ci'na). This is a somewhat larger species, the male measuring one fourth inch in length; the female, one third inch or more. The whole body is covered with black hairs which obscure the few light markings. The lighter median band of the cephalothorax is sometimes indistinct; it begins between the posterior eyes, where it is narrow, and is abruptly widened between the eyes and the median furrow, and is constricted in front of this furrow; on each side of the cephalothorax there is a row of irregular light spots. On the abdomen there is a black-margined lanceolate mark at base, followed on each side by a row of irregular pale spots; in some individuals there are irregular light chevrons near the hind end of the abdomen. The epigynum is represented in Fig. 731, *g*; the male palpus, in Fig. 732, *a*.

This is a common species in the Northern States, but it is

found also in the West and in the South. It occurs among stones along streams and also in hot and dry places.

Pardosa xerampelina (P. xer-am-pel'i-na).— The cephalothorax is deep brown or black above, with a median reddish brown band, which is widest about the median furrow; there are no distinct lateral stripes, but sometimes a few obscure light spots above the margin on each side posteriorly. The abdomen is black above or nearly so; there is a lanceolate basal mark of brown; the basal stripe is joined at two points on each side near its apex by the ends of a V-shaped mark, the apex of which is directed laterally; on the hinder part there are several, more or less distinct chevrons. The epigynum of the female (Fig. 731, *a*) differs from that of our other species it being widest in front. The female measures one third inch in length; the male is smaller.

This is a Northern species which occurs especially in mountainous regions.

Pardosa grœlandica (P. grœn-land'i-ca).— The cephalothorax is black or nearly so, with a lighter, brown, median band beginning only a little in front of the dorsal furrow and narrowing to a line on the posterior declivity of the carapace; from the front of the median band a horn-shaped yellow mark extends outward and forward on each side toward the corresponding eye; these marks are sometimes obscure or absent; there is a row of three or four curved light marks above the margin on each side. The abdomen is covered above with brown hair, with bunches of white hair forming a row of white spots along each side. The form of the epigynum is shown in Fig. 731, *c*. The female measures a little more than four tenths of an inch; the male, a little less.

This species was first described from Greenland; but its range extends south to the White Mountains and to Colorado and Utah. In Oregon and Washington it is the dominant species of Lycosa.

Pardosa modica (P. mod'i-ca).— The cephalothorax is dark brown, with a reddish brown median stripe, which is widest just behind the third row of eyes, constricted at the front end of the dorsal groove, and again immediately behind it; this stripe is divided in front of the groove by a dark middle line, widest in front and extending back as far as the dorsal groove; there is a light stripe on each side extending under the eyes to the clypeus. The abdomen is brownish black above, with a reddish brown basal lanceolate stripe, and a series of more or less distinct chevrons on the hinder half. The male measures one third inch in length; the female a little more. This is a variable species; Fig. 731, *d* represents the epigynum of the typical form and Fig. 731, *e* that of the variety *P. modica brunnea*. Figure 732, *b* represents the palpus of *P. modica brunnea*.

This is a Northern species whose range extends from Greenland south into the northern part of the United States from New England to Oregon.

Pardosa emertoni (P. em-er-to′ni).—The cephalothorax is yellow with two brown stripes which unite and become black between the middle eyes and there is a fine black line near the edge of the thorax on each side. The abdomen has a light middle band not much widened in front, where it includes a light stripe with dark brown edges, which tapers to a point about the middle of the abdomen. The hinder part of the middle stripe is indistinctly divided into four or five segments; at the sides of the middle stripe the abdomen is dark brown or black in small irregular spots and becomes gradually lighter toward the sides. The epigynum is represented in Fig. 731, *f*. The length of the body is about one fourth inch.

This is a Northern species whose range extends south to the District of Columbia. It was originally described by Emerton under the name *Pardosa pallida;* but as the name *pallida* was preoccupied, Chamberlin has proposed the name *emertoni* for the species.

Family OXYOPIDÆ (Ox-y-op′i-dæ)

The Lynx-spiders

The lynx-spiders are so called because some species chase their prey with great rapidity over herbage and the foliage of trees and shrubs; they even jump from branch to branch like the attids; but other species lie in wait near flowers and spring upon insects that visit the flowers.

The legs are long, with three tarsal claws but without scopulæ; the trochanters are not notched as in the two preceding families or but slightly so. The eyes are eight in number, dark in colour, and unequal in size; the anterior median being very small. The anterior row of eyes is strongly recurved; the posterior row, procurved; so that there may appear to be four rows of eyes of two each (Fig. 733). The abdomen tapers to a point behind.

Superfamily Argiopoidea

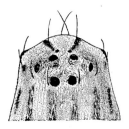

Fig. 733.
EYES OF OXYOPES

These are diurnal, hunting spiders, which make no use of webs for capturing their prey. They do not carry their egg-sacs about as do the members of the two preceding families; but fasten them to a branch or to a leaf or suspend them in a little web which they spin for this purpose. This family is feebly represented in the North; but some of the species are quite common in the South.

Three genera are represented in our fauna; these can be separated by the following table:

TABLE OF GENERA OF THE OXYOPIDÆ

A. Posterior row of eyes slightly procurved; posterior lateral eyes much farther from the anterior lateral eyes than from the posterior median eyes; lower margin of the cheliceræ unarmed. P. 666.   PEUCETIA

AA. Posterior row of eyes strongly procurved; posterior lateral eyes about as far from the anterior lateral eyes as from the posterior median eyes; lower margin of the chelicera with one tooth.

    B. Four posterior eyes equidistant; quadrangle limited by the posterior median eyes and the anterior lateral eyes longer than wide. P. 667.   OXYOPES

    BB. Four posterior eyes not equidistant. P. 668.

        HAMATALIVA

Genus PEUCETIA (Peu-cet'i-a)

The members of this genus can be recognized by the slight procurvature of the posterior row of eyes (Fig. 734). They are quite large spiders and are usually of a beautiful green colour, with red spots and black spines. The following is our only species.

Peucetia viridans (P. vir'i-dans). — This is a common species in the Southern States, where it often attracts attention on account of its

Fig. 734. EYES OF PEUCETIA

Superfamily Argiopoidea

beautiful bright transparent green colour marked with red spots and black spines.

The adult female measures three fourths of an inch in length, the male a little less; the eye-space is red, and there are irregular red spots, which vary in number and size, along the middle of the cephalothorax, and on each side; on the abdomen there are two rows of red spots which are sometimes united into a stripe, and

Fig. 735. PEUCETIA VIRIDANS AND EGG-SAC

a variable number of pairs of yellowish, oval, oblique spots, edged with brown or red. The legs are pale and conspicuously marked with black dots, and sometimes with red lines and dots.

The spiders reach maturity in the latter part of the summer. At this time the female makes its egg-sac. This is hemispherical in outline, with small projecting tufts (Fig. 735); it is suspended by threads among twigs or leaves. After the egg-sac is completed, the female remains near it guarding it. The young emerge from the egg-sac in the autumn.

Genus OXYOPES (Ox-y-o'pes)

In this genus the posterior row of eyes is strongly procurved and the eyes of which it is composed are equidistant from each

Superfamily Argiopoidea

other (Fig. 733). These are smaller spiders than those of the preceding genus, and the colour and markings are very variable. The egg-sac is very different from that of *Peuticia;* it is discoid, very flat, and, according to Simon, is strongly fastened like that of *Philodromus,* but according to F. O. P. Cambridge it is spun up in an irregular web amongst the leaves and twigs. Several species have been described from the United States. The following are most commonly met with and may be taken as examples for the genus.

Oxyopes salticus (O. sal'ti-cus).—This pretty species is about one third of an inch in length. The cephalothorax is pale yellow, clothed with white scales, and with rows of dark scales in the middle which form bands. The clypeus is marked with a narrow line from each anterior median eye which passes over it and down each chelicera. The legs are pale, with a narrow black line beneath the femora. The abdomen is mostly white above, with a basal darker marking. The venter has a dark median band. The male resembles the female but has black palpi and a dusky abdomen which is covered with iridescent purple scales. The tarsus of the male palpus is angled on the outer side. The epigynum of the female is a forwardly directed process which is acute at the distal end. This is a southern species which is found as far north as Long Island, N. Y. and Connecticut.

Oxyopes scalaris (O. sca-lar'is).—This species is almost uniform dark brown in colour, but the body is covered with gray hairs which make the spider appear paler when alive. The epigynum is a forwardly directed process which is rounded at the end. The tarsus of the male palpus is not angled on the sides.

This species is widely distributed.

Genus HAMATALIVA* (Ham-a-tal'i-va)

In addition to the characters given in the table of genera above it may be noted that the clypeus is very wide; in other words the eyes are remote from the front edge of the clypeus. These spiders are found in the extreme southern part of our territory.

Hamataliva grisea (H. gris'e-a).—The cephalothorax is a yellowish red-brown, with the eye area darker. The abdomen is thickly clothed with brown and white hairs. In the original

---

*The original spelling of this name as *Hamataliwa* (Keyserling '87 p. 458) is obviously a typographical error; on page 489 of the same paper it is spelled as above.

Superfamily Argiopoidea

description the only locality given is North America. I collected the species at Miami, Fla. This is the only species reported from the United States; but I collected what appears to be a distinct species at Austin, Tex.

Family ATTIDÆ (At'ti'dæ)

The Jumping Spiders

The jumping spiders are of medium or small size, with a short body and stout legs furnished with two tarsal claws. They are common on plants, logs, fences, and the sides of buildings. They are apt to attract attention by their peculiar appearance; their short, stout legs, bright colours, conspicuous eyes, quick jumping movements being very different from those of other spiders.

The members of this family can be easily recognized by the characteristic arrangement of their eyes and the relative size of the different pairs of eyes (Fig. 736). The eyes occupy the whole length of the head-part of the cephalothorax, and limit a quadrilateral area, which is termed the *ocular quadrangle*. The anterior eyes are situated on the vertical face and are large; the anterior median eyes are very large. The posterior median eyes, which are usually designated as *the small eyes*, are very small, often difficult to see, and

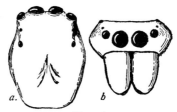

Fig. 736.
EYES OF A JUMPING SPIDER
a, cephalothorax from above
b, face and chelicerae

are situated in front of the posterior lateral eyes, sometimes called the *dorsal eyes*, forming a transverse row distinct from them; the eyes of these spiders are therefore in three rows. This arrangement recalls that of the eyes of the Lycosidæ, but in that family the eyes of the second row are large. The eyes of the jumping spiders are all of the diurnal type.

In a small group of genera, represented in our fauna only by *Lyssomanes*, the anterior lateral eyes are situated behind the anterior median eyes, thus forming a second row of eyes (Fig. 737); the eyes of these spiders are therefore in four rows, each row consisting of two eyes.

The body is usually thickly covered with hair or with scales.

Superfamily Argiopoidea

and many species are brightly coloured, even iridescent; but unfortunately the appearance of a specimen is usually greatly changed when it is put in alcohol. In certain members of this family the body is longer than in the typical forms, and ant-like in appearance.

The sexes differ little in size; but often they differ much in colour and in the form of the clothing of hairs and scales. In many cases the males have peculiar bunches of hairs on the front legs. At the mating time, the males of some species have been observed to dance before the females, and to assume singular attitudes, holding their legs extended sidewise or over their heads in such a way as to display their ornaments (Fig. 738), or moving them about so as to attract attention. These curious habits have been carefully described by Mr. and Mrs. Peckham ('89 and '90).

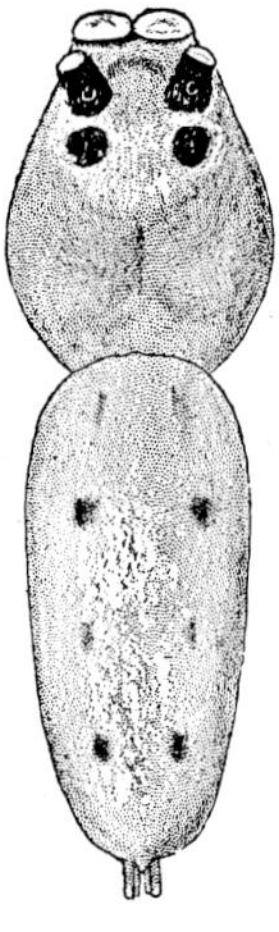

The jumping spiders are hunters, pursuing their prey or springing upon it when it comes near them. They move sidewise or backward with great ease, and can jump a long distance. They spin a dragline. I have seen them jump away from the side of a building, to catch an insect flying near, and quickly regain their position by means of the dragline.

Fig. 737. LYSSOMANES VIRIDIS

Fig. 738. MALE OF PELLENES VIRIDIPES DISPLAYING ORNAMENTS (after Peckham)

They make no webs except nests in which they hide in winter or when moulting or laying eggs. These nests are sac-like in form, composed of several envelopes, and usually furnished with two openings.

The egg-sacs are frail; as they are made within the sac-like nest, there is not the necessity for a dense cocoon that there is with most other spiders. The cocoon is usually lens-shaped and suspended, like a hammock, from the walls of the nest. There may be several cocoons within a single nest; but usually there is only one. With the species that we have observed, the eggs are laid early in the season, and the young soon hatch. They are guarded by the parent female until they disperse. On the approach of winter the young make the sac-like nests in which to pass the winter.

The Attidæ is one of the largest and most widely distributed of the families of spiders; representatives of it are found in all parts of the world except in the polar regions and several thousand species have been described. They are especially abundant in the tropics; and nearly three hundred species, representing more than forty genera, are known to occur in America north of Mexico.

A revision of the American species of the family has been published by Mr. and Mrs. Peckham ('09) in which are given detailed descriptions of all of our known species; this work is indispensable to one making an exhaustive study of this family, and has been freely used in a revision of the following account, which was mostly written before the appearance of the work of the Peckhams.

I have omitted descriptions of some Mexican genera, representatives of which are found in this country only along our southern border, also of some genera represented only by one or two rare species; and in case of the larger genera, it has seemed best to include accounts of only the more common species.

The genera of this family are grouped by Simon into three, somewhat artificial sections, which are distinguished by the nature of the armature of the lower margin of the furrow of the chelicerae, as follows:

I. With the lower margin of the furrow of the chelicerae armed with several, isolated teeth forming a series (Fig. 739, a).   Attidæ Pluridentati

II. With the lower margin of the furrow of the chelicerae armed with a single tooth or unarmed (Fig. 739, c).

Attidæ Unidentati

III. With the lower margin of the furrow of the chelicerae armed with a large tooth which is compressed so as to form a keel, and divided into two points by a notch, rarely truncate or furnished with a serrula (Fig. 739, b).   Attidæ Fissidentati

Superfamily Argiopoidea

Nearly all of our genera belong to the Attidæ Unidentati; and of those that belong to the other two sections according to Simon there are several that do not clearly exhibit the distinctive character of the section in which they are placed.

The following North American genera are placed by Simon in the Attidæ Pluridentati: *Lyssomanes, Thiodina, Ballus,* and *Synemosyna*. In our species of *Lyssomanes* (Fig. 740), the lower margin of the chelicera is armed with three isolated teeth and a large fissidentate tooth. In the examples of *Thiodina* that I have studied the chelicera is armed with a single fissidentate tooth (Fig. 741, and Fig. 742). Of this genus the Peckhams state: "The teeth on the lower border of the falces are exceedingly variable; in some the tooth is compound, in others single and in a few there are several teeth. Even the two falces in the same specimen are occasionally unlike." Of the genus *Synemosyna*, in *S. formica* the cheliceræ of both sexes are fissidentate (Fig. 743, and Fig. 744). Of the genus *Ballus* I have not seen specimens; but the Peckhams state that here the cheliceræ are pluridentate.

Among the Attidæ Unidentate I found that *Wala palmarum* is fissidentate; and the Peckhams state that this is also true of *Icius wickhamii*.

Under the section Attidæ Fissidentati are included by Simon the following North American genera: *Mævia, Fuentes, Zygoballus,* and *Peckhamia*. These are all truly fissidentate except *Fuentes tæniola*, which is unidentate, and which has been made the type of the genus *Metacyrba*. *Fuentes lineatus* has been made the type of the new genus *Onondaga*, and there is left no species of *Fuentes* in our fauna.

For the reasons above given, it has not seemed wise to attempt to separate in the table of genera the three sections based on the nature of the armature of the lower margin of the cheliceræ, but the sequence of genera adopted in the text following the table follows closely that used by Simon.

TABLE OF GENERA OF THE ATTIDÆ

A. The posterior margin of the cephalothorax and the pedicel covered by the abdomen; the second and third coxæ of each side contiguous. Not ant-like spiders.
  B. Eyes in four rows. P. 677.   LYSSOMANES
  BB. Eyes in three rows.

Superfamily Argiopoidea

Fig. 739.
THE THREE TYPES OF CHELICERÆ
a, pluridentate   *b*, fissidentate
c, unidentate (after Simon)

Fig. 740.
CHELICERA OF LYSSOMANES

Fig. 741.
CHELICERA OF THIODINA
PUERPERA, MALE

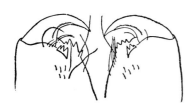

Fig. 742.
CHELICERÆ OF THIODINA PUERPERA
FEMALE

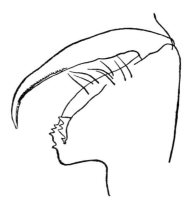

Fig. 743.
CHELICERA OF SYNEMOSYNA
FORMICA, MALE

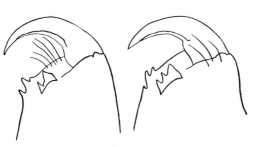

Fig. 744.
CHELICERÆ OF SYNEMOSYNA FORMICA, FEMALE

673

Superfamily Argiopoidea

C. Tibia and patella of the third leg shorter than tibia and patella of the fourth leg.
D. Small eyes situated midway between the anterior lateral and the posterior lateral eyes, or nearly so.
E. Sternum not greatly narrowed in front, the anterior coxæ being separated by a distance greater than the width of the labium.
F. Quadrangle of eyes occupying less than one half of the length of the cephalothorax.
G. Quadrangle of eyes narrower behind than before.
H. Cheliceræ unarmed below. P. 683.
EUOPHRYS
HH. Cheliceræ with one tooth below. P. 682.
STOIDIS
GG. Quadrangle of eyes as wide behind as before or wider.
H. Abdomen marked with longitudinal white stripes.
I. Abdomen with two white stripes (*Wala palmarum*, male). P. 687. WALA
II. Abdomen with three white stripes. P. 696.
PHLEGRA
HH. Abdomen not marked with longitudinal bands.
I. Abdomen marked with transverse white bands.
J. Cephalothorax much longer than wide.
K. Quadrangle of eyes occupying much less than one half the cephalothorax. Posterior margin of anterior eyes in a straight or slightly procurved line. P. 684. SALTICUS
KK. Quadrangle of eyes occupying about one half the cephalothorax; posterior margin of anterior eyes in a recurved line. (*Dendryphantes capitatus*.) P. 692. DENDRYPHANTES
JJ. Cephalothorax slightly longer than wide. P. 679. BALLUS

II. Abdomen not marked with transverse white bands, except sometimes a whitish band at base.
    J. Posterior three pairs of legs white, translucent, without marks. P. 687.
                                              WALA
    JJ. Posterior three pairs of legs not uniform white. P. 681.   ICIUS
FF. Quadrangle of eyes occupying more than one half of the length of the cephalothorax.
    G. Thorax sloping from just behind the posterior eyes; posterior eyes prominent; total length 2.5 mm. P. 683.   NEON
    GG. Thorax nearly flat in the first half then sloping abruptly; posterior eyes not prominent. Total length 5 + mm. P. 684.   SITTICUS
EE. The sternum narrowed in front so that the anterior coxæ are separated by a distance less than the width of the labium.
    F. Tibia of the first legs with four pairs of spines beneath.
        G. Lower margin of the furrow of the chelicera armed with a single tooth. P. 685. HYCTIA
        GG. Lower margin of the furrow of the cheliceræ armed with a compound tooth.
            H. Abdomen with four longitudinal white lines on a dark ground. P. 703.
                                ONONDAGA
            HH. Abdomen without white lines. P. 702.
                                MÆVIA
    FF. Tibia of the first legs with less than four pairs of spines beneath.
        G. Tibia of the first legs with two pairs of spines beneath or with less. P. 702.
                                METACYRBA
        GG. Tibia of the first legs with three pairs of spines beneath and some anterior lateral spines.
            H. First pair of legs stouter than the second pair. P. 686.   MARPISSA

Superfamily Argiopoidea

        HH. First two pairs of legs similar. P. 701.
                                            PLEXIPPUS
DD. Small eyes much more remote from the posterior eyes than from the anterior eyes.
    E. Ocular quadrangle much wider behind than before.
        F. Cephalothorax very wide, the thoracic part* shorter than the head-part or at all events not longer.
            G. Abdomen marked with white in addition to the basal band. P. 704.   ZYGOBALLUS
            GG. Abdomen not marked with white except a basal band. P. 695.   AGASSA
        FF. Thorax longer than head. Pass to F. and FF. under EE. immediately below.
    EE. The sides of the ocular quadrangle nearly parallel.
        F. Cephalothorax more or less depressed behind the posterior eyes and furnished with a median furrow situated behind the eyes but before the middle of the thoracic part.
            G. Cephalothorax high, heavy, and convex; first legs heavy and very hairy, often fringed; large species, rarely less than .28 inch in length. P. 683.   PHIDIPPUS
            GG. Cephalothorax not heavy; first legs not especially hairy; small species, rarely more than .24 inch in length.
                H. Third leg as long as or longer than fourth. P. 682.   TALAVERA
                HH. Fourth leg longer than the third.
                    I. Posterior three pairs of legs white, translucent, without marks. P. 687.   WALA
                    II. Posterior three pairs of legs not uniform white. P. 692.   DENDRYPHANTES
        FF. Cephalothorax not depressed behind the eyes; median furrow remote from the eyes, very small, obsolete.
            G. Cephalothorax narrow and long; no spines on the tibia of the first legs. P. 687.
                                          ADMESTINA

---

*In descriptions of Attidæ that part of the cephalothorax behind the posterior eyes is regarded as the thoracic part, and the remainder as the cephalic part.

Superfamily Argiopoidea

      GG. Cephalothorax short and thick; two or three pairs of spines on the tibia of the first legs. P. 694.   SASSACUS

  CC. Tibia and patella of the third leg as long as or longer than the tibia and patella of the fourth leg.

    D. Quadrangle of eyes wider in front than behind. P. 695.   HABROCESTUM

    DD. Quadrangle of eyes with sides parallel or wider behind than in front.

      E. Tibia of first pair of legs with four bulbous setæ. P. 678.   THIODINA

      EE. Tibia of first pair of legs without four bulbous setæ. P. 696.   PELLENES

AA. The posterior margin of the cephalothorax and the pedicel of the abdomen visible from above. Ant-like spiders.

  B. Hind part of thorax narrow and with parallel sides, thus adding to the apparent length of the pedicel. P. 679.   SYNEMOSYNA

  BB. Hind part of thorax not narrowed so as to add to apparent length of pedicel.

    C. Quadrangle of eyes more than one half the length of the cephalothorax. P. 707.   PECKHAMIA

    CC. Quadrangle of eyes less than one half the length of the cephalothorax. P. 680.   MYRMARCHNE

Genus LYSSOMANES (Lys-som'a-nes)

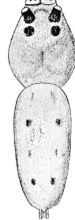

Fig. 745. LYSSOMANES VIRIDIS

This genus is sharply distinguished from all others represented in our fauna by the position of the anterior lateral eyes behind the anterior median eyes and occupying an area but little if any wider than that occupied by the anterior median eyes, the eyes being in four rows, each row consisting of two eyes (Fig. 745). Only a single species has been found in the United States.

Lyssomanes viridis (L. vir'i-dis).— This is a light green spider common in the Southern States. The male is one fourth inch in length; the female one third. There are four black tubercles on the head; the first pair of tubercles bear the second and third rows of eyes; and the second pair, the fourth row. There are usually four pairs of black dots on the abdomen; but these are sometimes wanting. The armature of the chelicera of the female is represented in Fig. 740.

Hentz, who first described this spider, says that it is very active, and apparently fearless, jumping on the hand that threatens it.

Genus THIODINA (Thi-o-di'na)

This genus is the only representative in our fauna of a group of genera distinguished by the presence of four bulbous setæ on the tibiæ of the first pair of legs (Fig. 746). The function of these setæ is not known; but they are believed to be sense-organs. Two very closely allied species have been found in the United States.

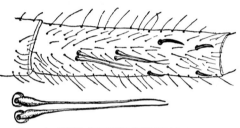

Fig. 746. TIBIA OF FIRST LEG OF THIODINA

Thiodina sylvana (T. syl-va'na). — In the male "the cephalothorax varies from light to dark reddish brown and has on the cephalic-plate a large oval spot of white just in front of the dorsal eyes, two white lines on each side near the posterior border, running upward from the lower margin, and just behind and below each dorsal eye three short parallel lines; the abdomen varies from light testaceous to brown and has on the dorsum two longitudinal white bands; on each side of the bands are some scattered black dots." In the female, "the cephalothorax is yellow, darkest in the eye-region; the eyes are on black spots, and there is a brown spot just above the anterior middle eyes, the abdomen is yellow with three longitudinal white bands, the middle often less distinct, and many black dots; the other parts are yellow" (Peckham). The male measures from .36 to .44 inch in length; the female from .36 to .48 inch. This species is found in the South and is distributed from the Atlantic to the Pacific.

Thiodina puerpera (T. pu-er'pe-ra).— "The general colour of the male is like sylvana, but instead of the oval white spot there is a white band running from between the dorsal eyes down the thoracic slope; under the dorsal eyes there is a white band instead of the three lines, and there is only one white line on each side coming up from the lower margin." "The female can be distinguished from sylvana only by its smaller size, the slight differences in the spines of the second tibia and the epigynum" (Peckham). The male measures from .20 to .24 inch in length; the female from .26 to .40 inch. The armature of the chelicera is represented in Figs. 741 and 742. See also the statement on page 664. This species is found in the Gulf States and extends north to Pennsylvania.

Genus BALLUS (Bal'lus)

The spiders of this genus are small, short, and stocky, with the abdomen not much larger than the cephalothorax. They jump little, but run over plants and build their sacs under stones and under bark. The following is our only known species.

Ballus youngii (B. young'i-i).— This spider measures less than one eighth inch in length; it is black, thinly covered with short yellow hairs; on the abdomen the thickening of these hairs form two yellow spots on the anterior part, and three transverse yellow bands. The species has been found in Pennsylvania and Wisconsin.

Genus SYNEMOSYNA (Sy-nem-o-sy'na)

This is one of three genera of ant-like spiders that occur in our fauna north of Texas. It is distinguished from *Peckhamia* by the greater length of the thorax, the quadrangle of eyes occupying only about one third of the length of the cephalothorax; and from *Myrmarachne* by the form of the hind part of the cephalothorax which is narrow, with parallel sides, thus adding to the apparent length of the pedicel.

The cephalothorax is strongly constricted a short distance behind the posterior eyes but is inflated behind this constriction, so that the middle portion of the thorax is about as high as the head. There is a deep dorsal depression near the middle of the length of the abdomen, opposite which there is usually, but not always, a lateral constriction.

A single species occurs in our fauna.

Superfamily Argiopoidea

Synemosyna formica (S. for-mi'ca).— This is the most common of our ant-like spiders, and the one in which the mimicry is most perfect (Fig. 747). It is about one fourth inch in length and very slender. The cephalothorax and basal part of the abdomen is brown; behind the constriction the abdomen is black. There is a pale white spot in front of the constriction of the cephalothorax, and one on each side widening downward under the posterior eyes. There is also a white stripe on the abdomen extending downward from the dorsal depression on each side and uniting in a large white patch underneath. The chelicerae differ in the two sexes; those of the female are cylindrical with the inferior margin of the furrow armed with one tooth subdivided by notches into two or three unequal points (Fig. 744); those of the male are much longer and more angular, and armed on the lower margin of the furrow with a large tooth divided by notches into four unequal points (Fig. 743).

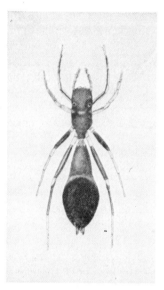

Fig. 747.
SYNEMOSYNA FORMICA

This species lives on plants and runs like an ant.

Genus MYRMARACHNE (Myr-ma-rach'ne)

The genus *Myrmarachne* is a large one and is widely distributed; but it is doubtful if any species occur in our fauna. It includes ant-like spiders, in which the sternum is long and narrow, and in which the lower margin of the furrow of the chelicerae is armed with five or six teeth. In this genus is placed by the Peckhams the following species although the sternum is wider than is typical of *Myrmarachne* and the lower margin of the furrow of the chelicerae is armed with a single tooth. It seems to me that the species in question is more closely allied to the tropical genus *Zuninga* than to this one; but to avoid confusion I leave it where it is placed by the Peckhams.

Myrmarachne albocinctus (M. al-bo-cinc'tus).— This is an ant-like spider which differs from the species of *Peckhamia* in that the quadrangle of eyes occupies less than half of the length of the cephalothorax, and from *Synemosyna* in that the hind part of the thorax is not narrowed so as to resemble a pedicel. In the palpi the tibia and tarsus are very much thickened and covered with stiff hairs on the under side. The cephalothorax is yellowish brown with an indistinct light mark across the middle. The abdomen has, at the constriction, a white band, which slopes backward along the sides and nearly to the spinnerets. The length of the body is from .20 to .28 inch.

This species is widely distributed throughout the eastern half of the United States.

Genus ICIUS (Ic'i-us)

The cephalothorax is not very high and only slightly convex, with the sides usually nearly parallel. The quadrangle of eyes occupies less than half of the length of the cephalothorax; the small eyes vary in position from midway between the anterior laterals and posterior laterals to nearer the anterior laterals. The sternum is not greatly narrowed in front. In our common species the abdomen is iridescent or metallic.

Ten species have been found in the United States; of these the following are the more common.

Icius elegans (I. el'e-gans).— This is a small spider measuring from one sixth to one fourth inch in length. It is bronze-green, being clothed with brilliant iridescent scales. The legs are yellow, with a black longitudinal line above; but in the female the femur of the first is nearly all black; and in the male there is on the inner side of the tibia of the first legs near the distal end an inky black spot, from which grows a fringe of black hairs. In the male there are two tufts of long hairs pointed forward. In both sexes there is a white band around the lower margin of the cephalothorax. The abdomen of the male is without bands or spots; but in the female there is a white basal band.

This species is widely distributed in the eastern half of the United States.

Icius similis (I. sim'i-lis).— This species very closely resembles the preceding; but the male lacks the dark spot at the

Superfamily Argiopoidea

end of the tibia of the first leg; and in the female there is no white basal band on the abdomen. It occurs throughout the greater part of the United States.

Icius hartii (I. hart'i-i).— This species is unusually low and flat. In the male there are no tufts over the anterior eyes; and the femur, patella, and tibia of the first legs are enlarged. In both sexes the cephalothorax is brown with a covering of gray hairs and the abdomen is dark gray with a white border broken into bars; sometimes the abdomen has a metallic lustre. It occurs from Massachusetts to Nebraska.

Genus TALAVERA (Tal-a-ve'ra)

This genus includes very small spiders, with the cephalothorax moderately high and with the sides parallel and vertical. The eye-region occupies nearly half of the length of the cephalothorax, is about one fourth wider than long, and is wider in front than behind. The second row of eyes is about midway between the others. The front coxæ are separated by a little more than the width of the labium.

A single known species occurs in our fauna.

Talavera minuta (T. mi-nu'ta).— This is a small species, the male measuring only one twelfth inch in length; the female, one tenth. "The cephalothorax is reddish brown, with the eye-region blackish. The abdomen, in the male, is black, in the female gray with indistinct pale chevrons. The legs are white banded with black, except the femur of the first in the female, which is more of less darkened, and the femur and tibia of the first in the male, which are entirely dark. The palpi are white, the falces yellowish" (Peckham).

This is not a common species; but it is distributed from the Atlantic to the Pacific.

Genus STOIDIS (Sto'i-dis)

In our species of this genus the small eyes are situated midway between the first and third rows; the sternum is not greatly narrowed, but is nearly round; and the ocular quadrangle is much wider in front than behind.

A single Floridian species represents this genus in this country.

Stoidis aurata (S. au-ra′ta).— The cephalothorax is black, with a few white scales in front of the posterior lateral eyes, and a white band on the sides of the thoracic part. The abdomen is grayish brown with four black spots, the two of each side more or less connected. The adult measures from one fifth to one fourth inch in length.

Genus NEON (Ne′on)

To this genus belong the smallest species of the Attidæ known. The quadrangle of eyes occupies more than one half of the length of the cephalothorax; the cephalothorax is flat, for a short distance inclined cephalad, and abruptly sloping behind the eyes. The anterior eyes are large, close together, and in a straight line; the posterior lateral eyes are very large and projecting; they form a row as wide as the cephalothorax.

Neon nellii (N. nel′li-i).— The adult is only one tenth inch in length. The cephalothorax is brown, darkest in the eye-region. The abdomen is brown with pale spots and chevrons. This minute species is common under stones and leaves at all seasons.

Genus EUOPHRYS (Eu-oph′rys)

The small eyes are halfway between those of the first and third rows or a little nearer the third row; the quadrangle of eyes occupies less than one half of the length of the cephalothorax; the sternum is oval, widely separating the coxæ of the first legs; and the lower margin of the furrow of the cheliceræ is unarmed.

Two uncommon species occur in our fauna.

Euophrys monadnock (E. mo-nad′nock).— The male measures one sixth inch in length; the female, one fifth. The colour of the male is black; the cephalothorax of the female is dark brown in front and lighter behind, marked with dark radiating lines, the abdomen is pale gray with light chevrons along the middle and irregular oblique lines on the sides. Although not common, this species is distributed from New Hampshire to California.

Euophrys cruciatus (E. cru-ci-a′tus).— Of this species only the male is known; this was found in New Hampshire. The abdomen is black with a distinct white cross in the middle. The length of the body is one fifth inch.

Superfamily Argiopoidea

Genus SITTICUS (Sit'ti-cus)

The lower margin of the furrow of the cheliceræ is unarmed except in a Californian species. The cephalothorax is high, convex, and rather wide; the abdomen is short and wide. The quadrangle of the eyes is nearly twice as wide as long, and occupies less than two fifths of the length of the cephalothorax. The coxæ of the first legs are widely separated.

The following is our best known and most widely distributed species. Several other species have been described, most of them from the western and southwestern portions of our country.

Sitticus palustris (S. pa-lus'tris).— The adult measures from one fifth to one fourth inch in length. The cephalothorax is light brown in the female and dark in the male; there is a narrow white stripe in the middle, and one on each side, as high as the lateral eyes. The abdomen has a wide, white transverse marking, just behind the middle, and several angular marks behind it. On the front half are two white spots. In the male the large middle marking is usually divided into two white spots. This species lives on plants and makes nests among leaves (Emerton); it is distributed from New York to Colorado and from Anticosti to Alberta.

Fig. 748. SALTICUS SENICUS

Genus SALTICUS (Sal'ti-cus)

The cephalthorax is much longer than wide, narrow in front and a little wider behind; the thorax is nearly flat in the first half and then slopes abruptly. The anterior eyes are very unequal in size; and the posterior eyes are not prominent. The quadrangle of eyes occupies much less than one half of the length of the cephalothorax. The abdomen is long and narrow and the sides are nearly parallel.

Four species have been found in our fauna; but three of them are known only by a few specimens from the Southwest.

Salticus senicus (S. sen'i-cus).— This species is one of the commonest members of the Attidæ in our fauna (Fig. 748). It is

Superfamily Argiopoidea

gray and white, about one fourth inch long, and lives on the sides of houses and on fences. The front of the head around and above the eyes is white. There is a white band across the anterior end of the abdomen, and two or three oblique white bands on the sides. In some cases, according to the Peckhams, the oblique bands meet on the back, and a longitudinal white band passes down the middle of the abdomen, widening, at the spinnerets.

Genus HYCTIA (Hyc'ti-a)

This genus includes long, slender spiders, with the first pair of legs much stouter than the other legs, and with the tibia of the first legs with four pairs of spines beneath. The sternum is narrowed in front so that the anterior coxæ are separated by a distance less than the width of the labium. Three species have been found in the United States.

Hyctia pikei (H. pi'ke-i).— This species can be easily distinguished from all other attids occurring in our fauna by the shape of the abdomen, which is more than three times as long as wide. The male measures from .28 to .35 inch in length; the female from .32 to .38 inch. In the male there is a wide black stripe, with four notches on each side, on the middle of the abdomen, and a white stripe on each side. In the female the central band is dark brown and is less definite and the sides are not so white as in the male.

This species is widely distributed in the eastern half of the United States.

Hyctia bina (H. bi'na).— In this species the abdomen is only twice as long as wide. In the male the abdomen is bronze-brown, with a white basal band, extending back about one third of the length of the abdomen on the sides and a broken white band on each side of the middle. In the female there is a white, median band extending the entire length of the abdomen, on each side of which there is a black band which does not reach either end of the abdomen.

This species is also widely distributed in the eastern half of the United States.

Hyctia robusta (H. ro-bus'ta).— This species has been found only in the Far West. The abdomen is three times as long as wide. It is a larger and hairier species than *H. pikei* and has white in

Superfamily Argiopoidea

the middle of the abdomen where *H. pikei* has dark brown. Only the female is known.

Genus MARPISSA (Mar-pis'sa)

The first pair of legs are slightly stouter than the second pair and are armed with three pairs of spines beneath the tibia. The abdomen is marked by an angular central band. The genus is represented by three species in this country.

Marpissa undata (M. un-da'ta).— This is a common species in the eastern half of the United States and is found as far west as Utah. It measures from .38 to .42 inch in length. The dorsal aspect of the cephalothorax is of a uniform gray; and there is a wide gray band with indented margins along the middle of the abdomen (Fig. 749); this band is bordered with black.

Fig. 749.
MARPISSA UNDATA

Fig. 750.
MARPISSA BIVITTATUS, MALE

Marpissa californica.—This species occurs in the Far West. It is smaller than *M. undata* and has a wide gray band, bounded by dark brown, on the cephalothorax.

Marpissa bivittatus (M. bi-vit-ta'tus).—This form differs from the other two species in having a white band bounded by dark rufus on the cephalothorax. In the male the abdomen has bright rufus bands in the middle and on the sides, alternating with two longitudinal white bands (Fig. 750). In the female

(Fig. 751) the whole upper surface of the abdomen is covered with mixed gray and rufus hairs.

This species occurs in Florida.

Genus ADMESTINA (Ad-mes-ti'na)

The cephalothorax is long, narrow in front and wider in the thoracic part; the median furrow is remote from the eyes, very small, almost obsolete. There are no spines on the tibia of the first legs.

Admestina tibialis (A. tib-i-a'lis).— This is a small species measuring only one sixth inch in length. The cephalothorax is black or brown, with the cephalic part often marked with two obscure spots. The abdomen is pale and marked with one or two black bands which are often notched; the first legs are brown, the others are pale, spotted or lined in black. It is found in the Eastern United States and west to Wisconsin and Texas.

Fig. 751.
MARPISSA BIVITTATUS, FEMALE

Fig. 752.
WALA PALMARUM, MALE

Genus WALA (Wa'la)

The cephalothorax is nearly flat, longer than wide, margins rounded near the middle. The ocular quadrangle occupies less than one half of the length of the cephalothorax, is wider than long, and is wider behind than in front; the anterior eyes are in a straight or slightly procurved line, with the median eyes sub-contiguous. The abdomen is long and narrow. The first legs in the males are much longer than the others.

Two species are widely distributed in the eastern half of the United States; a third, a South American species, *Wala grenada*, has been found by Mr. Banks in southern Florida.

Wala palmarum (W. pal-ma'rum).— The male is reddish, with a broad white band on the sides extending the whole length of the body (Fig. 752). The clypeus is marked with white hairs and yellow scales in the eye-region; and there are dark spots at the base of the eyes. The front legs are dark, all others are white. In the female the cephalothorax is reddish, and the eyes are on dark spots. The legs and palpi are white. The abdomen is light, with large triangular spots in the centre and small dark spots at the sides. The length of the body is about one fifth inch.

This species is common on trees and bushes throughout the summer months in the Eastern United States.

Wala mitrata (W. mi-tra'ta).— This species closely resembles the preceding one; but the legs are all white in both sexes, and the cheliceræ of the male are white and not long and horizontal as in *W. palmarum*. The distribution of the species is similar to that of *W. palmarum*.

Genus PHIDIPPUS (Phi-dip'pus)

The cephalothorax is high, heavy, and convex; the first legs are heavy and very hairy, and are often fringed. This genus includes spiders which are above the medium size, and is represented by thirty-eight known species in our fauna. The following are those most likely to be taken in the more thickly settled portions of our country. These can be separated by the following table. A table to all of our species is given by the Peckhams ('09), from which this is compiled.

A. Abdomen red or marked with red.
  B. Males.
    C. Cephalothorax red above. P. 690.   *P. whitmanii*
  CC. Cephalothorax black above.
    D. Abdomen black spotted with red. P. 691.
                                 P. insolens (in part)

DD. Abdomen red, or red banded with black.
  E. Abdomen banded with black. P. 691.
  P. clarus
  EE. Abdomen uniform red. P. 691.
  P. insolens (in part)
BB. Females.
  C. Cephalic plate with two black patches between which the colour is orange or red; abdomen with seven spots red, edged with black. P. 692. *P. mineatus*
  CC. Markings unlike those of *P. mineatus*
    D. Abdomen marked with black.
      E. Abdomen with two black bands. P. 691.
      P. clarus
      EE. Abdomen with one black band. P. 691.
      P. insolens (in part)
    DD. Abdomen all red.
      E. Cephalic plate with a bare black region back of first row of eyes. P. 690. *P. whitmanii*
      EE. Cephalic plate all red. P. 691.
      P. insolens (in part)
AA. Abdomen neither red nor marked with red.
  B. Abdomen black marked with white.
    C. Cephalothorax dark, distinctly marked with white. P. 690. *P. variegatus*
    CC. Cephalothorax all black. P. 689. *P. audax*
  BB. Abdomen brown or gray and spotted.
    C. Cephalic plate with a transverse row of white spots. P. 692. *P. mystaceus*
    CC. Cephalic plate without a row of white spots. P. 691.
    P. purpuratus

Phidippus audax (P. au′dax).—The cephalothorax and abdomen are black with many long white hairs. The abdomen is marked by a white basal band, a large more or less triangular central white spot, behind which are two pairs of white bars. In front of the large white spot is a pair of indistinct white spots, in some individuals there are traces of oblique lateral stripes (Fig. 753). In the middle of the back, behind the large white spot, there is a metallic band.

This is a common and widely distributed species throughout

Superfamily Argiopoidea

the East and as far west as Texas and Colorado. It varies in size from one third to one half inch in length; it lives under sticks and stones and passes the winter half-grown in a silken bag.

Phidippus variegatus (P. va-ri-e-ga'tus).— This is a large black species measuring from .44 to .60 inch in length. The cephalothorax is black with two white bands on the sides which do not meet behind (Fig. 754). The abdomen is black marked

Fig. 753.
PHIDIPPUS AUDAX

with white; there is a basal band which extends back on the sides, a large central triangular spot, and farther back on the sides there are two pairs of bars. The tibiæ of the first legs are heavily fringed.

Fig. 754. PHIDIPPUS VARIEGATUS

This is a Southern species occurring throughout the Gulf States.

Phidippus whitmanii (P. whit-man'i-i).— The body in this species is almost entirely red above. In the male the cephalothorax is red except for a black hairless region extending from the first to the second row of eyes; and in the female a light band comes up from between the anterior median eyes, nearly crossing the black hairless region. The abdomen is red with a more or less distinct white band on the base and sides, and sometimes there are two pairs of white bars on the posterior part.

This is a Northern species; it has been taken from New England to Wisconsin.

Phidippus clarus (P. cla′rus).— This is a red and black species. In the male the cephalothorax is black and the abdomen red with a central longitudinal black band, the margins of which are notched by three pairs of red or white spots, and there is a basal white band and oblique white bands on the sides. The first pair of legs are stout and long, exceeding the second pair by the length of the tarsus and a part of the tibia. In the female both the cephalothorax and abdomen are red; upon the abdomen there are two longitudinal black stripes spotted with white, and a basal white band, and oblique white bands at the sides. The male measures from .20 to .34 inch in length; the female from .32 to .52 inch.

This is a widely distributed species, occurring from the Atlantic to the Pacific.

Phidippus insolens (P. in′so-lens).— This is a Southern red and black species, which is dimorphic in both sexes. "The male has a black cephalothorax covered with inconspicuous brownish hairs. The abdomen may be black with red bands and spots, or may be uniform red, in which case the pattern shows in deeper spots of colour when the spider is under alcohol. In the black form the red marks consist of a basal band, an oblique band on each side, a pair of spots directly behind the basal band, a central triangular spot, and two pairs of bars farther back." "In the female the cephalothorax is red on the upper surface, the sides and thoracic slope being black. The abdomen, as in the male, has two forms, being sometimes of a uniform red on the dorsum, or, in some cases, with a small black V pointing up, just above the spinnerets, a faint white basal band, and one white diagonal on each side; while others, as in the type, have the basal band and diagonals yellowish and marked with a wide central black band which reaches from the spinnerets to a point in front of the middle." (Peckham.) The male measures .34 inch in length; the female from .40 to .56 inch.

Phidippus purpuratus (P. pur-pu-ra′tus).— This is a common widely distributed species, thickly clothed with light gray hairs; in alcohol it appears brown. The abdomen is light gray with a broad dark central band, which is more or less distinctly divided along the middle line, and which is marked with four pairs of white spots (Fig. 755).

This spider is usually found under stones or other objects lying on the ground and often in a silken nest.

Superfamily Argiopoidea

Phidippus mystaceus (P. mys-ta′ce-us).— I took this fine species at Austin, Tex.; it was originally described by Hentz from North Carolina; it is probable, therefore, that it is distributed throughout the Gulf States. The female only is known; this sex measures from .38 to .44 inch in length. The body is clothed with gray hairs and is spotted and banded with white as shown in Fig. 756. Instead of the large central white spot on the abdomen there is, in some individuals, a pair of spots more or less coalesced, or entirely distinct.

Phidippus miniatus (P. min-i-a′tus).— This is another

Fig. 755. PHIDIPPUS PURPURATUS   Fig. 756. PHIDIPPUS MYSTACEUS

Southern species of which only the female is known (Fig. 757). It is a large hairy spider, marked with red. The female measures from .52 to .72 inch in length. The cephalothorax is clothed with yellowish gray hairs and has two black patches in the eye-region, between which the colour is orange or red. The abdomen is covered with short bright red and long whitish hairs; there is a basal band of gray hairs, and seven red spots edged with black, a central larger spot and three pairs of spots, one in front of the central spot and two behind; these are not well shown in the figure.

Genus DENDRYPHANTES (Den-dry-phan′tes)

The genus *Dendryphantes* as recognized by American writers is much less extended than that of Simon, who includes in it the species placed by these writers in *Phidippus* as well as those classed

Superfamily Argiopoidea

by them in *Dendryphantes*. In the more restricted sense, *Dendryphantes* includes spiders of moderate size, rarely measuring more than one fourth inch in length. The cephalothorax is rather high and convex, widest in the thoracic part, highest at the third row of eyes. The anterior eyes are large, slightly separated; the second row is halfway between the other rows; and the eye-space occupies two fifths of the length of the cephalothorax.

Even as restricted here the genus is a very large one, twenty species having been described from our fauna alone; as these resemble each other closely, it is often difficult to determine them.

The two following are the most common species:

Dendryphantes marginatus (D. mar-gi-na'tus) The two sexes of this species differ greatly in appearance. The male (Fig. 758) is

Fig. 758.
DENDRYPHANTES
MARGINATUS, MALE

Fig. 757.
PHIDIPPUS MINIATUS

Fig. 759.
DENDRYPHANTES
MARGINATUS, FEMALE

yellowish brown or bronze-brown marked with white. There is a white band on each side of the cephalothorax below the eyes, and extending back nearly to the abdomen, and there is a white band around the edge of the abdomen; sometimes there are two rows of white spots on the dorsum of the abdomen. The female (Fig. 759) is brown or bronze-brown; the abdomen is marked with a basal white band, four pairs of dorsal white spots, and, on the sides, several oblique marks. The male measures one fourth inch in length, the female, one third.

This is a very common species, over a large part of our territory.

Superfamily Argiopoidea

Dendryphantes capitatus (D. cap-i-ta'tus).— In the male the cephalothorax is dark brown with a white stripe on each side under the eyes and extending back over the thorax; on the posterior part of the thorax, these stripes converge but do not meet; there are also white hairs above the anterior eyes. The abdomen is white in front and around the sides; it is dark brown in the middle; the brown area is often notched at the sides and sometimes indistinctly divided into four pairs of spots as in the female.

"The females are of two varieties, which run into each other. The light variety has the light parts white or light yellow and the dark parts dark brown covered with white hairs and scales. The cephalothorax is dark brown, thinly covered with scales, so that the dark colour shows between them in places. The abdomen is brighter yellow than the thorax, with four pairs of purplish brown spots, the second pair largest, connected with a paler brown middle marking. The dark variety is generally smaller and covered with longer hairs and scales. The dark spots on the abdomen are larger and more closely connected, so that the markings appear as light spots on a dark ground" (Emerton). The males measure from one sixth to one fifth inch in length; the females are a little larger.

This is a very common and variable species; it is found on shrubs and small trees over a large part of the United States.

Genus SASSACUS (Sas'sa-cus)

The cephalothorax is short and thick; it is flat in both sexes in the first two thirds, the posterior third slopes a little more abruptly in the male. The ocular quadrangle is wider behind than in front, and occupies one half of the length of the cephalothorax. The tibia of the first legs is armed with two or three pairs of spines.

Sassacus papenhoei (S. pap-en-ho'e-i).— The spiders of this species are small, dark, and covered with iridescent scales. The male is a little less, the female a little more than one fifth inch in length. The palpi are brown, and are covered with white hairs. The legs are reddish, with the last two segments white, with black rings at the distal end of each. The abdomen has four punctate spots on the dorsum and is encircled by a wide, snowy white band.

This species is found in the Southwest and in California.

Superfamily Argiopoidea

Genus AGASSA (A-gas'sa)

The cephalothorax is thick, about as wide as long, flat above, and hollowed behind to receive the abdomen. The ocular quadrangle occupies two thirds of the length of the cephalothorax and is much wider behind than before; the second row of eyes is much more remote from the posterior eyes than from the anterior eyes. The abdomen is oval, short, and truncate in front.

Agassa cyanea (A. cy-a'ne-a).—This is a small species, measuring a little less than one fifth inch in length. The whole body is covered with iridescent scales which give it a coppery green colour. The abdomen is marked with a narrow, yellowish white basal band. The species is distributed over a large portion of the region east of the Rocky Mountains.

Genus HABROCESTUM (Ha-bro-ces'tum)

The cephalothorax is high with the cephalic part convex, and the thoracic part steeply inclined. The ocular quadrangle occupies less than one half of the length of the cephalothorax and is wider in front than behind; the second row of eyes is halfway between the others. The abdomen is elongate oval in outline.

Four species are known to occur in our fauna; but only one of them is common and widely distributed.

Habrocestum pulex (H. pu'lex).— This is a small species, the male measuring only one sixth inch in length, the female a little more. In the male the cephalothorax is reddish, dark in the eye-region, with a narrow white triangle pointing forward, and lighter behind the eyes; the abdomen is dark brown, with two longitudinal, nearly parallel, light lines on the basal half and a broad transverse white mark just behind the middle. In the female the cephalothorax is dark, with a large triangular light spot reaching from the eyes to the hind end. The abdomen is dark, with either two light longitudinal lines on the basal half, as in the male, or with several irregular light spots in this region, and with a transverse light band just behind the middle. Around and behind the transverse band are other irregular light markings.

This species is widely distributed in the eastern half of the United States.

Superfamily Argiopoidea

Genus PHLEGRA (Phle'gra)

The cephalothorax is long and narrow, a little the widest behind the middle. The thorax is level for more than one half its length, then sloping abruptly. The anterior eyes are in a much recurved line; the ocular quadrangle is short, occupying only one third the length of the cephalothorax, with the sides parallel. The abdomen is elongate oval in outline.

Phlegra leopardus (P. le-o-par'dus).— The leopard-spider is easily recognized by its stripes of which there are two on the cephalothorax and three on the abdomen; these stripes are white on a dark ground. In the female the white stripes on the cephalothorax extend back from the anterior eyes to the hind end of the abdomen; in the male they begin at the posterior eyes. This species is widely distributed in the region east of the Rocky Mountains.

Genus PELLENES (Pel-le'nes)

The cephalothorax is high, convex, and a little longer than wide. The ocular quadrangle is wider than long, and usually wider behind than in front; the second row of eyes is about midway between the first and the third.

"The males of *Pellenes* have usually some peculiar modification of form, colour or ornament, appearing in the first and third legs. These fringes, enlargements and markings are used to attract and delight the female during courtship, the posturing and dancing being such as to show off every beauty to the greatest advantage. They make the identification of the males comparatively easy, while the females resemble each other so closely as to make it difficult to distinguish them. The young males are like the females until one or two moults from maturity" (Peckham).

In their "Revision of the Attidæ" the Peckhams describe forty species of *Pellenes* and give keys for their separation. But of many of these species only a few individuals exist in collections and of a considerable number of them only one sex is known. The following species are those that are most common; they can be separated by the following keys, compiled from those of the Peckhams.

KEY TO MALES OF PELLENES

A. First or third legs modified or fringed.
  B. Enlargements or fringes on both first and third legs.

Superfamily Argiopoidea

    C. Clypeus red; spatulate spines on tibia of first legs. Occurs in the Southern States and in Colorado. P. 699.   *P. coronatus*
    CC. Clypeus not red. P. 700.   *P. peregrinus*
  BB. Only first leg fringed or enlarged. P. 697.   *P. agilis*
AA. Legs not modified nor fringed.
  B. Abdomen with iridescent scales. P. 700.   *P. splendens*
  BB. Abdomen without iridescent scales.
    C. Abdomen dark marked with white. P. 698.   *P. borealis*
    CC. Abdomen light golden on the back with an encircling white band. P. 699.   *P. hoyi*

KEY TO FEMALES OF PELLENES

A. Clypeus marked with white and dark bands. P. 697.   *P. agilis*
AA. Clypeus not banded.
  B. Abdomen plainly banded above.
    C. Abdomen with at least one transverse band behind the basal band.
      D. Colour dark gray, banded with light gray or white; occurs in the Northern States. P. 698.   *P. borealis*
      DD. Colour yellowish with whitish bands; occurs in the Southern States and Colorado. P. 699.   *P. coronatus*
    CC. Only longitudinal bands behind the basal band on the abdomen.
      D. Abdomen with a central light band reaching a light basal band. P. 700.   *P. peregrinus*
      DD. Abdomen without a central light band reaching a light basal band. P. 700.   *P. splendens*
  BB. Abdomen not plainly banded above. P. 699.   *P. hoyi*

Pellenes agilis (P. ag'i-lis).—In the male, the cephalothorax is covered with short black hairs mixed with others of a yellowish brown colour, these latter being more numerous in the eye-region; there is a short, median, white stripe in the eye-region, and on each side a white stripe extending the whole length of the abdomen; these stripes are bowed outward at the posterior eyes and inward on the thorax. The abdomen is black, marked with white as follows: a basal band, a notched median band, and three bars on each side. The first legs are the stoutest and are heavily fringed with white, both above and below. In the female the markings of the cephalothorax are much less distinct than in the male, and gray hairs take the place of the black ones of the male, the cephalothorax being nearly all gray; the clypeus is marked with white and dark bands. The central band on the abdomen is broken up into several spots or pairs of spots. The male measures one fifth inch in length; the female, one fourth inch. The species is widely distributed throughout the region east of the Rocky Mountains.

Pellenes borealis (P. bor-e-a'lis).— Two types of the male have been described, the differences being due perhaps to the markings of the one having been rubbed away. "In the one, the cephalothorax, including the clypeus, is all black, and the abdomen is black with a white basal and a white encircling band which sends up two bars on each side, the anterior pair of which is continued in a straight transverse band across the back, in front of the middle. This transverse band is connected with the base by an indistinct stripe of white. On the posterior part of the back is a good-sized central white spot, and behind this are two white dots." "In the second male the cephalothorax has whitish bands from the dorsal eyes to near the hind margin, where they turn and pass forward along the lower side, and the upper sides are covered with gray hairs. On the abdomen the spaces between the bands and spots are filled with gray hairs. In front of the transverse band are two narrow transverse black lines, one on each side of the middle. Behind, instead of the white spot, there is a band of tiny white chevrons bordered by short oblique black bars, reaching the spinnerets. The white dots above the spinnerets are present." "In the female the ground colour of the whole body is dark gray the effect being produced by a thin covering of gray hairs on a black integument. The cephalothorax has light gray dorsal and marginal bands, and the abdomen has light gray basal and transverse bands, the two being connected by two light lines instead of by a solid band. Just behind the transverse band are two fawn dots, and farther back are four chevrons and two dots of the same colour. On the posterior sides are several oblique gray bands. The hair on the clypeus and above the front eyes is fawn" (Peckham).

Mr. Emerton discovered that the young male of this species has a red clypeus which is lacking in the adult.

The male measures a little more than one fifth inch in length; the female, one fourth inch.

The species is widely distributed in the North from New England to Wisconsin.

Pellenes coronatus (P. cor-o-na′tus).— This is a Southern species closely allied to *P. borealis*, which is a Northern species. In the male the clypeus is covered with short brilliant red hair (in *P. borealis* only immature males have the clypeus red), and on the anterior face of the tibia of the first legs there are two long spatulate spines; the third legs are also abnormal in form. The cephalothorax varies in colour from dark to pale; it has a marginal white line on each side, these bow inward till they meet at the posterior eyes and then spread outward on the thorax. The abdomen is black or brown; marked with white as follows: a basal band, a transverse stripe near the middle, a central spot behind this, and a pair of spots above the spinnerets. In the female the colour is yellowish, with whitish bands, which, however, are much less distinct than in the male. The male measures about one sixth inch in length; the female, one fifth.

This is a Southern species, the range of which extends north to Long Island and west to Texas and Colorado.

Pellenes hoyi (P. hoy′i).— "The male is a brilliant spider with variable markings. The upper part of the cephalothorax is bright yellowish red, marked, above the front eyes, by a snowy band which curves back to the eyes of the second row, where it merges in the white side region, and by a small white spot between the dorsal eyes. Below the eyes are wide white bands which occupy the entire sides in front, but are limited to the upper half farther back, the lower sides being black." "The abdomen is light golden on the back with an encircling white band, scalloped behind the middle, bordered with red around the base and front sides, and with black toward the spinnerets. Two short white lines run back from the middle of the basal band. On the posterior part, above, are one or two pairs of oblique bars."

"The female is less brilliant but not less variable than the male. The cephalothorax is covered with a mixture of orange, black and white hairs, the sides being lighter than the back. White angular lines, less distinct in front than behind, begin at the front lateral eyes, point up over the small eyes, down under the dorsals and then up again, but do not meet behind. Many specimens have a scalloped black line between the dorsal eyes. The pale golden region down the middle of the abdomen may or may not be marked in the second half with fine white chevrons. The bands on either side of the posterior central region are black, or red mixed with black, and the white basal band is mixed with red. In other females the surface is of a uniform pale brown with an encircling band gray. The white marks behind the middle consist of two long oblique lines, commonly bordered with black, which meet at an acute angle in front, and some small black-bordered chevrons further back." "The epigynum is unique, having two large anterior openings, and two other openings behind, near the edge" (Peckham). The males measure about one fifth inch in length; the females one fourth.

This species is distributed from the Atlantic to the Pacific.

Pellenes peregrinus (P. per-e-gri'nus).— On each side of the cephalothorax, there is a white stripe, which extends from the anterior lateral eyes backward under the dorsal eyes, and then curves inward on the first half of the thorax and then outward on the second half; at the posterior lateral angle of the thorax it turns forward and following the margin of the cephalothorax extends to the clypeus. On the abdomen there are three white longitudinal stripes. In the male the first pair of legs is fringed and the third pair is modified in form, the patella being widened and somewhat triangular in shape.

The male measures about one fifth inch in length; the female, one fourth inch.

This species has been taken in the Atlantic region from Maine to Florida.

Pellenes splendens (P. splen'dens).— "The male, in life, is one of the most brilliant of our Attidæ, the cephalothorax, including clypeus, being covered with iridescent blue scales, and the abdomen, above and below, with iridescent pinkish red. In alcohol, two white scallops appear behind the dorsal eyes, and the abdomen shows light bands at the base and sides and a fleur-de-lis mark in the middle, this being the pattern of the young male and of the female."

Superfamily Argiopoidea

"The female is variable. In the typical form the eye-region is mottled with short fawn and black hairs, with a darker transverse band between the dorsal eyes, behind which are two white scallops. The hairs behind and on the upper sides are yellowish brown, and those on the lower sides and clypeus are white. The abdomen is velvety black with a white band at base, a wide scalloped white band on each side behind the middle, and a central, irregular, white band which begins in front of the middle in a broad arrow head, and which may or may not reach the apex. At the end is a pair of white dots. The legs are medium brown. Another form has a white abdomen with a black band behind the base and four large black spots on the back" (Peckham). The male measures .26 inch in length; the female, .30 inch.

The range of this species extends from the Atlantic to the Pacific.

Genus PLEXIPPUS (Plex-ip'pus)

The cephalothorax is high and convex, with the sides parallel in the cephalic part and rounded on the thoracic. The ocular quadrangle occupies about one third of the length of the cephalothorax and has the sides nearly parallel. The tibia of the first legs is armed with three pairs of spines beneath; and the first two pairs of legs are similar. Only one species occurs in our fauna.

Plexippus paykullii (P. pay-kul'li-i).—This is a dark spider with a median light band extending the whole length of the body; this band is wider on the posterior half of the abdomen. In the female (Fig. 760), the median band is divided on the first half of the abdomen by a dark line, and the wider portion of this band, on the second half of the abdomen, is indistinctly marked with dark chevrons, and has a pair of transverse white bars extending from it. The male measures from .36 to .40 inch in length; the female from .36 to .48 inch.

Fig. 760.
PLEXIPPUS PAYKULLII. FEMALE

This species is found in all of the warm regions of the world, including the extreme southern portions of our territory.

Superfamily Argiopoidea

Genus METACYRBA (Met-a-cyr'ba)

The cephalothorax is low and flat; it is broadest in the middle and narrowed both in front and behind. The tibiæ of the first legs are armed with one or two short spines, or none at all, beneath. The sternum is much narrowed in front between the anterior coxæ and widened behind. The ocular quadrangle is wider than long.

The following species is widely distributed.

Metacyrba tæniola (M. tæ-ni'o-la).— This is a flat, nearly black species. The males are a little less than one fifth inch in length; the females, a little more than one fourth inch. The cephalothorax is black, smooth, and without markings or with a white line around the margin. The abdomen is dark gray, with two longitudinal narrow lines of white hairs more or less broken into short bars. The femora of the first legs are flattened and much stouter than those of the other legs, especially in the males.

This species is distributed throughout the United States and is found also in Mexico.

Genus MÆVIA (Mæ'vi-a)

The cephalothorax is rather high, with the sides nearly vertical, and nearly parallel in the head region and slightly rounded in the thoracic. The ocular quadrangle is slightly wider in front than behind and occupies less than one half of the length of the cephalothorax; the anterior eyes are in a straight or slightly recurved line; the second row of eyes is nearly midway between the other two rows. The legs are slender. The sternum is long, narrow in front, the coxæ of the first legs being nearly contiguous. The lower margin of the furrow of the chelicera is armed with a compound tooth (Fig. 761).

Fig. 761. CHELICERA OF MÆVIA VITTATA

As now restricted only two species of this genus occur in our fauna, *M. poultonii* from Texas and Arizona, and the following.

Mævia vittata (M. vit-ta'ta).— This species is of especial interest on account of a striking dimorphism of the male; it is a widely distributed and very common one.

Superfamily Argiopoidea

In the female the ocular quadrangle is either entirely black or with the three sides occupied by the eyes black; the remainder of the cephalothorax is light brown with three, more or less distinct, longitudinal dark bands (Fig. 762); on the lateral and posterior edge of the thorax there is a narrow black line. On the abdomen there are two longitudinal red bands and indistinct chevrons of the same colour in the middle of the hinder half. In alcohol the red marks change to brown.

The typical male resembles the female; but the red bands on the abdomen are broken up into two rows of spots which are connected with the chevrons (Fig. 763).

In the black male both cephalothorax and abdomen are black;

Fig. 762.
MÆVIA VITTATA, FEMALE

Fig. 763.
MÆVIA VITTATA, TYPICAL MALE

but there is a pale spot in the centre of the thorax which is divided by a longitudinal black line. In the middle of the ocular quadrangle there is a transverse row of three prominent tufts of hairs.

Genus ONONDAGA (On-on-da'ga)

The cephalothorax is low and nearly flat, with a transverse groove behind the dorsal eyes; the sides are a very little dilated behind the dorsal eyes, nearly vertical in front, rounded behind; the cephalic part is inclined and the thoracic part falls slightly in the first half, then more steeply. The eye-region occupies less than half of the length of the cephalothorax, and is slightly wider behind than in front; the second row of eyes is a little nearer the

Superfamily Argiopoidea

third than the first. The tibia of the first legs is thickened and armed with four pairs of spines below. The lower margin of the furrow of the cheliceræ is armed with a wide tooth which is nearly of the fissidentate type (Fig. 764).

Fig. 764. ONONDAGA LINEATA CHELICERA OF THE FEMALE

The following is our only species:

Onondaga lineata (O. lin-e-a'ta).— The ocular quadrangle is black; the remainder of the cephalothorax is light brown; the abdomen, in alcoholic specimens, is dark brown with light markings. In fresh specimens the abdomen is marked with four longitudinal lines of white hairs. The length of the body is about one fifth inch.

This species is widely distributed in the Atlantic region and as far west as Wisconsin.

Genus ZYGOBALLUS (Zyg-o-bal'lus)

These spiders are easily recognized by the characteristic form of the cephalothorax. The ocular quadrangle occupies more than one half of the length of the cephalothorax, and the thorax slopes very abruptly from behind the posterior eyes. The first pair of legs are much stouter than the others; and the cheliceræ of the males are greatly enlarged. This genus belongs to the fissidentate series; Fig. 765 represents the cheliceræ of a female; and Fig. 766 one of a male.

Four species are known to occur in the United States; they are all small, measuring from one

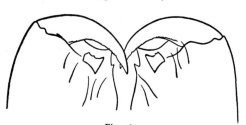

Fig. 765.
CHELICERÆ OF ZYGOBALLUS BETTINII, FEMALE

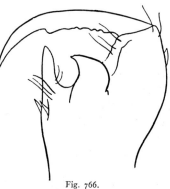

Fig. 766.
CHELICERA OF ZYGOBALLUS BETTINII, MALE

eighth to one fifth inch in length, and they vary so much in colour, size, and markings that it is difficult to distinguish them. One of the four *Z. rufipes* is a Mexican species, which has been found in Texas; a second, *Z. sexpunctatus*, is found only in the Southern States.

The males of these four species are separated as follows by the Peckhams; the females must be distinguished by differences in the epigynum.

KEY TO MALES OF ZYGOBALLUS

A. Face and sides of cephalothorax, beyond the second row of eyes, covered thickly with white scales. Tibia of the first legs about four times as long as wide.
- B. Cephalothorax with a large white spot at the beginning of the thoracic slope. Bulb of palpus with a longitudinal division. P. 706. *Z. sexpunctatus*
- BB. Cephalothorax with no white spot on thorax. Bulb of palpus with a transverse division. P. 705. *Z. bettinii*

AA. Sides of cephalothorax dark, or covered with rather inconspicuous whitish scales. Tibia of the first legs either from five to six times as long as wide or only two and one half times as long as wide.
- B. Patella and tibia of the first legs long and slender, the tibia from five to six times as long as wide. Found only in Texas and farther south. *Z. rufipes*
- BB. Joints of the first legs comparatively short. Tibia two and one half times as long as wide. Found in New England and as far south as Virginia. P. 706.
Z. nervosus

Zygoballus bettinii (Z. bet-ti′ni-i).— When living this is a very beautiful spider, having spots of white hairs and shining bronze and copper-coloured scales. The male is bronze-brown, with the face and sides of the cephalothorax beyond the second row of eyes covered thickly with white scales, and with a basal and two transverse bands of white on the abdomen. The female is also bronze-brown; it has a white basal band on the abdomen, and two short, longitudinal, angular, white bands on the front part of the abdomen; and several whitish chevrons on the hind part. The form of the epigynum is shown in Fig. 767. This is a common and

Superfamily Argiopoidea

widely distributed species; it occurs throughout the eastern half of the United States.

Zygoballus nervosus (Z. ner-vo'sus).— "In both sexes the cephalothorax is brown, thinly covered with whitish scales. The narrow clypeus is white. In the male the abdomen is brown, slightly metallic, with a very bright white basal band extending two thirds of the way along the sides, a nearly longitudinal white bar edged with black, on each side at the posterior end, and a white spot at the spinnerets. The female abdomen, of a lighter brown, is marked much like that of *Z. bettinii*, with two short

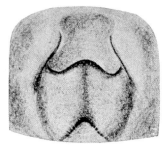

Fig. 767. EPIGYNUM OF ZYGOBALLUS BETTINII

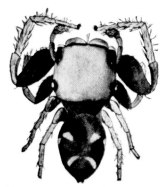

Fig. 768. ZYGOBALLUS SEXPUNCTATUS, MALE

Fig. 769. EPIGYNUM OF ZYGOBALLUS SEXPUNCTATUS

curved bands just back of basal band, followed by two large white spots with black spots behind them, and farther back a series of indistinct whitish chevrons, with a second pair of black spots a little in front of the spinnerets" (Peckham). This species has been taken from Maine to Illinois and south to Virginia.

Zygoballus sexpunctatus (Z. sex-punc-ta'tus).— This species closely resembles *Z. bettinii;* but the male differs in having a large white spot at the beginning of the thoracic slope, and in having the bulb of the palpus with a longitudinal division. In the typical form the two transverse bands of the abdomen are so broken that the first consists of four spots and the second of two (Fig. 768), giving the six spots which suggested the specific name, but these spots may be wanting. The female can be best distinguished by

Superfamily Argiopoidea

the form of the epigynum (Fig. 769). This species has been taken only in the Southern States.

Genus PECKHAMIA (Peck-ham'i-a)

These spiders are ant-like in form, having the pedicel of the abdomen visible from above; they differ from the other ant-like spiders found in our fauna in that the quadrangle of eyes occupies more than one half of the length of the cephalothorax. The thoracic part of the cephalothorax is short and marked behind the eyes with a transverse and slightly recurved furrow, behind which it is slightly convex, and then inclined to the posterior border, which is a little turned up and obtusely truncate. This genus belongs to the fissidentate series, the lower margin of the furrow of the cheliceræ being armed with a compound tooth (Fig. 770).

The three following species are our better known ones.

Fig. 770. CHELICERÆ OF PECKHAMIA PICATA
a, male   *b*, female

Peckhamia picata (P. pi-ca'ta).— This is our most common species. The ocular quadrangle is black with violet reflections; the thorax reddish brown, with two white spots in the depression behind the posterior lateral eyes; the anterior part of the abdomen is reddish brown, the posterior part black; there is a white band on each side in the constriction in the abdomen. The sternum is dark brown or black; the coxæ are light. The first pair of legs are considerably enlarged. The male is .13 inch in length; the female .19 inch.

Peckhamia scorpiona (P. scor-pi-o'na).— This is the smallest of our three species, the male measuring .09 inch in length; the female .14 inch. The first pair of legs are only slightly enlarged. "In the *male*, the cephalothorax is brown; the abdomen is brown

Superfamily Argiopoidea

anteriorly, encircled by a white line in front of the middle, behind which it is blackish; other parts brown, excepting a pale spot on the anterior part of the venter. In the *female*, the cephalothorax is brownish white, with the eyes on black spots; the abdomen is pale, with two short, curved dark bands near the spinnerets; cheliceræ brownish; venter pale, with a dark region near the spinnerets; legs brown above, pale beneath; other parts all pale" (Peckham, '92). This species is distributed from the Atlantic to the Pacific.

Peckhamia americana (P. a-mer-i-ca′na).— This is a larger and heavier species than the preceding, the female measuring one fifth inch in length. It differs from *P. picata* in being much lighter in colour and in having the sternum and coxæ of a uniform light reddish brown; while in *P. picata* the dark brown or black sternum makes a contrast with the light coxæ (Peckham, '92). This is a Southern species occurring in the Gulf States and in the Southwest.

INDUSTRY, BY P. VERONESE, DUCAL PALACE, VENICE

BIBLIOGRAPHY

The following list includes only the titles of the books and papers to which references have been made in the preceding pages. For a more complete list the student should refer to the annual volumes of the *Zoological Record*, published by the Zoological Society of London, and sold at their house in Hanover Square, London. The part of the *Zoological Record* treating of the Arachnida can be purchased separately. Citations of original generic and specific descriptions of North American spiders can be readily obtained by reference to the *Synonymic Index-Catalogue of Spiders of North, Central and South America, etc.* by Dr. A. Petrunkevitch ('11). This is published as Volume XXIX in the Bulletin of the American Museum of Natural History.

In the following list the number in parenthesis indicates in each case the year in which the work was published. When reference is made to more than one work published by an author in the same year, the different works are distinguished by an added letter.

The abbreviations of the titles of journals are those adopted by the editors of the *International Catalogue of Scientific Literature* of which the *Zoological Record* is now a part.

APSTEIN, C. ('89). "Bau und Function der Spinndrüsen der Araneida." *Arch. Naturg.*, Berlin. Vol. 55, pp. 29–74.

ATKINSON, G. F. ('86). "Descriptions of Some Trap-door Spiders; their Nests and Food Habits." *Ent. Amer.* Vol. 2, pp. 109–137.

BALZAN, L. ('91). "Voyage de M. E. Simon au Venezuela 16e Mémoire. Arachnides. Chernetes (Pseudoscorpiones)." *Paris Ann. Soc. Ent.*, 1891, pp. 496–552.

BANKS, N. ('92 a). "The Spider Fauna of the Upper Cayuga Lake Basin." Philadelphia, Pa., *Proc. Acad. Nat. Sci.*, 1892, pp. 11–81.

BANKS, N. ('92 b). "A Classification of the North American Spiders." *Canad. Entom.* London, Can., Vol. 24, pp. 88–97.

BANKS, N. ('92 c). "On *Prodidomus rufus* Hentz." Washington, D. C. *Proc. Ent. Soc.*, Vol. 2, pp. 259–261.

BANKS, N. ('93 a). "The Phalangida Mecostethi of the United States." Philadelphia, Pa. *Trans. Amer. Ent. Soc.*, Vol. 20, pp. 149–152.

BANKS, N. ('93 b). "The Phalanginæ of the United States." *Canad. Entom.*, London, Can., Vol. 25, pp. 205–211.

Bibliography

BANKS, N. ('93 c). "Notes on Spiders." New York, N. Y., *J. Ent. Soc.*, Vol. 1, pp. 123-134.

BANKS, N. ('94 a). "The Nemastomatidæ and Trogulidæ of the United States." *Psyche.* Cambridge, Mass., Vol. 7, pp. 11-12; 51-52.

BANKS, N. ('94 b). "The Phalangida of New York." New York, N. Y., *J. Ent. Soc.*, Vol. 2, pp. 40-41.

BANKS, N. ('94 c). "Some New American Acarina." Philadelphia, Pa. *Trans. Am. Ent. Soc.*, Vol. 21, pp. 209-222.

BANKS, N. ('94 d). "Two Families of Spiders New to the United States." *Ent. News*, Philadelphia, Pa. 1894, pp. 298-300.

BANKS, N. ('94 e). "Notes on *Larinia* and *Cercidia*." *Ent. News*, Philadelphia, Pa. 1894, pp. 8-9.

BANKS, N. ('95 a). "On the Oribatoidea of the United States." Philadelphia, Pa. *Trans. Am. Ent. Soc.* Vol. 22.

BANKS, N. ('95 b). "Notes on the Pseudoscorpionida." New York, N. Y., *J. Ent. Soc.*, Vol. 3, pp. 1-13.

BANKS, N. ('95 c). "A List of the Spiders of Long Island, N. Y., with descriptions of New Species." New York, N. Y., *J. Ent. Soc.*, Vol. 3, pp. 76-93.

BANKS, N. ('00). "Synopses of North American Invertebrates. IX. The Scorpions, Solpugids, and Pedipalpi." *Amer. Nat.*, Boston, Mass. Vol. 34, pp. 421-427.

BANKS, N. ('01). "Synopses of North American Invertebrates. XVI. The Phalangida." *Amer. Nat.*, Boston, Mass., Vol. 35, pp. 669-679.

BANKS, N. ('04). "A Treatise on the Acarina or Mites." Washington, D. C., Smithsonian Inst., *U. S. Nation. Mus. Proc.*, Vol. 28, pp. 1-114.

BANKS, N. ('05). "Synopses of North American Invertebrates. XX. Families and Genera of Araneida." *Amer. Nat.*, Boston, Mass., Vol. 39, pp. 293-323.

BANKS, N. ('07). "A Catalogue of the Acarina, or Mites, of the United States." Washington, D. C., Smithsonian Inst., *U. S. Nation. Mus. Proc.*, Vol. 32, pp. 595-625.

BANKS, N. ('08). "A Revision of the Ixodoidea or Ticks, of the United States." Washington, D. C., *Tech. Ser. U. S. Dep. Agric. Bur. Ent.*, No. 15.

BANKS, N. ('10). "Catalogue of Nearctic Spiders." Washington, D. C., Smithsonian Inst., *U. S. Nation. Mus. Bull.*, No. 72.

BARROIS, J. ('96). "Mémoire sur le développement des Chelifer." *Revue Suisse Zool. Genève.* Vol. 3, pp. 461-498.

BARROWS, W. M. ('19). "The Taxonomic Position of Mysmena bulbifera (Glenognatha bulbifera) Banks with Some Observations on its Habits." *Ohio Journ. Sci.*, Vol. 20, pp. 210-212.

BEIER, M. ('32). "Pseudoscorpionidea." Das Tierreich, *Preuss. Akad. Wiss. Berlin*, Schulze und Kukenthal; 57 Lieferung (I. Suborders Chthoniinea and Neobisiinea, pp. 1-258); 58 Lieferung (II. Suborder Cheliferinea, pp. 1-294).

BERG, C. ('93). "Pseudoscorpionidenkniffe." *Zool. Anz.*, Leipzig. Vol. 16, pp. 446-448.

BERNARD, H. M. ('95). "On the Spinning Glands in *Phrynus*." London, *J. Linn. Soc. Zool.*, Vol. 25, pp. 272-278.

Bibliography

BERNARD, H. M. ('96). "The Comparative Morphology of the Galeodidæ." London, *Trans. Linn. Soc.*, Second Series, Vol. 6, pp. 305–417.

BERTEAUX, L. ('89). "Le Poumon des Arachnides." *Cellule*, Louvain, Vol. 5, pp. 200–317.

BERTKAU, P. ('72). "Ueber die Respirationsorgane der Araneen." *Arch. Natg.*, Berlin, Vol. 38, pp. 208–233.

BERTKAU, P. ('75). "Ueber den Generationsapparat der Araneïden." *Arch. Natg.*, Berlin. Jahrgang 41, Band 1, pp. 235–262.

BERTKAU, P. ('78). "Versuch einer natürlichen Anordnung der Spinnen." *Arch. Natg.*, Berlin. Jahrgang 44, Band 1, pp. 351–410.

BERTKAU, P. ('86). "Beiträge zur Kenntniss der Spinnesorgane der Spinner. I. Die Augen der Spinne." *Arch. mik. Anat.*, Bonn, Vol. 27, pp. 589–631.

BISHOP, S. C. ('24). "A Revision of the Pisauridæ of the United States." *New York State Mus.* Bull. No. 252, pp. 1–140.

BISHOP, S. C. and CROSBY, C. R. ('30). "Studies in American Spiders: Genera Ceratinopsis, Ceratinopsidis and Tutaibo." *Journ. New York Ent. Soc.*, Vol. 38, pp. 15–33.

BISHOP, S. C. and CROSBY, C. R. ('32). "Studies in American Spiders: The Genus Grammonota." *Journ. New York Ent. Soc.*, Vol. 40, pp. 393–421.

BISHOP, S. C. and CROSBY, C. R. ('35). "Studies in American Spiders: Miscellaneous Genera of Erigoneæ, Part I." *Journ. New York Ent. Soc.*, Vol. 43, pp. 217–281.

BISHOP, S. C. and CROSBY, C. R. ('38). "Studies in American Spiders: Miscellaneous Genera of Erigoneæ, Part II." *Journ. New York Ent. Soc.*, Vol. 46, pp. 55–107.

BLACKWALL, J. ('55). "Experiments and Observations on the Poison of Animals of the Order Araneida," London, *Trans. Linn. Soc.*, Vol. 21, pp. 31–37.

BLAUVELT, H. H. ('36). "The Comparative Morphology of the Secondary Sexual Organs of Linyphia and Some Related Genera, Including a Revision of the Group." *Festschrift E. Strand*, Riga, Vol. 2, pp. 81–172.

BÖRNER, C. ('04). "Ein Beiträg zur Kenntniss der Pedipalpen." *Zoologica*, Stuttgart, Heft 42.

BOUVIER, E. L. ('96). "Sur la ponte et le développement d'un Pseudoscorpionide." Paris, *Bull. Soc. Ent.* 1896, pp. 342–343.

BRYANT, E. B. ('35). "A Rare Spider." *Psyche*, Vol. 42, pp. 163–166.

CAMBRIDGE, F. O. P. See Pickard-Cambridge, F. O.

CAMBRIDGE, F. P. ('02). "On the Spiders of the Genus *Latrodectus*." London, *Proc. Zool. Soc.*, 1892, Vol. 1, pp. 247–261.

CAMBRIDGE, O. P. See Pickard-Cambridge, O. P.

CANESTRINI, G. ('85–'99). *Prospetto dell' Acaro-fauna Italiana.* Padova.

CAUSARD, M. ('96). "Recherches sur l'appareil circulatoire des Araneides." Paris, *Bull. Sci. France Belgique*, Vol. 29, pp. 1–109.

CHAMBERLIN, J. C. ('31). "The Arachnid Order Chelonethida." *Stanford Univ. Publ. Biol. Sci.*, Vol. 7, pp. 1–284.

CHAMBERLIN, R. V. ('04). "Notes on Generic Characters in the Lycosidæ." *Canad. Entom.*, London, Can., Vol. 36, pp. 145–148; 173–178.

Bibliography

CHAMBERLIN, R. V. ('08). "Revision of North American Spiders of the Family Lycosidæ." Philadelphia, Pa., *Proc. Acad. Nat. Sci.*, 1908, pp. 157–318.

CHAMBERLIN, R. V. ('22). "The North American Species of the Family Gnaphosidæ." *Proc. Biol. Soc. Washington*, Vol. 35, pp. 145–171.

CHAMBERLIN, R. V. and IVIE, W. ('32). "North American Spiders of the Genera Cybæus and Cybæina." *Bull. Univ. Utah*, Vol. 23, pp. 1–42.

CHAMBERLIN, R. V. and IVIE, W. ('33). "A New Genus in the Family Agelenidæ." *Bull. Univ. Utah*, Vol. 24, pp. 1–7.

CHAMBERLIN, R. V. and IVIE, W. ('37). "New Spiders of the Family Agelenidæ from Western North America." *Ann. Ent. Soc. America*, Vol. 30, pp. 211–241.

COMSTOCK, J. H. ('10). "The Palpi of Male Spiders." *Ann. Ent. Soc. Amer.*, Columbus, O. Vol. 3, pp. 161–185.

COMSTOCK, J. H., and KOCHI, C. ('02). "The Skeleton of the Head of Insects." *Amer. Nat.*, Boston, Mass., Vol. 36, pp. 13–45.

COOK, O. F. ('99). "*Hubbardia*, A new genus of Pedipalpi." Washington, D. C., *Proc. Ent. Soc.*, Vol. 4, pp. 249–261.

CROSBY, C. R., and BISHOP, S. C. ('25). "Studies in New York Spiders: Genera Ceratinella and Ceraticelus." *New York State Mus.* Bull. No. 264, pp. 5–71.

CROSBY, C. R., and BISHOP, S. C. ('28). "Revision of the Spider Genera Erigone, Eperigone and Catabrithorax (Erigoneæ)." *New York State Mus.* Bull. No. 278, pp. 3–98.

CROSBY, C. R., and BISHOP, S. C. ('33). "American Spiders: Erigoneæ, Males with Cephalic Pits." *Ann. Ent. Soc. America*, Vol. 26, pp. 105–182.

DUFOUR, L. ('62). "Anatomie, Physiologie et Histoire naturelle des Galéodes." Paris, *Mem. Acad. Sci.*, Vol. 17, pp. 338–446.

DUGES, A. ('36). "Observations sur les Araneides." *Ann. Sci. Nat.* (Zool.), Paris, Vol. 6, pp. 159–218.

EMERTON, J. H. ('78). *The Structure and Habits of Spiders*. Salem: S. E. Cassino.

EMERTON, J. H. ('82). "New England Spiders of the Family Theridiidæ." New Haven, Conn., *Trans. Acad. Arts Sci.*, Vol. 6, pp. 1–86.

EMERTON, J. H. ('84). "New England Spiders of the Family Epeiridæ." New Haven, Conn., *Trans. Acad. Arts Sci.*, Vol. 6, pp. 295–342.

EMERTON, J. H. ('85). "New England Lycosidæ." New Haven, Conn., *Trans. Acad. Arts Sci.*, Vol. 6, pp. 481–505.

EMERTON, J. H. ('88). "New England Spiders of the Family Ciniflonidæ." New Haven, Conn., *Trans. Acad. Arts Sci.*, Vol. 7, pp. 443–458.

EMERTON, J. H. ('90). "New England Spiders of the Families Drassidæ, Agalenidæ, and Dysderidæ." New Haven, Conn., *Trans. Acad. Arts Sci.*, Vol. 8, pp. 2–42.

EMERTON, J. H. ('91). "New England Spiders of the Family Attidæ." New Haven, Conn., *Trans. Acad. Arts Sci.*, Vol. 8, pp. 220–251.

EMERTON, J. H. ('92). "New England Spiders of the Family Thomisidæ." New Haven, Conn., *Trans. Acad. Arts Sci.*, Vol. 8, pp. 359–381.

EMERTON, J. H. ('94). "Canadian Spiders." New Haven, Conn., *Trans. Acad. Arts Sci.*, Vol. 9, pp. 400–429.

Bibliography

EMERTON, J. H. ('02). *The Common Spiders of the United States.* Boston: Ginn & Co.

EMERTON, J. H. ('09). "Supplement to the New England Spiders." New Haven, Conn., *Trans. Acad. Arts Sci.*, Vol. 14, pp. 173-236.

ENGELHARDT, V. v. ('10). "Beiträge zur Kenntniss der weiblichen Copulationsorgane einiger Spinnen." *Zs. wiss. Zool.*, Leipzig, Vol. 96, pp. 32-117.

EWING, H. E. ('09). "The Oribatoidea of Illinois." Urbana, *Bull. Ill. Lab. Nat. Hist.*, Vol. 7, pp. 337-389.

EWING, H. E. ('23). "Holosiro Acaroides, New Genus and Species, the only New World representative of the Mite-like Phalangids of the Suborder Cyphophthalmi." *Ann. Ent. Soc. America*, Vol. 16, pp. 387-391.

EWING, H. E. ('29). "A synopsis of the American Arachnids of the primitive order Ricinulei." *Ann. Ent. Soc. America*, Vol. 22, pp. 583-600.

FABRE, J. H. ('07). *Souvenirs Entomologiques.* Paris: Ch. Delagrave.

GERTSCH, W. J. ('34). "Some American Spiders of the Family Hahniidæ." *American Mus. Novit.*, No. 712, pp. 1-32.

GERTSCH, W. J. ('37). "New American Spiders." *American Mus. Novit.*, No. 936, pp. 6-7.

GERTSCH, W. J. and MULAIK, S. ('39). "Report on a New Ricinuleid from Texas." *American Mus. Novit.*, No. 1037, pp. 1-5.

GERTSCH, W. J., and WALLACE, H. K. ('36). "Notes on New and Rare American Mygalomorph Spiders." *American Mus. Novit.*, No. 884, pp. 1-12.

GRABER, V. ('80). "Ueber das unicorneal Tracheaten—und Special das Arachnoideen—und Myriopoden—Auge." *Arch. mikr. Anat.*, Bonn, Vol. 17, Heft 1, pp. 58-93.

GRENACHER, H. ('79). *Untersuchungen über das Sehorgan der Arthropoden, insvesondere der Spinnen, Insecten und Crustaceen.* Göttingen: Vandenhöck und Ruprecht.

HANCOCK, J. L. ('99). "The Castle-building Spider." *Ent. News*, Philadelphia, Pa., Vol. 10, pp. 23-29.

HANSEN, H. J. and SORENSEN, W. ('04). "On Two Orders of Arachnids." Cambridge.

HASSELT, A. W. M. VAN ('89). "Le Muscle Spiral et la Vésicule du Palpe des Araignées Mâles." *s. Gravenhage, Tijdschr. Ent.*, Vol. 32, pp. 161-203.

HENKING, H. ('88). "Biologische Beobachtungen an Phalangiden." *Zool. Jahrb.* Jene (Syst) Vol. 3, pp. 319-335.

HENTZ, N. M. ('75). *The Spiders of the United States.* A collection of the Arachnological Writings of Nicholas Marcellus Hentz, M.D. Edited by Edward Burgess, with Notes and Descriptions by James H. Emerton. Boston: Boston Society of Natural History.

HUTTON, T. ('43). "Observations on the Habits of a large species of Galeodes." *Ann. Mag. Nat. Hist.*, London, Vol. 12, pp. 81-85.

KEYSERLING, G. E. and MARX, G. ('80-'93). *Die Spinnen Amerikas.* Vols. 1-4. Nürnberg: Baner & Raspe.

KORSCHELT, E. and HEIDER, K. ('99). *Text-book of the Embryology of Invertebrates* (English ed.). Vol. 3. New York: The Macmillan Co.

KRAEPELIN, K. ('99). "Scorpiones and Pedipalpi." *Das Tierreich*, 8 *Lieferung*, pp. 1-265. Berlin: R. Friedländer und Sohn.

Bibliography

Krohn, A. ('67). "Ueber die Anwesenheit Zweier Drüsensacke im Cephalothorax der Phalangiden." *Arch. Natg.*, Berlin, Vol. 33, pp. 79–83.

Lang, A. ('91). *Text-book of Comparative Anatomy*. (English Edition.) New York: Macmillan & Co.

Latreille, P. A. (1806). *Genera Crustaceorum et Insectorum*, Vol. 1. Parisiis et Argentorati: König.

Laurie, M. ('94). "On the Morphology of the Pedipalpi." London, *J. Linn, Soc. Zool.*, Vol. 25, pp. 20–48.

Loman, J. C. C. ('96). "On the Secondary Spiracles on the Legs of Opilionidæ." *Zool. Anz.*, Leipzig, Vol. 19, pp. 221–222.

Marx, G. ('81). "On Some New Tube-constructing Spiders." *Amer. Nat.*, Boston, Mass., 1881, pp. 396–400.

Marx, G. ('88). "On a New and Interesting Spider." *Entom. Amer.*, Brooklyn, N. Y., Vol. 4, pp. 160–162.

Marx, G. ('89). "On a New Species of Spider of the genus *Dinopis* from the Southern United States." Philadelphia, Pa., *Proc. Acad. Nat. Sci.*, Vol. 41, pp. 341–343.

Marx, G. ('90). "Catalogue of the Described Araneæ of Temperate North America." Washington, D. C., Smithsonian Inst., *U. S. Nation. Mus. Proc.*, Vol. 12, pp. 497–594.

Marx, G. ('91). "A Contribution to the Knowledge of North American Spiders." Washington, D. C., *Proc. Ent. Soc.*, Vol. 2, pp. 28–37.

McCook, C. ('89–'93). *American Spiders and their Spinning-Work*. 3 Vols. Philadelphia: The Author.

Megnin, P. ('80). *Les Parasites et les Maladies Parasitaires. Insectes, Arachnides, Crustacés*. Paris: Libraire de l'Académie de Medécine.

Meijere, J. C. H. de ('01). "Ueber das letze Glied der Beine bei den Arthropoden." *Zool. Jahrb.*, Jene, Vol. 14, (Anat.), pp. 417–476.

Menge, A. ('66). *Preussische Spinnen*. Danzig: A. W. Kafemann.

Metschnikoff, E. ('71). "Entwicklungsgeschichte des Chelifer." *Zs. wiss. Zool.*, Leipzig, Vol. 21, pp. 513–524.

Moggridge, J. T. ('73). *Harvesting Ants and Trap-door Spiders*. London: L. Reeve & Co.

Moggridge, J. T. ('74). *Supplement to Harvesting Ants and Trap-door Spiders, with Specific Descriptions of the Spiders*, by O. Pickard-Cambridge. London: L. Reeve & Co.

Montgomery, T. H. ('03). "Studies on the Habits of Spiders, Particularly Those of the Mating Period." Philadelphia, Pa., *Proc. Acad. Nat. Sci.*, 1903, pp. 59–149.

Montgomery, T. H. ('08). "Further Studies on the Activities of Araneads." *Amer. Nat.*, Boston, Mass., Vol. 42, pp. 697–709.

Montgomery, T. H. ('10). "Further Studies on the Activities of Araneads." II. Philadelphia, Pa., *Proc. Acad. Nat. Sci.*, 1909, pp. 548–569.

Murray, A. ('77). "Economic Entomology." *South Kensington Museum Handbooks*. London: Chapman and Hall.

Nelson, J. A. ('09). "Evolution and Adaptation in the Palpus of Male Spiders." *Ann. Ent. Soc. Am.*, Columbus, O., Vol. 2, pp. 60–64.

Packard, A. S. ('05). "Change of Colour and Protective Colouration in a Flower-spider (*Misumena vatia* Thorell)." New York, N. Y., *J. Ent. Soc.*, Vol. 13, pp. 85–96.

Bibliography

Parrott, Hodgkiss, and Schoene ('06). "The Apple and Pear Mites." Geneva, N. Y., *Bull. N. Y. Agr., Exp. Sta.*, No. 283.

Peckham, G. W. and E. G. ('89). "Observation on Sexual Selection in Spiders of the Family Attidæ." Milwaukee: Cramer, Aikers & Cramer.

Peckham, G. W. and E. G. ('90). "Additional Observations on Sexual Selection in Spiders of the Family Attidæ." Milwaukee, *Occ. Papers Wis. Nat. Hist. Soc.*, Vol. 1, pp. 117–151.

Peckham, G. W. and E. G. ('09). "Revision of the Attidæ of North America." Madison. *Trans. Wis. Acad. of Sci.*, Vol. 16, pp. 355–646.

Petrunkevitch, A. ('10 b). "Ueber die Circulationsorgane von *Lycosa carolinensis* Walk." *Zool. Jahrb.*, Jene, Vol. 31, (Anat.), pp. 161–170.

Petrunkevitch, A. ('11). "A Synonymix Index-Catalogue of Spiders of North, Central and South America With All Adjacent Islands, etc." *Bull. American Mus. Nat. Hist.*, Vol. 29, pp. 1–791.

Petrunkevitch, A. ('23). "On Families of Spiders." *Ann. New York Acad. Sci.*, Vol. 29, pp. 145–180.

Petrunkevitch, A. ('28). "Systema Aranearum." *Trans. Connecticut Acad. Sci.*, Vol. 29, pp. 1–270.

Petrunkevitch, A. ('33). "An Inquiry into the Natural Classification of Spiders, Based on a Study of their Internal Anatomy." *Trans. Connecticut Acad. Sci.*, Vol. 31, pp. 299–389.

Pickard-Cambridge, O. ('74). *Specific Descriptions of Trap-door Spiders.* In *Harvesting Ants and Trap-door Spiders*, by J. T. Moggridge. Supplement. pp. 254–301.

Pocock, R. I. ('98). "The Nature and Habits of Pliny's Solpuga." *Nature*, London, Vol. 57, pp. 618–620.

Putnam, J. D. ('83). "The Solpugidæ of America." Davenport, Ia., *Proc. Acad. Nat. Sci.*, Vol. 3, pp. 249–310.

Roewer, C. F. ('23). "Die Weberknechte der Erde." Jena, Verlag von Gustav Fischer.

Roewer, C. F. ('34). "Solifugae, Palpigradi." *Bronns Klassen und Ordnungen des Tierreichs*, Leipzig, 5 Band, IV Abt., 8 Buch, pp. 550–577 (Eremobatidæ), pp. 578–608 (Ammotrechidæ).

Rössler, R. ('82). "Beiträge zur Anatomie der Phalangiden." *Zs. wiss. Zool.*, Leipzig, Vol. 36, pp. 671–700.

Rucker, A. ('01). "The Texan Kœnenia." *Amer. Nat.*, Boston, Mass., Vol. 25, pp. 615–630.

Rucker, A. ('03 a). "Further Observations on Kœnenia." *Zool. Jahrb.*, Jene, Vol. 18, pp. 401–437.

Rucker, A. ('03 b). "A New Kœnenia from Texas." *Q. J. Micros. Sci.*, London, Vol. 47, pp. 215–231.

Schimkewitsch, W. ('84). "Etude sur l'anatomie de l'epeire." *Ann. Sci., Nat.* (Zool.), Paris, Vol. 17, pp. 1–94.

Scudder, S. H. ('77). "The Tube-constructing Ground Spider of Nantucket." *Psyche*, Cambridge, Mass. Vol. 2, pp. 2–9.

Seeley, R. M. ('28). "Revision of the Spider Genus Tetragnatha." *New York State Mus.* Bull. No. 278, pp. 99–150.

Silvestri, F. ('13). "Novi generi e specie de Koeneniidæ." *Boll. Lab. Zool. Gen., Portici*, Vol. 7, pp. 211–217.

Simon, E. ('79). Les Arachnides de France. Vol. 7, Paris: *Libraire Encyclopédique de Roret*.

Bibliography

SIMON, E. ('92–'03). Histoire Naturelle des Araignées. Deuxième Edition, Paris: *Libraire Encyclopédique de Roret.*

SMITH, C. P. ('08). "A Preliminary Study of the Araneæ Theraphosæ of California." *Ann. Ent. Soc. Amer.*, Columbus, O., Vol. 1, pp. 207–250.

THORELL, T. ('88). "Pedipalpi e Scorpioni dell' Arcipelago Malese." Genova, *Ann. Museo Civ. st. nat.*, Vol. 6.

VOGT, C. et YUNG, E. ('94). *Traité d'Anatomie Comparée Pratique.* Vol. II. Paris: C. Reinwald & Cie.

WAGNER, W. ('87). "Copulationsorgane des Männchens als Criterium für die Systematik der Spinnen." St. Petersburg, *Horae Soc. Ent. Russ.*, Vol. 22, pp. 3–132.

WAGNER, W. ('88). "La Mue des Araignées." *Ann. Sci. Nat.* (Zool.), Paris, Vol. 6, pp. 281–393.

WALCKENAER et GERVAIS ('37–'47). *Histoire Naturelle des Insectes. Apteres.* (4 Vols.) Paris: *Libraire Encyclopédique de Roret.*

WARBURTON, C. ('90). "The Spinning Apparatus of Geometric Spiders." *Q. J. Micros. Sci.*, London, Vol. 31, pp. 29–39.

WARBURTON, C. ('09). "Arachnida Embolobranchiata (Scorpions, Spiders, Mites, etc.)." *The Cambridge Natural History.* Vol. 4, pp. 297–473. London: Macmillan & Co.

WEED, C. M. ('87). "The Genera of North American Phalangiinæ." *Amer. Nat.*, Boston, Mass., Vol. 21, p. 935.

WEED, C. M. ('89 *a*). "A Descriptive Catalogue of the Phalangiinæ of Illinois." Urbana, *Bull. Ill. Lab. Nat. Hist.*, Vol. 3, pp. 79–97.

WEED, C. M. ('89 *b*). "A Partial Bibliography of the Phalangiinæ of North America." Urbana, *Bull. Ill. Nat. Hist.*, Vol. 3, pp. 99–106.

WEED, C. M. ('92 *a*). "Descriptions of New or Little-known North American Harvest Spiders (Phalangiidæ)." Philadelphia, Pa., *Trans. Amer. Ent. Soc.*, Vol. 19, pp. 187–194.

WEED, C. M. ('92 *b*). "A Preliminary Synopsis of the Harvest Spiders (Phalangiidæ) of New Hampshire." Philadelphia, Pa., *Trans. Amer. Ent. Soc.*, Vol. 19, pp. 261–272.

WEED, C. M. ('92 *c*). "The Striped Harvest Spider: A Study in Variation." *Amer. Nat.*, Boston, Mass., Vol. 26, pp. 999–1008.

WEED, C. M. ('92 *d*). "Notes on Harvest Spiders." *Amer. Nat.*, Boston, Mass., Vol. 26, pp. 528–530.

WEED, C. M. ('93). "A Descriptive Catalogue of the Harvest Spiders (Phalangiidæ) of Ohio." Washington, D. C., Smithsonian Inst., *U. S. Nation. Mus. Proc.*, Vol. 16, pp. 543–563.

WERNER, F. ('35). "Scorpiones, Pedipalpi." *Bronns Klassen und Ordnungen des Tierreichs*, Leipzig, 5 Band, IV Abt., 8 Buch, pp. 1–318.

WESTRING, N. ('61). *Araneæ Svecicæ Descriptæ.* Gothoburgi: Sumter et Litteris, D. F. Bonnier.

WHEELER, W. M. ('00). "A Singular Arachnid (*Kœnenia mirabilis* Grassi) occurring in Texas." *Amer. Nat.*, Boston, Mass., Vol. 34, pp. 837–850.

WIDMAN, E. ('07). "Der feiner Bau der Augen einiger Spinnen." *Zool. Anz.*, Leipzig, Vol. 31, pp. 755–762.

WILDER, B. G. ('66). "On the *Nephila plumipes* or Silk Spider of South Carolina." Boston, Mass., *Proc. Soc. Nat. Hist.*, Vol. 10, pp. 200–210.

WILDER, B. G. ('68 *a*). "On the *Nephila plumipes* or Silk Spider." Boston, Mass., *Proc. Amer. Acad. Arts Sci.*, Vol. 7, pp. 52–57.

WILDER, B. G. ('68 b). "The Harmlessness of the Bite of *Nephila plumipes.*" Boston, Mass., *Proc. Soc. Nat. Hist.*, Vol. 11, p. 7.
WILDER, B. G. ('68 c). "How Spiders Begin their Webs." *Amer. Nat.*, Boston, Mass., Vol. 2, pp. 214–215.
WILDER, B. G. ('73 a). "The Habits and Parasites of *Epeira (Argiope) riparia*, with a Note on the Moulting of *Nephila plumipes.*" *Proc. Amer. Ass. Adv. Sci.*, Vol. 22, Pt. 2, pp. 257–263.
WILDER, B. G. ('73 b). "The Nets of *Epeira (Argiope), Nephila* and *Hyptiotes (Mithras).*" *Proc. Amer. Ass. Adv. Sci.*, Vol. 22, Pt. 2, pp. 264–274.
WILDER, B. G. ('75). "The Triangle Spider." New York, N. Y., *Popular Science Monthly*, Vol. 6, pp. 641–655.
WOLCOTT, R. H. ('05). "A Review of the Genera of the Water Mites." Lancaster, Pa., *Trans. Amer. Micros. Soc.*, Vol. 26, pp. 161, 243.
WOOD, H. C. ('68). "On the Phalangeæ of the United States of America." Salem, Mass., *Proc. Essex Inst.*, Vol. 6, pp. 10–40.

INDEX

Names of genera and species and references to generic and specific descriptions are in italics.

Abdomen, 4, 95, 126
Abdomen, parts of the, 126
Abdominal endosternites, 142
Abdominal sclerites, 128
Acacesia, 461, *522*
　foliata, 523
Acanthophyrnus coronatus, 20
Acarina, 13, 81
Accessory claws, 124
Acrosoma, 527
Actinoxia, 231, *235*
　versicolor, 235
Adelosimus, 367
　studiosum, 367
Admestina, 676, *687*
　tibialis, 687
Aeronautic spiders, 215
Afterspinnen, 53
Agassa, 676, *695*
　cyanea, 695
Agelena, 167, 599, *601*
　nævia, 119, 210, *601*
　pennsylvanica, 602
　potteri, 602
Ageleneæ, 599, 601
Agelenidæ, 255, 597; table, 599
Agelenids, the, 597
Ageleninæ, 599, 600
Agræca, 585, *588*
　pratensis, 588
Alæ of the guide, 131
Alary muscles, 151
Aliatypus, 237, *239*
　californicus, 239
Alimentary canal, 152
Alimentary tubules, 156
Alveolus, 108
Amaurobiidæ, 254, 273; table, 274
Amaurobiids, the, 273
Amaurobius, 164, 167, *274*, 275
　bennetti, 276
　ferox, 277
Amblyocarenum, 231, *235*
　talpa, 235
Ambushing spiders, 186
Ammotrecha, 39
　californica, 39
　peninsulana, 39
Ammotrechidæ, 36, 38
Ammotrechona, 39
　cubæ, 39
　texana, 39
Ammotrechula, 39
Anahita, 569, *570*
　punctulata, 570
Anal sternite, 54
Ancylorrhanis, 351, *382*
　hirsutum, 382
Anelli of the subtegulum, 119
Anelosimus, 367
Anthrobia, 389, *391*
　mammouthia, 391
Antistea, 614, *615*
　brunnea, 615
Antrodiætinæ, 218, 224; table, 237

Antrodiætus, 220, 223, 226, 237
Anuroctonus, 30
　phæodactylus, 30
Anus, 136
Anyphæna, 575, *577*
　celer, 577
　fraterna, 577
　pectorosa, 577
Anyphænella, 575, *576*
　saltabunda, 576
Anyphæninæ, 574; table, 574
Aorta, 149, 150
Apical division of the bulb, 109, 112
Apical sclerite of the embolus, 120
Apodemes, 141
Aponeurotic plates, 142
Apollophanes, 555, *558*
　texana, 558
Apostenus, 585, *588*
　cinctipes, 588
Aptostichus, 231, *236*
　atomarius, 236
　clathratus, 236
　stanfordianus, 236
Aquatic spider, *Argyroneta aquatica*, 597
Arachnida, 3, 8, 9
Aranea, 460, 461, *481*; type of palpus, 114
　angulata, 483
　cavatica, 483, *484*
　corticaria, 486
　displicata, 508
　frondosa, 114, 499, *501;* palpus of, 115
　gemma, 132, 482, *486*
　gigas, 190, *489;* palpus of, 118
　gigas conspicellata, 490
　miniata, 488
　nordmanni, 483, *484*
　ocellata, 500, *503;* palpus of, 115
　pegnia, 508
　sericata, 499, *500*
　silvatica, 484
　thaddeus, *504;* retreat of, 206
　trifolium, 210, 489, *493*
　trifolium candicans, 498
Araneæ, 39
Araneæ theraphoseæ, 221
Araneæ veræ, 221, 252
Araneas, the large angulate, 482'
　the large round-shouldered, 482, 488
　the smaller angulate, 482, 486
　the smaller round-shouldered, 482, 504
　the three house, 482, 498
Araneida, 13, 39, 218
Araneinæ, 415, 457; table, 458
Arctosa, 638, *656*
　funerea, 656
　littoralis, 656
　rubicunda, 656
Areas of the head, 98
Argenna, 280, *288*
　aktia, 288
Argiope, 448, *456*
　argentata, 457
Argiopidæ, 255, 414
Argiopinæ, 415, 447

718

Index

Argiopoidea, 221; table to families, 221
Argyrodes, 187, 349, 353
   nephilæ, 355
   trigonum, 168, 210, 353
Ariadna, 306, 307
   bicolor, 307
Arolium, 124
Artema, 341, 344
   atlanta, 345
Arteries, lateral abdominal, 149, 151
Artery, caudal, 149, 151
Arthropoda, 3
Arthropods, table of classes of, 8
Articular sclerite, 102
Asagena, 349, 377
   americana, 379
Atriolum, 132
Attachment disks, 188
Attidæ, 173, 192, 208, 256, 669; table, 672
Attidæ Fissidentati, 671
Attidæ Pluridentati, 671
Attidæ Unidentati, 671
Atypidæ, 225, 226, 247
Atypoides, 237, 238
   riversi, 238
Atypus, 159, 160, 220, 223, 248
   abbotii, 248
   bicolor, 110, 251
   milberti, 251
   niger, 251
   piceus, 251
Avicularia, 240, 244
   avicularia, 185
   californica, 244
Aviculariidæ, 226, 239
Aviculariinæ, 240
Avicularioidea, 221, 222, 225
Aysha, 574, 575
   gracilis, 575
Azilia, 431, 439
   vagepicta, 439

Babesia bovis, 90
Bacilli, 162
Ballus, 674, 679
   youngii, 679
Banana Spider, the, 565
Banded Garden Spider, 452
Basal division of the bulb, 109
Basal spot of the chelicera, 100
Basement membrane, 137, 139
Basilica Spider, the, 431
Bathyphantes, 390, 392
   concolor, 393
   nigrinus, 393
Bdella peregrina, 81
Bdellidæ, 87
Beak, of mites, 83
Beautiful Leucauge, the, 436
Beetle mites, 91
Bird-spider, the, 185
Bird-spiders, 224
Blabomma, 611
Blackheads, 94
Body-wall, 137
Bolyphantes bucculenta, 404
Book-lungs, 145, 146
Boophilus, 89
   annulatus, 89
Bothriocyrtum, 230, 233
   californicum, 233
Bowl and Doily Spider, the, 400
Brachybothrium, 237
   accentuatum, 238
Brachythele, 245, 246
   longitarsis, 246
   theveneti, 246
Brain, 160
Bridge, the, 196

Broteas, 29
   alleni, 29
Bryobia pratensis, 88
Buthidæ, 25
Buthinæ, 25, 26

Caddo, 65, 68
   agilis, 68
   boopis, 68
Cæca, 156
Cæcal ring, 156
Calamistrum, 125
Callilepis, 321, 335
   imbecilla, 336
Callioplus, 274, 278
   tibialis, 278
Calymmaria, 599, 607
   cavicola, 607
Camerostoma, of mites, 83
Capitulum, of mites, 83
Caponiidæ, 145, 254, 303
Caponiids, the, 303
Carapace, 95
Cardiac ligaments, 149, 151
Carepalxis tuberculifera, 462
Castianeira, 592
   cingulata, 593
   descripta, 593
   longipalpus, 594
Cattle tick, 89
Centipedes, 6
Central nervous system, 160
Centruroides, 25, 27
   exilicaudata, 27
   gracilis, 27
   hentzi, 27
   sculpturatus, 27
   vittatus, 27
Centrurus, 27
Cephalothorax, 4, 95
Ceraticelus fissiceps, 385
Ceraticelus lætabilis, 385
Ceratinopsis interpres, 388
Cercidia, 459, 469
   funebris, 469
   prominens, 469
Cervical groove, 95
Cesonia, 323, 333
   bilineata, 334
Chactidæ, 25, 29
Chela, 12
Chelate, 12
Chelate cheliceræ, 102
Cheliceræ, 11, 99
Chelifer cancroides, 48
Cheliferinea, 47
Cheridiidæ, 39
Cheridiodea, 48
Chernetidæ, 48
Chilopoda, 6, 8
Chiracanthium, 579, 583
   inclusum, 583
Chitin, 138
Chorizomma, 611
Chorizommoides, 611
Chthoniidæ, 46
Chthoniinea, 45
Chthonius, 46
   cæcus, 46
   crosbyi, 46
   packardi, 46
Cicurina, 612
   robusta, 612
Circulatory system, 148
Citharoceps, 306, 307
   californica, 307
Clavis, 112
Claw of the chelicera, 100
Claws, 124

Index

Cleobis, 38
Clover mite, 88
Clubiona, 579, 580
  abbotii, 580
  canadensis, 581
  obesa, 581
  pallens, 581
  riparia, 581
  tibialis, 583
Clubionidæ, 256, 571; table, 573
Clubionids, the, 571
Clubioninæ, 579; table, 579
Clypeus, 99
Cochlear, 132
Cocoon, 182
Coleosoma, 349, 376
  floridana, 377
Collecting, methods of, 178
Collecting outfit, 177
Colulus, 136
Comb-footed spiders, the, 345
Comb-like organs, 22
Comedones, 94
Commensal spiders, 187
Complete orbs, 193
Conductor of the embolus, 110, 112
Connate type of embolus, 119
Copulatory apparatus, of Ricinulei, 50
Coras, 599, 608
  fidelis, 610
  medicinalis, 610
Coriarachne, 537, 544
  brunneipes, 544
  floridana, 544
  utahensis, 544
  versicolor, 544
Corinninæ, 573, 595; table, 596
Cork type of trap-door, 228
Cornea, 161
Corneal hypodermis, 161, 162
Cornicularia directa, 386
Cosmetidæ, 59
Coupling device, of Ricinulei, 49
Coxa, 104
Coxal glands, 170
Coxal spur, 114
Crab-spiders, the, 534
Cribellum, 132, 136
Crosbycus, 78
  dasycnemon, 78
Crustacea, 4, 8
Crustulina, 351, 374
  guttata, 375
Cryphœca, 613
  montana, 613
Cryphœceæ, 599, 611
Cryptocellus, 49, 52
  dorotheæ, 52
Cryptostemmatidæ, 52
Ctenidæ, 256, 568; table, 569
Ctenizidæ, 224, 226
Cteninzinæ, 226; table, 230
Ctenus, 569
  byrrhus, 569
  captiosus, 569
  hibernalis, 569
Cucullus, 49
Cupiennius, 569, 571
  sallei, 571
Curculioides, 52
Cuticula, 137, 138, 139
  moulting of the, 183
Cuticular appendages, 139
Cybæina, 600
Cybæinæ, 599
Cybæota, 600
Cybæozyga, 600
Cybæus, 600
  giganteus, 600

Cyclocosmia, 230, 232
  truncata, 232
Cyclosa, 458, 464; table, 464
  bifurca, 464, 467; egg-sacs of, 208
  caroli, 464, 467
  conica, 465
  turbinata, 465, 468
  walckenæri, 464, 468
Cymbium, 108, 112
Cynorta, 59
Cyphophthalmi, 58

Daddy-long-legs, 53
Dancing madness, 620
Deinopidæ, 254, 272
Deinopis, 272
  spinosus, 272
Demodecidæ, 94
Demodex, 94
  folliculorum, 94
Demodicoidea, 86, 93
Dendrolasma, 78, 80
  mirabilis, 81
Dendryphantes, 674, 692
  capitatus, 694
  marginatus, 693
Dermanyssus gallinæ, 90
Development of spiders, 182
Dictyna, 280, 281
  foliacea, 285
  sublata, 282
  volucripes, 285
  volupis, 285
Dictynidæ, 254, 279; table, 280
Dictynids, the, 279
Dictynina, 279
Diguetia, 313, 314
  canities, 315
Diplocentridæ, 24, 28
Diplocentrus, 28
  keyserlingi, 28
  lesueurii, 28
  whitei, 28
Diplopoda, 4, 8
Diplosphyronida, 42, 45, 46
Diplothele, 223
Dipluridæ, 226, 245; table, 245
Dipœna, 349, 371
  buccalis, 371
  crassiventris, 371
  lascivula, 371
  nigra, 371
Distal hæmatodocha, 115
Dithidæ, 46
Diurnal eyes, 96, 166
Diverticula of the heart, 149, 150
Dolichognatha, 431, 439
  tuberculata, 439
Dolomedes, 620, 622, 625
  albineus, 627
  fimbriatus, 620
  ideoneus, 627
  scriptus, 627
  sexpunctatus, 631
  tenebrosus, 626
  triton, 631
  urinator, 627
  vittatus, 630
Domed orb, 195
Domestic-spider, the, 360
Double door type, 228
Dragline, 188
Drapetisca, 390, 394
  alteranda, 394
  socialis, 394
Drassidæ, 321
Drassodes, 322, 324
  neglectus, 324
Drassus neglectus, 325

Index

Drassyllochemmis, 586, *591*
  captiosus, *591*
Drassyllus, 323, *330*
  frigidus, *331*
  rufulus, *330*
Dysdera, 159, *303*; palpus of, 109
  crocata, *303*
Dysderidæ, 173, 192, 254, 303
Dysderids, the, 303

Ebo, 555, *560*
  latithorax, *560*
  mexicana, *560*
  oblongus, *560*
Egg-sac, 182
  in rolled leaf, 213
  life within the, 182
  maker unknown, 211
Egg-sacs, silk of the, 192
Eilica bicolor, *333*
Ejaculatory duct, 108
Embolic subdivision, 113
Embolus, 108, 110
  different types of, 119
Empodium, 124
Endite, 11, 104
Endoskeleton, 141
Endosternite of the cephalothorax, 141
Enoplognatha, 350, *379*
  marmorata, *379*
  rugosa, *379*
  tecta, *379*
Epeira, 414, *481*
  alba, *464*
  caudata, *468*
  cinerea, *486*
  cophinaria, *450*
  domiciliorum, *511*
  foliata, *503*
  globosa, *509*
  hortorum, *438*
  insularis, *493*
  mitrata, *530*
  parvula, *525*
  patagiata, *504*
  prompta, *525*
  riparia, *450*
  sclopetaria, *501*
  scutulata, *488*
  spinea, *527*
  strix, *503*
  triaranea, *509*
  trivittata, *511*
  vertebrata, *515*
Epeiras, the three house, 499
Epeiridæ, 414
Epicardiac ligaments, 151
Epigastric furrow, 128
Epigastric plates, 132
Epigastrium, 128
Epigyna, different types of, 130
Epigynum, 129
Epipharynx, 102, 153
Episinus, 348, *357*
  amœnus, *357*
Epistoma, of mites, 83
Erebomaster, *63*
  flavescens, *63*
Eremobatidæ, 36
Eremobatinæ, 37, 38
Eremohax, *37*
  magnus, *37*
Eremohaxinæ, 37
Ergatis, 279
Erigone autumnalis, *387*
Erigonids, the, 384
Erigoninæ, 384
Erineum, 94
Eriophora, 459, 460, 461, *516*

Eriophora, bivariolata, *517*
  circulata, 115, *517;* palpus of, 117
Eriophyes pryi, *93*
Eriophyidæ, 83, 93
Ero, 532, *533*
  furcata, 210, *533*
Eucteniza, 235, *321*
Eucynortella bimaculata, 60
Euophrys, 674, *683*
  cruciatus, *683*
  monadnock, *683*
Eupodoidea, 87
Eurybunus, 67, *72*
  brunneus, *72*
  formosus, *72*
  spinosus, *72*
Euryopis, 349, *357*
  argentea, *359*
  funebris, *359*
  quinquemaculata, *359*
  scriptipes, *359*
Eurypelma, 240, 241
  californicum, 242, *243*
  helluo, 242, *243*
  hentzii, *243*
  marxi, 242, *243*
  rusticum, 242, *243*
  steindachneri, 242, *243*
Eustala, 461, *523*
  anastera, *524*
Evagrus, 245, *246*
  comstocki, *246*
Exocardiac ligaments, 151
External anatomy of spiders, 95
Eye-capsule, 161, 163
Eye-muscle, 167
Eye-space, 98
Eyes, 10, 96, 161

Face, 99
Falces, 100
Fasciculi unguiculares, 125
Fat-body, 155, 157
Faucheurs, 53
Fecal crystals, 157
Femoral groove, 114
Femur, 104
Filistata, 292, *294;* type of palpus, 108
  hibernalis, 108, *294*
Filistatidæ, 253, 292
Filistatids, the, 292
Filistatinella, 292, *301*
  crassipalpus, *301*
Filistatoides, 292, *301*
  insignis, *301*
Filmy Dome Spider, the, 404
Flagellum, 33, 41
Flagellum complex, of Solpugida, 37
Folding-door tarantulas, 236
Foliate spider, the, 501
Folium, 128
Food of spiders, 185
Fore-intestine, 152, 153
Foundation lines, 197
Four-lunged true spiders, 256
Free zone, 203
Front, 99
Frontinella communis, 402
Fuentes, *664*
  lineatus, *664*
Fundus, 108
Funnel-web spiders, the, 597
Funnel-web tarantulas, 225, 245
Funnel-webs, 193
Furrow of the chelicera, 100
Furrow of the posterior spiracle, 128

Galea, 40
Galeodea, 32

Index

Galeodes arabs, 35
Gall-mites, 93
Gamasid mites, 90, 91
Gamasidæ, 91
Gamasoidea, 86, 90
Gamasomorpha floridana, 309
Gamasus, 83
Garden spiders, the, 447
Garypidæ, 47
Garypoidea, 47
Gasteracantha, 525
   cancriformis, 526
Gasteracanthinæ, 525
Gayennina, 574, 576
   britcheri, 576
Gea, 448, 457
   heptagon, 457
Generalized types of palpi, 106
Genital bulb, 108
Genital sternite, 54
Geodrassus, 322, 325
   auriculoides, 325
Geolycosa, 638, 651
   fatifera, 651
   pikei, 651
Giant crab-spiders, the, 564
Glands of spiders, 169
   aciniform or berry-shaped, 171
   aggregate or treeform, 174
   ampullate or bellied, 172
   cribellum, 174
   cylindrical, 173
   lobed, 174
   of the spermathecæ, 160
   poison, 170
   pyriform or pear-shaped, 172
Glenognatha, 421, 422
   emertoni, 422
Globipes, 66, 71
   spinulatus, 71
Glyptocranium, 458, 462
   bisaccatum, 462
   cornigerum, 462
Gnaphosa, 321, 334
   gigantea, 334
   sericata, 335
Gnaphosidæ, 256, 321
Gnaphosids, the, 321
Gnathonagrus unicorn, 386
Gosiphrurus, 585, 587
   scleratus, 588
Grab for gray-beards, 53
Grass-spiders, the, 601
Grate-form tapetum, 165
Grocers' itch, 92
Guide, 131

Habrocestum, 677, 695
   pulex, 695
Hackled bands, 190
   of *Amaurobius*, 275
   of *Filistata*, 299
Hadrobunus, 67, 75
   grandis, 76
   maculosus, 76
Hadrurus, 30, 32
   hirsutus, 32
Hæmatodocha, 112
Hahnia, 614
   cinerea, 614
Hahniidæ, 225, 613; table, 614
Hahniids, the, 613
Hairs, 139
Halacaridæ, 88
Hamataliva, 666, 668
   grisea, 668
Hammock Spider, the, 408
Hand, 23
Haplodrassus, 323, 326

Haplodrassus, *signifer*, 326
Harvestmen, 53
Harvest-mites, 87, 88
Hay-makers, 53
Head, 95
Heart, 147, 148, 149
Hentzia, 430, 431
   basilica, 431
Heptathela, 218, 223
Herpyllus, 323, 332
   ecclesiasticus, 332
   vasifer, 332
Hersiliidæ, 255, 633
Hersiliids, the, 633
Hesperauximus, 274, 279
   sternitzkii, 279
Hesperocosa, 638, 652
   unica, 653
Heteropoda, 565
   venatoria, 565
Heteropodidæ, 256, 564; table, 565
Heterosphyronida, 45
Hexapoda, 6, 8
Hexura, 220, 223, 245, 247
   picea, 247
Hexurinæ, 245, 247
Hind-intestine, 152, 156
Histagonia, 381
Holosiro, 58
   acaroides, 58
Homalonychidæ, 256, 338
Homalonychids, the, 338
Homalonychus, 338
   selenopoides, 339
Homolophus, 66, 71
   biceps, 71
Hood, 49
Hub, 199
Hyctia, 675, 685
   bina, 685
   pikei, 685
   robusta, 685
Hydrachnidæ, 88, 89
Hydrachnoidea, 86, 88
Hypocardiac ligaments, 151
Hypochilidæ, 253, 256
Hypochilus, 258
   thorelli, 258
Hypodermis, 137, 138
   corneal, 162
Hypopharynx, 153
Hypopodium, 125
Hypopus, of mites, 85
Hyptiotes, 262, 269
   cavatus, 269
   gertschi, 272

Icius, 675, 681
   elegans, 681
   hartii, 682
   similis, 681
   wickhami, 672
Incomplete orb, 195
Intermediate types of palpi, 110
Internal anatomy of spiders, 137
Intima, 152
Iris, 165
Irregular nets, 193
Irregular webs with hackled bands, 196
Ischnothyreus, 310, 311
   barrowsi, 312
Ischyropsalidæ, 65, 76
Isometrus, 25, 26
   maculatus, 26
Itch-mites, 81, 92
Ixodes, 83
Ixodoidea, 86, 89

Jumping-spiders, the, 669

Index

Kaira, 458, *464*
  alba, *464*
Katadysas pumilus, 587
Kibramoa, 313, *314*
  suprenans, *314*
Kœnenia, 9, *15*
  florenciæ, *15*
  wheeleri, 15
Koeneniidæ, 15
Krohn's glands, openings of, 53

Labium, 102
  of mites, 83
Laboratory equipment, 179
Labulla, 390, *397*
  altioculata, *397*
Labyrinth Spider, the, 462
Lacinius, 66, 70
  ohioensis, 70
  texanus, 70
Laniatores, 59
Larinia, 460, 520
  borealis, *521*
  directa, *521*
  famulatoria, *522*
Laronia, 321, *333*
  bicolor, *333*
Lateral condyle of the chelicera, 100
Lateral pores, 55
Lateral sclerites, 132
Lateral subterminal apophysis, 113, 115
Lathys, 280
  foxii, *281*
Latrodectus, 348, 349, *372*
  geometricus, *374*
  mactans, *373*
  mactans bishopi, *374*
  mactans hesperus, *374*
  mactans texanus, *374*
Lattice-spider, the, 504
Lauricius, 579
  hemiclæinus, *580*
Leg formulæ, 126
Leiobunum, 67, 73
  bicolor, 75
  bimaculatum, *74*
  calcar, 73
  crassipalpis, 74
  exilipes, 73
  flavum, 75
  formosum, 56, 74
  longipes, 74
  nigripes, 74
  nigropalpi, 73
  politum, 57, 75
  speciosum, 75
  townsendi, 74
  ventricosum, 75
  verrucosum, 75
  vittatum, 74
Lentigen, 162
Lephthyphantes, 390, *393*
  leprosus, *394*
  nebulosus, *394*
Leptobunus, 67, *71*
  borealis, *72*
  californicus, *72*
Leptoneta, *318*
  californica, *318*
Leptonetidæ, 255, 318
Leptonetids, 318
Leucauge, 430, *435*
  venusta, *436*
Leuronychus, 67, 73
  pacificus, 73
  parvulus, 73
Libitioides sayi, 60
Life of spiders, 177
Limulus, 7

Limulus, *polyphemus*, 7
Linyphia, 174, 390, *398*
  clathrata, *398*
  coccinea, *399*
  communis, *400*
  insignis, *402*
  lineata, *402*
  litigiosa, *406*
  marginata, *404*
  phrygiana, *408*
  pusilla, *412*
  variabilis, *399*
Linyphia type of palpus, 112
Linyphiidæ, 174, 255, 382; table, 389
Liocraninæ, 573, *585*; table, *585*
Liocranoides, 586, *591*
  unicolor, *591*
Liodrassus, 323, *333*
  utus, *333*
Liphistiidæ, 95, 218, 221, 223
Liphistius, 134, 218, 221, 223
Lithyphantes, 349, 377
  corollatus, *377*
  fulvus, *377*
Litopyllus, 323, *332*
  rupicolens, *333*
Lorum of the pedicel, 127
Lost Atypus, 256
Loxosceles, 108, 313, *315;* palpus of, 109
  unicolor, *316*
Lunate plate, 115
Lung-slits, 128, 129
Lutica, 337, *338*
  maculata, *338*
Lycosa, 167, 635, *638*, *639*; table, *643*
  arenicola, 651
  aspersa, *643*, *646*
  avara, *644*, *650*
  avida, 635, 636, *644*, *648*
  baltimoriana, *644*
  carolinensis, *644*, *647*
  communis, *648*
  frondicola, *644*, *649*
  gulosa, *644*, *650*
  helluo, *643*, *645*
  inhonesta, *647*
  kochii, *650*
  lenta, *643*
  nidicola, *645*
  nidifex, 651
  pikei, 651
  pratensis, *644*, *649*
  punctulata, *643*
  rabida, *643*, *648*
  riparia, *643*, *645*
  tarentula, 222
  tigrina, *647*
  vulpina, *647*
Lycosidæ, 255, 635; table, 637
Lynx-spiders, the, 665
Lyriform organs, 125, 168
Lyssomanes, 672, *677*
  viridis, *677*

Macrothelinæ, 245
"Made to order" webs, 180
Mævia, 675, *702*
  vittata, *702*
Mallos, 279
Malpighian vessels, 157
Mandibles, 100
Mangora, 460, *517*
  gibberosa, *518*
  maculata, 520
  placida, *518*
Marcellina, 579
  piscatoria, *579*
Margaropus annulatus, 90
Marpissa, 675, *686*

Index

Marpissa, bivittatus, 686
   californica, 686
   undata, 686
Marxia, 459, *469*
   mæsta, 471
   stellata, 470
Masticatory ridges, 11
Mastigoproctus giganteus, 19
Maxillæ, 104
Maypacius floridanus, 624
Mazax, 592
   spinosa, 592
Mecostethi, 58
Median apophysis, 112
Median ocular area, 98
Megamyrmecion, 322, 327
   californicum, 327
Menthidæ, 47
Meriola, 596, *597*
   californica, 597
   decepta, 597
Mesal subterminal apophysis, 115
Meshed hub, 199
Mesosoma, 66, *76*
   nigrum, 76
Mesothele, 218
Meta, 167, 431, *433*
   menardii, 210, *433*
Metacynorta ornata, 60
Metacyrba, 672, 675, *702*
   tæniola, 702
Metargiope, 448, *452*
   fasciata, 455
   transversa, 455
   trifasciata, 198, *452*
Metazygia, 459, *475*
   wittfeldæ, 475
Metepeira, 460, *476*
   labyrinthea, 476
Methods of study, 177
Metids, the, 429
Metinæ, 415, 429
Micaria, 592, *595*
   aurata, 595
Micariinæ, 573, 592
Micrathena, 525, *527*
   gracilis, 529
   maculata, 530
   reduviana, 536
   sagittata, 527
Microcreagris cavicola, 47
Microcreagris rufulus, 48
Microhexura, 246
   montivagus, 246
Micrommata undata, 625
Microneta, 390, *392*
   viaria, 392
Microthelyphonida, 13
Micro-whip-scorpions, 13
Middle division of the bulb, 109, 112
Middle hæmatodocha, 115
Millipedes, 4
Mimetidæ, 254, 531; table, 532
Mimetids, the, 531
Mimetus, 532
   interfector, 532
Mimognatha, 421, *422*
   foxi, 423
Miranda, 187, *448*
   aurantia, 182, 209, 210, *448*
Misumena, 167, 224, 537, *538*
   vatia, 539
Misumeninæ, 536; table, 537
Misumenoides, 537, *540*
   aleatorius, 540
Misumenops, 537, *542*
   asperatus, 543
Mites, 13, 81
Mitopus, 66, *69*

Mitopus californicus, 69
   dorsalis, 70
   montanus, 70
Modisimus, 340, *341*
   texanus, 341
Monosphyronida, 42, 45, 47
Motherhood of spiders, 208
Mouth, 104
Muscle impressions, 127
Muscles of the abdomen, 144
   of the appendages, 143
   of the body-wall, 143
   of the sucking stomach, 144
Muscular system, 142
Mygale, 222
   unicolor, 237
Mygalomorphæ, 222
Myriapoda, 6
Myrmarachne, 677, *680*
   albocinctus, 681
Myrmeciophila, 231, *234*
   fluviatilis, 234
   torreya, 234
Myrmecotypus, 592
   cubanus, 592
Mysmena, 351, *381*
   guttata, 381

Nemastoma, 78
   inops, 78
   modestum, 78
   troglodytes, 78
Nemastomatidæ, 65, 78
Neoanagraphis, 322, 327
   chamberlini, 327
Neoantistea, 614, *615*
   agilis, 615
Neobisiinea, 46
Neobisiodea, 47
Neobisium brunneum, 47
Neobisium carolinense, 47
Neon, 675, *683*
   nelli, 683
Neoscona, 461, *509*
   arabesca, 510
   benjamina, 511
   nautica, 513
   oaxacensis, 515
   pratensis, 515
Neoscotolemon, 62
   pictipes, 63
   spinifer, 62
Nephila, 173, *440*
   clavipes, 440
   plumipes, 442
Nephilinæ, 415, 440
Nervous system, 160, 161
Nesticus, 431, *438*
   pallidus, 439
Nests of spiders, 206, 234
Net-building spiders, 187
Nidivalvata marxii, 237
Nocturnal eyes, 96, 166
Nodocion, 323, *328*
   barbaranus, 328
Notched zone, 199
Nursery-web Weavers, the, 615

Ochyrocera, 318, *319*
   pacifica, 319
Ocular quadrangle, 99
Ocular tubercle, 99
Odiellus, 66, *70*
   pictus, 70
Œclus, 28
Œcobiidæ, 253, 289
Œcobius, 290
   isolatus, 290
   parietalis, 290, *291*

724

Index

Œcobius, texanus, 291
Œdothorax montiferus, 386
Œsophagus, 154
Ogre-faced spiders, 272
Olios, 565, 567
  fasciculatus, 567
Oniscidæ, 4
Onondaga, 675, 703
  lineata, 704
Onychium, 124
Oonopidæ, 254, 309
Oonopids, the, 309
Oonops, 310, 311
  floridanus, 311
Openings of the epigynum, 130
Opiliones, 53
Opisthacanthus, 28, 29
  elatus, 29
Opisthothelæ, 220
Opopæa, 310, 312
  meditata, 312
Optic rod, 162
Oral tube, of mites, 83
Orange Garden Spider, the, 448
Orb and irregular net, 196
Orb-weavers, the, 414
Orb-webs, 193, 196
Orchestina, 310
  saltitans, 311
Organs of taste and smell, 169
Organs of touch, 169
Oribatid mites, 91
Oribatidæ, 91
Oribatoidea, 85, 86, 91
Origanates rostratus, 386
Orodrassus, 323, 326
  coloradensis, 326
Ortholasma, 79
  pictipes, 79
  rugosa, 80
Orthonops, 304
  gertschi, 305
Ostia of the heart, 149, 150
Ovaries, 158, 159
Ovipositor, 132
Oxyopes, 666, 667
  salticus, 668
  scalaris, 668
Oxyopidæ, 255, 665; table, 666
Oxyptila, 537, 543
  conspurcata, 543

Pachygnatha, 110, 159, 421
  brevis, 422
Pachyliscus, 63
  californicus, 63
Pachylomerides, 230, 231
  audouinii, 232
Paidisca, 351, 382
  marxi, 382
Paired claws, 124
Pairing of spiders, 207
Palæostracha, 7, 8
Palmula, 124
Palpal organ, 158
Palpatores, 65
Palpigradi, 13
Palpus, 105, 106, 121
Paracymbium, 112
Pardosa, 167, 638, 660
  albopatella, 662
  annulata, 662
  emertoni, 665
  grœnlandica, 664
  lapidicina, 165, 663
  milvina, 662
  modica, 664
  modica brunnea, 665
  nigropalpis, 663

Pardosa, pallida, 665
  saxatilis, 661
  sternalis, 663
  xerampelina, 664
Parmula, 132, 375
Pars cephalica, 95
Pars pendula, 109, 120
Pars thoracica, 95
Patella, 104
Pear-leaf blister, 93
Peckhamia, 677, 707
  americana, 708
  picata, 707
  scorpionea, 707
Pectines, 22
Pedicel, 127
Pediculoides, 92
Pedipalpi, 16
Pedipalpida, 13, 16
Pedipalps, 11, 82, 104
Pellenes, 677, 696; table, 696
  agilis, 697
  borealis, 698
  coronatus, 699
  hoyi, 699
  peregrinus, 700
  splendens, 700
Pelopatis, 622, 624
  undulata, 624
Pericardial cavity, 150, 151
Pericardium, 150, 151
Pericurus, 319, 321
  pallida, 321
Peritoneal layer, 155
Petiole, 109
Peucetia, 666
  viridans, 666
Phalangida, 13, 53
Phalangides, 53
Phalangiidæ, 65
Phalangita, 53
Phalangium, 66, 69
  opilio, 69
  parietinus, 69
Phalangodes, 63
  armata, 64
  brunneus, 64
  californicus, 64
Phalangodidæ, 59, 62
Phanetta, 389, 391
  subterranea, 391
Pharyngeal gland, 154
Pharynx, 153
Phidippus, 676, 688
  audax, 689
  clarus, 689, 691
  insolens, 688, 689, 691
  mineatus, 689, 692
  mystaceus, 689, 691
  purpuratus, 689, 691
  variegatus, 689, 690
  whitmanii, 688, 690
Philodrominæ, 536, 554; table, 555
Philodromoides, 555, 558
  pratariæ, 558
Philodromus, 555, 556
  minutus, 556, 557
  ornatus, 558
  pernix, 557
Phlegmacera cavicolens, 77
Phlegmacera occidentale, 77
Phlegra, 674, 696
  leopardus, 696
Pholcidæ, 253, 339; table, 340
Pholcids, the, 339
Pholcophora, 341, 343
  americana, 343
Pholcus, 167, 340, 342
  phalangioides, 342

725

Index

Photographing, methods of, 180, 181
Phrurolithus, 585, 590
  formica, 590
Phrurotimpus, 585, 589
  alarius, 589
Phrynidæ, 19
Physocyclus, 341, 343
  globosus, 343
Piabuna, 585, 590
  longispina, 591
Pigment cells, 165
Pirata, 638, 653
  insularis, 655
  marxi, 655
  minutus, 654
  montanus, 655
Pisaurid type of palpus, 118
Pisauridæ, 255, 615
Pisaurina, 622, 625
  mira, 625
Plagiostethi, 58, 64
Plagula, 127
Plantula, 124
Platform Spider, the, 412
Platæcobius, 290, 291
  floridanus, 292
Plectana, 470
Plectreurys, 313
  castanea, 314
  tristis, 314
Plesiometa, 430, 438
  argyra, 438
Plexippus, 676, 701
  paykulli, 701
Pœcilochroa, 322, 329
  montana, 329
  variegata, 330
Poison glands, 170
Postabdomen, 128
Postbacillar eyes, 97
Post-retinal membrane, 163
Poultry, mite infesting, 90
Prey, means of destroying the, 187
Process of the finger, 40
Procurved, 98
Prodalia foxii, 281
Prodidomidæ, 256, 319
Prodidomus, 319
  rufus, 319
Prokœnenia californica, 15
Prokœnenia wheeleri, 15
Protolophus, 65, 67
  singularis, 68
  tuberculatus, 68
Pseudogarypidæ, 48
Pseudoscorpiones, 39
Pseudoscorpionida, 13, 39
Pseudoscorpions, 39
Pseudo-spiders, 53
Pseudo-stigmatic organs of mites, 91
Psilochorus, 341, 344
  pullulus, 344
Pteropus, 83
Pulmonary veins, 150, 151
Pulvillus, 124
Purse-web Spider, the, 248

Quadrangle of the posterior eyes, 99

Rachodrassus, 322, 328
  echinus, 328
Racquet-organs, 34
Radii, 197
Radix, 113
Rake of the chelicera, 100
"Ram's-horn organs," 43
Ray-formed orb-web, 195
Ray-spider Family, the, 415
Receptaculum seminis, 108

Rectum, 157
Recurved, 98
Red-bugs, 88
Red-spiders, 87
Reproduction of lost organs, 185
Reproductive organs, 10, 158
Reservoir, 108
Respiratory organs, 10, 145
Retina, 161, 162
Retinacula, 113
Rhechostica, 240
  texense, 241
Rhomphaea, 349, 351
  fictilium, 352
Ricinodidæ, 52
Ricinoides, 49, 52
Ricinulei, 13, 48
Ricinulids, the, 48
Robertus, 351, 380
  riparius, 380
Rostrum, of mites, 83
  of spiders, 102
Runcinia, 542

Sabacon, 77
  crassipalpe, 77
Salticus, 674, 684
  scenicus, 684
Sarcoptes ovis, 93
Sarcoptes scabei, 92
Sarcoptidæ, 92
Sarcoptoidea, 87, 92
Sassacus, 677, 694
  papenhoei, 694
Scape, 132, 375
Scaphiella, 310, 312
  hespera, 312
Schizocosa, 638, 657; table, 658
  bilineata, 658
  ocreata, 658
  saltatrix, 658
Schizomidæ, 17
Schizomus, 17
Schizonotidæ, 17
Sclerite, 128, 139
Sclerobunus, 61
  brunneus, 61
  robustus, 61
Scoloderus, 458, 461
  tuberculiferus, 461
Scopodes, 322, 328
  catharius, 328
Scopula, 100, 125
  of the chelicera, 100
  of the pedipalp, 105
  of the tarsi, 125
Scopus, 112
Scorpionida, 12, 21
Scorpionidæ, 24, 28
Scorpions, 21
  table of families of, 24
Scotolathys, 280, 289
  pallidus, 289
Scytodes, 313, 316
  intricata, 316
  longipes, 316
  thoracica, 316
Scytodidæ, 253, 313
Scytodids, the, 313
Sea of gossamer, 216
Second antenna, 100
Second trochanter, 41
Segestria, 159, 160, 306
  pacifica, 306
Segestriidæ, 254, 306
Segestriids, the, 306
Selenopidæ, 256, 563
Selenops, 563
  aissus, 563

Index

Semichelate, 12
Seminal vesicle, 158
Serrula, 41, 105
   exterior, 41
   interior, 41
Setæ, 139
Shamrock spider, the, 493
Sheep-dip, 93
Sheep-scab, 92
Sheet and irregular net webs, 196
Sheeted hub, 199
Sheet-webs, 193
Sheet-web weavers, 382
Shepherd-spiders, 53
Sicariidæ, 313
Side eyes, 97
Sierra dome spider, the, 406
Silk glands, 43, 176
Silk of spiders, 187
Silk spiders, the, 440
Silvered Garden Spider, the, 457
Singa, 459, 475
   pratensis, 476
   truncata, 476
   variabilis, 475
Sitalces, 61, 64
   californicus, 64
Sitticus, 675, 684
   palustris, 684
Smell, 169
Solifugæ, 32
Solpugida, 13, 32
Solpugides, 32
Solpugids, 32
Sosilaus, 638, 659
   spiniger, 659
Sosippus, 637, 639
   californicus, 639
   floridanus, 639
   mimus, 639
Sosticus, 322, 329
   continentalis, 329
Specialized types of palpi, 111
Specimens, preservation of, 178
Spermathecæ, 158, 159, 160
Spermophora, 340, 341
   meridionalis, 342
Spigot, 135
Spine formulæ, 126
Spinneret, 40, 132, 134
Spinning field, 134
Spinning organs, 132
Spinning tubes, 135, 176
Spintharus, 349, 355
   flavidus, 359
Spiny-bellied spiders, the, 525
Spiracles, 128
Spiral guy-line, 202
Spiral muscle, 112
Stabilmentum, 203
   of *Metargiope*, 400
Steatoda, 174, 349, 375
   borealis, 375
Stenoonops, 310, 311
   minutus, 311
Stercoral pocket, 157
Sternophora, 40
Sternophoridæ, 48
Stink glands of Phalangida, 58
Stipes, 113
Stoidis, 674, 682
   aurata, 683
Stomodæum, 153
Storena, 337
   americana, 337
Stridulating organs, 136
   of the mouth-parts, 106
Stylet, 40
Subœsophageal ganglion, 160

Subtegulum, 112
Sucking stomach, 153, 154
Superciliary ridge, 23
Suture, 139
Swathing band, 189
Swathing film, 189
Syarinidæ, 47
Synema, 537, 553
   bicolor, 554
   obscura, 554
   parvula, 553
Synemosyna, 677, 679
   formica, 680
Syrisca, 585, 586
   affinis, 586
Syspira, 585, 586
   eclectica, 586

Table of families of true spiders, 252
   of mygalomorph spiders, 225
Tailed whip-scorpions, 18
Tailless whip-scorpions, 19
Talavera, 676, 682
   minuta, 682
Tama, 634
   mexicana, 634
Tapinauchenius, 240, 244
   cærulescens, 244
   texensis, 244
Tapinopa, 390, 413
   bilineata, 413
Taracus, 76
   packardi, 77
   pallipes, 77
   spinosus, 76
Tarantism, 222
Tarantula, 20
   fusimana, 20
   marginemaculata, 2.
   palmata, 20
   whitei, 20
Tarantula type of palpus, 109
Tarantulas, 222
Tarantulidæ, 17, 19
Tarentula, 638, 652
   pictilis, 652
Tarsonops, 304, 305
   systematicus, 305
Tarsus, 104, 114
Taste, organs of, 169
Tegenaria, 599, 607
   derhami, 608
Tegulum, 112
Tenent hairs, 125
Tentabunda, 565, 568
   cubana, 568
Tergum of the thorax, 121
Terminal apophysis, 113, 115
Terminal tenent hairs, 125
Testes and their ducts, 158
Tetrablemma, 95
Tetragnatha, 159, 160, 210, 212, 420, 423
   caudata, 429
   elongata, 425
   extensa, 425
   grallator, 425
   laboriosa, 426
   lacerta, 429
   pallescens, 429
   straminea, 428
   vermiformis, 428
Tetragnathids, 419
Tetragnathinæ, 415, 419; table, 420
Tetranychidæ, 87
Tetranychus telarius, 88
Teutana, 350, 376
   triangulosa, 376
Thalamia parietalis, 291
Thanatidius, 622

Index

Thanatidius, dubius, 624
  tenuis, 624
Thanatus, 556, 561
  coloradensis, 561
  lycosoides, 561
Thaumasia, 631
Thelyphonidæ, 17, 18
Thelyphonus, 144
Theonæ, 351, 381
  stridula, 381
Theridiidæ, 254, 345; table, 348
Theridion, 167, 174, 350, 359
  differens, 367
  eximius, 367
  foliacea, 285
  fordum, 361
  frondeum, 362
  globosum, 363
  kentuckyense, 364
  murarium, 368
  punctosparsum, 365
  quadripunctatus, 186
  rupicola, 362
  spirale, 369
  studiosum, 365
  tepidariorum, 210, 330, 360, 435
  unimaculatum, 364
  zelotypum, 365
Theridiosoma, 210, 416
  radiosa, 416
Theridiosomatinæ, 414, 415
Theridula, 350, 369
  opulenta, 369
  quadripunctata, 369
  sphærula, 369
Thiodina, 677, 678
  puerpera, 678
  sylvana, 678
Third claw, 124
Thomisid type of palpus, 119
Thomisidæ, 256, 534
Thomisus caudatus, 538
  vulgaris, 557
Thorax, 121
Tibellus, 556, 562
  duttonii, 562
  oblongus, 563
Tibia, 104
Ticks, 89
Tick-fever, 89
Tinus, 622, 631
  peregrinus, 633
Titanœca, 274, 278
  americana, 278
Titiotus, 569, 570
  californicus, 570
Tityus, 25, 27
  floridanus, 27
  tenuimanus, 27
Tmarus, 537
  angulatus, 338
Tomicomerus, 78
  bryanti, 78
Touch, organs of, 169
Trabea, 638, 659
  aurantiaca, 131, 659
Tracheal spiracles, 128, 129
Tracheal sternite, 54
Trachelas, 167, 596
  tranquilla, 596
Trachyrhinus, 67, 72
  favosus, 72
  marmoratus, 73
Tracks of orb-weavers, 204
Transformation, 184
Trap-door spiders, 226
Trapline, 204
Triænonychidæ, 59, 61
Triangle Spider, the, 269

Triangular web, 196
Trichogen, 138
Tricholathys, 280, 288
  spiralis, 288
Trichopore, 138
Trilobites, 8
Trithyreus pentapeltis, 17, 18
Trochanter, 104
Trochantin, 42
Troglohyphantes, 390, 396
  cavernicolus, 397
  incertus, 397
Trogulidæ, 65, 79
Trombidiidæ, 88
Trombidium sericeum, 88
Trombidoidea, 87
True spiders, the, 252
Trunk of the embolus, 120
Tubular tracheæ, 147
Tutaculum, 119, 536
Typical Orb-weaver, the, 457
Tyroglyphidæ, 92

Ulesanis, 351, 371
  americana, 371
Uloboridæ, 254, 261
Uloborids, the, 261
Uloborus, 261, 262
  americanus, 262, 263
  geniculatus, 267
  plumipes, 263
Unca, 112
Uncate, 12
Urate cells, 158
Uroctonus, 30
  mordax, 30
Uroplectes, 25, 26
  mexicanus, 26
Uropoda, 83
Usofila, 318
  gracilis, 318
Uterus, 158, 159

Vagina, 158, 159
Vasa deferentia, 158
Veins, pulmonary, 151
Vejovidæ, 25, 29
Vejovis, 30, 31
  boreus, 31
  carolinus, 32
  flavus, 31
  mexicanus, 31
  punctipalpi, 31
  spinigerus, 32
Venom of spiders, 213
Venous circulation, 152
Verrucosa, 459, 479
  arenata, 479
Visceral nervous system, 161
Viscid silk of *Aranea,* 181
Viscid spiral, 202
Viscid thread, 189
Visual cell, 162
Vitreous layer, 162
Vitreum, 162
Vulva, 132

Wadotes, 599, 611
  calcaratus, 611
Wagneriana, 459, 474
  tauricornis, 474
Wala, 675, 687
  palmarum, 688
  mitrata, 688
Wandering spiders, the, 186, 568
Warp, 192
Water-mites, 88
Webs of spiders, the types of, 193
Whip-scorpions, 13, 16

Index

Willibaldia, 397
Wind-scorpions, 35
Wixia, 460, 480
  ectypa, 481
Wolf-spiders, the, 635
Woof, 192

Xysticus, 199, 537, *545*
  elegans, 546
  emertoni, 547, *549*
  ferox, 547
  formosus, 547
  gulosus, 547
  limbatus, 547
  luctans, 549
  nervosus, 551
  quadrilineatus, 551
  triguttatus, 551

Zelotes, 323, *330*

Zelotes, subterraneus, 330
Zilla, 459, *471*
  atrica, 473
  californica, 474
  montana, 474
  x-notata, 474
Zilla type of orb-web, 195
Zodariidæ, 255, 326
Zodariids, the, 336
Zora, 586, *587*
  pumilus, 587
Zorocrates, 302
  æmulus, 302
Zoropsidæ, 254, 302
Zoropsids, the, 302
Zuninga, 680
Zygoballus, 676, *704;* table, 705
  bettinii, 705
  nervosus, 705, *706*
  rufipes, 705
  sexpunctatus, 705, *707*